Genetic Engineering of Crops

Genetic Engineering of Crops

Edited by Harvey Parker

SYRAWOOD
PUBLISHING HOUSE
New York

Published by Syrawood Publishing House,
750 Third Avenue, 9th Floor,
New York, NY 10017, USA
www.syrawoodpublishinghouse.com

Genetic Engineering of Crops
Edited by Harvey Parker

International Standard Book Number: 978-1-68286-400-5 (Hardback)

Cataloging-in-publication Data

Genetic engineering of crops / edited by Harvey Parker.
p. cm.
Includes bibliographical references and index.
ISBN 978-1-68286-400-5
1. Crops--Genetic engineering. 2. Plant genetic engineering. I. Parker, Harvey.
SB123.57 .G46 2017
631.5233--dc23

Printed in China.

TABLE OF CONTENTS

PREFACE

Genetic engineering of crops are methods of modification and engineering to alter the traits of crops. This book on genetic engineering of crops discusses topics that deal with plant characteristics and disease resistance. This book outlines the progresses and applications of genetic engineering in crops in detail. The topics included in this book are of utmost significance and bound to provide incredible insights to readers. Those in search of information to further their knowledge will be greatly assisted by this text as it presents researches contributed by acclaimed experts from across the globe. Academicians and students alike would find this book on genetic engineering of crops especially helpful.

This book unites the global concepts and researches in an organized manner for a comprehensive understanding of the subject. It is a ripe text for all researchers, students, scientists or anyone else who is interested in acquiring a better knowledge of this dynamic field.

I extend my sincere thanks to the contributors for such eloquent research chapters. Finally, I thank my family for being a source of support and help.

Editor

1

Comparative Proteomic Analysis of Aluminum Tolerance in Tibetan Wild and Cultivated Barleys

Huaxin Dai[1], Fangbin Cao[1], Xianhong Chen[1], Mian Zhang[1], Imrul Mosaddek Ahmed[1], Zhong-Hua Chen[2], Chengdao Li[3], Guoping Zhang[1]*, Feibo Wu[1]*

1 Department of Agronomy, College of Agriculture and Biotechnology, Zijingang Campus, Zhejiang University, Hangzhou, P.R. China, **2** School of Science and Health, Hawkesbury Campus, University of Western Sydney, New South Wales, Australia, **3** Department of Agriculture, Government of Western Australia, South Perth, Western Australia, Australia

Abstract

Aluminum (Al) toxicity is a major limiting factor for plant production in acid soils. Wild barley germplasm is rich in genetic diversity and may provide elite genes for crop Al tolerance improvement. The hydroponic-experiments were performed to compare proteomic and transcriptional characteristics of two contrasting Tibetan wild barley genotypes Al- resistant/tolerant XZ16 and Al-sensitive XZ61 as well as Al-resistant *cv.* Dayton. Results showed that XZ16 had less Al uptake and translocation than XZ61 and Dayton under Al stress. Thirty-five Al-tolerance/resistance-associated proteins were identified and categorized mainly in metabolism, energy, cell growth/division, protein biosynthesis, protein destination/storage, transporter, signal transduction, disease/defense, etc. Among them, 30 were mapped on barley genome, with 16 proteins being exclusively up-regulated by Al stress in XZ16, including 4 proteins (S-adenosylmethionine-synthase 3, ATP synthase beta subunit, triosephosphate isomerase, Bp2A) specifically expressed in XZ16 but not Dayton. The findings highlighted the significance of specific-proteins associated with Al tolerance, and verified Tibetan wild barley as a novel genetic resource for Al tolerance.

Editor: Tianzhen Zhang, Nanjing Agricultural University, China

Funding: This study was financially supported by the National 863 program (2012AA101105) and the National Natural Science Foundation of China (30630047). The funders had no role in study design, data collection and analysis, decision to publish, or preparation of the manuscript.

Competing Interests: The authors have declared that no competing interests exist.

* E-mail: wufeibo@zju.edu.cn (FBW); zhanggp@zju.edu.cn (GPZ)

Introduction

Ionic aluminum (Al^{3+}), highly toxic to plant growth, is a major factor limiting crop productivity on acid soils [1]. The strategies for maintaining production on acid soils include lime application to raise soil pH and use of plants with high tolerance to acid soils. Development and planting of Al tolerant cultivars is a cost-effective and practically acceptable approach for full utilization of acid soil [2].

Barley (*Hordeum vulgare* L.) is one of the most Al-sensitive species among small grain cereals [3]. Al toxicity limits the growth and productivity of barley on acid soils and its expansion as a crop into many agricultural areas in the world [4]. In order to breed barley cultivars tolerant to Al toxicity, it is especially important to identify genetic resources with Al tolerance. Wild barley germplasm is rich in useful genes for crop improvement [5]. Tibetan annual wild barley from Qinghai-Tibet Plateau is regarded as one of the progenitors of cultivated barley and is rich in genetic diversity [6]. We successfully identified Tibetan wild annual barley genotypes with high tolerance to both low pH and Al stress [7]. However, their underlying physiological and molecular mechanisms in Al tolerance remain unclear.

Comparative proteomic analysis and bioinformatics techniques provide powerful tools to identify proteins expressed under abiotic stress [8]. Root proteomic analysis showed that proteins involved in stress defense, metabolisms and signal transduction were important for soybean [9], tomato [10] and Arabidopsis [11] plants survival under Al toxicity. However, only limited information is available on Al accumulation/translocation and Al tolerance mechanisms in barley. Moreover, physiological and proteomic responses to Al stress in Tibetan wild barley genotypes have never been investigated and compared with elite Al-tolerant barley cultivars. Thus, precise knowledge of the proteomic basis is required to dissect the mechanisms underlying acid/Al tolerance in wild barley. In the present study we examined stress-specific proteins for acid/Al tolerance in wild barley by comparing the proteomic responses of the two Tibetan wild barley genotypes XZ16 (high acid/Al tolerant), XZ61 (acid/Al sensitive) and Al-tolerant *cv.* Dayton using two-dimensional gel electrophoresis (2-D) and mass spectrometry (MS). These results are useful to better understand the mechanisms of Al tolerance in barley, and provide an effective pathway for the exploration of Al-tolerant genes in plants.

Materials and Methods

Plant Materials and Experimental Design

Hydroponic experiments were performed using two Tibetan annual wild barley XZ16 and XZ61 (*H. vulgare* L. *ssp. spontaneum*), acid/Al- tolerant and sensitive genotypes, respectively, and one Al-tolerant-cultivar Dayton. Seeds were surface sterilized in 1% H_2O_2 for 30 min, rinsed with distilled water, and then germinated

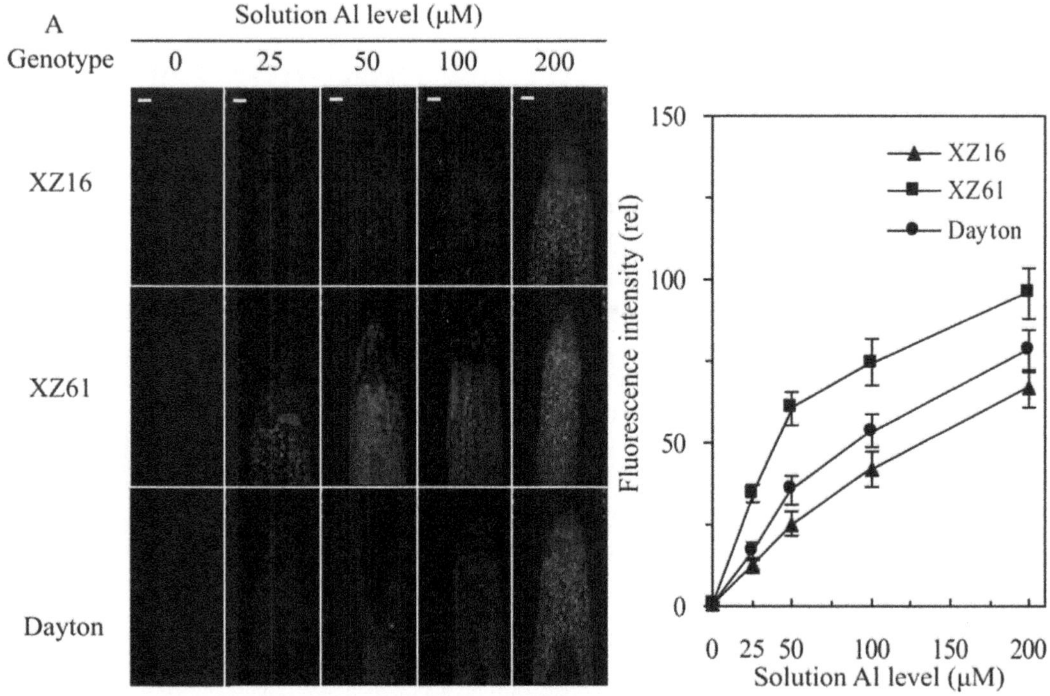

Figure 1. Al localization in barley roots exposed to different Al levels for 24 h. (A) Al was monitored by morin fluorescence using confocal laser scanning microscopy. Fluorescence intensity of image analysis was calculated using Image J software. (B) Seedlings were subjected to 0, 25, 50, 100 and 200 μM $AlCl_3$, respectively, in 0.5 mM $CaCl_2$ solution at pH 4.3 for 24 h, and then root tips (0–10 mm) were stained with morin. Images shown in the Figures are representative of more than fifteen seedlings per treatment. Bar = 100 μm. Data are means ± SD (n = 5).

in sterilized moist quartz sand in an incubator at 20±1°C. Seven-day-old uniform seedlings were transplanted into 5-L containers filled with 4.5 L basal nutrient solution (BNS). The composition of BNS was described in Wu *et al.* [12]. The container was covered with a polystyrol plate with 7 evenly spaced holes (2 plants per hole) and placed in a greenhouse. Solution was continuously aerated with pumps and renewed daily after Al addition. The solution pH was adjusted to 5.8±0.1 with NaOH or HCl, as required.

On the day 7 after transplantation, seedlings were cultured for 1 d in 0.5 mM $CaCl_2$ at pH 4.3, and then exposed to 0, 50 or 200 μM Al in 0.5 mM $CaCl_2$ at pH 4.3 for 24 h. The plantlets were kept in a growth room at 25/20°C with a 14/10 h (day/night) photoperiod and irradiance of 340 μmols m^{-2} s^{-1} light intensity. A split-plot design was adopted with treatment as the main plot and genotype as sub-plot with five replicates in each treatment. At 3 days after treatment, plants were harvested from each treatment, and roots were washed with distilled water thoroughly and collected for two 2-DE experiments and qRT-PCR analysis. Dry weights of the plants were determined and used for Al concentration.

Determination of Al Concentration and Root Al Distribution

Al concentrations was determined after digestion in an acid mixture (HNO_3:$HClO_4$ = 4:1, v/v) at 150°C for 10 h, using inductively coupled plasma atomic emission spectrometry (ICP/AES) (Thermo Jarrel Ash, San Jose, CA).

Morin staining for Al in the root tip region was determined according to Zheng et al. [13]. Briefly, 1-cm root tips were stained with 100 mM morin (Sigma-Aldrich, St. Louis, MO, USA) in 10 mM MES buffer (pH 5.5) for 30 min. Images were acquired using a Leica TCS SP2 confocal laser scanning microscope (Leica Microsystems, Heidelberg, Germany) with excitation at 420 nm and emission at 515 nm.

Protein Extraction, Quantification, Visualization and Image Analysis

Total root protein extracts were prepared essentially according to phenol extraction method [14] with minor modification. Root sample (3 g) of control and Al treated plants were grounded in a mortar separately to a fine powder in liquid nitrogen and homogenized in an extraction buffer containing 30 mg PVPP. The homogenate was suspended in 7 ml ice-cold phenol extraction buffer (0.7 M sucrose; 0.1 M KCl; 50 mM EDTA,0.5 M Tris-HCl, 1% w/v DTT, pH 7.5; complete protease inhibitor cocktail (Roche Applied Science)) and immediately added 7 ml ice-cold Tris buffered phenol and vortexed for 15 s. The sample was then vortexed for 10 s every 5 min and repeated for six times at 4°C. After centrifugation (30 min, 5000×g, 4°C) the phenolic phase was collected, and the sample was re-extracted with 14 ml of extraction buffer (added with the same volume of phenol extraction as collected items), and vortexed for 10 s every 5 min and repeated for six times at 4°C. After centrifugation (30 min, 5000×g, 4C) the phenolic phase was collected and precipitated overnight with five volumes 100 mM ammonium acetate in methanol at −20°C. After centrifugation at 5000×g for 30 min at 4°C, the supernatant was removed and the pellet was rinsed twice in ice-cold acetone/0.2% DTT. Between the two rinsing steps, the sample was incubated for 60 min at −20°C. The pellet was air-dried, resuspended in 200 μl lysis buffer (7 M urea, 2 M thiourea, 4% CHAPS, 20 mM Tris-HCl, pH7.4, containing 1% w/v DTT; Amersham Biosciences), and vortexed for 1 h at room temperature. Protein concentration was determined by

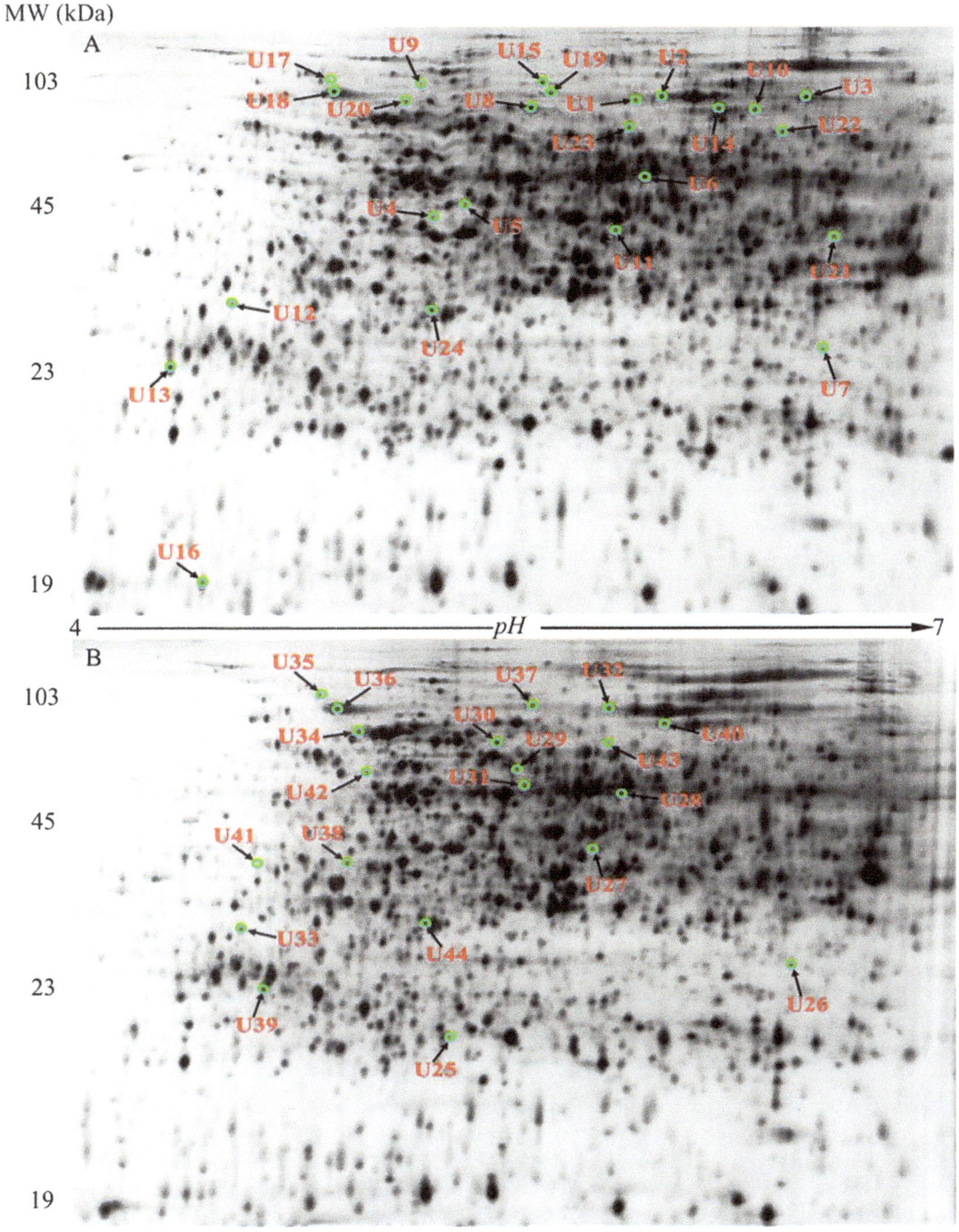

Figure 2. Representative 2-DE maps of root proteins in XZ16 exposed to Al for 24 h. The proteins were isolated from the root of XZ16 exposed to 50 μM (A, upper) and 200 μM (B, below) Al for 24 h. Total proteins were extracted and separated by 2-DE. In IEF, 100 μg proteins were loaded onto pH 4–7 IPG strips (24 cm, linear). SDS-PAGE was performed with 12.5% gels. The spots were visualized by silver staining. Differentially accumulated protein spots are indicated by green sashes. Arrows indicate the differentially expressed protein spots whose expressions were significantly induced (fold changes ≥1.5) or unchanged (−1.50< folds <1.5) in XZ16 but down-regulated (folds <−1.50) in XZ61; or up-regulated in XZ16 but unchanged in XZ61, under 50 μM Al (U1–U24) and 200 μM Al (U25–U44) stress.

standard Bradford assay using bovine serum albumin as standard (Bio-Rad, Hercules, CA, USA). All chemicals used were, if not further specified in the text, p.a. or electrophoresis grade. All electrophoresis units employed were from Amersham Biosciences.

Protein visualization, image analysis and quantification were determined according to Bah *et al.* [15]. For each sample, at least three independent protein extracts were prepared after each treatment and at least three 2-DE analyses were performed for each protein extract. To analyze the expressed protein patterns, stained gels were scanned and calibrated using a PowerLook1100 scanner (UMAX), followed by analysis of protein spots using GE HealthCare Software (Amersham Biosciences). Spot detection was

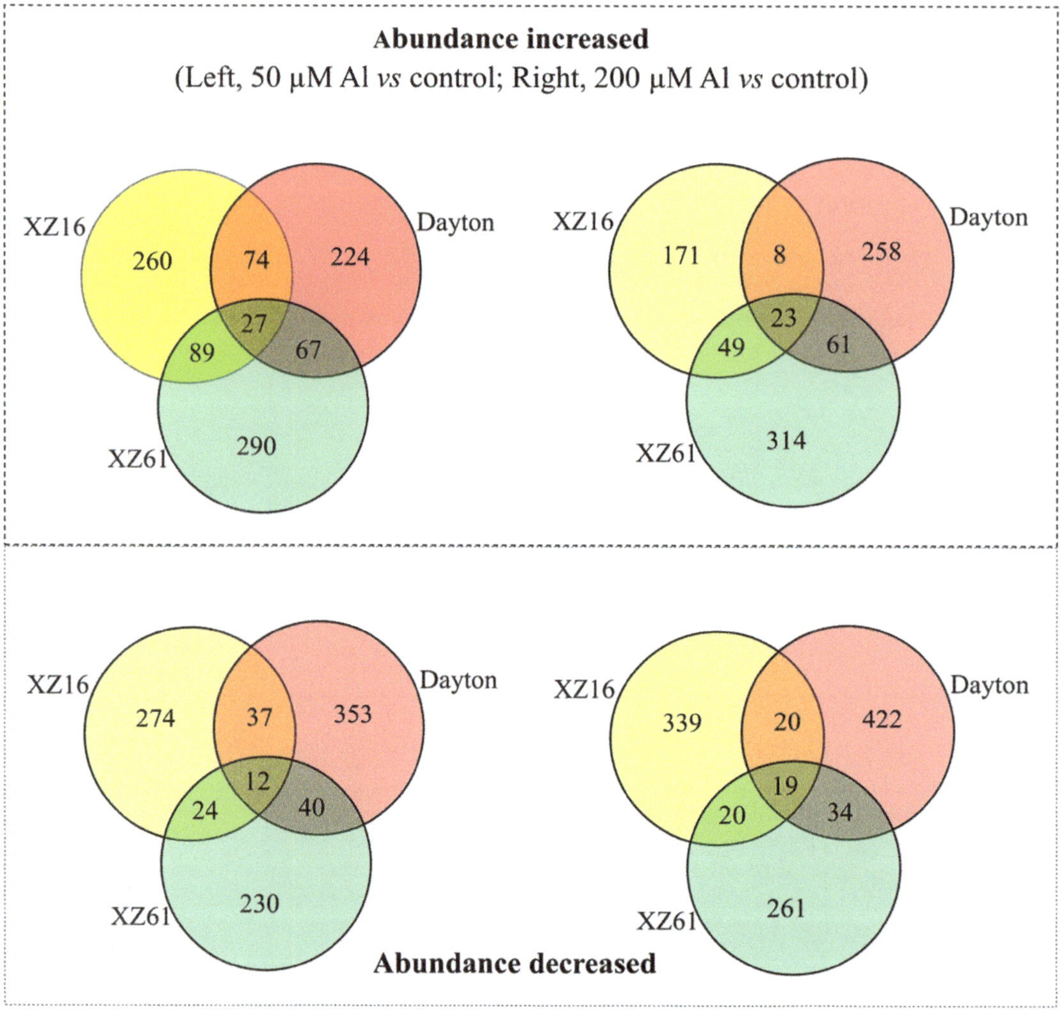

Figure 3. Venn diagram illustrating the expression patterns of Al stress-responsive proteins in roots of XZ16, XZ61 and Dayton. The numbers of differentially expressed spots up- or down-regulated are shown in the different segments. As to protein spots altered by 50 and 200 μM Al stress, 450, 473, 392 spots (50 μM Al vs control) and 251, 447, 350 spots (200 μM Al vs control) were up-regulated in XZ16, XZ61, Dayton, respectively; while 347, 306, 442, and 398, 334, 495 spots down-regulated. The abundance of 101, 31 spots increased in both of XZ16 and Dayton under 50, 200 μM Al stress, respectively, and that of 49, 39 decreased.

realized without spot editing. The protein spots were quantified using the % volume criterion. Only those with significant and reproducible changes ($P<0.05$) were considered to be differentially accumulated proteins. The target protein spots were automatically excised from the stained gels and digested with trypsin using a Spot Handling Workstation (Amersham Biosciences). Peptides gel pieces were placed into the EP tube and washed with 1:1 mixture of 50 μL of 30 mM $K_3Fe(CN)_6$ and 100 mM NaS_2O_3 for 10–15 min until completely discolored then washed with 200 μL bi-distilled water (two times for 5 min each). The washed solution was drained and washed with 50% ACN (acetonitrile, Fisher A/0626/17) and 100% ACN rotationally, and then incubated in 25 mM NH_4HCO_3 (ammonium bicarbonate, Sigma A6141) for 5 min at 37°C. After leaching out of the incubation solvent, 50% ACN and 100% ACN was rotationally added and dried at 40°C for 5 min respectively. Trypsin digestion was carried out as follows: sequencing-grade porcine trypsin (Promega, Madison, WI, USA) was suspended in 25 mM NH_4HCO_3 at a concentration of 12.5 ng per μl to rehydrate the dried gel pieces. The trypsin digestion was carried out for 16 h at 37°C. Peptides were extracted from the digest as follows for three times: 10 μL of 50% ACN containing 0.1% TFA (trifluoroacetic acid, GE HealthCare) was added to each tube and incubated for 5 min at 37°C and the supernatants were transferred to new EP tube. The extracts were pooled and then vacuum concentrated for about 2 h. A solution of peptides was filtrated via Millipore (Millipore ZTC18M096) and mixed with the same volume of a matrix solution consisting of saturated α-cyano-4-hydroxycinnamic acid (CHCA) in 50% ACN containing 0.1% TFA. After the peptides were co-crystallized with CHCA by evaporating organic solvents, tryptic-digested peptide masses were measured using a MALDI-TOF-TOF mass spectrometer (ABI4700 System, USA). All mass spectra were recorded in positive reflector mode and generated by accumulating data from 1000 laser shots. The following threshold criteria and settings were used: detected mass range of 700–3200 Da (optimal resolution for the quality of 1500 Da), using a standard peptide mixture (des-Argl-Bradykinin Mr904.468, Angiotensin I Mr1296.685, Glul-Fihrinopeptide B Mr1570.677, ACTH (1–17) Mr2093.087, ACTH (18–39) Mr2465.199; ACTH (7–38) Mr3657.929) as an external standard calibration, with laser

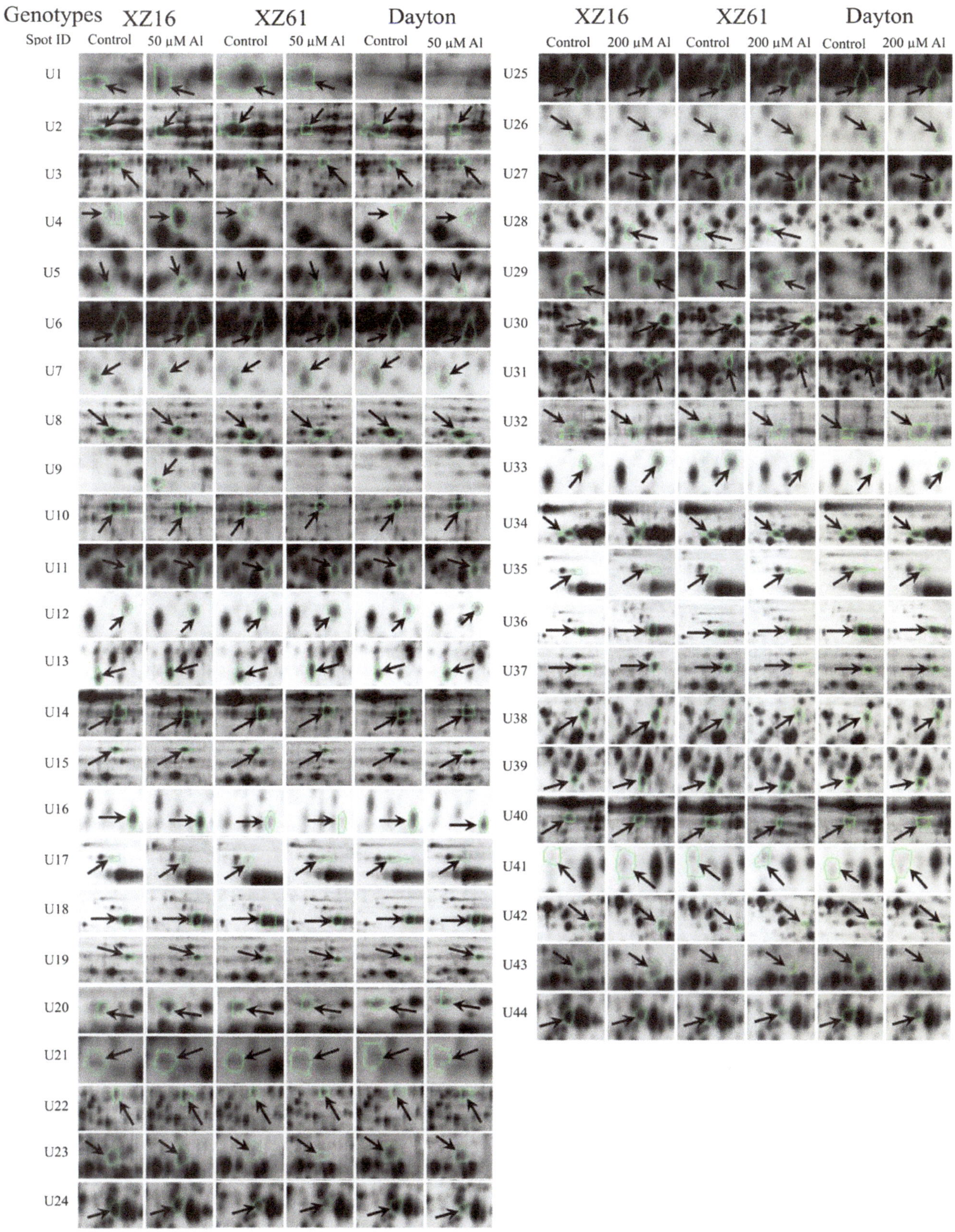

Figure 4. 'Spot view' of the abundance of differentially expressed proteins in roots of barley seedlings under 50 or 200 μM Al for 24 h. The images of three genotypes: XZ16, XZ61 and Dayton (indicated with green circles) from control and 50 or 200 μM Al (pH 4.3) treated plants. Protein spot ID refers to numbers in Figure 4 and Tables 1, 2.

Table 1. Proteins whose expression were significantly induced (+) in XZ16 roots but down-regulated (−)/unchanged in XZ61, or unchanged in XZ16 but down-regulated in XZ61 under 50 μM Al stress.

Spot ID	Protein name	Accession number	MW Da	pI	AASC %	MP	Folds (Al *vs* control)		
							XZ16	XZ61	Dayton
	Metabolism								
U1	S-adenosylmethionine synthase 3 [*Hordeum. vulgare* subsp. *vulgare*]	gi|122220777	43138	5.51	42	11	+2.39	−2.44	0.00
U2	Methionine synthase [*H. vulgare* subsp. *vulgare*]	gi|50897038	84794	5.68	35	19	−1.34	−3.39	−2.23
U3	Methionine synthase [*H. vulgare* subsp. *vulgare*]	gi|50897038	84794	5.68	22	12	−1.24	−2.68	+1.02
U4	Glutamine synthetase [*Arabidopsis thaliana*]	gi|228456	47123	6.73	15	7	+6.50	-10^6	−1.09
U5	γ-glutamylcysteine synthetase [*Triticum aestivum*]	gi|57903694	43079	5.30	20	9	+4.25	−2.74	−3.50
U6	Putative asparate aminotransferase [*H. Vulgare* subsp. *vulgare*]	gi|89511843	45377	5.75	14	4	+1.80	−1.34	+1.36
U7	Predicted pirin-like protein [*Brachypodium distachyon*]	gi|357114735	41337	9.37	17	8	+1.32	−1.72	+1.36
	Energy								
U8	Os06g0133800 [*Oryza sativa japonica*]	gi|115466224	73973	5.44	16	10	+1.07	−1.60	+1.04
U9	ATP synthase beta subunit [*Triticum. monococcum*]	gi|525291	59326	5.56	19	7	$+10^6$	0.00	0.00
U10	Aconitate hydratase 3 [*Citrus clementina*]	gi|285309967	98669	5.89	10	9	−1.03	−2.53	+1.04
U11	Fructose-bisphosphate aldolase [*H. vulgare*]	gi|226316443	39071	6.08	29	7	+1.71	+1.10	+1.16
	Cell growth/division								
U12	Predicted proliferating cell nuclear antigen-like [*Brachypodium distachyon*]	gi|357137519	29514	4.61	14	5	+1.69	+1.33	+1.04
	Protein biosynthesis								
U13	Putative elongation factor 1 beta [*H. vulgare*]	gi|7711024	24716	4.52	34	7	+1.86	+1.29	+1.18
	Protein destination/storage								
U14	HSP organizing protein/stress-inducible protein [*Dactylis glomerata*]	gi|281399029	64793	6.11	10	5	+1.07	−1.81	−1.10
U15	Heat shock protein 93-V [*A. Lyrata* subsp. *lyrata*]	gi|297795893	103611	6.36	35	30	+1.06	−1.94	+1.12
U16	Os01g0839700 [*O. sativa japonica*]	gi|115440951	19059	5.34	21	3	−1.03	−1.93	−1.12
U17	Heat shock protein [*Spinacia oleracea*]	gi|425194	71231	5.15	22	12	+2.08	+1.01	+1.10
U18	Cytosolic heat shock protein 90 [*H. vulgare*]	gi|32765549	80654	4.96	27	15	+1.21	−1.99	+1.63
U19	Cytosolic heat shock protein 90 [*H. vulgare*]	gi|32765549	80654	4.96	17	9	+1.01	−1.81	+1.03
	Signal transduction								
U20	RNA-binding Ras-GAP SH3 binding protein [*T. aestivum*]	gi|290579509	45493	4.95	18	7	+1.60	−1.51	−1.11
	Unknown								
U21	Hypothetical protein SORBIDRAFT_10g022570 [*Sorghum bicolor*]	gi|242096224	47231	6.08	13	7	+1.80	−1.57	−1.03
U22	Hypothetical protein LOC100383520 [*Z. mays*]	gi|293336836	77345	6.03	16	11	+1.59	+1.29	+1.26
U23	Predicted protein [*H. vulgare* subsp. *vulgare*]	gi|326513540	58531	6.05	22	40	+2.19	+1.23	−1.72
U24	Predicted protein [*H. vulgare* subsp. *vulgare*]	gi|326513418	35914	5.06	22	9	+1.08	−2.60	−1.36

AASC, Amino acid sequence coverage; MP, Matched peptides.
Protein spot ID refers to numbers in Fig. 7. Accession number of top database match from the NCBInr database. 'Al *vs* control' referred to fold variation of Al exposed *vs* unexposed plants. Fold increase and decrease were calculated as Al/control, and –control/Al for up and down -regulated proteins respectively. All ratios shown are statistically significant ($p<0.05$).

frequency of 50 Hz, repetition rate of 200 HZ, UV wavelength of 355 nm, and accelerated voltage of 20,000 V. Peptide mass fingerprint data were matched to the NCBInr database using Profound program under 50 ppm mass tolerance.

Peptide and Protein Identification by Database Search

Data were processed using the Data Explorer software (Applied Biosystems) and proteins were unambiguously identified by searching against a comprehensive non-redundant sequence database (NCBInr) using the MASCOT software search engine (http://www.matrixscience.com/cgi/search form.pl?FORMVER = 2&SEARCH = MIS). Folds of increase and decrease in Al exposed *vs* unexposed roots were calculated as treated/control and -control/treated for up- and down-regulated proteins, respectively. For single-peptide identified proteins, up- and down-regulation were assigned when the regulation factors were above 1.5 ($p<0.05$).

Table 2. Proteins whose expression were significantly induced (+) in XZ16 roots but down-regulated (−)/unchanged in XZ61, or unchanged in XZ16 but down-regulated in XZ61 under 200 μM Al stress.

Spot ID	Protein name	Accession number	MW, Da	pI	AASC %	MP	Folds (Al *vs* controls)		
							XZ16	XZ61	Dayton
	Metabolism								
U25 (U6*)	Putative asparate aminotransferase [*H. vulgare* subsp. *vulgare*]	gi|89511843	45377	5.75	14	4	+1.71	−1.34	+1.20
U26 (U7)	Predicted: pirin-like protein [*Brachypodium distachyon*]	gi|357114735	41337	9.37	17	8	−1.25	−2.05	+1.10
	Energy								
U27 (U11)	Fructose-bisphosphate aldolase [*H. vulgare*]	gi|226316443	39071	6.08	29	7	+2.06	−1.22	+1.11
U28	Os09g0535000 [*O. sativa* japonica]	gi|115480367	32715	6.96	34	9	$+10^6$	+1.29	0.00
U29	Bp2A protein [*T. turgidum* subsp. *dicoccoides*]	gi|133872436	25759	5.86	27	5	+1.71	−1.72	0.00
U30	Phosphoglycerate mutase [*T. aestivum*]	gi|32400802	29615	5.43	52	14	+1.94	+1.38	+1.74
U31	Enolase (2-phosphoglycerate dehydratase) [*O. sativa* japonica]	gi|780372	48299	5.42	18	7	+1.25	−1.72	+2.53
U32	atp1 [*Secale strictum*]	gi|166165274	53979	6.01	31	14	+1.15	−4.10	+1.29
	Cell growth/division								
U33 (U12)	Predicted: proliferating cell nuclear antigen-like [*B. distachyon*]	gi|357137519	29514	4.61	14	5	+2.16	+1.08	−1.15
	Protein destination/storage								
U34	Heat shock 70 kda protein [*Z. mays*]	gi|226500540	72989	5.62	10	4	+1.21	−1.97	+1.07
U35 (U17)	Heat shock protein [*Spinacia oleracea*]	gi|425194	71231	5.15	22	12	+1.92	−1.25	−1.01
U36 (U18)	Cytosolic heat shock protein 90 [*H. vulgare*]	gi|32765549	80654	4.96	27	15	+1.12	−2.91	+1.15
U37 (U19)	Cytosolic heat shock protein 90 [*H. vulgare*]	gi|32765549	80654	4.96	17	9	+1.05	−1.76	−1.14
	Transporters								
U38	Vacuolar proton-atpase D subunit [*T. aestivum*]	gi|108925894	41321	4.89	46	10	+1.05	−1.66	+1.12
	Signal transduction								
U39	14-3-3D protein [*H. vulgare* subsp. *vulgare*]	gi|83271056	28742	4.80	44	9	+1.31	−1.58	+1.10
	Disease/defense								
U40	Phenylalanine ammonia-lyase [*Phyllostachys edulis*]	gi|224998176	76278	6.15	19	12	+1.21	−3.50	+1.96
	Unknown								
U41	Predicted protein [*H. vulgare* subsp. *vulgare*]	gi|326496891	38099	4.83	25	7	+2.08	−1.79	+1.22
U42	Hypothetical protein Sb01g000380 [*S. bicolor*]	gi|242032147	60974	5.20	21	9	+1.09	−2.03	+1.36
U43 (U23)	Predicted protein [*H. vulgare* subsp. *vulgare*]	gi|326513540	58531	6.05	22	40	+1.84	−1.11	−3.17
U44 (U24)	Predicted protein [*H. vulgare* subsp. *vulgare*]	gi|326513418	35914	5.06	22	9	−1.32	−3.34	−1.27

*, Spot U25 is the same protein spots as U6 identified in 50 μM Al stress.
Protein spot ID refers to numbers in Fig. 7. Accession number of top database match from the NCBInr database. 'Al *vs* control' referred to fold variation of Al exposed *vs* unexposed plants. Fold increase and decrease were calculated as Al/control, and −control/Al for up and down -regulated proteins respectively. All ratios shown are statistically significant ($p<0.05$).

qRT-PCR Analysis

Total RNA was isolated from roots with the TRIzol reagent following manufacturers' recommendation (Invitrogen, Karlsruhe, Germany). cDNA samples were assayed by quantitative real time PCR (qRT-PCR) in the iCycler iQTM Real-time PCR Detection System (Bio-Rad, Hercules, CA, USA) using the SYBR Green PCR Master Mix (Applied Biosystems). The PCR conditions consisted of denaturation at 95°C for 3 min, followed by 40 cycles of denaturation at 95°C for 30 s, annealing at 58°C for 45 s and extension at 72°C for 45 s. Gene-specific primers (Table S1) were designed using the Primer Express software (Applied Biosystems). Barley *actin* gene was used as control (AY145451) fw-5′-GACTCTGGTGATGGTGTCAGC-3′, rv-5′-GGCTGGAAGAGGACCTCA-3′.

Statistical Analysis

Statistical analysis were performed using the Data Processing System (DPS) Software Package [16]. Statistical significance of the data was evaluated by two-way ANOVA using Duncan's multiple range test (SSR).

Results

Tibetan Wild Barley XZ16 Is Highly Tolerant to Al Toxicity

Time of appearance and severity of Al toxicity symptoms differed greatly among the three genotypes (Figure S1). XZ16 was less affected by 24 h exposure to 50 or 200 μM Al (pH 4.3), whereas XZ61 was obviously affected, as reflected by severe root growth inhibition. No significant difference between control and 50 or 200 μM Al stressed plants was found in root DW (dry weight) of XZ16, and the whole plant DW of XZ16 and Dayton. However, root and the whole plant DW of XZ61 decreased by

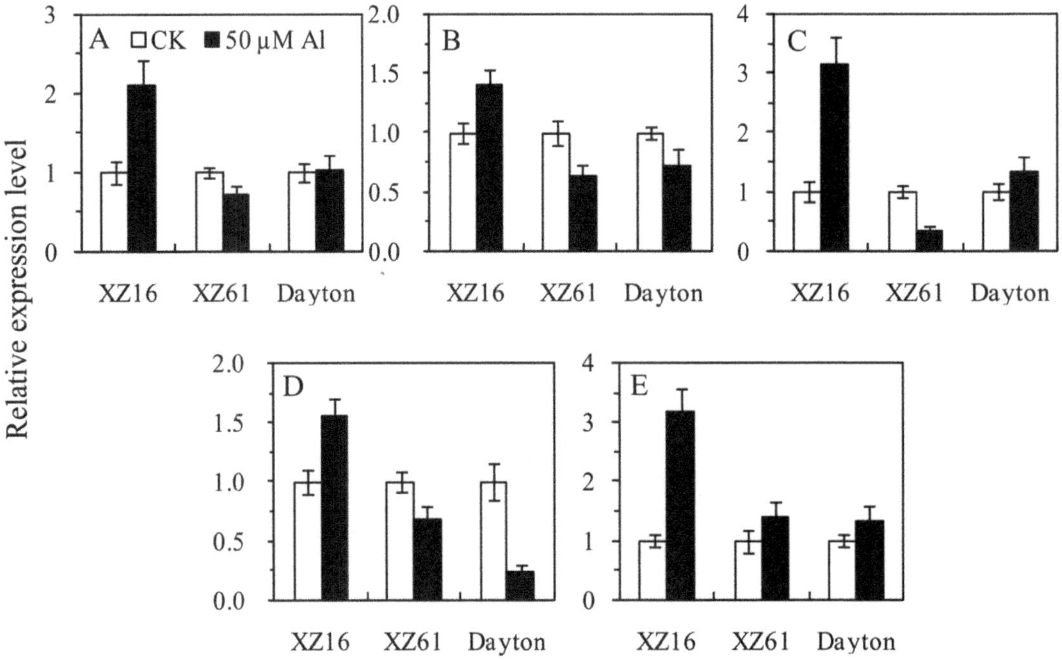

Figure 5. Three root proteins associated with low pH isolated from the root of XZ16. Representative 2-DE maps of root proteins isolated from XZ16 exposed to pH 4.3 for 24 h (A). 'Spot view' of the abundance of differentially expressed proteins that were significantly higher expressed (+) in XZ16 compared with Dayton roots under control condition at pH 4.3 (XZ16 *vs* Dayton) but suppressed (−) at pH 6.0 (B).

10.7% and 7.8% (50 μM Al *vs* control) and by 19.1% and 13.5% (200 μM Al *vs* control), respectively.

Al localization in barley roots exposed to different Al levels for 24 h was monitored by morin fluorescence using confocal laser scanning microscopy (Figure 1A), and fluorescence intensity of image analysis was calculated using Image J software (Figure 2B). Root Al fluorescence showed that root Al concentration increased with increasing external Al levels. XZ16 exhibited similar fluorescent signal in root tips with Dayton, being significantly ($p<0.05$) less than that of XZ61 in both Al levels.

As to root Al accumulation, there was no significant difference between XZ16 and Dayton, but both exhibited significantly ($p<0.05$) less accumulation than that of XZ61 in the two Al levels (Figure S2A). Shoot Al accumulation, however, was significantly lower ($p<0.05$) in XZ16 than in Dayton and XZ61 in response to Al treatments (Figure S2B). The transferring rates from roots to shoots among them were not differed significantly except that Dayton showed 18.2% and 13.6% higher than XZ16 and XZ61 when exposed to 50 and 200 μM Al stress, respectively (Figure S2C). The above results indicate that Tibetan wild barley XZ16 is highly tolerant to Al stress.

Differential Al-induced Protein Expression in Roots of the Three Genotypes

Approximately 2268 spots, ranging from 2038 to 2642, were resolved in each of two reproducible SDS-polyacrylamide gels (Figure 2). Protein spots altered by Al stress in Al-tolerant genotypes XZ16 and Dayton and Al-sensitive XZ61 or differentially accumulated among the three genotypes under Al stress were further analyzed and shown in Figure 3. For the protein spots altered by 50 and 200 μM Al stress, 398, 334, 495 spots (50 μM Al *vs* control) and 251, 447, 350 spots (200 μM Al *vs* control) were up-regulated in XZ16, XZ61, Dayton, respectively; while 347, 306, 442, and 398, 334, 495 spots were down-regulated. In XZ16 and Dayton, the abundance of 101 and 31 spots increased and that of 49 and 39 decreased in response to 50 and 200 μM Al, respectively,.

Al-responsive protein spots (44 spots, U1–U24, U25–U44; Figure 4) were analyzed by MALDI-TOF/TOF MS, and identified by MS/MS data with significant probability ($p<0.05$). Twenty-four and 20 spots were up-regulated (fold change >1.50) in XZ16 but unaltered/down-regulated in XZ61, or unaltered in XZ16 but down-regulated in XZ61 under 50 (U1–U24, Figure 2A, Table 1) and 200 μM Al (U25–U44, Figure 2B, Table 2), respectively. The resulting spectra of the 44 protein spots (nine spots overlapped at 50 and 200 μM Al, as shown in brackets in Table 2, were identified using MASCOT software search engine against *H. vulgare* and homologous proteins of other green plants in the NCBI non-redundant (nr) protein database and barley ESTs databases (Tables 1 and 2). These proteins were classified into nine groups based on their biochemical functions [17]. The majority of the protein profile was energy (spots U8–U11, U27–U32) and metabolism (U1–U7), followed by protein destination/storage (U14–U19 and U34–U37) and unknown (U21–U24, U41 and U44). The other six minor groups included signal transduction (U20 and U39), cell growth/division related proteins (U12), protein biosynthesis (U13), transporter (U38) and disease/defense (U40). Further comparison of the 44 identified spots with that of Dayton revealed that 16 (Spots U1, U4, U5, U6 (U25), U9, U11 (U27), U12 (U33), U13, U17 (U365), U20, U21, U22, U23 (U43), U28, U29 and U41; Tables 1 and 2) proteins up-regulated in XZ16 were surprisingly down-regulated or unaltered in both Dayton and XZ61. There were a protein uniquely expressed (U9, ATP synthase beta subunit) in XZ16 and three proteins not expressed in Dayton [U1, S-adenosylmethionine synthase 3 (SAMS); U28, a homologue of triosephosphate isomerase (TPI); U29, Bp2A protein]. On the contrary, three proteins (U18, U31,

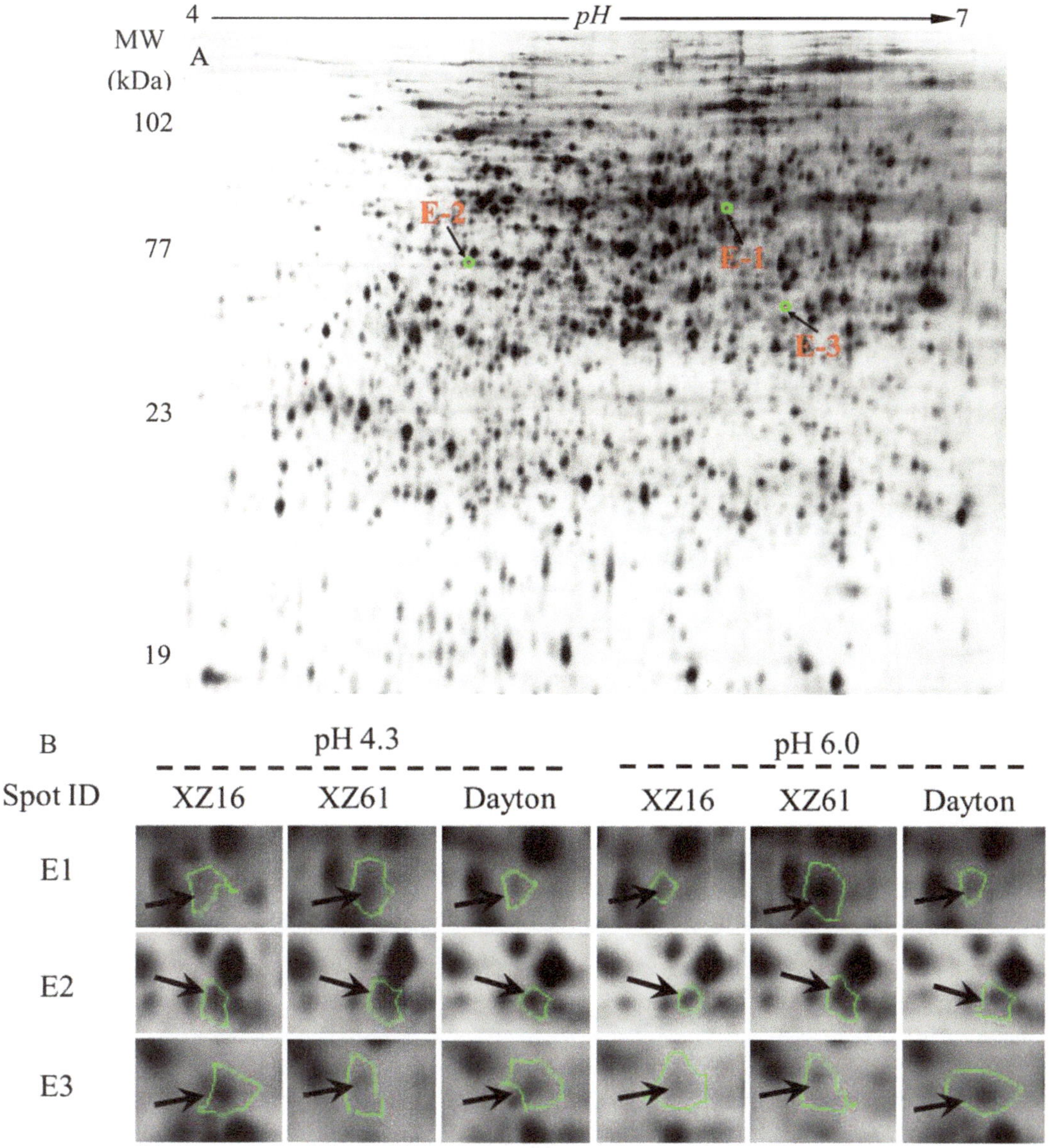

Figure 6. Effect of 50 μM Al on the transcript levels of gene expression encoding selected proteins of three barley genotypes exposed to 50 μM Al for 24 h. (A) S-adenosylmethionine synthase 3 (SAMS3, spot U1); (B) methionine synthase (MeSe, spot U2); (C) glutamine synthetase (GS, spot U4); (D) γ-glutamylcysteine synthetase (γ-GCS, spot U5) and (E) ATP synthase beta subunit proteins in roots (spot U9) of three barley genotypes exposed to 50 μM Al for 24 h. Error bars represent SD values (n = 3).

and U40) were slightly up-regulated in Dayton under Al stress, but unaltered in XZ16.

Six chaperone-related proteins were identified: heat shock protein (HSP, U17), HSP 93-V (U15), cytosolic HSP 90 (U18 and U19), HSP organizing protein (U14) and HSP 70 kDa (U34). These spots showed normal expression in XZ16 and Dayton, but were down-regulated in XZ61 under Al stress (Tables 1 and 2), indicating that these proteins are involved in Al detoxification in both XZ16 and Dayton.

In addition, expression of three proteins, classified as the energy category (Figure 5 and Table S2) was significantly higher in XZ16 *vs* Dayton under control condition at pH 4.3, and simultaneously suppressed at pH 6.0. They are predicted to be 6-phosphogluconate dehydrogenase decarboxylating-like isoform 1 (6-PGDH) (E1), plastid glutamine synthetase isoform GS2b (E2) and iron deficiency specific clone no. 3 (IDS3) (E3). These proteins are potentially responsible for the low pH tolerance in XZ16 compared with Dayton.

XZ16 Shows Higher Expression of Genes Corresponding to Al Up-regulated Proteins

To determine whether the changes in protein abundance detected by 2-DE were correlated with changes at the transcriptome level, quantitative RT-PCR was performed using RNA isolated from the roots of a separate set of plants treated with 0 or 50 μM Al for 24 h (Figure 6 and Table S2). Transcript levels of five Al-regulated genes including *SAMS3*, *MeSe*, *GS*, *γ-GCS* and *ATP synthase beta subunit* (U1, U2, U4, U5 and U9) were chosen and successfully detected. Among them, *SAM3*, *MeSe*, *GS* and *γ-GCS* were up-regulated in XZ16 and down-regulated in XZ61,

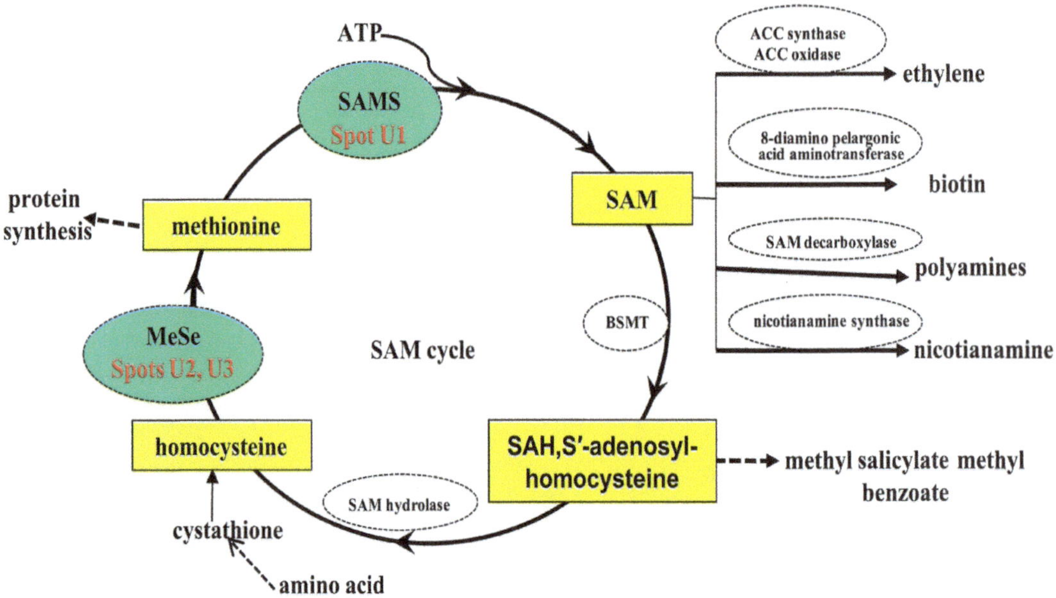

Figure 7. Two of indentified SAM cycle-related enzyme proteins SAMS and MeSe (green oval-shaped box) associated with Al-tolerance in Tibetan wild barley XZ16. SAM regeneration and utilization reactions of the SAM cycle are depicted. Enzymes of SAM cycle are: MeSe, methionine synthase; SAMS, SAM synthetase; BSMT, benzoic acid/salicylic acid: S-adenosyl-methionine carboxyl methyltransferase; SAM hydrolase. SAH, S'-adenosyl-homocysteine; ACC, 1-aminocyclopropane-1-carboxylate. ACCS, 1-aminocyclopropane-1-carboxylic acid synthase.

following the expression trend detected by 2-DE (Table 1). Fifty micromolar of Al significantly up-regulated *ATP synthase beta subunit* by 3.2 folds in XZ16 compared to the control, but no change was detected in XZ61 and Dayton (Figure 6).

For genome mapping, BLASTn and CAP3 were used to blast the *Brachypodium distachyon* protein databases. The top hits were then mapped to barley chromosome based on barley genome zipper and sequence [18], [19]. Finally, 32 (including 30 of the 35 Al-tolerance related proteins, and 2 of the 3 low pH responsive spots) out of the above 38 proteins were linked to their corresponding candidate genes in barley genome (Table S3). The mapping of these candidate genes provides a short cut for the identification and transformation of Al responsible genes in future.

Discussion

The current study showed that Al-tolerant Tibetan wild barley XZ16 is characterized by less Al accumulation both in roots and shoots (Figure 1 and Figure S2). Our current data is the first study to identify Al-responsive proteins in Tibetan wild barley (XZ16) using a proteomic approach. Barley is one of the most Al sensitive cereal species and responds to Al toxic ion immediately through releasing organic acid in roots [3,35]. It is important to successfully induce Al-tolerant-specific-proteins before visible morphological-stress-symptoms for plants to achieve tolerance to Al toxicity. In addition, the micro- and minor- elements, containing in the nutrient solution for the long term Al exposure of 15 days, trends to complicatedly react with Al ion compared to simplified adding Ca^{2+} solution. Furthermore, after long term Al treatment, barley roots (especially in sensitive genotype) is severely damaged, out of vigorousness and partly death, thus not suitable for protein extraction. Therefore, in this study, Al-tolerant-specific-proteins in wild genotype XZ16 were verified under 50 μM or 200 μM Al condition for 24 h. The treatment of 200 μM Al on the plants could provide us the special mechanism (proteins) of Al tolerance in XZ16 in comparison to Dayton or XZ61 under high Al concentration condition. Because Tibetan wild barley XZ16 and XZ61 are the two contrasting genotypes with different Al tolerances, the Al-regulated proteins identified using comparative proteomics will provide a good foundation to elucidate the mechanisms involved in Al- tolerance/resistance in Tibetan wild barley. Here we identified 35 proteins associated with Al-tolerance in wild barley XZ16 (Figures 2, 5 and Tables 1, 2). There were 16 proteins, up-regulated in roots of XZ16 but down-regulated or unaltered in both Dayton and XZ61, indicating their specificity and importance for Al tolerance in XZ16. Among them, four proteins (SAMS3, U1; ATP synthase beta subunit, U9; Os09g0535000, U28; Bp2A protein, U29), were markedly induced by Al stress in XZ16 but not in Dayton, while repressed in XZ61. Obviously, XZ16 has different stress response and defense mechanisms against Al stress as Dayton did. Further examination of these proteins may elucidate the mechanism of Al tolerance in XZ16 and provide new genetic materials for developing Al-tolerant crops. The selected stress-responsive proteins are discussed below according to their function.

Metabolism Category

In the root proteomic analyses, seven of the identified proteins are involved in metabolism: SAMS3 (U1), methionine synthase (MeSe, U2 and U3), glutamine synthetase (GS, U4), γ-glutamyl-cysteine synthetase (γ-GCS, U5), putative aspartate aminotransferase (AST, U6) and predicted pirin-like protein (U7). SAMS, as an essential enzyme in cellular metabolism, has been long regarded as a 'housekeeping' function and it catalyzes the nucleophilic substitution reaction between methionine and ATP into SAM (S'-adenosyl-L-methionine) (Figure 7). SAM serves as an aminopropyl and methyl donor for ethylene and polyamine (PAs) [20], [21], [22]. It is well documented that ethylene and PAs are involved in response to biotic and abiotic stresses in plants [23], [24]. Thus SAMS plays an important role in Spd and Spm biosynthesis and stress response in plants [25], [26]. Qi *et al.* [27]

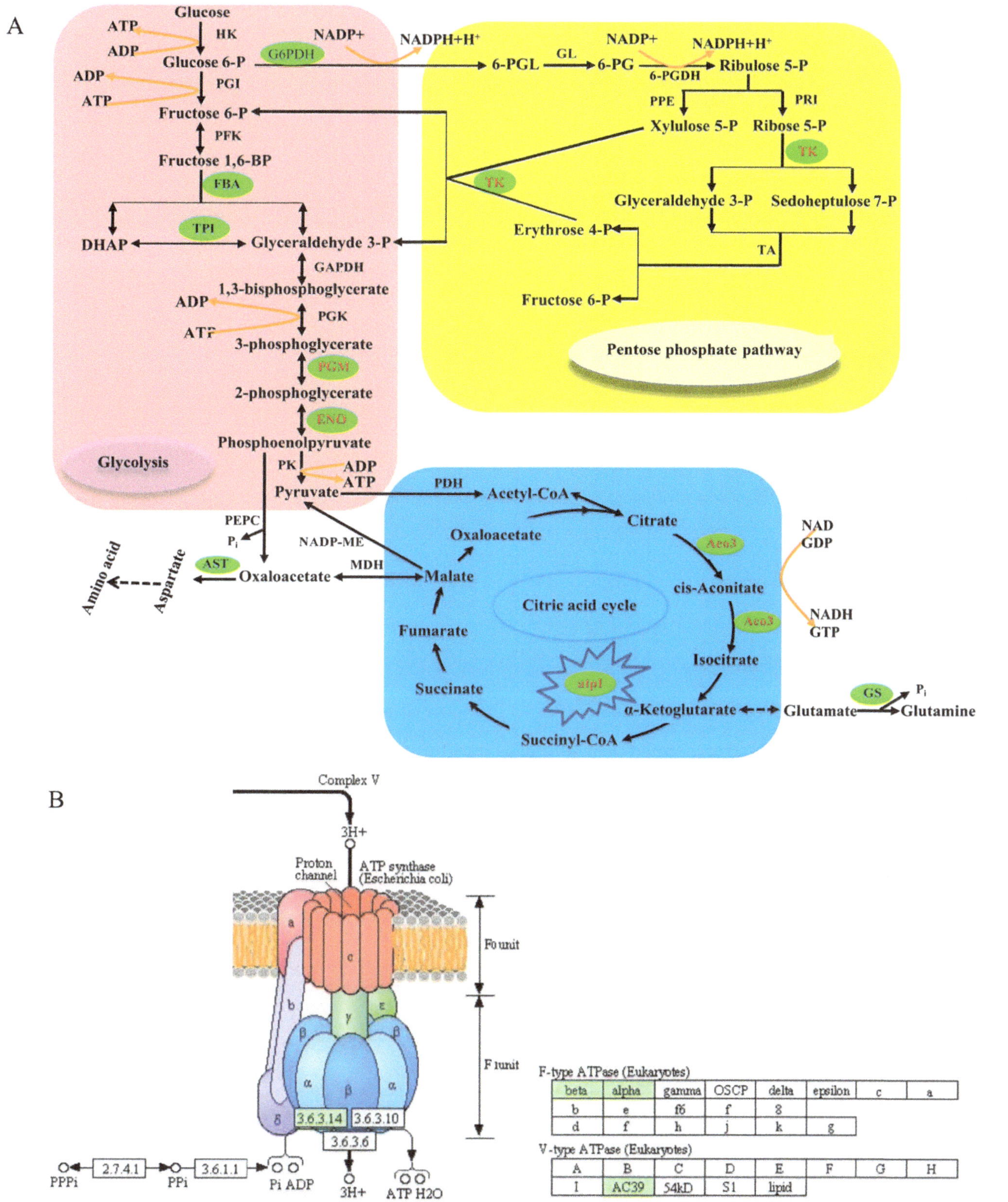

Figure 8. Some of indentified proteins (green oval-shaped box, cycle-related enzymes) associated with Al-tolerance involving in the metabolic pathways of glycolysis (purple box), pentose phosphate pathway (yellow box), and citric acid cycle (blue box) in Tibetan wild barley XZ16 (A). Searching via the KEGG (Kyoto Encyclopedia of Genes and Genomes), ATP synthase beta subunit (spot U9), atp1 (F-tyep ATP synthase α subunit, spot U32) and V-ATPase (V-type ATPase AC39, spot 38) were represented by "3.6.3.14" in oxidative phosphorylation (B). Glucose 6-P either proceeds through glycolysis (participated by FBA, TPI, PGM and ENO) or the PPP (participated by TK) to produce pyruvate, which is converted to acetyl-CoA, entering the TCA cycle and producing citrate and malate (participated by Aco3). atp1 belongs to mitochondrial membrane ATP synthase generates ATP from ADP in the presence of a proton gradient across the membrane.

Glutamine and other amino acids also feed by the TCA cycle; for example, α-ketoglutarate can converte glutamate, which is converted to glutamine by GS; malate can be converted oxaloacetate, which is coverted to aspartate through AST. Abbreviation: HK, hexokinase; PGI, phosphateisomerase; PFK, phosphofructokinase; BP, bisphoshate; FBA, fructose-bisphosphate aldolase; TPI, triose phosphate isomerase; GAPDH, glyceraldehyde-3-phosphate dehydrogenase; PGK, phosphoglycerate kinase; PGM, phosphoglycerate mutase; ENO, enolase; PK, pyruvate kinase; G6PDH, glyceraladehyde 6-phosphate dehydrogenase; 6-PDL, 6-phosphogluconolactone; GL, gluconolactonase; 6-PG, 6-phosphogluconate; 6-PGDH, 6-phosphogluconate dehydrogenase; PPE, pentose phosphate epimerase; PRI, phosphoriboisomerase; TK, transketolase; TA, transaldolase; PDH, pyruvate dehydrogenase; CoA, coenzyme A; Aco3, aconitate hydratase 3; GS, glutamine synthetase; PEPC, phosphoenolpyruvate carboxylase; MDH, malate dehydrogenase; NADP-ME, NADP-malic enzyme; AST, asparate aminotransferase.

found that over-expression of *Suadea salsa SAMS2* in transgenic tobacco plants leads to an increase in polyamine (PAs) content and enhancement of salt tolerance. In the present study, SAMS3 was up-regulated in XZ16 but not expressed in Dayton and repressed in XZ61 under 50 μM Al stress, highlighting the function of this protein in Al tolerance of XZ16. Although SAM is the precursor of ethylene, the key enzyme of ethylene synthesis is ACCS [28] (Figure 7). Since ACCS was not identified in this study, further research is required to determine whether up-regulation of SAMS is involved in ethylene-mediated inhibition of root growth and/or the alteration of cell wall and polymer structures in roots under Al stress. On the other hand, *SAMS3* was expressed normally in Dayton under Al stress as determined by qRT-PCR (Figure 6). Therefore, Al-induced expression of *SAMS3* is likely to be regulated at the post-transcriptional or translational level in Dayton.

In addition, we also observed that MeSe was depressed in XZ61 (U2 and U3), but not altered in XZ16 under Al stress. Higher MeSe levels would increase the amount of methionine for SAM synthesis (Figure 7). Enhanced accumulation of MeSe contributes to increase PAs biosynthesis, which is critical to plant survival in many environmental stresses [29]. The decreased accumulation of MeSe in XZ61 indicates that Al stress triggered damages in barley roots may be associated with the modification of amino acid metabolism and synthesis of other amino acids derived metabolites. The next important questions is how SAMS3 and MeSe elaborate to cope with Al toxicity and which metabolism factors, e.g., SAM, ethylene, PAs (spermidine or spermine), are responsible for Al detoxification in conjunction with SAMS3 and MeSe. Therefore, it is necessary to investigate the accumulation of related metabolite and enzymatic activities of the SAMS family for a better understanding of their functions in the Al stress response.

Two glutathione (GSH) synthesis proteins γ-GCS and GS (U4 and U5), play a pivotal role in various metabolic processes involved in plant growth and development and stress responses including Al toxicity [30], [31]. The pirin-like protein (U7 and U26) plays important roles in a number of different biological processes, however, its physiological role in plants remains unclear [32]. Aspartate aminotransferase (AST, U6 and U25) is a key antioxidative enzyme for detoxifying reactive oxygen species (ROS) under abiotic stresses [33]. Similarly, up-regulation of AST expression was reported in Arabidopsis in response to Al stress [34]. Under Al stress, γ-GCS (U5), GS and AST were significantly increased in XZ16, whereas dramatically decreased in XZ61 and Dayton (Table 1), again indicating that there a unique mechanism in Al tolerance for the Tibetan wild barley genotype.

Energy Category

Similar to the proteomic results of Zhou *et al.* [10], ATP synthase beta subunit (U9) was only detected in XZ16 after 24 h of 50 μM Al stress (Table 1), indicating that the energy provided to ATPase for active Al efflux and detoxification in XZ16 was increased when exposed to Al stress. qRT-PCR of *ATP synthase beta subunit* confirmed a remarkable increase in XZ16 but not in X61 and Dayton under 50 μM Al (Figure 6). Therefore, we conclude that ATP synthase beta subunit significantly contributes to the ATPase-mediated active Al efflux and detoxification, and is regulated at both transcriptional and translational level in XZ16 in response to Al.

In addition to ATP synthase beta subunit, we also identified another type of ATPase, atp1 (U32), which was markedly down-regulated in XZ61 after Al stress, but remained unchanged in XZ16 and Dayton (Table 2). The efflux of Al-induced citrate and malate is usually mediated by anion channels and ATPase-driven active co-transporters [35], [36]. Interestingly, a significant increase in ATP synthase beta subunit and unchanged levels of atp1 in XZ16 were consistent with higher root citrate and malate efflux, whereas the lack of ATP synthase beta subunit expression and the Al-induced severe inhibition of atp1 in XZ61 were associated with lower citrate and malate secretion. These results firmly suggest that high level of ATP synthase beta subunit and atp1 in XZ16, unlike that in Dayton, may partly contribute to the active OA transport and secretion during protection from Al-toxicity. Thus, we may speculate that a novel protein synthesis of ATP synthase beta subunit may be involved only in Al inducing organic acid secretion in Tibetan wild barley XZ16. Obviously, it deserves a more detailed investigation in the future.

A BLAST search revealed that Os09g0535000 (U28) is a homologue of triosephosphate isomerase (TPI). TPI was reported to be induced in rice [37] and maize [38] under drought stress, indicating the importance of cellular homeostasis maintenance and emphasizing the role of this protein in energy production. Enolase (ENO, U31) is responsive to salt, low and high temperature and anaerobic stresses [8], [39]. Spot U10, identified as aconitate hydratase 3 (Aco3), plays a role in regulating resistance to oxidative stress and cell death in *Arabidopsis* and *Nicotiana benthamiana* [40]. Spot U29, the Bp2A protein, was first identified in the wheat genome [41]. The present study is the first to examine Bp2A (U29) expression under Al stress. However, the function of this protein and its direct involvement in Al tolerance are poorly understood. Therefore, the mechanisms underlying the differential expression of this protein in different barley genotypes should be further explored.

In the glycolytic pathway, the levels of other Al stress responsive proteins fructose-bisphosphate aldolase (FBA, U11 and U27) and phosphoglycerate mutase (PGM, U30) were increased in XZ16 under Al treatments, consistent with the findings of Fukuda *et al.* [42] in rice under Al stress and Yan *et al.* [43] in the rice response to chilling. Down-regulation of Os06g0133800 (U8), a homologue to transketolase (TK), in XZ61 may reversibly inhibit ribose 5-phosphate, which is a substrate for nucleic acid synthesis associated with the synthesis of RNA under Al stress [42].

Taken together, up-regulation of FBA, PGM, TPI, Bp2A, and ATP synthase beta subunit (U11, U27, U30, U28, U29 and U9) and maintaining the normal expression levels of the other four energy proteins in XZ16 might help to produce more energy needed in the defense processes under Al stress conditions. The higher abundance of all these enzymes catalyzing various reactions in glycolysis, pentose phosphate pathway, and citric acid cycle in roots of Al resistant genotypes, suggest less Al-triggered disruption of energy metabolism in XZ16 (Figure 8). Up-regulation of ATP synthase beta subunit, Bp2A, TPI and FBA is of particular importance to XZ16 relative to Dayton.

Cell Growth/Division and Protein Biosyntheses Category

Proliferating cell nuclear antigen (PCNA) is an essential component in eukaryotic DNA synthesis [44]. PCNA interacts with many proteins and participates in a variety of metabolic processes, such as cell cycle control, nucleotide excision repair and post-replication mismatch repair [45]. Translation elongation factor 1β (EF1B, U13) is a highly conserved protein that catalyzes the exchange of bound GDP for GTP on EF-1α, a required step to ensure continued protein synthesis. In our study, PCNA (U12 and U33) and EF1B (U13) were up-regulated in in roots of XZ16, but were unchanged in XZ61 and Dayton. It suggests that XZ16 might possess a high actitity of DNA replication machinery in response to Al stress, and EF1B may play an important role in maintaining the root elongation rate of XZ16 under Al stress.

In addition, the other specific protein up-regulated in XZ16 is the RNA-binding Ras-GAP SH3 binding protein (G3BP, signal transduction category, U20). Zhou *et al.* [46] reported that the expression of the *G3BP* gene was associated with fertility conversion in male-sterile wheat. However, the exact function of G3BP in RasGAP-dependent signaling remains to be defined. These results provide a starting point for further investigation into the functions of these proteins using genetic and other approaches.

In conclusion, the response and defense mechanisms of Al stress in XZ16 appear different from those of Dayton, as reflected by the different expressions of these specific proteins associated with Al tolerance under Al stress between XZ16 and Dayton or XZ61. There are four proteins (i.e. SAMS3, ATP synthase beta subunit, TPI, Bp2A protein), which are exclusively expressed in XZ16 not in Dayton and XZ61 under Al stress, indicating their crucial role in development of Al stress tolerance in XZ16, and novelty of genetic resource for Al-tolerance. In addition, as the functions of some differentially expressed proteins and their direct involvement in stress tolerance are poorly understood, further studies are warranted to elucidate the underlying molecular and metabolic pathways for better understanding the mechanisms involved in Al-tolerance of wild barley and providing new genetic resources in Al-tolerant crop breeding.

Supporting Information

Figure S1 Effect of Al stress on root (A) and individual plant dry weight (B) of barely seedlings of XZ16, XZ61 and Dayton.

Figure S2 Al accumulation in roots (A) and shoots (B) and transferring rate from root to shoot (C) of barley seedlings of XZ16 (□), XZ61 (■) and Dayton (░).

Table S1 Oligonucleotides used as primers for real-time RT-PCR.

Table S2 Proteins whose expression were significantly higher expressed (+) in XZ16 compared with Dayton roots under control condition at pH 4.3 (XZ16 *vs* Dayton) but suppressed (−) at pH 6.0.

Table S3 Thirty-two proteins (including 2 low pH responsive proteins E1 and E2) corresponding candidates were mapped to chromosome based on barley genome zipper information.

Author Contributions

Conceived and designed the experiments: FBW GPZ HXD. Performed the experiments: HXD XHC FBC MZ IMA. Analyzed the data: HXD ZHC. Wrote the paper: HXD FBW GPZ CDL ZHC.

References

1. Kochian LV, Pineros MA, Hoekenga OA (2005) The physiology, genetics and molecular biology of plant aluminum resistance and toxicity. Plant Soil 274: 175–195.
2. Foy CD (1996) Tolerance of barley cultivars to an acid, aluminum-toxic subsoil related to mineral element concentrations in their shoots. J Plant Nutr 19: 1361–1380.
3. Zhao ZQ, Ma JF, Sato K, Takeda K (2003) Differential Al resistance and citrate secretion in barley (*Hordeum vulgare* L.). Planta 217: 794–800.
4. Kochian LV, Hoekenga OA, Pineros MA (2004) How do crop plants tolerate acid soils? - Mechanisms of aluminum tolerance and phosphorous efficiency. Annu Rev Plant Biol 55: 459–493.
5. Pickering R, Johnston PA (2005) Recent progress in barley improvement using wild species of *Hordeum*. Cytogenet Genome Res 109: 344–349.
6. Wang AH, Yu ZY, Ding Y (2009) Genetic diversity analysis of wild close relatives of barley from Tibet and the Middle East by ISSR and SSR markers. Cr Biol 332: 393–403.
7. Dai HX, Shan WN, Zhao J, Zhang GP, Li CD, et al. (2011) Difference in response to aluminum stress among Tibetan wild barley genotypes. J Plant Nutr Soil Sc 174: 952–960.
8. Xu CP, Sibicky T, Huang BR (2010) Protein profile analysis of salt-responsive proteins in leaves and roots in two cultivars of creeping bentgrass differing in salinity tolerance. Plant Cell Rep 29: 595–615.
9. Zhen Y, Qi JL, Wang SS, Su J, Xu GH, et al. (2007) Comparative proteome analysis of differentially expressed proteins induced by Al toxicity in soybean. Physiol Plantarum 131: 542–554.
10. Zhou S, Sauve R, Thannhauser TW (2009) Proteome changes induced by aluminium stress in tomato roots. J Exp Bot 60: 1849–1857.
11. Liu TW, Fu B, Niu L, Chen J, Wang WH, et al. (2011) Comparative proteomic analysis of proteins in response to simulated acid rain in Arabidopsis. J Proteome Res 10: 2579–2589.
12. Wu FB, Zhang GP, Dominy P, Wu HX, Bachir DML (2007) Differences in yield components and kernel Cd accumulation in response to Cd toxicity in four barley genotypes. Chemosphere 70: 83–92.
13. Zheng SJ, Yang JL, He YF, Yu XH, Zhang L, et al. (2005) Immobilization of aluminum with phosphorus in roots is associated with high aluminum resistance in buckwheat. Plant Physiol 138: 297–303.
14. Carpentier SC, Witters E, Laukens K, Deckers P, Swennen R, et al. (2005) Preparation of protein extracts from recalcitrant plant tissues: An evaluation of different methods for two-dimensional gel electrophoresis analysis. Proteomics 5: 2497–2507.
15. Bah AM, Sun HY, Chen F, Zhou J, Dai HX, et al. (2010) Comparative proteomic analysis of *Typha angustifolia* leaf under chromium, cadmium and lead stress. J Hazard Mater 184: 191–203.
16. Tang QY, Feng MG (1997) *Practical statistics and its DPS statistical software package.* Beijing, China: Agriculture Press.
17. Bevan M, Bancroft I, Bent E, Love K, Goodman H, et al. (1998) Analysis of 1.9 Mb of contiguous sequence from chromosome 4 of *Arabidopsis thaliana.* Nature 391: 485–488.
18. Mayer KF, Martis M, Hedley PE, Simkova H, Liu H, et al. (2011) Unlocking the barley genome by chromosomal and comparative genomics. Plant Cell 23: 1249–1263.
19. Mayer KF, Waugh R, Langridge P, Close TJ, Wise RP, et al. (2012) A physical, genetic and functional sequence assembly of the barley genome. Nature.
20. Tiburcio AF, Kaur-Sawhney R, Galston AW (1990) Polyamine metabolism. In: Miflin BJ, Lea PJ, eds. *The biochemistry of plants: a comprehensive treatise, intermediary nitrogen metabolism.* Orlando, USA: Orlando Academic Press, 283–325.
21. Galston A, Kaur-Sawhney R (1995) Polyamines as endogenous growth regulators. In: Davies PJ, eds. *Plant hormones: physiology, biochemistry and molecular biology.* Dordrecht, Netherlands: Kluwer Academic Publishers, 158–178.
22. McKeon TA, Fernandez-Maculet JC, Yang SF (1995) Biosynthesis and metabolism of ethylene. In: Davies PJ, eds. *Plant hormones: physiology, biochemistry and molecular biology.* Dordrecht, Netherlands: Kluwer Academic Publishers, 118–139.
23. Heby O, Persson L (1990) Molecular-genetics of polyamine synthesis in eukaryotic cells. Trends Biochem Sci 15: 153–158.

24. Espartero J, Pintortoro JA, Pardo JM (1994) Differential accumulation of S-Adenosylmethionine synthetase transcripts in response to salt stress. Plant Mol Biol 25: 217–227.
25. Ma XL, Wang ZL, Qi YC, Zhao YX, Zhang H (2003) Isolation of S-adenosylmethionine synthetase gene from *Suaeda salsa* and its differential expression under NaCl stress. Acta Bot Sin 45: 1359–1365.
26. Gupta P, Duplessis S, White H, Karnosky DF, Martin F, et al. (2005) Gene expression patterns of trembling aspen trees following long-term exposure to interacting elevated CO_2 and tropospheric O_3. New Phytol 167: 129–142.
27. Qi YC, Wang FF, Zhang H, Liu WQ (2010) Overexpression of suadea salsa S-adenosylmethionine synthetase gene promotes salt tolerance in transgenic tobacco. Acta Physiol Plant 32: 263–269.
28. Wang KLC, Li H, Ecker JR (2002) Ethylene biosynthesis and signaling networks. Plant Cell 14: S131–S151.
29. Flores HE, Protacio CM, Signs MW (1989) Primary and secondary metabolism of polyamines in plants. Recent advances in phytochemistry 23: 329–393.
30. Noctor G, Arisi ACM, Jouanin L, Kunert KJ, Rennenberg H, et al. (1998) Glutathione: biosynthesis, metabolism and relationship to stress tolerance explored in transformed plants. J Exp Bot 49: 623–647.
31. Ruiz J, Rivero R, Romero L (2006) Boron increases synthesis of glutathione in sunflower plants subjected to aluminum stress. Plant Soil 279: 25–30.
32. Kim MC, Chung WS, Yun DJ, Cho MJ (2009) Calcium and calmodulin-mediated regulation of gene expression in plants. Mol Plant 2: 13–21.
33. Valderrama R, Corpas FJ, Carreras A, Gomez-Rodriguez MV, Chaki M, et al. (2006) The dehydrogenase-mediated recycling of NADPH is a key antioxidant system against salt-induced oxidative stress in olive plants. Plant Cell Environ 29: 1449–1459.
34. Goodwin SB, Sutter TR (2009) Microarray analysis of Arabidopsis genome response to aluminum stress. Biol Plantarum 53: 85–99.
35. Ma JF, Ryan PR, Delhaize E (2001) Aluminium tolerance in plants and the complexing role of organic acids. Trends Plant Sci 6: 273–278.
36. Shen H, He LF, Sasaki T, Yamamoto Y, Zheng SJ, et al. (2005) Citrate secretion coupled with the modulation of soybean root tip under aluminum stress. Up-regulation of transcription, translation, and threonine-oriented phosphorylation of plasma membrane H^+-ATPase. Plant Physiol 138: 287–296.
37. Salekdeh GH, Siopongco J, Wade LJ, Ghareyazie B, Bennett J (2002) Proteomic analysis of rice leaves during drought stress and recovery. Proteomics 2: 1131–1145.
38. Riccardi F, Gazeau P, de Vienne D, Zivy M (1998) Protein changes in response to progressive water deficit in maize. Plant Physiol 117: 1253–1263.
39. Forsthoefel NR, Cushman MAF, Cushman JC (1995) Posttranscriptional and posttranslational control of enolase expression in the facultative crassulacean acid metabolism plant *Mesembryanthemum crystallinum* L. Plant Physiol 108: 1185–1195.
40. Moeder W, del Pozo O, Navarre DA, Martin GB, Klessig DF (2007) Aconitase plays a role in regulating resistance to oxidative stress and cell death in Arabidopsis and *Nicotiana benthamiana*. Plant Mol Biol 63: 273–287.
41. Haudry A, Cenci A, Ravel C, Bataillon T, Brunel D, et al. (2007) Grinding up wheat: a massive loss of nucleotide diversity since domestication. Mol Biol Evol 24: 1506–1517.
42. Fukuda T, Saito A, Wasaki J, Shinano T, Osaki M (2007) Metabolic alterations proposed by proteome in rice roots grown under low P and high Al concentration under low pH. Plant Sci 172: 1157–1165.
43. Yan SP, Zhang QY, Tang ZC, Su WA, Sun WN (2006) Comparative proteomic analysis provides new insights into chilling stress responses in rice. Molecular & Cellular Proteomics 5: 484–496.
44. Kosugi S, Ohashi Y (1997) PCF1 and PCF2 specifically bind to *cis* elements in the rice proliferating cell nuclear antigen gene. Plant Cell 9: 1607–1619.
45. Daimon K, Kawarabayasi Y, Kikuchi H, Sako Y, Ishino Y (2002) Three proliferating cell nuclear antigen-like proteins found in the hyperthermophilic archaeon *Aeropyrum pernix*: Interactions with the two DNA polymerases. J Bacteriol 184: 687–694.
46. Zhou LL, Song GQ, Li HY, Hu YG, He BR (2010) Cloning and expression analysis of *TaG3BP* gene related to fertility conversion in male sterile lines of wheat. Acta Agronomica Sinica 36: 1683–1690.

2

Genome Re-Sequencing of Semi-Wild Soybean Reveals a Complex *Soja* Population Structure and Deep Introgression

Jie Qiu[1], Yu Wang[1], Sanling Wu[1], Ying-Ying Wang[1], Chu-Yu Ye[1], Xuefei Bai[1], Zefeng Li[1], Chenghai Yan[1], Weidi Wang[1], Ziqiang Wang[1], Qingyao Shu[2], Jiahua Xie[3], Suk-Ha Lee[4], Longjiang Fan[1]*

1 Department of Agronomy & James D. Watson Institute of Genome Sciences, Zhejiang University, Hangzhou, China, **2** Institute of Nuclear Agricultural Science, Zhejiang University, Hangzhou, China, **3** Department of Pharmaceutical Sciences, North Carolina Central University, Durham, North Carolina, United States of America, **4** Department of Plant Science and Research Institute for Agriculture and Life Sciences, Seoul National University, Seoul, Korea

Abstract

Semi-wild soybean is a unique type of soybean that retains both wild and domesticated characteristics, which provides an important intermediate type for understanding the evolution of the subgenus *Soja* population in the *Glycine* genus. In this study, a semi-wild soybean line (Maliaodou) and a wild line (Lanxi 1) collected from the lower Yangtze regions were deeply sequenced while nine other semi-wild lines were sequenced to a 3-fold genome coverage. Sequence analysis revealed that (1) no independent phylogenetic branch covering all 10 semi-wild lines was observed in the *Soja* phylogenetic tree; (2) besides two distinct subpopulations of wild and cultivated soybean in the *Soja* population structure, all semi-wild lines were mixed with some wild lines into a subpopulation rather than an independent one or an intermediate transition type of soybean domestication; (3) high heterozygous rates (0.19–0.49) were observed in several semi-wild lines; and (4) over 100 putative selective regions were identified by selective sweep analysis, including those related to the development of seed size. Our results suggested a hybridization origin for the semi-wild soybean, which makes a complex *Soja* population structure.

Editor: Matthew E. Hudson, University of Illinois, United States of America

Funding: This work was supported by the National Basic Research Program of China (2011CB109306) (to L.F.), the Next Generation BioGreen 21 Program (PJ008117), Rural Development Administration, Republic of Korea (to S.-H.L.) and National Science Foundation of China (31301046) and China Postdoctoral Science Foundation (2013M531467) (to Y.W.). The funders had no role in study design, data collection and analysis, decision to publish, or preparation of the manuscript.

Competing Interests: The authors have declared that no competing interests exist.

* Email: fanlj@zju.edu.cn

These authors contributed equally to this work.

Introduction

The genus *Glycine* has two subgenera: *Glycine* and *Soja*. The latter one consists of the cultivated soybean (*Glycine max*) and its progenitor wild soybean (*G. soja*). *G. max* is an important cash crop for dietary protein and oil world-wide. It is generally believed that *G. max* was domesticated from its annual wild relative *G. soja* in China around 5,000–6,000 years ago [1,2]. After domestication, *G. max* displays distinct differences in several traits from *G. soja*. For example, *G. soja* has much smaller seeds (<3.0 g per 100 seeds) and a darker seed coat, whereas *G. max* (generally >9.0 g per 100 seeds) has a yellow seed coat [3]. Besides the above two species with distinct morphological characters in the subgenus *Soja*, an intermediate type can be found in accessions of landrace soybean germplasm collections or wild line collections in China [4–6]. For example, an intermediate type known as *G. gracilis* has been described as a semi-wild soybean, which usually has an intermediate seed weight (>3.0 g per 100 seeds) with a dark seed coat [6,7]. In China, this form is also a popular type of soybean cultivated in the northeast and the Yangtze regions, because it usually has a very high seed germination rate and a short growth period as well as a robust adaptability to various environments [4,8]. For example, Maliaodou (refers to "beans for horse fodder" in Chinese) is a popular growing semi-wild soybean in the Yangtze regions. In the modern Chinese Soybean Breeds [9], Maliaodou (No. 495) was listed as a landrace in the Jinhua, Zhejiang province. It has a dark seed coat with a mud film, which in general is a typical phenotype for wild species (Fig. 1). However, it possesses an erected plant architecture with a main stem and large leaves, which are similar to cultivated soybean lines. In brief, the intermediate type, i.e. semi-wild soybean, provides an important population for the subgenus *Soja* and may be beneficial in shedding light on its origin and the process of soybean domestication.

As a new species, *G. gracilis* was first proposed by Skvortzow [10]. However, its origin has been a subject of intense debate. Some early studies denied its novelty as a new species and proposed that it be incorporated into *G. max* [11]. Two original

Figure 1. Phenotypes of cultivated, semi-wild and wild soybeans used in the present study. Seed weight (per 100 seeds) and plant architecture are shown.

views have been projected for *G. gracilis*: an intermediate evolutionary type between *G. soja* and *G. max* [12] and a hybridization origin from *G. max* and *G. soja* [13]. The latter hypothesis was supported by several studies based on the analyses of the frequency/distribution of alleles [14] and molecular markers [15–17]. A recent observation of inter-species gene flow, which captured the natural occurrence of introgression between cultivated and wild accessions, provided further phenotypic evidence to support the hybridization hypothesis [6,18,19]. However, the origin of this semi-wild type of soybean has not clearly been established.

Gene flow between *G. max* and the wild relative *G. soja* has been observed [7,19,20]. Despite significant phenotypic differences between the two species, no reproductive isolation has been found yet. Introgressions between wild and cultivated soybeans were revealed by RFLP [15] and SSR [3,6,17,19,21–23]. It has been proven that the introgression between wild and cultivated soybean is bidirectional rather than unidirectional, i.e. from wild to cultivated soybean populations [21,24,25], and also cultivated soybean into the wild population [18,19].

The *Soja* population has been investigated using diverse molecular markers such as SSR [3,6,17,19,21–23]. Based on 111 fragments from 102 soybean genes, the genetic bottleneck associated with artificial selection in soybean was first illustrated [26]. Later, SNPs based on genome-resequencing [25,27,28] or SNP chips [23,29] were identified in the wild and cultivated soybeans, and this provided the first investigation of population structure and the estimation of loci under domestication and genetic improvement from the whole soybean genome. With the low cost and high-throughput sequencing platforms, genome resequencing of a representative set of semi-wild soybean accessions is now possible, therefore providing us with a unique opportunity to investigate a more comprehensive subgenus *Soja* population structure and the origin of the semi-wild soybean at the genomic level.

In this study, we sequenced 10 semi-wild soybeans with a wild line and then analyzed our sequence data together with other available genomic data from wild and cultivated soybean lines. Our results demonstrate the hybridization origin of semi-wild soybean and a high rate of genetic introgression among the members of the subgenus *Soja*, which resulted in a mixed population structure including wild, semi-wild and cultivated soybean.

Materials and Methods

Plant materials

The semi-wild soybean Maliaodou and wild soybean (*G. soja*) Lanxi 1 were collected from the Jinhua basin in Zhejiang province, China (N29°04′, E119°38′). Other semi-wild soybeans were collected from wide geographical locations in China and kindly provided by Institute of Crop Sciences, Chinese Academy of Agricultural Sciences (CAAS) (Table S1). No specific permissions were required for the location, and the study did not involve endangered or protected species.

Genomic sequencing

Green leaves from a single plant of each accession were used for DNA extraction following previously described protocol [30]. Two libraries with 500 bp and 2 kb insertion sizes for Lanxi 1 and one library with a 500 bp insertion size for each of other lines were generated for Illumina Hiseq2000 sequencing platform. Paired-end (PE) reads with 100 bp were determined and a clean data set was collected from raw reads, which were pre-processed to remove adaptors to filter out low quality reads (≥50% of its nucleotides with <Q20). E-corrections were performed with the program "Correction" to reduce the low frequent K-mer for better assembly [31].

Genome assembly

To get the draft genome sequences of Maliaodou and Lanxi1, different K-mer sizes ranging from 29-mer to 55-mer were tried to perform *de novo* assembly by SOAPdenovo v1.05 with clean paired-end reads [31]. The best assembly draft data (i.e., its contigs with the longest N50) was achieved at the 49-mer parameter for both Maliaodou and Lanxi1. Scaffold construction was performed based on the paired-end information of reads, and the gaps between the scaffolds were then closed by GapCloser v1.12 [31].

For the assembly of the Lanxi 1 chloroplast genome, the clean paired-end reads of Lanxi 1 that could be mapped to the known *G. max* chloroplast genome (NC_007942 [32]) by Bowtie2 v2.0.5 [33] were collected. The average mapping depth reached above 4000×, which is ~80 times of the whole genome mapping depth. Collected reads with low mapping depth (<250x) was filtered in our assembly effort. The remaining reads were used for *de novo* assembly using Velvet v1.2.07 [34] with 51-mer length. Considering the possibility that some reads from the chloroplast genome may be missed due to the used 250× threshold, we closed the gaps using all the clean reads by the software GapCloser v1.10 [35]. The overlap-based CAP3 [36] was utilized to merge redundant sequences for assembly refinement.

Genome annotation

Repeat regions in the assembled genome of Lanxi 1 were first identified using *de novo* methods implemented in RepeatScout v1.0.5 [37] and further masked by homology-based RepeatMasker v3.3.0 [38]. Genes were predicted using Augustus v2.5.5 [39] with Arabidopsis as a species parameter. Functions of the predicted genes were annotated by the BLASTP search against nr [40] and the annotated gene set of Williams 82 genome (ftp://ftp.jgi-psf.org/pub/compgen/phytozome/v8.0/Gmax_v1.0/annotation/Gmax_109_peptide.fa.gz) [41] with e-value<1e-5 as a threshold value.

SNP calling and structural variations

The clean reads of each soybean line were aligned to the Williams 82 reference genome (ftp://ftp.jgi-psf.org/pub/compgen/phytozome/v8.0/Gmax_v1.0/assembly/Gmax_109.fa.gz) using Bowtie2 v2.0.5 with default settings [33]. Consecutive steps were applied for SNPs detection. Samtools v0.1.18 [42] was used for SNP calling with the parameter –C = 50, which aims to reduce the effect of reads with excessive mismatches. In order to avoid paralogue inference, –q = 1 as threshold was used to filter reads that could align to multiple regions. Besides, to alleviate the false positive calling result due to relative low sequencing depth of some semi-wild soybean lines, we combined the variant calling information of all the 43 soybean accessions (31 by Lam *et al.* [27], one by Kim *et al.* [46] and 11 by this study), and selected SNPs present in at least two accessions or present in only one accession but with above 10 reads supported. The bam file produced from the mapping procedure was further analyzed for structural variations detection by BreakDancer v1.1 [43] with default parameters. Structural variations were displayed using Circos v0.62 [44].

Genetic diversity estimation

The average pairwise divergence (π, [45]) within a population was estimated for the wild, semi-wild and cultivated soybean populations. An in-house custom PERL script was applied for the estimation. Based on SNP calling results of each line, all variant sites across the whole genome were identified for each population and the number of nucleotide substitutions per site was estimated. The whole length of the reference genome was taken as a total number of nucleotides for π estimation.

Identification of cultivated and wild soybean-specific sequences

Nineteen wild lines (17 by Lam *et al.* [27] and two by Kim *et al.* [46] and this study) and 15 cultivars (14 by Lam *et al.* [27] and one Williams 82 by Schmutz *et al.* [41]) were used to identify the cultivated and wild soybean-specific sequences. Any sequences in the reference genome Williams 82 (ftp://ftp.jgi-psf.org/pub/compgen/phytozome/v8.0/Gmax_v1.0/assembly/Gmax_109.fa.gz), which could not be mapped by any reads from Lanxi 1 and other wild lines, were identified and defined as cultivated-specific sequences. For wild-specific sequences, the reads from the wild line Lanxi 1, which were not mapped into Williams 82, were collected and assembled by *De novo* into a 10 Mb set using Velvet v1.2.07 [34] with optimal K-mer size of 43 bp. The set was further mapped by the reads from the other 14 cultivars (Lam *et al.* [27]). Those sequences that could not be mapped by any reads from the 14 cultivated lines were considered as wild-specific sequences. Moreover, for identification of wild/cultivated common sequences, the genomic regions mapped by the all 14 cultivated lines but were not wild Lanxi 1 were considered as to be common sequences of cultivated soybeans, while the common-mapped regions in the *de novo* assembly of Lanxi 1 (as a wild reference genome) by the 18 wild lines, but not the Williams 82, were selected as common sequences of wild soybeans. In the process of identification specific or common regions, BWA v0.6.1 [54] was applied for mapping with default parameters. Mapping coverage and location information was obtained from the Samtools v0.1.18 using 'Samtools depth' parameter. The unmapped regions with at least 100 bp in length were extracted by in-house PERL scripts.

Phylogenetic tree and population structure

To collect a solid SNP dataset for constructing a *Soja* phylogenetic tree, the loci with a minimum coverage of three bases for each line across the genome were first selected, and a total number of 3,794,973 loci were found in all soybean lines. Heterozygous SNPs and the adjacent SNPs within 50 bp detected in each individual were further excluded for further genotyping. Finally, a total of 7,424 SNPs were used to construct the phylogenetic tree and population structure. The neighbor-joining tree was constructed by MEGA5.2 [47] with bootstrap support (1,000 replicates). The program STRUCTURE v2.3.4 [48] was applied for population structure; and the length of burn-in period was set to 100,000 with 100,000 MCMC reps afterwards. The number of genetic clusters was assigned using the Delta*K*-method described by Evanno *et al.* [49]. The number of clusters (*K*) was tested ranging from 2 to 9 with 8 replicates per *K*.

Selection analysis

To detect signals of recent selection, a method based on reduced pooled heterozygosity [50,51] was used. Considering the unbalanced sequencing data, in order to alleviate bias we preprocessed the sequences from each line by randomly choosing the fastq sequences to make all lines with 3~4× sequencing depth. Three subgroup pools (wild, semi-wild and cultivated) with pretreated sequence data from each line were created. By taking the genome of Williams 82 as a reference, variable sites in each pool were identified with the coverage ranging from 10 to 500. Reference and variant allele counts (n_{ref} and n_{alt}) at identified SNP positions from each pool were used to identify selections in 100 kb sliding windows with a step size of 50 kb. For a 100 kb window along the reference genome, the pooled heterozygous (H_p) was calculated by the formula: $H_p = 2\sum n_{ref} \sum n_{alt} / (\sum n_{ref} + \sum n_{alt})^2$, where $\sum n_{ref}$ and $\sum n_{alt}$ are the sums of n_{ref} and n_{alt} for all SNPs in the window. Z transformation ($ZH_p = (H_p - \mu H_p)/\sigma H_p$) was then applied to locate the putatively selected regions from the extreme tails by a threshold of 4 standard deviations as previously used [50,51].

GO enrichment analysis

The genes located in the selected regions were extracted from the soybean gene annotation file in Phytozome (ftp://ftp.jgi-psf.org/pub/compgen/phytozome/v8.0/Gmax_v1.0/annotation/Gmax_109_gene.gff3.gz), and the GO enrichment study conducted through AgriGO and '*Glycine max*' was set as the species setting (http://bioinfo.cau.edu.cn/agriGO/) [52]. The p-value and FDR criteria for the considered enrichment GO terms were <0.0001 and <0.05, respectively.

Experimental validation

To confirm the genomic variations in Maliaodou, 25 variations (10 SNPs and 15 indels) were selected for experimental validation. Primers were designed based on the flanking sequences of the variation sites using the on-line software Primer3 (http://frodo.wi.mit.edu/) (Table S2). The PCR reaction was carried out as described previously [53]. The PCR amplification products were checked by electrophoresis on 1.0% (w/v) agarose, then purified and sequenced directly or through clone sequencing using the

Sanger platform (ABI 3737XL) (Sunny, Shanghai). In the clone sequencing effort, at least five clones were sequenced for each sample.

Results

Genome re-sequencing

Most of the semi-wild soybeans possess an erected plant architecture with a main stem and/or large leaves, which are similar to cultivated soybean lines (Maliaodou is shown in Fig. 1). The ten semi-wild soybeans have an intermediate seed size (4.34–12.32 g), which is between the wild line (e.g. Lanxi 1, 1.22±0.04 g) and the cultivated line (e.g. Williams 82, 22.44±1.28 g) (Fig. 1; Table S1).

All semi-wild soybeans were sequenced into a ~3-fold genome coverage by high-throughput sequencing; Maliaodou was further sequenced into a 41-fold coverage (Table S3). In order to obtain a reference for the wild soybean genome for comparative analysis, a wild line Lanxi 1 was collected from the same location as Maliaodou and was sequenced into a 55-fold genome coverage by two libraries with different insertion sizes. After *de novo* assembly, a 929.9 Mb genome with contig/scaffold N50 sizes of 21.7/51.0 kb was obtained and used as a wild-type reference in the following studies. A total of 56,298 genes were further annotated, of which 48,240 were able to find their orthologs (hits) in the cultivated reference genome Williams 82 (Table S3). For the assembly of the Lanxi 1 chloroplast genome, we finally obtained complete chloroplast genome of Lanxi 1 with length of 152,199 bp, which harbors 139 protein coding genes. The assembly information is available in GenBank under the accession number KC779227.

Genetic diversity

Variation calling for 43 soybean accessions was carried out using the Williams 82 genome as a reference (Table S4ab). After SNP calling by Samtools v0.1.18 [42] and further steps to alleviate false positive calling by combining the variant calling information of all 43 soybean accessions, a total number of 7,704,637 SNPs were identified. Summary of shared (supported by at least two accessions) and unique variations (supported by at least 10 reads) of each accession is listed in Table S4a. In general, relatively lower variations were observed in the semi-wild soybeans than in the wild lines. The average variation of semi-wild lines had 910,373 SNPs and 38,258/32,907 insertion/deletions. Based on the available genomic data of 19 wild soybean genomes from previous studies [27,46] and the current study (Lanxi 1), the wild lines had an average of 1,628,253 SNPs and 110,181/94,292 insertion/deletions (Table S4b). For Maliaodou, 1,587,320 SNPs and 111,399/92,112 small insertions/deletions (indels) (<5 nt) were observed compared to the reference (Fig. 2). About 14,157 large regions (>1 kb) or 27.8 Mb cumulated sequences (about 2.8% of the reference genome) in the reference genome were not mapped by any reads of Maliaodou (Table S5a). Among 14,157 regions, 257 were over 10 kb in size (the largest one with 83.5 kb); they are shown in Fig. 2. The results were similar to those obtained by another approach based on information of paired-ends reads using BreakDancer [43], which detected 2,983 deletion events with 8.7 Mb cumulated sequences (score >60). Meanwhile, limited inversion (total number of 60) and translocation (160 and 419 intra- and inter-chromosomal) events were observed in the semi-wild (Table S5b). We also measured genetic divergence of wild, semi-wild and cultivated soybeans using the parameter π [45]. The π value was 1.416×10^{-3} for semi-wild, which is intermediate between the wild (2.173×10^{-3}) and cultivated soybeans (1.332×10^{-3}).

Unusual high heterozygous rates (0.19–0.49) were observed in several semi-wild soybeans (Table S4b). For example, a heterozygous rate of 0.49 was observed among the 1,587,320 SNPs in Maliaodou. Using the same pipeline, Lanxi 1 and most other wild lines were found to have lower heterozygous rates (<0.05) except for three lines with 0.12–0.15 (Table S4b). Similar higher heterozygous rates (>0.35) were also observed for indels in Maliaodou but not in any wild lines.

To confirm the variations, especially the unexpected high heterozygosity of Maliaodou, an independent SNP calling a different mapping algorithm implemented in BWA v0.6.1 [54] was carried out for Maliaodou, which is less sensitive but more specific than Bowtie2 v2.0.5 [33]. The results showed 0.50, 0.35, and 0.36 heterozygous rates for SNP, insertion and deletion, respectively (Table S4b). When 25 variations (10 SNPs and 15 indels) were selected for experimental validation by traditional Sanger sequencing, 21 (84%) were confirmed (Table S2). Three indels caused by repetitive sequences were not validated, perhaps due to sequencing errors [55]. The high heterozygous rates were also consistent with our fragmented assembly result of the Maliaodou genome (contig N50 size = 0.5 kb). Previous studies have indicated that the heterozygous rate is one of major disruptive factors in the algorithm for genome assembly [56].

Phylogenetic relationship and population structure

A phylogenetic relationship of the 30 re-sequencing wild and cultivated soybean genomes has been analyzed. One of the cultivated soybean accessions (C19) was excluded due to its high admixture [27]. Based on their phylogenetic structure, 10 semi-wild soybean lines from this study and two additional wild soybeans (Lanxi 1 in this study and a line from South Korea by Kim *et al.* [46]) were added into the tree (Fig. 3a). The semi-wild soybeans were not grouped into an independent branch but scattered into wild subgroups.

Seed size is a key target trait of soybean domestication as cultivated soybeans usually have a bigger seed size. According to the phylogenetic tree (Fig. 3a), species with a large seed size (cultivated lines in green) are generally separated from those with a smaller seed size (wild lines in blue). However, in the mixed branch with semi-wild soybeans (in red), their seed weight seems to not be consistent with their evolutionary relationship (Fig. S1). The result indicated that seed size, the most important trait under domestication selection, could hardly be used as a single factor to estimate soybean divergence, especially for those in the intermediate evolutionary stage. This is consistent with the fact that seed size is merely one of the target traits during soybean domestication and is significantly affected by the environmental conditions [57].

Bayesian clustering (STRUCTURE) of soybean accessions was further carried out. Using the method described by Evanno *et al.* [49], ΔK reached a peak when K was set to 3 (Table S6), indicating that $K = 3$ is optimal for the population structure. When we ordered the accessions based on their genetic background, they could be divided into three subgroups: a cultivated, wild, and mixed subgroup (including both wild and semi-wild soybeans) (Fig. 3b). These results are consistent with our phylogenetic results (Fig. 3a). Similarly, semi-wild soybeans do not have an independent subpopulation and are mainly grouped with wild soybeans.

Selective sweep analysis

Crops usually experience two stages (domestication with subsequently genetic improvement) during the evolutionary

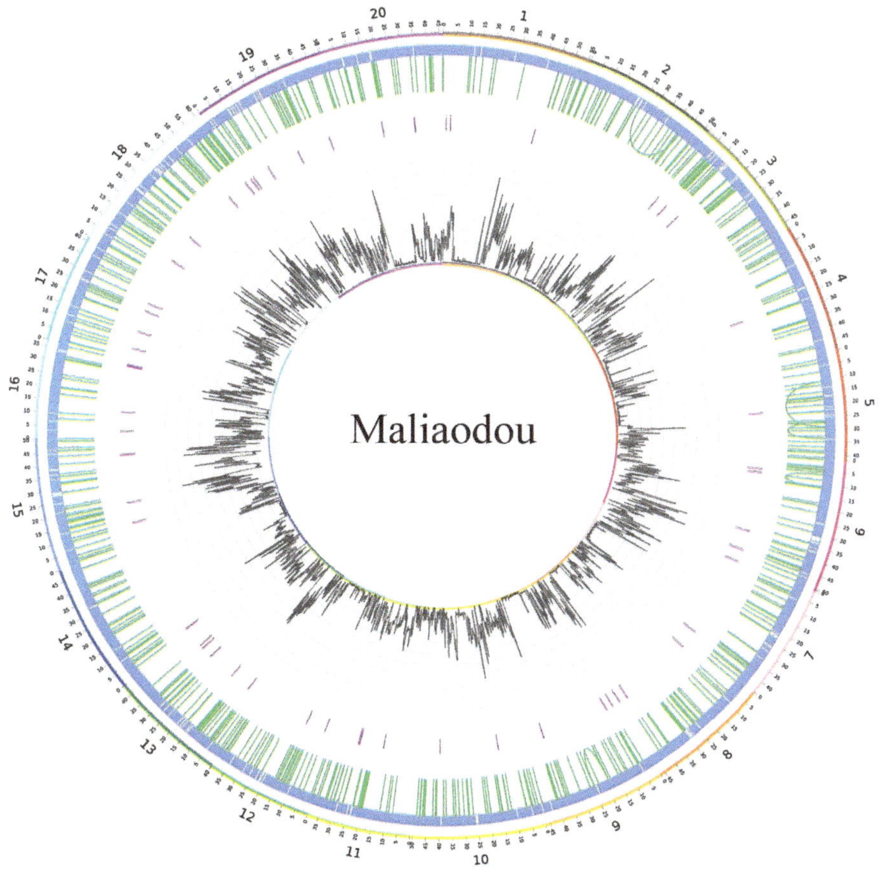

Figure 2. Genomic variations between the semi-wild genome Maliaodou and the reference genome Williams 82. Circles from outside: 20 chromosomes labeled with different colors; in the blue bars, each white vertical line presents a >10 kb un-mapped region in the reference genome; green and purple vertical lines present the intra-chromosomal translocations and inversion events, respectively. Distribution of the SNP density of the Maliaodou is labeled with grey lines.

process from wild-type, and semi-wild is a transitional point between these two stages [58]. To detect signals of artificial selection, we searched the wild, semi-wild, and cultivated soybean genome for selection regions with reduced pooled heterozygosity (H_p). Using autosomal 100 kb as the scanning window suggested by Axelsson *et al.* [51] and Chung *et al.* [28], several regions in cultivated and semi-wild soybean populations had extremely low heterozygosity rates (Fig. 4; Table S7a). By the threshold of at least four standard deviations away from the mean ($Z(H_p)<-4$), a total of 102 and 144 putatively selected regions could be detected in the semi-wild and cultivated populations, respectively. It is reasonable to believe that more selected loci could have been identified in the latter because cultivated population experience additional artificial selection in recent genetic improvements [26].

In the 144 selected loci, only 101 loci were detected in the cultivated soybeans, but not in the wild or semi-wild soybean were $Z(H_p)<-2$ was as a threshold, suggesting that they are the putative target loci under recent genetic improvements. The 101 loci include 282 protein-coding genes (Table S7bc), which could play a vital role in agronomy-related traits. We performed GO enrichment analysis on the 282 genes and identified 9 gene ontology terms with FDR (false discovery rate) <0.05 (Table S8). Interestingly, the GO categories were enriched in embryonic and embryo sac development as well as megagametogenesis, which are involved in the development of seed size. The variants within these genes may lay a genetic foundation to the seed size evolution.

Discussion

A mixed semi-wild group complicated population structure of the subgenus *Soja*

It is believed that the cultivated soybean (*G. max*) was domesticated from the wild *G. soja* [1,2]. Our results demonstrate a complex transitional stage from wild to cultivated soybean: a mixed population including a series of semi-wild types, which suggests no distinct or independent transitional period for soybean domestication. The complex evolutionary situation might be due to the hybridization events between domesticated and wild soybeans, which frequently occurred in natural field conditions (see next section for details). Many crops have had significant reproductive isolation from their wild progenitors after domestication, which made their phylogenetic relationship clearly (such as rice [59]). Our results are consistent with the observation by Wang *et al.* [6] who clearly revealed that there is no existence of a transitional intermediate ancestor between the wild and cultivated soybean. Furthermore, our results do not support the semi-wild soybean as an independent species (*G. gracilis*). Apparently, the mixture of semi-wild and wild accessions in the phylogenetic tree not as an independent sub-branch (Fig. 3a) strongly suggests that the semi-wild soybeans belong to the wild category and should be

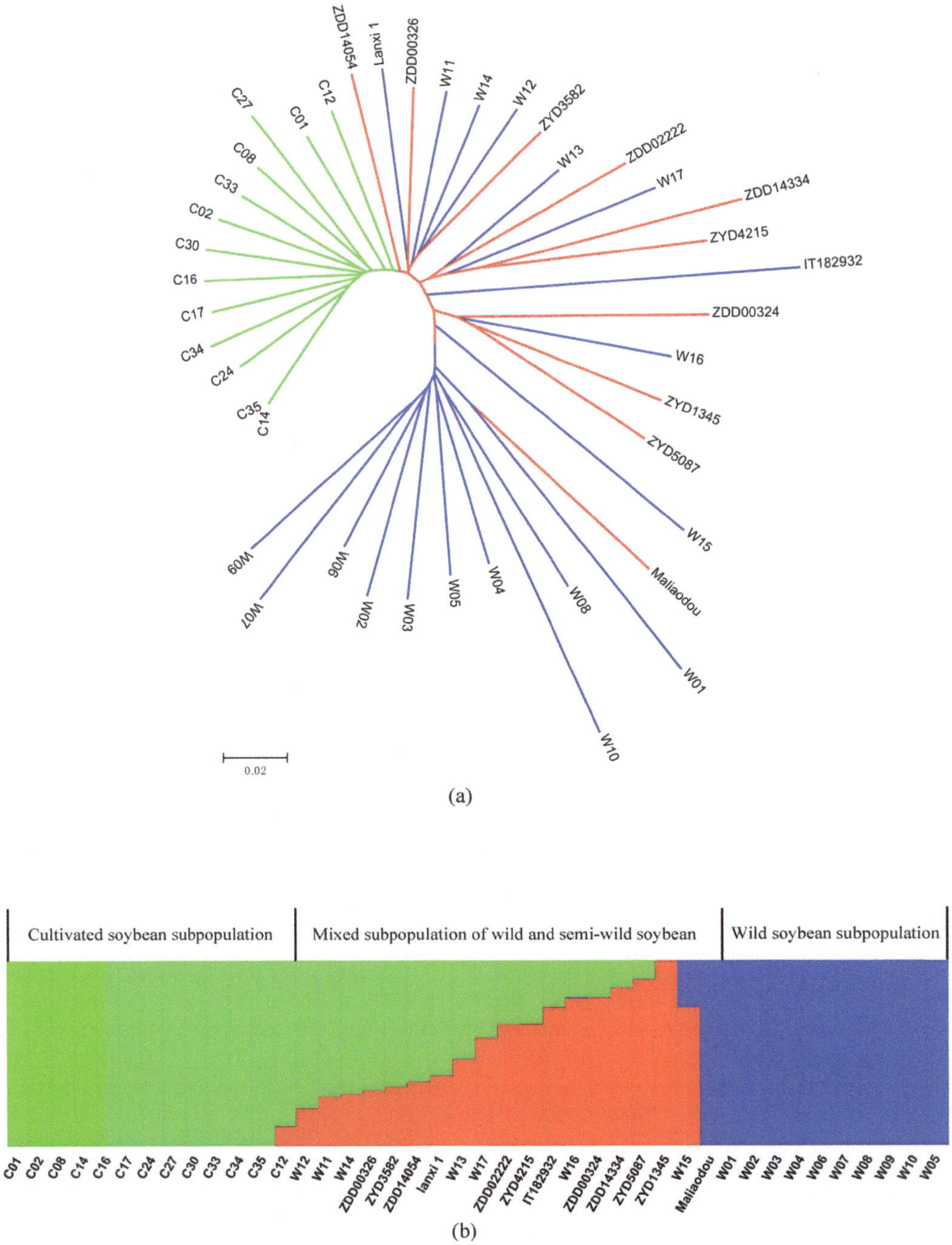

Figure 3. Phylogenetic tree and population structure of wild, semi-wild, and cultivated soybeans. (a) A neighbor-joining phylogenetic tree constructed using SNP data. Cultivated, semi-wild, and wild soybeans are labeled with green, red, and blue lines, respectively. (b) Bayesian clustering (STRUCTURE, $K=3$) of soybean accessions were grouped based on their species. Cultivated lines were designated with 'C' as prefix while wild lines were with 'W' as prefix except IT182932 (Korean) and Lanxi 1 (this study). Semi-wild lines were titled with 'Z' as prefix except Maliaodou.

considered a variant or subclade of *G. soja*. However, in order to fully understand the *Soja* population structure, more semi-wild lines from wide geographical locations and their sequence data need to be sampled or generated in future.

The hybridization origin of the semi-wild soybean

The hybridization origin of the semi-wild soybean has been proposed [13,60]. However, direct evidence in support of this hypothesis is lacking, although hybridization events between cultivated and wild soybean have been observed in natural field conditions [6,18,19]. In this study, several genomic evidences favor the above hypothesis.

First, the semi-wild soybeans were not grouped into an independent branch in the *Soja* phylogenetic tree (Fig. 3a). Instead, they were scattered into the wild soybean subgroup. As mentioned above, our results failed to support an independent speciation and domestication of an intermediate evolutionary type between *G. soja* and *G. max* [12]. Second, the semi-wild soybean hosts both cultivated and wild soybean novel sequences. Based on the available genomic sequences of the wild lines including Lanxi 1

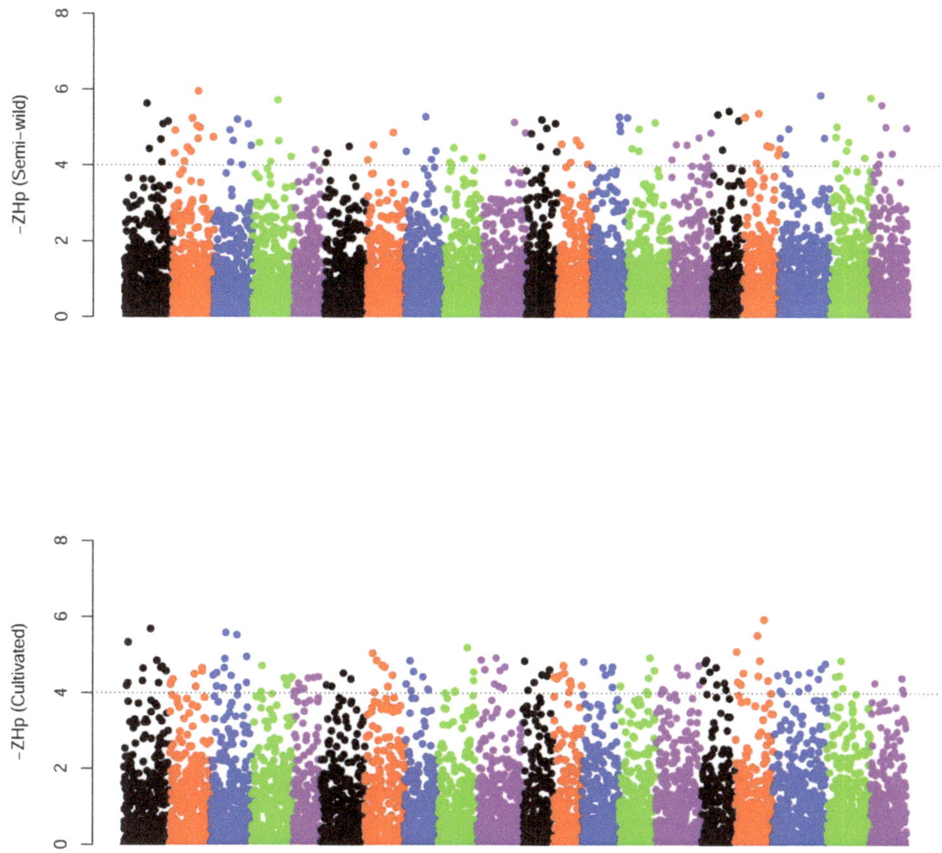

Figure 4. Summary of selective sweep analysis. Distribution of *Z*-transformed average pooled heterozygosity (H_p) in semi-wild (top) and cultivated soybeans (bottom), respectively. The negative end of the $Z(H_p)$ distribution plotted along soybean autosomes 1–20 are shown. A dashed horizontal line indicates the cut-off ($Z<-4$) used for extracting outliers.

and cultivars [27,46], we identified 3.07/0.25 Mb cultivated/wild soybean-specific sequences and 3.35/0.82 Mb cultivated/wild soybean-common sequences (for detailed definition see Methods section). Using the deeply sequenced Maliaodou as an example, more than 99% of 3.07 Mb cultivated-specific sequences were covered by the reads from the Maliaodou genome (Table S9). Meanwhile, most (>95%) of the 0.82 Mb wild-common sequences and part (>12%) of the 0.25 Mb wild-specific sequences were also covered by the Maliaodou genome. Again, these findings strongly support the above hypothesis that Maliaodou originated from the hybridization of the wild and domesticated soybean. Third, unexpectedly high heterozygous rates were observed in Maliaodou and three other semi-wild lines. Soybeans are predominantly self-pollinating and usually have low outcrossing rates of <3.0% in cultivated lines but relatively higher rates (2.4–13%) in wild lines [19,61]. This is the first time that the high heterozygous rate was observed and reported in the semi-wild soybean. Regarding the significantly higher heterozygous rate of the Maliaodou genome as compared with the 19 wild lines (*t*-test *P* value$<0.1\times10^{-3}$ for SNP and $P=0.4$–0.9×10^{-3} for indel), a reasonable explanation is that the semi-wild line may have a recent hybrid origin rather than an intermediate evolutionary type between *G. soja* and *G. max*.

In addition, the above hypothesis is also supported by chloroplast genome data. Xu *et al.* [17] investigated chloroplast DNA SSR in 326 wild and cultivated soybean accessions and identified 52 haplotypes. However, no cultivar-specific haplotypes were found and thus hybrid swarms between cultivated and wild forms were suggested. In this study, we collected green soybean leaves of Lanxi 1 for genome re-sequencing. This generated chloroplast genomic sequences and provided us with a wild soybean chloroplast genome (Accession number: KC779227) for comparison study. Using this and the known *G. max* chloroplast genome (NC_007942) as references, the wild and cultivated lines with available genomic data [27,46] were genotyped. No haplotype is shared by all cultivars, all wild lines or all semi-wild lines (i.e. within each group of material) used in this study (Table S10). Our results are similar to the findings by Xu and co-workers [17].

Taken together, we believe that introgressions from local wild soybeans to cultivated soybeans occurred frequently during the long domestication process of soybean, i.e. a kind of hybridization origin, which creates a complex genetic background for species in subgenus *Soja*. Several important crops are polyploidy and have originated from the hybridization of two or three ancient species, such as wheat, cotton and tobacco [62]. However, the hybridization origin of diploid crops from a domesticated line and its progenitor is rare, with only one case with genomic evidence coming from *indica* rice (*Oryza sativa*), which originated from a cross between the cultivated *japonica* rice and local wild rice [59]. For the semi-wild soybean, this should be another case of genomic evidence for the hybridization origin of diploid crops from cultivated and wild lines.

The extremely high heterozygous rates observed in Maliaodou and other semi-wild soybeans (Table S4b) suggest that their hybridization events occurred recently. This is in agreement with the fact that no record of "Maliaodou" could be found in the Chinese ancient literatures until last century (Hu & Tian 1993), although soybeans became one of the most important crops in China several thousand years ago. Wang *et al.* [18] also proposed a short creation history of semi-wild soybean based on their observation of some newly collected semi-wild lines. Additionally, the ancestry inference based on STRUCTURE result for Maliaodou indicated it is different from that of other semi-wild accessions. Based on the SNP calling result, we observed an unexpectedly high heterozygosity rate in the Maliaodou (~50%), which indicated that this semi-wild line might originate from a recent hybridization between cultivated and wild soybeans. Tracing the location in which the Maliaodou was collected, we found that many wild soybeans are also grown in the same field, and thus genetic introgressions from wild lines to Maliaodou probably frequently occurred.

Footprints of artificial selection in the soybean genome

Both natural and domestication selections target genes/loci controlling adaptive or agronomic traits and leave footprints of selection in the soybean genome. Several candidate domestication regions have been identified by recent genomic investigations of wild and cultivated soybeans [25,26,28]. Of the 282 genes located in the putative selective regions in this study (Table S7), at least 63 were also identified as putative selective genes by Li *et al.* [25] or Chung *et al.* [28] (Table S7). The gene coding for ABSCISIC ACID-INSENSITIVE 5-like protein 3 (Glyma13g03880) was detected by both of their studies as well as our study. It is one of the seed color related genes based on QTL mapping [63], and it suggests that this gene was targeted by the recent genetic improvement. Besides seed size, seed color is also a main target trait of domestication and subsequent improvement. In our study, a relatively strict criteria ($Z(H_p)<-4$) was applied, which may exclude some putative artificially selected genes. Moreover, as shown above, 14,157 large regions (>1 kb), or about 2.8% of the reference genome of cultivated Williams 82, were not mapped by any reads from the deep sequenced (~41×) semi-wild line Maliaodou (Table S5). These regions might be kept only in the cultivated soybean population during improvement. Taken together, the above-identified genes in the selective and unmapping regions in this study that may relate to the artificial selection and important agronomic traits provide some candidate targets for further functional investigation.

Supporting Information

Figure S1 A neighbor-joining phylogenetic tree of wild (blue), semi-wild (red) and cultivated soybeans (green) labelled with seed weight per 100 seeds.

Table S1 Detailed information of the Chinese accessions of subgenus *Soja* used in this study.

Table S2 Experimental validation results for the genomic variations and PCR primers used in this study.

Table S3 Global statistics of the wild soybean Lanxi 1 and semi-wild soybean genomes.

Table S4 Detailed information of genomic variations in the all soybean lines in this study relative to the reference genome Williams 82.

Table S5 Genomic structural variations between the semi-wild Maliaodou and domesticated reference genome Williams 82.

Table S6 Inference of best *K* for separating soybean subgroups using the delta *K* method.

Table S7 Position information for selective loci and genes in the selective loci in cultivated soybeans.

Table S8 GO enrichment analysis of the genes in selective regions of cultivated soybeans.

Table S9 Genomic coverage of the semi-wild soybean Maliaodou in the cultivated and wild-specific sequences of soybean.

Table S10 Chloroplast SNPs of cultivated, wild and semi-wild soybeans relative to the reference chloroplast genome (Accession no. NC_007942).

Acknowledgments

We thank Dr. Farooqahmed S Kittur (North Carolina Central University) for his critical reading of the manuscript.

Author Contributions

Conceived and designed the experiments: LF. Performed the experiments: YW. Analyzed the data: JQ YW SW CHY WW XB ZL CYY LF. Contributed reagents/materials/analysis tools: ZW. Wrote the paper: LF JQ. Contributed to the discussion: CYY ZW QS JX SHL.

References

1. Carter TE, Nelson RL, Sneller CH, Cui Z (2004) Soybean: improvement, production, and uses. Agronomy 16: 303–416.
2. Zhao TJ, Gai JY (2004) The origin and evolution of cultivated soybean (*Glycine max* (L.) Merr.) (in Chinese). Sci Agric Sin 37: 954–962.
3. Chen YW, Nelson RL (2004) Genetic variation and relationships among cultivated, wild, and semiwild soybean. Crop Sci 44: 316–325.
4. Li Y, Guan R, Liu Z, Ma Y, Wang L, et al. (2008) Genetic structure and diversity of cultivated soybean (*Glycine max* (L.) Merr.) landraces in China. Theor Appl Genet 117: 857–871.
5. Wang KJ, Li XH, Liu Y (2012) Fine-scale phylogenetic structure and major events in the history of the current wild soybean (*Glycine soja*) and taxonomic assignment of semi-wild type (*Glycine gracilis* Skvortz.) within the Chinese subgenus *Soja*. J Hered 103: 13–27.
6. Wang KJ, Li XH (2013) Genetic diversity and gene flow dynamics revealed in the rare mixed populations of wild soybean (*Glycine soja*) and semi-wild type (*Glycine gracilis*) in China. Genet Resour Crop Ev 60: 2303–2318.
7. Kuroda Y, Kaga A, Tomooka N, Vaughan D (2010) The origin and fate of morphological intermediates between wild and cultivated soybeans in their natural habitats in Japan. Mol Ecol 19: 2346–2360.
8. Wang KJ, Li FS (2000) General situation of wild soybean (*G. soja*) germplasm resources and its utilization of introgression into cultivated soybean in China. Rev Agric Sci Technol 2: 69–72.
9. Hu M, Tian P (1993) Chinese Soybean Breeds. Beijing: Agriculture Press.
10. Skvortzow BV (1927) The soybean–wild and cultivated in Eastern Asia. In: Proceedings of the Manchurian Research Society, Natural History Section Publication Series A, No. 22. Harbin, China.

11. Hermann FJ (1962) A revision of genus Glycine and its immediate allies. No. 1268. United States Department of Agriculture: Washington, DC.
12. Fukuda Y (1933) Cytogenetical studies on the wild and cultivated Manchurian soybeans. J Jpn Bot 6: 489–506.
13. Hymowitz T (1970) On the domestication of the soybean. Econ Bot 24: 408–421.
14. Broich SL, Palmer RG (1980) A cluster analysis of wild and domesticated soybean phenotypes. Euphytica 29: 23–32.
15. Abe J, Hasegawa A, Fukushi H, Mikami T, Ohara M, et al. (1999) Introgression between wild and cultivated soybeans of Japan revealed by RFLP analysis for chloroplast DNAs. Econ Bot 53: 285–291.
16. Close PS, Shoemaker RC, Keim P (1989) Distribution of restriction site polymorphism within the chloroplast genome of the genus *Glycine*, Subgenus *Soja*. Theor Appl Genet 77: 768–776.
17. Xu H, Abe J, Gai Y, Shimamoto Y (2002) Diversity of chloroplast DNA SSRs in wild and cultivated soybeans: evidence for multiple origins of cultivated soybean. Theor Appl Genet 105: 645–653.
18. Wang KJ, Li XH, Zhang JJ Chen H, Zhang ZL, et al. (2010) Natural introgression from cultivated soybean (*Glycine max*) into wild soybean (*Glycine soja*) with the implications for origin of populations of semi-wild type and for biosafety of wild species in China. Genet Res Crop Evol 57: 747–761.
19. Wang KJ, Li XH (2011) Interspecific gene flow and the origin of semi-wild soybean revealed by capturing the natural occurrence of introgression between wild and cultivated soybean populations. Plant Breed 130: 117–127.
20. Nakayama Y, Yamaguchi H (2002) Natural hybridization in wild soybean (*Glycine max* ssp. *soja*) by pollen flow from cultivated soybean (*Glycine max* ssp. *max*) in a designed population. Weed Biol Manag 2: 25–30.
21. Concibido VC, La Vallee B, McLaird P, Pineda N, Meyer J, et al. (2003) Introgression of a quantitative trait locus for yield from *Glycine soja* into commercial soybean cultivars. Theor Appl Genet 106: 575–582.
22. Wen Z, Ding Y, Zhao T, Gai J (2009) Genetic diversity and peculiarity of annual wild soybean (*G. soja* Sieb. et Zucc.) from various eco-regions in China. Theor Appl Genet 119: 371–381.
23. Li YH, Li W, Zhang C, Yang L, Chang RZ, et al. (2010) Genetic diversity in domesticated soybean (*Glycine max*) and its wild progenitor (*Glycine soja*) for simple sequence repeat and single-nucleotide polymorphism loci. New Phytol 188: 242–253.
24. Leroy AR, Fehr WR, Cianzio SR (1991) Introgression of genes for small seed size from *Glycine soja* into *G. max*. Crop Sci 31: 693–697.
25. Li YH, Zhao SC, Ma JX, Li D, Yan L, et al. (2013) Molecular footprints of domestication and improvement in soybean revealed by whole genome re-sequencing. BMC Genomics 14: 579.
26. Hyten DL, Song Q, Zhu Y, Choi IY, Nelson RL, et al. (2006) Impacts of genetic bottlenecks on soybean genome diversity. Proc Natl Acad Sci U S A 103: 16666–16671.
27. Lam HM, Xu X, Liu X, Chen W, Yang G, et al. (2010) Resequencing of 31 wild and cultivated soybean genomes identifies patterns of genetic diversity and selection. Nat Genet 42: 1053–1059.
28. Chung WH, Jeong N, Kim J, Lee WK, Lee YG, et al. (2013) Population Structure and Domestication Revealed by High-Depth Resequencing of Korean Cultivated and Wild Soybean Genomes. DNA Res.
29. Hao D, Cheng H, Yin Z, Cui S, Zhang D, et al. (2012) Identification of single nucleotide polymorphisms and haplotypes associated with yield and yield components in soybean (*Glycine max*) landraces across multiple environments. Theor Appl Genet 124: 447–458.
30. Sambrook J, Russell DW (2001) Molecular cloning: a laboratory manual. 3rd edn. New York: Cold Spring Harbor Laboratory Press.
31. Li R, Zhu H, Ruan J, Qian W, Fang X, et al. (2010) *De novo* assembly of human genomes with massively parallel short read sequencing. Genome Res 20: 265–272.
32. Saski C, Lee SB, Daniell H, Wood TC, Tomkins J, et al. (2005) Complete chloroplast genome sequence of *Glycine max* and comparative analyses with other legume genomes. Plant Mol Biol 59: 309–322.
33. Langmead B, Salzberg SL (2012) Fast gapped-read alignment with Bowtie 2. Nat Methods 9: 357–359.
34. Zerbino DR, Birney E (2008) Velvet: algorithms for *De novo* short read assembly using de Bruijn graphs. Genome Res 18: 821–829.
35. Boetzer M, Pirovano W (2012) Toward almost closed genomes with GapFiller. Genome Biol 13: R56.
36. Huang X, Madan A (1999) CAP3: A DNA sequence assembly program. Genome Res 9: 868–877.
37. Price AL, Jones NC, Pevzner PA (2005) *De novo* identification of repeat families in large genomes. Bioinformatics 21 Suppl 1: i351–358.
38. Smit AF (1996) The origin of interspersed repeats in the human genome. Curr Opin Genet Dev 6: 743–748.
39. Stanke M, Tzvetkova A, Morgenstern B (2006) AUGUSTUS at EGASP: using EST, protein and genomic alignments for improved gene prediction in the human genome. Genome Biol 7 Suppl 1: S11 11–18.
40. Coordinators NR (2013) Database resources of the National Center for Biotechnology Information. Nucleic Acids Res 41: D8–D20.
41. Schmutz J, Cannon SB, Schlueter J, Ma J, Mitros T, et al. (2010) Genome sequence of the palaeopolyploid soybean. Nature 463: 178–183.
42. Li H, Handsaker B, Wysoker A, Fennell T, Ruan J, et al. (2009) The Sequence Alignment/Map format and SAMtools. Bioinformatics 25: 2078–2079.
43. Chen K, Wallis JW, McLellan MD, Larson DE, Kalicki JM, et al. (2009) BreakDancer: an algorithm for high-resolution mapping of genomic structural variation. Nat Methods 6: 677–681.
44. Krzywinski M, Schein J, Birol I, Connors J, Gascoyne R, et al. (2009) Circos: an information aesthetic for comparative genomics. Genome Res 19: 1639–1645.
45. Nei M, Miller JC (1990) A simple method for estimating average number of nucleotide substitutions within and between populations from restriction data. Genetics 125: 873–879.
46. Kim MY, Lee S, Van K, Kim TH, Jeong SC, et al. (2010) Whole-genome sequencing and intensive analysis of the undomesticated soybean (*Glycine soja* Sieb. and Zucc.) genome. Proc Natl Acad Sci U S A 107: 22032–22037.
47. Tamura K, Peterson D, Peterson N, Stecher G, Nei M, et al. (2011) MEGA5: molecular evolutionary genetics analysis using maximum likelihood, evolutionary distance, and maximum parsimony methods. Mol Biol Evol 28: 2731–2739.
48. Pritchard JK, Stephens M, Donnelly P (2000) Inference of population structure using multilocus genotype data. Genetics 155: 945–959.
49. Evanno G, Regnaut S, Goudet J (2005) Detecting the number of clusters of individuals using the software STRUCTURE: a simulation study. Mol Ecol 14: 2611–2620.
50. Rubin CJ, Zody MC, Eriksson J, Meadows JR, Sherwood E, et al. (2010) Whole-genome resequencing reveals loci under selection during chicken domestication. Nature 464: 587–591.
51. Axelsson E, Ratnakumar A, Arendt ML, Maqbool K, Webster MT, et al. (2013) The genomic signature of dog domestication reveals adaptation to a starch-rich diet. Nature 495: 360–364.
52. Du Z, Zhou X, Ling Y, Zhang Z, Su Z (2010) agriGO: a GO analysis toolkit for the agricultural community. Nucleic Acids Res 38: W64–70.
53. Mullis K, Faloona F, Scharf S, Saiki R, Horn G, Erlich H (1992) Specific enzymatic amplification of DNA *in vitro*: the polymerase chain reaction. Biotechnol Ser, 24, 17–27.
54. Li H, Durbin R (2009) Fast and accurate short read alignment with Burrows-Wheeler transform. Bioinformatics 25: 1754–1760.
55. Nakamura K, Oshima T, Morimoto T, Ikeda S, Yoshikawa H, et al. (2011) Sequence-specific error profile of Illumina sequencers. Nucleic Acids Res 39: e90.
56. Kelley DR, Salzberg SL (2010) Detection and correction of false segmental duplications caused by genome mis-assembly. Genome Biol 11: R28.
57. Kumar V, Rani A, Solanki S, Hussain SM (2006) Influence of growing environment on the biochemical composition and physical characteristics of soybean seed. J Food Compos Anal 19: 188–195.
58. Yamasaki M, Tenaillon MI, Bi IV, Schroeder SG, Sanchez-Villeda H, et al. (2005) A large-scale screen for artificial selection in maize identifies candidate agronomic loci for domestication and crop improvement. Plant Cell 17: 2859–2872.
59. Huang X, Kurata N, Wei X, Wang ZX, Wang A, et al. (2012) A map of rice genome variation reveals the origin of cultivated rice. Nature 490: 497–501.
60. Hirota T, Sayama T, Yamasaki M, Sasama H, Sugimoto T, et al. (2012) Diversity and population structure of black soybean landraces originating from Tanba and neighboring regions. Breed Sci 61: 593–601.
61. Fujita R, Ohara M, Okazaki K, Shimamoto Y (1997) The extent of natural cross-pollination in wild soybean (*Glycine soja*). J Hered 88: 124–128.
62. Mallet J (2007) Hybrid speciation. Nature 446: 279–283.
63. Yang K, Jeong N, Moon JK, Lee YH, Lee SH, et al. (2010) Genetic analysis of genes controlling natural variation of seed coat and flower colors in soybean. J Hered 101: 757–768.

3

Distribution and Differentiation of Wild, Feral, and Cultivated Populations of Perennial Upland Cotton (*Gossypium hirsutum* L.) in Mesoamerica and the Caribbean

Geo Coppens d'Eeckenbrugge[1]*, Jean-Marc Lacape[2]

1 CIRAD, UMR 5175 CEFE, Campus du CNRS, Montpellier, France, **2** CIRAD, UMR AGAP, Montpellier, France

Abstract

Perennial forms of *Gossypium hirsutum* are classified under seven races. Five Mesoamerican races would have been derived from the wild race 'yucatanense' from northern Yucatán. 'Marie-Galante', the main race in the Caribbean, would have developed from introgression with *G. barbadense*. The racial status of coastal populations from the Caribbean has not been clearly defined. We combined Ecological Niche Modeling with an analysis of SSR marker diversity, to elucidate the relationships among cultivated, feral and wild populations of perennial cottons. Out of 954 records of occurrence in Mesoamerica and the Caribbean, 630 were classified into four categories cultivated, feral (disturbed and secondary habitats), wild/feral (protected habitats), and truly wild cotton (TWC) populations. The widely distributed three first categories cannot be differentiated on ecological grounds, indicating they mostly belong to the domesticated pool. In contrast, TWC are restricted to the driest and hottest littoral habitats, in northern Yucatán and in the Caribbean (from Venezuela to Florida), as confirmed by their climatic envelope in the factorial analysis. Extrapolating this TWC climatic model to South America and the Pacific Ocean points towards places where other wild representatives of tetraploid *Gossypium* species have been encountered. The genetic analysis sample comprised 42 TWC accessions from 12 sites and 68 feral accessions from 18 sites; at nine sites, wild and feral accessions were collected in close vicinity. Principal coordinate analysis, neighbor joining, and STRUCTURE consistently showed a primary divergence between TWC and feral cottons, and a secondary divergence separating 'Marie-Galante' from all other feral accessions. This strong genetic structure contrasts strikingly with the absence of geographic differentiation. Our results show that TWC populations of Mesoamerica and the Caribbean constitute a homogenous gene pool. Furthermore, the relatively low genetic divergence between the Mesoamerican and Caribbean domesticated pools supports the hypothesis of domestication of *G. hirsutum* in northern Yucatán.

Editor: Xianlong Zhang, National Key Laboratory of Crop Genetic Improvement, China

Funding: The authors have no support or funding to report.

Competing Interests: The authors confirm that co-author Jean-Marc Lacape is a PLOS ONE Editorial Board member, and that this does not alter his adherence to PLOS ONE Editorial policies and criteria.

* Email: geo.coppens@cirad.fr

Introduction

Cotton (*Gossypium* spp.) is unique among crop plants in that four species have been independently domesticated in four different regions of the world: two tetraploids, *G. hirsutum* L. in Mesoamerica, *G. barbadense* L. in South America, and two diploids, *G. herbaceum* L. in Arabia and Syria and *G. arboreum* L. in the Indus Valley of India and Pakistan [1]. In the process, they were transformed from photoperiod-sensitive perennial sprawling or upright shrubs into short, compact, annualized day-length-neutral plants; and their small impermeable seeds sparsely covered by coarse, poorly differentiated hairs became larger and covered with abundant and long, white lint. Simultaneously, their seeds lost their impermeability and dormancy. The wide diversity of cotton results from the successive waves of agronomic improvement and human-mediated germplasm diffusion [1,2].

Phylogenetic investigations in *Gossypium* distinguish 45 modern diploid species distributed among three major geographic lineages and eight genomes. The American tetraploid lineage originated within the last 1–2 million years from a single hybridization event between a maternal African A and an American D genome [1]. It diversified into five species, three wild endemic species, *G. darwinii* Watt native to the Galapagos, *G. tomentosum* Nutt. ex Seem. from the Hawaiian Islands, *G. mustelinum* Miers ex Watt restricted to Northeastern Brazil, and the two cultivated species *G. barbadense* and *G. hirsutum*. The latter provides over 90% of the world cotton, spreading North and South to subtropical and temperate latitudes well over 30° as an annual crop. Its indigenous (preindustrial) range encompasses most of Mesoamerica and the Caribbean, with two centers of morphological and genetic diversity, one in Southern Mexico-Guatemala, considered a primary center of diversity, and one in the Caribbean, where some introgression took place with *G. barbadense* [2–5].

In these two regions, *G. hirsutum* exhibits a diverse array of perennial forms, which Hutchinson [6] classified into seven geographical races. The primitive and highly variable race 'punctatum' is mostly found in Yucatán and round the coasts and islands of the Gulf of Mexico. Race 'latifolium' has a center of diversity in Guatemala and southern Mexico, but its range extends from most of Mexico to El Salvador and Nicaragua. Race 'Marie-Galante' is distinct both geographically and morphologically, with its pronounced apical dominance and tree-like habit. Its range includes the Antilles and Central America, South from El Salvador into northern to northeastern South America. Its origin and diffusion seems to be closely related to human migrations that would have resulted in the introduction of *G. barbadense* into Central America and the Antilles and its introgression with *G. hirsutum* in these areas [2,3,7,8]. Together, these three most widespread races, 'latifolium', 'punctatum' and 'Marie-Galante', encompass most of the morphological variation in *G. hirsutum*. The remaining four races present a more restricted geographic distribution, with race 'palmeri' in the Mexican states of Oaxaca and Guerrero, race 'morrilli' in the central Mexican plateau, race 'richmondi' along the Pacific side of the Isthmus of Tehuantepec, and race 'yucatanense' limited to the northern coast of Yucatán. The latter is known only as a small, highly branched, sprawling shrub forming a dominant constituent of undisturbed beach strand vegetation. Hutchinson [6] considered race 'yucatanense' an extreme case of feral populations derived from primitive 'punctatum' landraces.

The persistence of wild populations of *G. hirsutum* has been the subject of considerable debate. On one hand, most germplasm collections came from man-made habitats, such as field plots and house yards, or highly disturbed habitats, such as roadsides and secondary vegetation, indicating that spontaneous cotton plants were escapes from cultivation. Furthermore, morphological differentiation appears similar and parallel for both landraces and feral plants [6,9,10]. Testing materials from Yucatán, Hutchinson [6] observed no differences between progenies of 'punctatum' from plants cultivated in dooryards and plants established in natural vegetation. On the other hand, Sauer [11] observed that the northern Yucatán wild cottons are negatively associated with human settlements and form a dominant constituent of "a complex vegetation type occupying a coherent and extensive area with natural and edaphic and climatic boundaries." He maintained this interpretation in his study of the Cayman Islands shoreline vegetation [12].

In a study of the effects of domestication in *G. hirsutum*, Stephens [9] extended the question to the seemingly wild populations of race 'punctatum' observed on the dry leeward sides of some of the Greater Antilles, on Florida Cays [6,13], along the coasts of the Gulf of Mexico as well as in Venezuela. For a long time, he could not rule out the possibility that these forms are feral relics of pre-Columbian or early post-Columbian cultivation [9,14], even though they have retained their small impermeable seeds with an impressive capacity for long distance dispersal [14,15]. Only from 1967 did he abandon the views of Hutchinson et al. [16] and refer without restriction to coastal populations in the Caribbean and the Gulf of Mexico as wild [3,7]. In their extensive collecting travels, Ano et al. [17], Ano and Schwendiman [18], and Schwendiman et al. [19] went even further in underlining the similarity of these cotton populations with those of northern Yucatán shores, relating their distribution to sea currents, and classifying them in the same race 'yucatanense'.

Long-range seed dispersal also explains the presence of *G. hirsutum* in the Pacific Ocean. Fryxell and Moran [20] described a truly wild small 'punctatum' population in Socorro, an island of the Revillagigedo archipelago, some 600 km West of Mexico. Similar wild forms have diffused to even more distant Pacific islands (Tahiti, Marquesas, Samoa, Fiji, and Wake islands) [20–22]. Indeed, Fryxell [23] suggested a close relationship between the evolutionary history of the tetraploid cotton species and their particular adaptation to strand habitats along marine beaches, underlining the importance of oceanic seed diffusion and citing a dozen cases of such populations, eight of which concerned *G. hirsutum*. He presented a hypothesis relating this coastal adaptation and capacity for diffusion via ocean currents to the significant mobility of shorelines during the Pleistocene. In 1979, Fryxell further developed his views in his monograph on the Malvaceae [24], adding to his arguments those of Sauer [11]. Since then, the question of the natural dispersion of *G. hirsutum* has been further complicated by the recent description of wild populations of *G. hirsutum* in Paraguay [25], confirming an intuition of Stephens [7].

Despite its importance for cotton genetics and breeding, the question of truly wild cottons has spawned relatively few genetic studies. In their RFLP study, Brubaker and Wendel [8] observed three groups: (1) races 'yucatanense' and 'punctatum', (2) races 'latifolium' and 'palmeri', and (3) race 'Marie-Galante'. They refuted Hutchinson's views on the regressive status of race 'yucatanense', and proposed a model where "the morphological intergradation, geographical proximity, and genetic similarity of race 'yucatanense' to inland 'punctatum' populations – of Yucatán – reflects a relationship between the first domesticated form of *G. hirsutum* and its wild progenitor." Thus, the initial stages of cotton (*G. hirsutum*) domestication would have taken place in northern Yucatán and the human-mediated transfer of the first 'punctatum' cottons out of the species' natural range would have triggered the process of concomitant differentiation into new and improved races, agronomic developments, and long range germplasm diffusion. This process would explain the current distribution of *G. hirsutum* diversity. The SSR study of Lacape et al. [26] supported the racial classification [6], and the interracial relations appeared consistent with the model of progressive domestication, diffusion and differentiation proposed by Brubaker and Wendel [8], except for the geographically more distant 'Marie-Galante', which appeared closely related to 'punctatum'. Their three 'yucatanense' accessions from Guadeloupe (as classified by Ano et al. [27]) exhibited a high number of unique alleles. Similarly, in the study of Liu et al. [28], the unique representative of race 'yucatanense', from Yucatán, appeared highly divergent from the other accessions.

The views of Brubaker and Wendel [8], which explain the pre-Columbian *G. hirsutum* diversity by successive waves of diffusion of genetic and agronomic developments, from northern Yucatán to inland Yucatán (race 'punctatum'), then to southern Mexico and Guatemala (race 'latifolium'), and finally to all Mesoamerica and the Caribbean, imply an early cotton domestication. This is consistent with the contributions of historical linguistics and archaeology. Thus, words for cotton can be reconstructed in Proto-Otomangue, a language that was spoken in Central Mexico at least 6500 BP [29,30]. According to Smith and Stephens [31], the oldest remains of Mesoamerican cotton, found in the Tehuacán Valley and dated 5500 to 4300 BP, represent fully domesticated introductions, being comparable in form and size to the landraces currently existing in the same area.

The domestication and diffusion scenario proposed by Brubaker and Wendel [8] for *G. hirsutum* has been generally accepted and it

Table 1. Repartition among races of *G. hirsutum* of the 110 selected accessions per country of origin and type/race (Nota: 9 locations had both 'truly wild' (TWC) and feral specimens).

	Feral						Total feral	Total wild	Total
Country race:	**MG**	**MO**	**PA**	**RI**	**PU**	**unk**		**TWC**	
Antigua & Barbuda								1	1
Aruba	1						1		1
Bahamas	1						1		1
Barbados	1						1		1
Bonaire	2						2	2	4
BWI_Grand Cayman	1						1		1
Colombia	2						2		2
Costa Rica	1						1		1
Cuba	1						1		1
Curaçao	4						4	4	8
Dominican Rep	4						4	4*	8
Dominica	1						1		1
French Guiana	1						1		1
Guadeloupe	6						6	4	10
Guam Pacific					1		1		1
Guatemala	1						1		1
Haïti	1						1		1
Jamaica	3						3	1	4
Maldives					1		1		1
Martinique	1						1		1
Mexico		1	1	1	9		12	8^{Y}	20
Nicaragua	1						1		1
Puerto Rico	8						8	$7^{\dagger}$	15
Saint-Kitts & Nevis	2						2	3	5
Saint-Vincent & Grenadines	1						1		1
Samoa						$1^{\P}$	1		1
Trinidad & Tobago	2						2		2
USA-Florida								3	3
Venezuela	7						7	5	12
Total‡	**53**	**1**	**1**	**1**	**11**	**1**	**68**	**42**	**110**

Races are referred as MG for 'Marie-Galante', MO for 'morrilli', PA for 'palmeri', PU for 'punctatum', RI for 'richmondi', unk for 'unknown'. TWC collectively refers to truly wild cotton populations including those of race 'yucatanense' from Mexico (see text).

* = also referred-to as *G. ekmanianum*,

^{Y} = 8 accessions from Mexico include 7 accessions from Yucatán and one from Socorro Islands,

† = local name "algodon brujo",

‡ = one additional accession from Australia (FM966) as modern cultivated (total = 111),

¶ = accession initially of «unknown» race (further assigned as race 'punctatum').

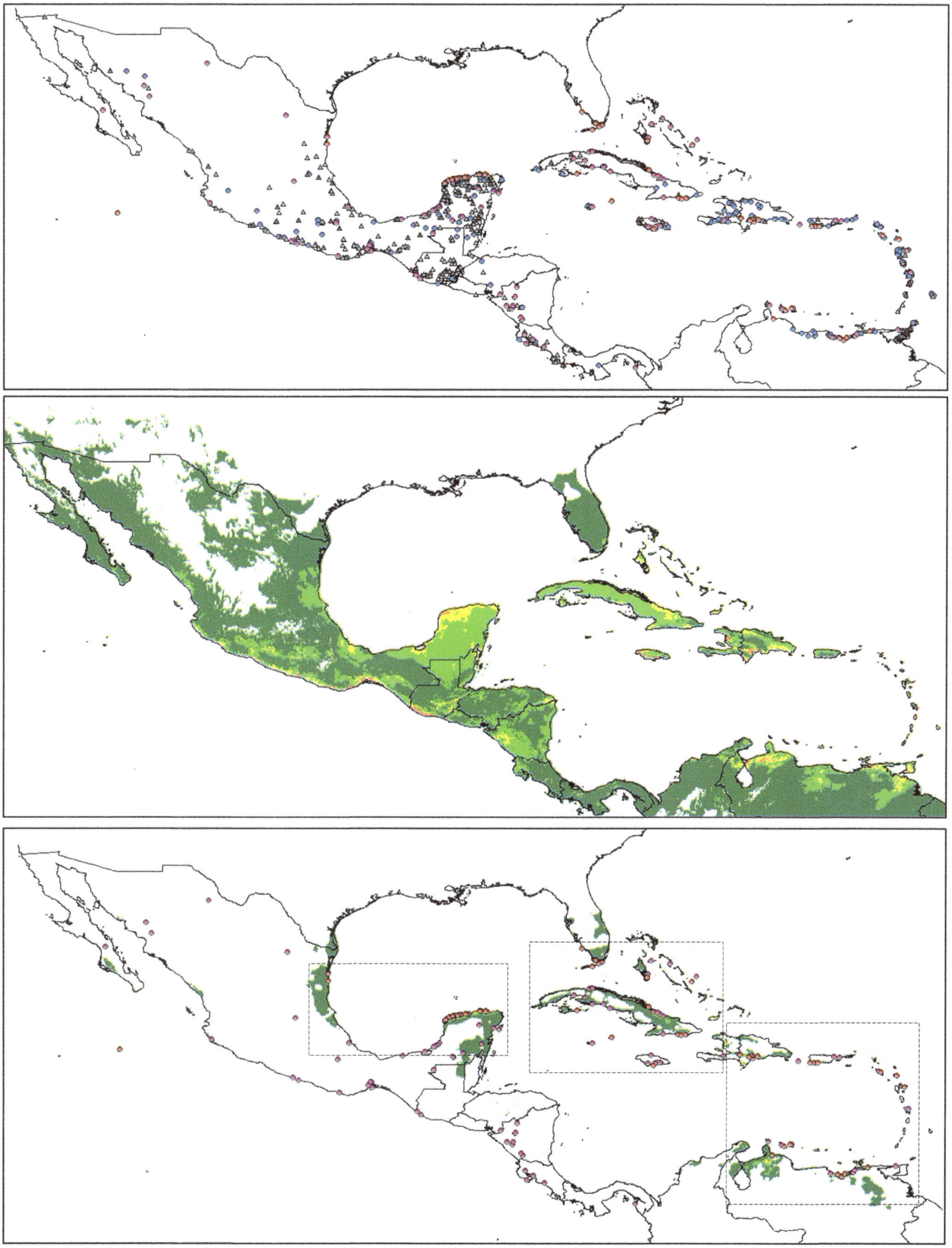

Figure 1. Distribution and climate model of perennial forms of *G. hirsutum* in Mesoamerica and the Caribbean. A. Distribution of 954 categorized datapoints for perennial forms of *G. hirsutum* in Mesoamerica and the Caribbean 'truly wild' (TWC) specimens/populations are represented by red dots, 'wild/feral' by purple dots, 'feral' (disturbed habitats) by blue dots, 'cultivated' by brown triangles, and unclassified plants by grey triangles. **B.** Climate model for distribution of both cultivated and spontaneous *G. hirsutum* in Mesoamerica and the Caribbean (complete set as presented in Figure 1A). Climate suitability is indicated by background color from unfavorable (no color) to marginal (dark green) or increasingly favorable (light green and warmer colors). **C.** Localization of the populations from categories TWC and 'wild/feral' and climate model for TWC populations. Red dots represent the datapoints used for the distribution model (TWC populations), whereas purple dots represent 'wild/feral' populations of uncertain status (truly or secondarily wild). Climate suitability is indicated as in Fig. 1B. Three dotted frames refer to map limits as magnified in Figure 2.

is found practically unchanged in the most recent syntheses [1,32]. The fact that it is based on only one wild population (from northern Yucatán) has not been challenged, and alternative scenarios have been overlooked. Nevertheless, as stated by Sauer [11], "if lint bearing cottons were naturally present in the New World as sea dispersed pioneers, they were not likely to be confined to Yucatán… The lint may have been widely gathered and perhaps traded long before regular cultivation began; the process of domestication may have been diffuse in space and time, involving wild cottons from Caribbean and Pacific coasts, as well as Yucatán." Indeed, if *G. hirsutum* is a perennial whose regressive forms thrive in disturbed human habitats and xerophytic secondary vegetation, domestication was not necessarily a linear process, moving from littoral strands to the agricultural field through the dooryard. Wendel et al. [4] questioned "whether *G. hirsutum* achieved widespread distribution and regional differentiation as a wild plant prior to domestication, or if it was widely distributed as a perennial semi-domestic by the pre-Columbian people from a much smaller native range". Casas et al. [33] have described how Mesoamerican societies have improved more than 200 plant species, through management practices that integrate cultivated areas, agroforestry systems and gathering from the wild, with or without conscious selection. As documented from many studies of cactus fruit species [34–36], the result of this *in situ* domestication process is a mosaic of habitats and useful plant populations with particular morphological, genetic, and even reproductive characteristics, according to management intensity. Similar management practices may have been used for cotton. Stephens [9] cites several accounts, from the first voyage of Columbus to much later periods in colonial times, mentioning the simultaneous exploitation of cultivated, feral and wild cottons, according to the quality objective.

If we recognize that Caribbean wild cotton populations may have been involved in the domestication process, we must also question the distinction between a primary centre of diversity in Mesoamerica and a secondary centre in the Caribbean. The strong dominance of race 'Marie-Galante' in the latter region, as well as in southern Central America and northern South America, poses the question of its origin and even of its possible separate domestication [7,32].

The question of the natural distribution of *G. hirsutum* is not only crucial for understanding the biogeography of tetraploid cottons, and their evolution and diffusion under domestication, but also for the continuation of the domestication process. Further improvement of the crop requires both a better exploitation of the available germplasm and better genetic tools to manipulate important economic traits [32]. For example, studies on the effects of domestication on such essential traits as fiber development [37] and the corresponding genetic transformations, with an altered expression of about 25% of the genes at transcriptome level [38], depend on the comparison of well-defined and representative samples of wild and domesticated germplasm.

We present here a double approach to investigate this question, combining Ecoclimatic Niche Modeling (ENM) and neutral genetic markers to assess whether coastal cotton populations are "truly wild," and investigate their relationship with inland perennial cottons. ENM methods derive an envelope for the environmental requirements of a taxon from a set of its occurrence localities. They have provided a powerful tool for investigating the ecology and distribution of both plant and animal species. An ENM study on *G. hirsutum* was recently published by Wegier et al. [39] aiming to understand not only the distribution of "wild" cotton populations from Mexico, but also the spatial organization of genetic diversity and potential gene flow from genetically modified cultivars using molecular markers. However, as compared to the wild cotton studies cited above, Wegier et al. [39] used much more permissive criteria to distinguish feral and wild cottons. In our approach, we have used ENM to document the relationship between perennial cotton domestication and distribution in Mesoamerica, the Caribbean and the Gulf of Mexico, by mapping and comparing the potential tropical/subtropical distributions of domesticated *G. hirsutum* populations, feral cotton (escaped from cultivation), and presumably or truly wild populations of races 'yucatanense' and 'punctatum'. Potential distribution of *G. hirsutum* was predicted for modern climatic conditions as well as for climatic parameters modeled for the Last Glacial Maximum (LGM). The underlying idea is that the original distribution of wild cotton in the early Holocene was necessarily related to its distribution during the Pleistocene, following an approach validated by several studies [40,41]. The identification of potential climatic refuges for the species should help in distinguishing natural and human factors in its dispersal.

As the ENM study confirmed the particular ecology and "truly wild" status of a number of coastal cotton populations, SSR neutral genetic markers were then used to characterize them and investigate their relationship with neighboring feral cottons.

Materials and Methods

Climatic modeling and analysis

Our ENM study focused on the centers of diversity of *G. hirsutum*, i.e., Mexico, Central America, and the Eastern Caribbean (from the coasts of Venezuela to Florida through the Antilles). From now on, we will collectively refer to this region as Mesoamerica and the Caribbean. Geographical and ecological information was extracted from the CIRAD cotton germplasm database and records [17–19,27,42,43] and related collecting reports by French and US scientists (collections in the 80s under the aegis of the former IBPGR), the scientific literature on wild cotton, regional floras, herbarium-label and germplasm-passport data obtained from the Global Biodiversity Information Facility (GBIF) portal, the Mexican Red Mundial de Información sobre Biodiversidad (REMIB), and relevant Mexican administrative documents. All geographic coordinates have been assigned or verified against associated geographic information with gazetteers (mostly Google Earth and Geonames). Incomplete or imprecise records were discarded, as were redundant data (dataset available upon request).

The information associated with the collections/observations was also used to classify cotton occurrences according to their status on a wild to cultivated scale, using four categories: 'cultivated' (fields and dooryards), 'feral' (plants found in disturbed habitats, such as roadsides and secondary vegetation), 'wild/feral' (plants found in preserved habitats and/or forming persistent populations), and 'truly wild' (populations described as such by experts, based on ecological and morphological grounds). This categorization is partially analogous to that used by Stephens [9], whose "wild forms" would include both our 'truly wild' and 'wild/feral' categories, whereas Stephens' "semiferal" and "commensal/cultivated" forms correspond to our 'feral' category and cultivated categories, respectively. The objective was also analogous: Stephens tested his categories on domestication traits (fiber and seeds) while we aimed at testing them on eco-climatic grounds.

For each occurrence record, 19 bioclimatic variables were extracted from WorldClim, a package consisting of global surfaces of climate, with a 2′30″ grid resolution (corresponding roughly to 4.4×4.6 km) [44]. These variables are: 1) annual mean temperature; 2) mean diurnal range (mean of monthly (max temp - min temp); 3) isothermality (Bio2/Bio7); 4) temperature seasonality; 5) maximal temperature of warmest month; 6) minimal temperature of coldest month; 7) temperature annual range; 8) mean temperature of wettest quarter; 9) mean temperature of driest quarter; 10) mean temperature of warmest quarter; 11) mean temperature of coldest quarter; 12) annual precipitation; 13) precipitation of wettest month; 14) precipitation of driest month; 15) precipitation seasonality; 16) precipitation of wettest quarter; 17) precipitation of driest quarter; 18) precipitation of warmest quarter; and 19) precipitation of coldest quarter.

For ENM, we chose the widely used Maxent machine learning method. It estimates the probability distribution of maximum entropy (i.e. closest to uniform) subject to the constraint that the expected value of each environmental variable (or its transform and/or interactions) under this estimated distribution matches its empirical average [45]. Maxent was run twice, firstly on the whole dataset, and secondly only on points in the 'truly wild' category. A logistic threshold value equivalent to the 10 percentile training presence was retained to separate climatically favorable areas from marginally fit areas. Maxent output provides measures of the contribution of each bioclimatic variable (percent contribution and permutation importance) and proposes a jackknife test to quantify the contribution of each variable from the gain when it is used in isolation and the gain loss when it is omitted from the model. However, the strong correlations among bioclimatic variables do not allow an easy interpretation of their relative importance. Therefore, we performed a principal component analysis (PCA) to characterize and compare the climatic envelopes of our categories of *G. hirsutum* observations, discarding those variables whose contribution appeared marginal. The factors with an eigenvalue above 1 were retained and a normalized varimax rotation was applied to maximize the sum of the variances of the squared loadings, simplifying the interpretation of the results. The different categories of populations were then plotted on the principal components plane to visualize and compare their ecoclimatic range.

To predict the potential distribution of *G. hirsutum* at LGM, the MIROC climatic model [46] derived from the PMIP2 database Paleoclimate Modelling Intercomparison Project Phase II for 21,000 BP was downloaded from the Worldclim website (http://www.worldclim.org/) and used on a dataset restricted to the 'truly wild' category.

Genetic analyses

The panel of accessions of perennial *G. hirsutum* cotton populations used for SSR genotyping comprised 110 feral and wild accessions supplemented by a modern cultivar, 'FM966' (Table 1). One hundred and eight accessions originated from the CIRAD seed bank, and three from USDA. Twenty-nine countries/provinces of Mesoamerica and the Caribbean were represented (Table 1). Particular attention was paid to geographic locations where both truly wild and feral populations could be identified in close proximity (such as for the populations of Pointe des Châteaux in Guadeloupe), or slightly more distant (such as for the populations from Yucatán sea-shores versus inland). Such sites with both truly wild and nearby feral specimens were identified in nine cases (Mexico/Yucatán, Jamaica, Dominican Republic, Puerto Rico, St Kitts & Nevis, Guadeloupe, Venezuela, Bonaire, and Curaçao). Three localities were represented only by wild specimens, Florida (one feral specimen discarded due to missing data), Antigua, and Socorro Islands of Mexico; and 18 additional localities were only represented by feral populations. A few additional locations where truly wild cotton (further abbreviated as TWC) populations had been reported (visible as red dots in Figure 1) could not be included in the genetic study due to lack of plant material, such as in Cuba, the western coast of the Gulf of Mexico (Tamaulipas), Bahamas and Grand Cayman. Detailed geographic information of the 110 accessions is available in Table S1; Figure S1 presents their localizations on the sites with TWC populations.

Five seeds per accession were sown in small pots in the greenhouse in Montpellier and DNA was extracted from pooled samples (1–3 different plants) of young leaves using the MATAB protocol [47]. Thirty-seven SSR markers were selected for genotyping based on previous experience [26], in order to optimize information and quality. They were mostly derived from non-coding genomic DNA sequences (majority from series 'BNL' and 'CIR'), preferably to the more frequent EST-derived SSRs, with presumptive neutrality (no evidence of having been targeted during domestication). They had shown in previous experiments the amplification of a single PCR product in tetraploid cotton, thus avoiding the ambiguity generated by homoeolog loci. SSRs were genotyped in multiplex panels of 8 SSRs (four dyes and two SSRs per dye). Simultaneous PCR amplifications in a final volume of 10 µl contained 5 ng of genomic DNA, 200 µM of each dNTP, 0.5 mM MgCl2, 1 U Taq polymerase, 0.08 µM of M13-tailed 'F' primer, 0.1 µM of both the 'R' primer and of an M13 oligonucleotide tailed with the ad hoc fluorochrome. PCR reactions were performed on an Eppendorf microcycler (Eppendorf, Madison, WI)) using the following profile, a hot start of 94°C for 5 min, 35 cycles of 30 sec at 94°C, 1 min at 55°C and 1 min at 72°C, and a final extension step of 30 min at 72°C. PCR products were pooled with 10 µl of GeneScan 600-LIZ size standard. PCR products were denaturated and size fractionated using capillary electrophoresis on an ABI 3500 Genetic Analyzer (Applied Biosystems). Subsequently, GeneMapper 4.1 (Applied Biosystems) software was used for allele size estimation.

Twenty-six SSRs showing strict and unambiguous bi-allelic patterns (coded as homozygote when a single peak/allele and heterozygote with 2 peaks/alleles) were selected. The 26 SSRs were mapped on 18 of the 26 chromosomes (Table S2). Expected heterozygosity at each locus was calculated as $He = 1 - \Sigma pi^2$ where pi is the frequency of the ith allele.

The data matrix of bi-allelic codings for the 26 SSRs and 111 genotypes was imported into the DARWin5 software [48] to calculate genetic dissimilarities. Bootstrap dissimilarity matrices were calculated by drawing 10 000 entries. A Principal Coordinate Analysis based on the similarity matrix was conducted also with DARWin package. In complement to this factorial analysis, unweighted trees without topological constraints were constructed using a neighbor joining (NJ) approach [49] to represent individual relations. Lastly, the methods implemented in the STRUCTURE software [50] were used to infer population clusters and estimate admixture (quantitative clustering). The number of clusters, K, was chosen based on 20 independent runs for K values ranging between 1 and 5 with a burn-in length of 500,000 followed by 750,000 MCMC iterations. The ΔK method [51] was then applied using Structure Harvester [52], and estimated membership for each genotype, in each cluster, was read from the STRUCTURE output.

Results and Discussion

Dataset composition and distribution for climatic modeling

A total of 954 datapoints were gathered, of which 630 could be ascribed to our four categories (Table 2). Figure 1A shows no clear differences in the distributions of the different categories, except for 'truly wild' cotton (TWC) populations, which only occur along the coasts of the Eastern Caribbean and the Gulf of Mexico. The sample is well balanced between Central America and Mesoamerica, on one hand, and the islands and shores from the Eastern Caribbean to Florida on the other hand. Feral and wild specimens are better represented than cultivated germplasm, which can be explained by a collecting bias of botanists, most often interested by spontaneous plants, and germplasm collectors, motivated by the rusticity expected from primitive and spontaneous materials. The poor representation of 'cultivated' cotton also reflects the decline of its cultivation in Mexico [10] and in the Caribbean [18].

Among the 544 datapoints from Central and Mesoamerica, few have been assigned to a geographical race: 2 for race 'morrilli' (state of Guerrero), 8 for 'palmeri' (Guerrero), 5 for 'richmondi' (Oaxaca), 41 for 'punctatum' (Yucatán peninsula and Socorro Island), and 32 for 'yucatanense' (state of Yucatán). Albeit poor, this information is consistent with their original description by Hutchinson [6] and, with the exception of 'yucatanense', all races are found in both 'cultivated' and 'feral' categories, illustrating the absence of morphological differentiation between cotton landraces and feral cottons within a same region, as reported by several collectors [9,10,24]. 'Punctatum' is the only race with important spontaneous populations classified as 'wild/feral', one in the state of Yucatán, around Celestún, and several ones on the southern coast of Campeche state, between Champotón and Isla del Carmen. The only 'truly wild' Mexican population of race 'punctatum' is the one described by Fryxell and Moran [20] in the Socorro Island (Revillagigedo archipelago).

For the Eastern Caribbean (410 accessions from Venezuela to Florida), most observations were from breeders, so the racial composition is much better documented. It shows a strong dominance of race 'Marie-Galante' (278 acc.). The only other identified race is 'punctatum', ascribed to the TWC category (64 datapoints) or, exceptionally, to the 'wild/feral' category (one datapoint). In our dataset, these TWC are classified as 'punctatum', following the early views of Hutchinson [13], author of the original classification, although the same materials collected by Ano et al. [27] and Schwendiman et al. [19] were later reclassified under race 'yucatanense'.

Ecoclimatic niche models for cultivated, feral, and wild *G. hirsutum*

Figure 1B presents the potential distribution extrapolated by the Maxent software for the whole dataset. Along the coasts of Mexico, climatically favorable lowland areas correspond to those identified by Wegier et al. [39], i.e., the Yucatán peninsula, the regions of Veracruz and Tamaulipas along the western shores of the Gulf of Mexico, and the tropical Pacific coast. The latter area appears particularly favorable. The state of Tabasco (southern shores of the Gulf of Mexico) is better represented than in the study of Wegier et al. [39]. Other favorable areas are found much further inland.

Given the relative over-representation of wild and feral materials in our sample, Figure 1B gives a likely picture of the Mesoamerican distribution of perennial *G. hirsutum* for the last three millennia at least, i.e. a period of very active agricultural development, during which modern climatic conditions were already established [53]. The distribution of favorable areas corresponds quite well with those areas where several of Hutchinson's geographic races were developed: Yucatán to Mexican shores of the Gulf of Mexico for race 'punctatum', Yucatán to Guatemala for race 'latifolium', Pacific regions and the southern side of the isthmus of Tehuantepec for races 'palmeri' and 'morrilli', and even regions of the central Mexican plateau for race 'richmondi'. In Central America, the pre-Columbian distribution of *G. hirsutum* appears related to the diffusion of race 'Marie-Galante', as the favorable areas close to the Guatemalan-Salvadoran border and in western Nicaragua show good correspondence with the distribution of this race, presented by Stephens [7]. As suggested by this author, these races probably differentiated under relative geographical, ecological and cultural isolation, the latter term covering "the combined effects of human selection, migration and diffusion."

Figure 1C presents the geographical distribution of 'wild/feral' and TWC populations, together with a distribution model based only on 'truly wild' populations (100 datapoints). The areas suitable for TWC populations (Figure 1C) cover a very small part of the favorable areas for the whole sample (Figure 1B). They are mostly found in three sub-regions: (i) Gulf of Mexico and northern Yucatán, (ii) Florida and western Greater Antilles and (iii) Venezuela and eastern Caribbean, as detailed in Figures 2A, B and C, respectively.

Table 2. Dataset composition and distribution among domestication status categories of perennial *G. hirsutum* as defined for the present study.

	Total	Uncategorized	cultivated	feral	wild/feral	truly wild
Meso- & Central America	544	308	61	80	59	36
Eastern Caribbean to Florida	410	16	96	188	46	64
Total	954	324	157	268	105	100

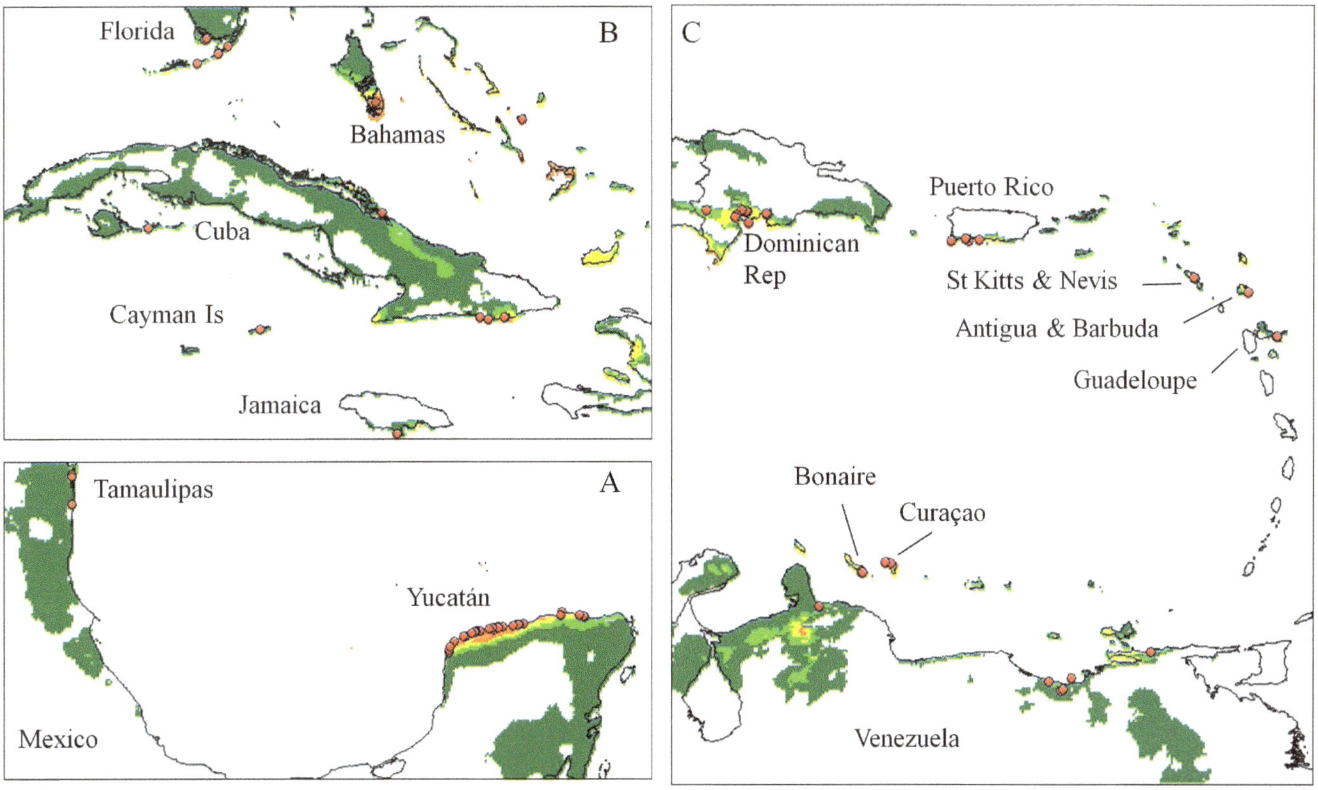

Figure 2. Localization of the truly wild cotton (TWC) populations and corresponding climate model. Map frames indicated as rectangles in Figure 1C. Climate suitability as indicated in Fig. 1. **A.** Gulf of Mexico. **B.** Florida and western Greater Antilles. **C.** Venezuela and eastern Caribbean.

The great majority of 'wild/feral' populations (purple dots on Figure 1C) fall in areas that are marginal (dark green areas) or unsuitable for TWC populations, validating our a priori categorization.

The 'yucatanense' population along the northern coast of Yucatán (Figure 2A), certainly constitutes the most extensive TWC population [43,54]. Our model confirms that its distribution is clearly limited by ecological parameters, as stated by Sauer [11]. Within this well-delimited area, a few specimens classified as 'wild/feral' are very probably incompletely documented representatives of race 'yucatanense'. Extensive spontaneous populations also exist on the western coast of the Yucatán peninsula, but we have found no indications that these are 'truly wild'. On the contrary, the model indicates that they have developed under climatic conditions that are not even marginally fit for TWC populations. West of the Gulf of Mexico, along the coast of Tamaulipas, 'truly wild' *G. hirsutum* was observed by Lukefahr cited in Stephens [7]. However, favorable areas are small and sparse in this region, and we could trace only three specimens whose labels mention that they were parts of natural coastal vegetation. Confirming the statement of Stephens [7], no population that could be classified as TWC has been documented for the Pacific coast of Mexico, where a very few small coastal areas appear climatically marginal for sustaining such populations. Thus, while the model confirms highly favorable climatic conditions in the Revillagigedo Islands, it gives no clear indication about areas where wild *G. hirsutum* could have developed on the western coast of Mexico before diffusing to islands in the Pacific Ocean.

In northern South-America and the southern Caribbean (Figure 2C), TWC populations are scattered along the coasts of Venezuela, between the Gulf of Venezuela (Saco de Maracaibo; state of Falcón) and the North of the state of Sucre, and on the shores of many islands along these coasts: Curaçao, Bonaire, Isla de Piritú. We have found only ambiguous information for the Chacachacaré Island. Mentions of colonial cotton plantation cast doubt on the only report of wild cotton populations in this area by Stephens [9]. On the other hand, the surroundings of Chacachacaré village in the Island of Margarita offer excellent conditions for TWC populations, suggesting that the homonymy of these neighbor sites may have created confusion. To the West, the shores of Colombia only offer marginal conditions for TWC (Figure 1C), which explains why Stephens [9] was not successful in his search for wild cotton in this area. To the Northeast of Venezuela, there seems to be another gap in the natural distribution of *G. hirsutum*, as no TWC populations have been identified in Trinidad and Tobago or in the southern half of the Lesser Antilles (Figure 2C), which is consistent with the descriptions of Hutchinson [13,55]. In the northern Lesser Antilles (Figure 2C), only three TWC populations have been described, in Guadeloupe [27], in Antigua and in Saint Kitts [9,18,19], and the model confirms favorable climatic conditions at these sites.

In the Greater Antilles (Figure 2B and 2C), the modeled distribution also agrees well with the wealth of previous reports of TWC populations of race 'punctatum', indicating favorable climatic conditions for the "algodón brujo" of southern Puerto Rico [9,13,19], for the populations around the Yaquí Valley of the Dominican Republic [9,19,56], in Haiti [13], Jamaica [19,57,58] and in the Cayman Islands [12,59,60]. In southern Cuba, similar populations exist around Guantánamo (specimen labels refer to the morphological type described by Britton in 1908 [57]). Further North, the modeled distribution is consistent with the observations of TWC in Florida [19,23] and in the Bahamas [61,62]. For Bermudas, much further North, the model indicates unfavorable conditions for TWC, which is consistent with the statement by Britton [63] about the absence of native cotton in these islands.

Climatic requirements of cultivated, feral and wild populations of perennial cotton

Seven variables were discarded for PCA on climatic variables, because of their poor specific contribution to the Maxent model obtained from the whole sample: isothermality (Bio3), maximal temperature of the warmest period (Bio5), precipitation of the wettest and driest periods (Bio13 and Bio14), precipitation seasonality (Bio15), and precipitation of the driest and warmest quarters (Bio17 and Bio 18).

The analysis on the remaining twelve variables produced three factors with an eigenvalue superior to 1 (Table 3). The first one is strongly associated with mean temperatures at all periods of the year (Bio1, and Bio8-11), with correlations between 0.82 and 0.95; the second one is associated with precipitation (Bio12, 16 and 19), with correlations between 0.80 and 0.95; and the third one is associated with variables related to latitude (Bio2-7: diurnal temperature range, temperature seasonality, minimal temperature of coldest period and temperature annual range). The third factor shows no clear differences among our categories, which is consistent with their similar latitudinal dispersion, from tropical Venezuela to subtropical northern Mexico and Florida. In contrast, the categories and origins present different patterns of dispersion in the plane formed by the two first principal component factors (Figure 3). On the continent (Central and Mesoamerica, Figure 3A), part of the observations come from cooler regions (along the x-axis of factor 1, to the left) or from wetter regions (along the y-axis of factor 2, upwards), while cotton-associated climates appear more uniform in the eastern Caribbean (Figure 3B). *G. hirsutum* was not observed in regions that are both cooler and wetter (upper left area in Figure 3), which gives the general shape of an inverted 'L' to the Mesoamerican dot cloud.

When considering domestication status, no clear distinction can be made between 'cultivated', 'feral', and 'wild/feral' materials (Figure 3C), as these categories share the same general inverted 'L' pattern of dispersion in the principal components plane. In contrast, TWC populations are clearly characterized by very uniform climatic conditions; thus the environment of both 'yucatanense' and truly wild 'punctatum' (Figure 3D) is clearly among the hottest and driest in our sample. The best represented geographical race, 'Marie-Galante', which is highly dominant throughout the Antilles, logically presents the same climatic dispersion as the general Caribbean sample, with occurrences under extremely arid conditions too (not shown). Indeed, several reports mention spontaneous 'Marie-Galante' populations in the vicinity of TWC populations, as in Puerto Rico [13], Saint Kitts [18], and Guadeloupe [27].

Potential distribution of native *G. hirsutum* in America and the Pacific

Both the ENM and factorial analyses clearly show that TWC populations of *G. hirsutum* present an exceptional combination of a narrow environmental niche and a highly geographically scattered distribution. Stephens [64] has related the capability for long distance dispersal of tetraploid cotton seeds to their buoyancy and tolerance to prolonged immersion in salt water. It is therefore interesting to extend the TWC climatic model derived from occurrences in Mesoamerica and the Caribbean to a larger area in South America and the Pacific. Figure 4 presents the results of this extrapolation in South America. Four areas offer favorable climatic conditions, two inland areas, Bolivia/Paraguay and Northeastern Brazil, and two coastal areas, Ecuador/Peru and Pacific islands. Strikingly, all of them are validated by the existence of wild populations of tetraploid cottons. The favorable area in Bolivia and Paraguay was suggested long ago by Stephens [7] and, indeed, a wild form of *G. hirsutum* has been reported there recently [25]. Its inland situation renews the question of tetraploid cotton dispersal, as it implies non-oceanic diffusion. A bird-related mechanism is the likely explanation [15]. The other potential inland area, in Northeastern Brazil, corresponds well to the distribution of *G. mustelinum*, a wild tetraploid endemic to the region [65–67]. The third area, in the arid coastal regions of southern Ecuador and northern Peru and in the Galapagos Islands, corresponds with the distribution of 2 other wild tetraploid *Gossypium species*: (i) the wild populations of *G. barbadense* (North and South of the Guayas estuary) and (ii) the wild tetraploid species *G. darwinii*, a close relative of *G. barbadense*, endemic to the Galapagos islands [68].

In the fourth favorable area (not shown), further west into the Pacific, Worldclim coverage is incomplete, particularly for small atolls, so all climatically suitable sites could not be detected. Among those cases where the extrapolation results can be compared to data from the literature, worth mentioning are the Hawaiian Islands (with marginal climatic conditions in leeward coastal areas of Honolulu, Lana'i, Kaua'i and Hawai'i), Wake Island, the Republic of Kiribati, Fiji, Samoa, and French Polynesia. Indeed, Hawaiian Islands are home of the endemic wild tetraploid *G. tomentosum*, while an unusual wild form of *G. hirsutum* is locally common in Wake Island [21,69]. The information available on the presence of wild cotton in Kiribati is less clear, with mentions of *G. tomentosum* [70–72], and/or another *Gossypium* species (probably *G. hirsutum*) [73]. Among the Pacific islands cited for wild populations of *G. hirsutum*, only Fiji and Samoa do not appear climatically fit for this species according to our extrapolation; however, this can be related to the rarity of *G. hirsutum* var. *taitense* Roberty in both archipelagos [74].

The excellent correspondence between areas potentially favorable to wild forms of *G. hirsutum* and the actual distributions of wild tetraploid species (*G. hirsutum* itself, *G. mustelinum*, *G. barbadense*, *G. darwinii* and *G. tomentosum*) provides a very interesting example of ecological niche conservatism in evolution [75]. In the present case, it constitutes a further confirmation that the model derived from our Caribbean and TWC population sample is accurate, and indicates that the main driver of tetraploid cotton radiation was geographic isolation, not environmental specialization.

Potential distribution of *Gossypium hirsutum* in Mesoamerica and the Caribbean at the Last Glacial Maximum

Figure 5 presents the potential distribution of 'truly wild' *G. hirsutum* for LGM climates, i.e. about 21,000BP. Sea level was ca. 125 m lower at that time, and rose markedly from 17,000 to 7,000 BP [76]. According to the MIROC model, most areas where 'truly wild' cotton populations are found under modern climates were only slightly less favorable at LGM. A few very small favorable areas, such as the one along the shores of Tamaulipas, were at best marginally fit for *G. hirsutum*. In contrast, three areas show a considerable extension at LGM, with many more favorable

Table 3. Principal component analysis (Varimax normalized rotation) on a set of bioclimatic variables retained for their contribution to the Maxent ecoclimatic model of distribution: factor loadings (values higher than 0.70 in bold characters).

Variable	Factor 1	Factor 2	Factor 3
1-Annual mean temperature	**0.93**	0.04	0.33
2- mean diurnal range	−0.06	0.04	**−0.82**
4- temperature seasonality	−0.01	−0.29	**−0.80**
6- minimal temperature of coldest month	0.54	0.12	**0.83**
7- temperature annual range	−0.07	−0.09	**−0.97**
8- mean temperature of wettest quarter	**0.90**	−0.18	−0.03
9- mean temperature of driest quarter	**0.82**	0.17	0.37
10- mean temperature of warmest quarter	**0.95**	−0.11	−0.15
11- mean temperature of coldest quarter	0.68	0.18	0.63
12- annual precipitation	−0.07	**0.95**	0.14
16- precipitation of wettest quarter	−0.00	**0.95**	−0.03
19- precipitation of coldest quarter	0.02	**0.80**	0.21
Proportion of total variance	0.34	0.22	0.31

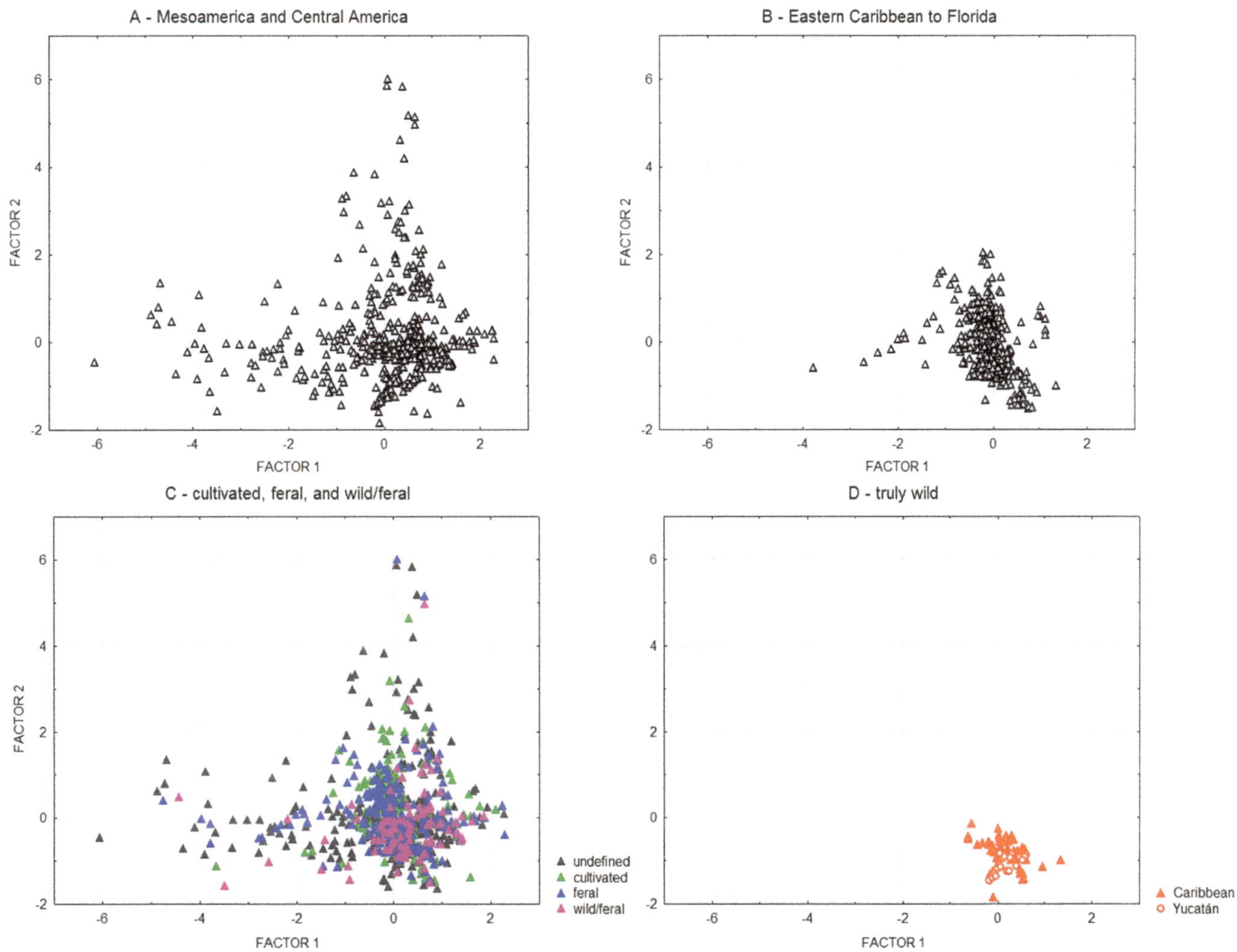

Figure 3. Principal component analysis of *G. hirsutum* climatic envelope. Climate variables are listed in Table 3. Comparison of different subsamples in Mesoamerica, the Eastern Caribbean and Florida, according to origin (A, B) and domestication status (C, D).

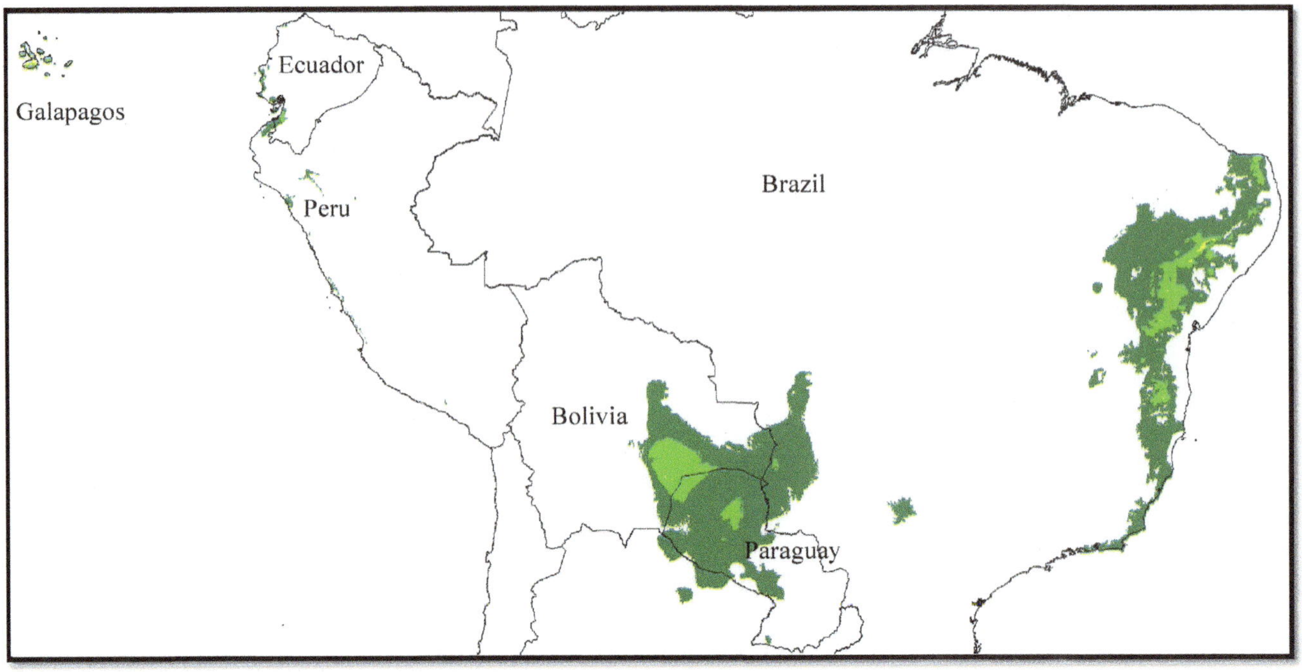

Figure 4. Potential distribution of truly wild *G. hirsutum* in South America. Distribution as extrapolated from the climate model presented in Figures 1C and 2.

emerged lands: (i) northern Yucatán, (ii) southern Florida, the Bahamas and Virgin Islands, and (iii) the western shore of Venezuela and a small area on the northeastern Colombian shores. On the whole, *G. hirsutum* distribution was probably much more extended in the Caribbean and in the Gulf of Mexico during late Pleistocene and early Holocene. The main picture is consistent with the hypothesis of Fryxell [24] that Pleistocene shoreline movements were decisive in the evolution and adaptation of tetraploid cottons.

Further south, in equatorial and southern America, LGM climatically favorable areas appear essentially similar to modern ones, except for the Brazilian Northeast, which was less favorable

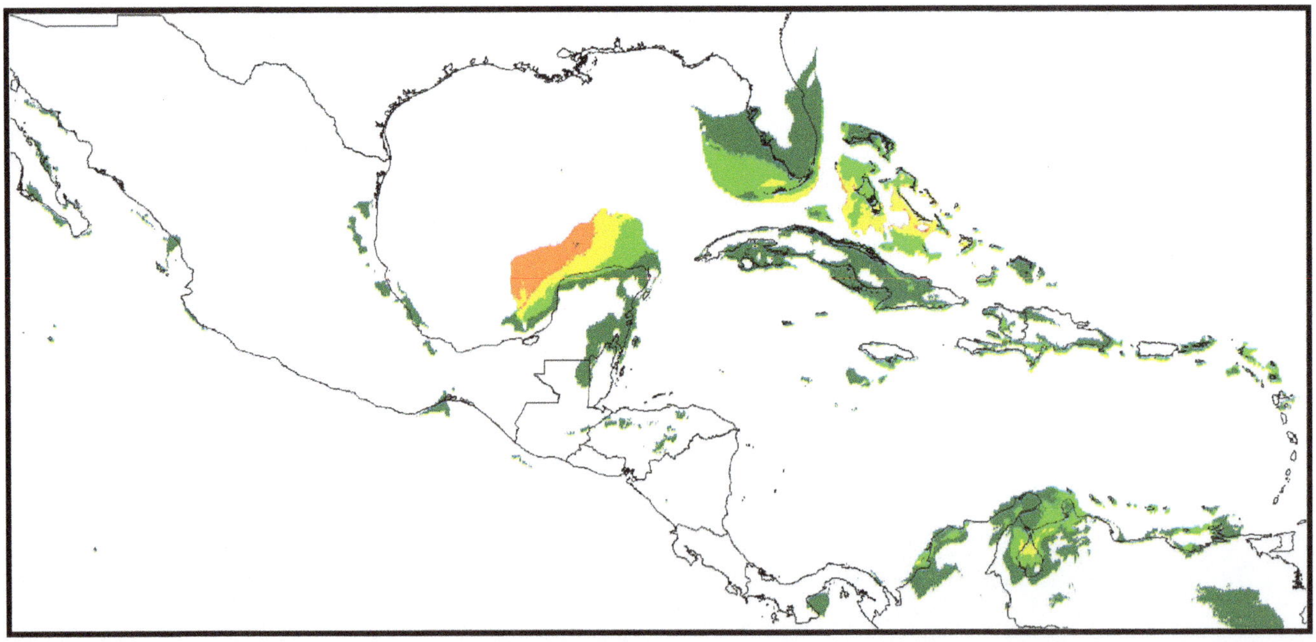

Figure 5. Potential distribution of *G. hirsutum* during the Last Glacial Maximum (21,000 BP). Potential distribution of *G. hirsutum* in the Caribbean and in the Gulf of Mexico, extrapolated according to the MIROC climatic model for LGM. (Note that sea level differences at LGM explain variation with modern sea shore delimitation).

(Figure S2) than in modern times (Figure 4). A few marginally favorable areas may have existed along the Mexican Pacific shores (Figure 5). In the Pacific Ocean, the situation appears similar to the modern one (not shown).

Genetic characterization of 'truly wild' cotton and their feral neighbors

The 42 TWC accessions from 11 different (2 for Mexico alone) countries (Figure 6, Table 1 and Table S1, Figure S1) ensure a good representation of the geographical range of truly wild *G. hirsutum* populations as described above; although several similar populations could not be sampled. It appeared very early in the analysis that these TWC populations, including the 'yucatanense' population of Yucatán as well as those from diverse places in the Caribbean, showed no racial or geographic differentiation, so we have pooled them in the following presentation. In our sample the 'feral' group was represented by 53 'Marie-Galante' accessions of northern South America and the Caribbean and by 15 other accessions ('punctatum' from Yucatán, other races from Mesoamerica, and un-ascribed material, Table 1).

SSR statistics are detailed in Table S2. In total, 204 alleles were coded over the 111 accessions and 26 SSR markers, ranging between 3 (HAU2861) and 19 (BNL3103) alleles per SSR. *He* values varied among markers, confirming previous results [26]. Unique alleles amounted to 37 in the feral 'Marie-Galante' group (53 accessions) and 43 in the TWC group (42 accessions). *He* shows only limited differences between the different races/categories (Table S3); globally it averages 24.2%, more than usually observed in cultivated cotton (between 5 and 15% under field conditions, but nil in the case of our cultivated control). *He* is slightly higher in wild accessions (28.2%) as compared to feral ones (22.0%). The genetic dissimilarity was also higher within the TWC group (D = 0.51) than within the feral group (D = 0.38) (Table S4).

Both distance-based methods implemented with DARwin, NJ classification (Figure S3) and principal coordinate analysis (Figure 7), separate TWC from feral accessions (first axis in the PCA, Figure 7, and basal branching in NJtree, Figure S3). Within the feral group the analyses further distinguished two subgroups. The first one includes 48 of the 53 accessions of race 'Marie-Galante' and the second one includes 17 accessions, 10 of race 'punctatum' from Mexico/Yucatán, 6 others (from races 'morrilli', 'palmeri' and 'richmondi' and 3 unassigned), as well as the modern cultivar. Thus, this clustering, which suffers only few exceptions, appears essentially to reflect domestication status (wild vs. feral), and secondarily race. In the nine locations that could be sampled for both feral and TWC accessions (Figure 6), the different analyses clearly indicated that wild/feral status was better than geographical distribution in determining genetic proximity among

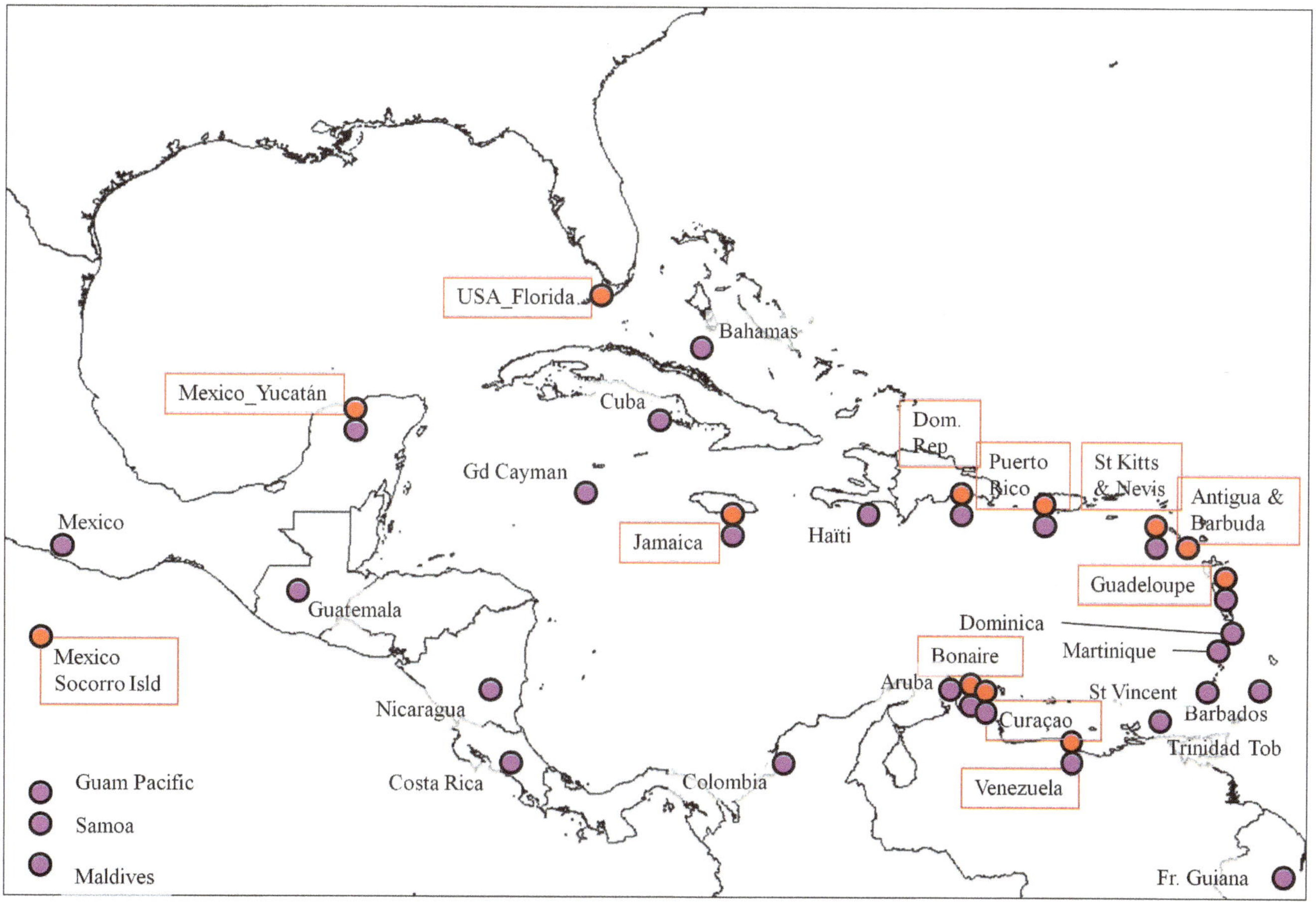

Figure 6. Distribution of the populations of perennial *G. hirsutum* sampled for the SSR-based genetic analysis. Samples include truly wild (TWC) and feral perennial populations in Mesoamerica and the Caribbean. TWC populations are shown as red dots and feral populations are shown as purple dots. Twelve locations where TWC were identified are labeled in red frame. All except USA/Florida, Antigua and Socorro Islands, are also represented by feral specimens, while 18 additional locations had only feral specimens. See Table S1 and Figures S1 for details and precise localizations of the accessions in the 11 sites with TWC populations (Socorro Islands not shown).

samples. For example, the TWC from the Atlantic shores of the Lesser Antilles (Guadeloupe, St Kitts and Antigua) were genetically much closer to TWC from northwestern Yucatán, distant by over 2,000 miles, than they were to the feral cottons of the same islands. The genetic relationship among feral cottons is not determined by geographical proximity either: for example, the six 'Marie-Galante' accessions from the island of Guadeloupe are not grouped in the same 'Marie Galante' branch of the dendrogram (Figure S3B).

The STRUCTURE analysis and ΔK method of Evanno [51] were fully consistent with the two previous ones, clustering the 111 accessions into either two or three clusters, both with high ΔK values (>1000) (Figure S4). Using K = 2 separated TWC from feral cottons (not shown). Using K = 3 further partitioned the feral group in two sub-groups. In Figure 8, we have organized our sample according to the same criteria inferred from both PCA and NJ analyses, but based on field observations: 'truly wild' vs. feral, and feral accessions assigned to 'Marie-Galante' vs. other feral accessions.

The 42 accessions from TWC populations form a fairly homogenous group (Figures 8 and S3) with an average 67% membership. Only few discrepancies were observed, whereby three accessions had very low (<5%) likelihood of membership to this cluster: W30 (acc. AS0340) from Venezuela, W102 (acc. BPS1240) and W103 (acc. BPS1247) from Puerto-Rico. These accessions were probably wrongly assigned due to an error in collection (although passport data are unambiguous) or a mixture at some stage of multiplication. For a few other TWC assignations, the possibility of *in situ* hybridization cannot be dismissed, as they show an important level of admixture (<50% membership to TWC): W105 (acc. INC035) from Socorro Island, W58 (acc. AS0653) from Yucatán, W86 (acc. BPS1157) from Bonaire, W148 (acc. BPS1239) from Puerto Rico, and W95 (acc. BPS1225) from Dominican Republic. For the latter, the collector mentioned a "different" phenotype with orange pollen and yellow petals [19].

The 53 accessions (22 countries) of race 'Marie-Galante' have an average membership of over 81%. This group encompasses the same geographical distribution as the TWC group except for Mexico (Figure 6). Four 'Marie-Galante' accessions present higher membership to the other feral group, probably because of wrong race assignation: W153 (acc. CR2000A) from Costa-Rica, W65 (acc. Texas184) from Guatemala, W150 (acc. BPS1243) from Puerto Rico, and W27 (acc. AS0335) from Venezuela. Two 'Marie-Galante' accessions, W98 (acc. BPS1230) from Puerto Rico and W59 (acc. AS0681) from St Kitts and Nevis, show high levels of admixture with the TWC cluster and both present unusually high rates of heterozygote SSR, of 64% and 58% respectively; they probably result from an hybridization. Of the two accessions sampled in Colombia, one (W32, acc. AS0435) presents 98% membership to the 'Marie Galante' cluster while the other one (W33, acc. AS0437) presents some admixture. It is noteworthy that the latter, W33, belongs to a series of 'Marie-Galante' from Northern Colombia, near Barraquilla, described by Ano and Schwendiman [42] as "híbrido nativo; offspring of ancient deliberate crossings between local spontaneous 'Marie-Galante' and commercial varieties of *G. hirsutum* or *G. barbadense*."

Lastly, the mostly 'punctatum' branch of feral cottons presents the lowest level of admixture (>97% membership). With 16 accessions, this group includes 12 'punctatum' accessions [9 from inland-Yucatán (as opposed to TWC from the northern shores of Yucatán, - 1 from Maldives in the Indian Ocean (W181, acc. KLM1872), - 2 from Pacific islands (W157, acc. TX-0997 from Guam, W108 acc. TX-1295 from Samoa)], one representative of Mexican races, 'morrilli', 'richmondi' and 'palmeri', and the modern cultivar (FM966) from Australia. The homogeneity of this group indicates that the genetic differentiation among Mexican races [26] is negligible as compared to their divergence from both 'Marie-Galante' and TWC populations.

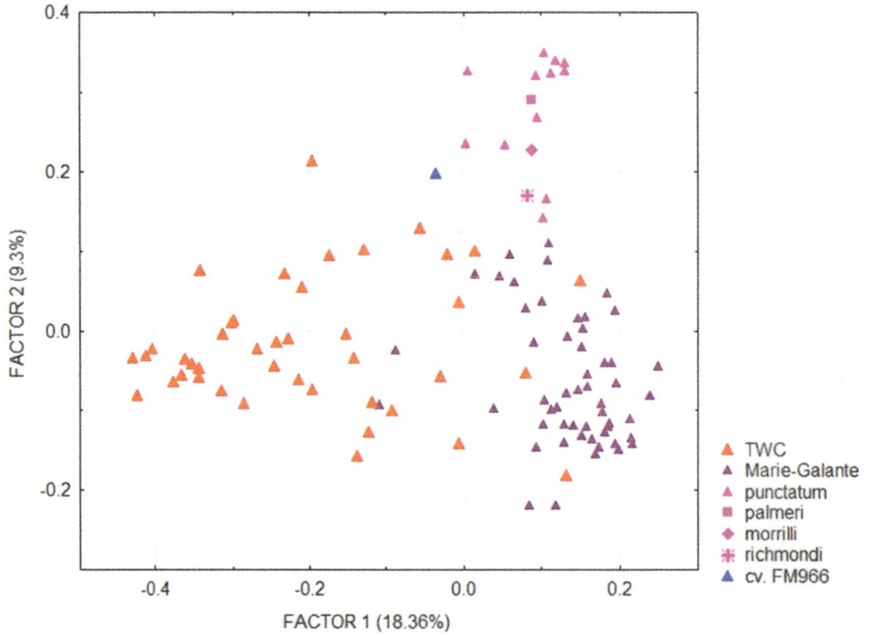

Figure 7. Principal coordinates analysis (PCA) on SSR data in truly wild (TWC) and feral *G. hirsutum*. PCA based on the similarity matrix for 26 SSR markers and 111 accessions represented according to their racial assignation. Factor 1 separates TWC from feral cottons and Factor 2 separates race 'Marie-Galante' from other feral cottons.

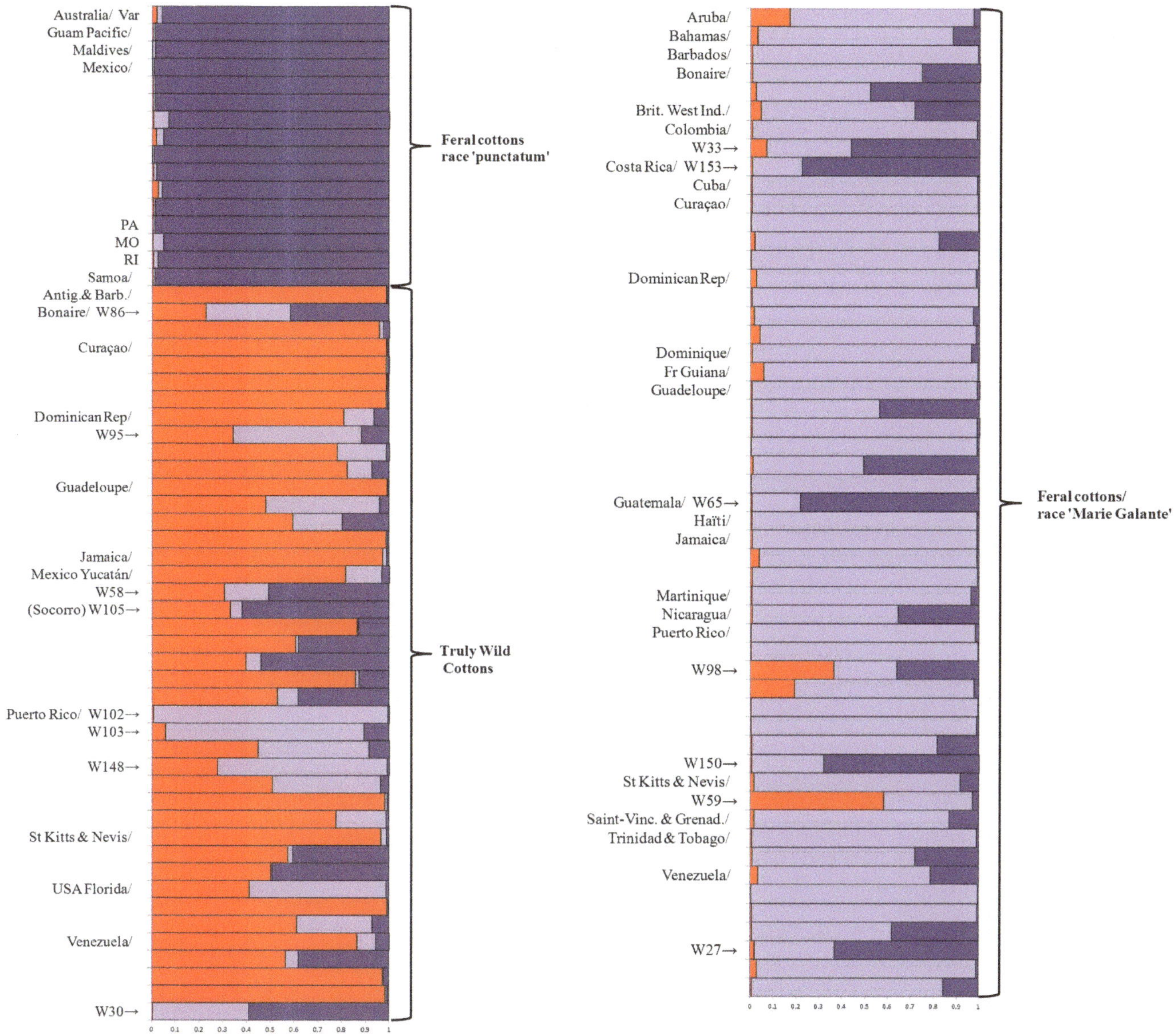

Figure 8. STRUCTURE plot of 111 perennial cottons of *Gossypium hirsutum* with K = 3 clusters. The y-axis shows the proportion membership to the cluster (three clusters depicted in light purple, deep purple and red). Each horizontal bar represents a single accession. The accessions are arranged according to their domestication status and, for feral accessions, their racial assignation, and then alphabetically per country of origin. Fourteen questionable cases (membership to cluster <33%) are indicated with their 'W' accession numbers as detailed in Table S1 (see also comments in the main text). Within cluster 'feral cotton/punctatum', MO, RI, PA, and Var refer to 'morrilli', 'richmondi', 'palmeri' and modern cultivar, respectively.

Distinctiveness of wild and feral populations

The genetic structure observed in a broad collection of cottons representing a vast region of Mesoamerica, Central America, the eastern Caribbean, and even Pacific islands (110 accessions, from 29 different countries or islands) demonstrates that the major driver organizing this collection is the status, feral or wild, of the cotton population, rather than any geographical factor. Thus, ENM and genetic analyses converge in discriminating TWC populations from feral populations, as assessed in our categorization exercise. We can conclude definitively that, not only do 'truly wild' populations of *G. hirsutum* still exist, but they are ecologically and genetically distinct, occupying a narrow and well defined habitat. Their genetic distinctiveness and homogeneity invalidate any racial or specific distinction among TWC populations of *G. hirsutum*, such as their classification into race 'yucatanense' in northern Yucatán and into race 'punctatum' in the Caribbean. *A fortiori*, our results do not support any particular status for the wild perennial cottons from the Dominican Republic, which were the most 'inland' collections among our TWC samples (see Figure S1). These wild cottons had been given racial status (*G hirsutum* race 'ekmanianum') or specific status, as *G. ekmanianum* Wittmack [56,77], and they had even been proposed as a new species by Wendel [78], and other authors, of genome AD_6 (other 5 tetraploid species being denoted as AD_1 to AD_5). Our results do not support such proposals as these specimens fall within the overall range of TWC accessions (Figure S3B). Instead, they unambiguously validate the opinion of Schwendiman and colleagues [19,27] who recognized their morphological unity, from Yucatán to Florida, the Antilles and Venezuela, grouping them under race 'yucatanense'.

The genetic and ecological divergence between race 'yucatanense' *sensu* Schwendiman and feral populations is clearly stronger than the splitting of the latter into two clusters corresponding to (i) races of pure *G. hirsutum* from Mesoamerica, and (ii) the Caribbean and Central American representatives of race 'Marie-Galante' resulting from an introgression with *G. barbadense*. This comparison indicates that domestication resulted in a major infraspecific division in *G. hirsutum*. In any case, the low level of admixture between neighboring TWC and feral populations shows the effects of surprisingly strong reproductive barriers and/or very strict ecological adaptation, resulting in very limited gene flow, despite their geographical proximity.

Distribution and domestication status in *Gossypium hirsutum*

The much stronger differentiation of TWC populations is reminiscent of the study of Stephens [9] who used a similar categorization approach to evaluate the effects of domestication on seed and fiber properties of perennial cottons, well before he formally admitted the existence of 'truly wild' populations of tetraploid cottons. As in our ecological and genetic analyses, his "wild" category was clearly the most distinct. Thus, there were highly significant differences in seed grade, seed index and lint index between the wild and feral categories, whereas differences among feral and cultivated categories were much less marked. The morphological, genetic, and ecological proximity between cultivated and feral cottons can be easily explained if they are closely related, i.e. if the latter are still part of the domesticated genepool. This is first suggested by the fact that feral plants show the same geographic patterns of morphological differentiation as cultivated materials [6,9,10]. Second, the correlation between the occurrence of feral populations and the cultivation of perennial cotton has been reported by most experts, including Ulloa et al. [10] who observed that feral populations are getting rarer as the cultivation of cotton declines in Mexico. This indicates that most feral populations depend on cultivation of ancient landraces for their perpetuation, following a sink-source dynamics model; in ecological terms, their realized niche is wider than their fundamental niche [79]. This double dependence on man, for their man-made habitat and for their reproduction from cultivated plants, contrasts with the long-term permanency of wild coastal populations of *G. hirsutum*. For example, the wild population of Portland Point in Jamaica was mentioned by Britton in 1908 [57], Schwendiman et al. in 1986 [19] and Stoddart and Fosberg in 1991 [58]. Such cases provide excellent illustrations of the fact that, in its original condition, *G. hirsutum* is a pioneer plant colonizing disturbed coastal habitats, but that "this (habitat) instability is in itself highly stable" and very ancient, so "that the pioneers are simultaneously old residents", as Fryxell [24] put it. The relationship between extreme aridity and the occurrence of wild cotton is obviously related to the fact that very few other plant species can compete under such conditions, suggesting that our TWC-specific climatic model is fairly representative of its realized niche. Indeed, as stated by Hutchinson [80], even the most mesophytic members of *Gossypium* are intolerant to competition, particularly at the seedling stage. Contrary to feral cotton, the realized niche of TWC populations is narrower than their fundamental niche.

Thus, the present study provides an opportunity to analyze the effect of domestication on the distribution of cultivated perennials, a rarely studied aspect of domestication. Miller and Knouft [81] have analyzed the case of the jocote or purple mombin (*Spondias purpurea* L.), a small fruit tree native from the dry forests of southern Mexico and Central America, and cultivated for its fruit and/or as a fence. They found that the climatic envelope of the wild populations is nested in that of the cultivated forms. In other words, domestication and cultivation mostly expanded the range of the species. Miller and Knouft [81] attributed this expansion to genetic adaptation, discarding the effect of tending cultivated trees, and, more surprisingly, neglecting the effect of the domestication syndrome itself. Indeed, the domesticated purple mombin produces mostly sterile fruits, so it is essentially reproduced from cuttings that grow much faster than seedlings [82], under much less intense competition.

As compared to purple mombin and the majority of perennial fruit crops, *G. hirsutum* differs in its relatively high level of autogamy and endogamy [83]. Domestication has considerably increased the diversity of the species [6,26] and apparently extended its ecoclimatic range (Figure 3C and D), well beyond the most peripheral and arid habitats of TWC populations. Thus the question remains fundamentally the same: have domestication and selection under cultivation widened the fundamental ecoclimatic envelope of perennial *G. hirsutum* through selection and genetic adaptation? As this envelope is common to feral and cultivated populations, and the feral populations depend on the permanent contribution of cultivated cotton, the most likely answer is that the much wider distribution of these two categories is essentially related to the reduction of competition in cultivated and neighboring disturbed habitats, not to a genetic effect. Furthermore, as the domestication syndrome involves seed characteristics (e.g. seed permeability, hardseededness, dormancy) that are essential for the survival of wild populations, most feral cottons are unable to re-colonize, and persist in, the original habitat of the species. Thus, the apparent paradox is that, although the geographical distribution of perennial *G. hirsutum* has been considerably widened by domestication and cultivation, its niche has been reduced by the loss of reproductive capacity in its natural habitat. In fact, there would have been a true paradox if cultivation, while reducing exposure to both extreme aridity and competition, had increased the competitive potential of *G. hirsutum* in secondary habitats.

Domestication of *Gossypium hirsutum*

Among the important reasons to study the natural distribution of perennial *G. hirsutum* in Mesoamerica and the Caribbean are the identification of potential areas for the early domestication processes and the comparative characterization of domesticated versus wild cottons. Our distribution maps do not contradict the hypothesis of Brubaker and Wendel [8] of an initial domestication of *G. hirsutum* in northern Yucatán, as this region, home of the most extensive wild populations, indeed corresponds to the largest favorable area (Figure 2A). This has been true not only for the last three millennia under modern climates [53] but very likely also for all the Holocene and even earlier, during the late Pleistocene (Figure 5). On the other hand, our maps also support the views of Sauer [11] on a more diffuse process in space and time, with early lint gathering and even trade preceding regular cultivation. Such a process is consistent with the descriptions of multiform exploitation of different cotton populations by Caribbean natives in early colonial chronicles [9], reminiscent of the domestication processes described by Casas et al. [33]. Clearly, *G. hirsutum* lends itself particularly well to such practices. It is naturally restricted to marginal habitats, where it does not suffer much from competition, but as a pioneer species it could have responded very fast and positively to disturbance by man. Its propensity to cross the boundaries between wild, disturbed and cultivated habitats is still obvious today. We can easily imagine how cotton may have invaded spontaneously the surroundings of fishing communities living close to a natural population. Some basic selection in this new habitat would have steadily brought some improvement, progressively providing the genetic basis for more intense management under managed cultivation, and thereby triggering the domestication process. Once the domestication syndrome was acquired, cultivated forms could not revert to the 'truly wild' condition, favoring spatial isolation between the two forms, and in turn further strengthening selection and domestication processes.

However, while the process described above may have taken place both in Mesoamerica and the Caribbean, our genetic data do not favor domestication in the latter area, as Caribbean feral populations appear more closely related to Mesoamerican cultivated and feral cottons than to local TWC populations. Thus, the most likely hypothesis remains that of Brubaker and Wendel [8], with a very early domestication of *G. hirsutum* in northern Yucatán, followed by its progressive diffusion and racial differentiation in all Mesoamerica, then Central America and northern South America. There, race 'Marie-Galante' would have developed through introgression with domesticated forms of *G. barbadense*, as hypothesized by Stephens [7], before reaching the Caribbean.

Conservation of the genetic diversity of *G. hirsutum* and potential interest of wild perennial cottons for breeding

Strategies for the conservation of cotton genetic resources must take into account the relationship between cultivated, feral and wild populations, and the risks of genetic erosion. In the case of the domesticated gene pool, Ulloa et al. [10] have underlined that in southern Mexico *G. hirsutum* perennial cottons survive only as curiosities in garden plots or dooryards, or as occasional feral plants; while attempts at commercial cotton production have been abandoned. In the case of wild cottons, their very ancient habitat is being increasingly threatened, as international tourism covets the same sea-and-sun ecoclimatic niche of dry tropical coasts [19].This point is important in considering the long term *in situ* conservation of perennial cotton *G. hirsutum* populations. Although not considered in this study, the cases of endangered wild *G. barbadense* populations of southern Ecuador/northern Peru, as well as of *G. mustelinum* from northeastern Brazil [84], are similar in ecology and climatic conditions. Only *G. darwinii* from Galapagos is not threatened [85]. The conservation and further plant exploration of wild cottons is important. As highlighted by the results of Liu et al. [28] and Lacape et al. [26], these cottons may have up to 70% unique alleles.

The ecological niche where these wild cotton populations are encountered in Mesoamerica clearly indicates that they represent a great reservoir for genes and alleles related to tolerance to abiotic stresses (water, high temperature or saline stresses). Even though these wild cottons are excellent sources for widening the genetic base for breeding because of their complete interfertility with modern cultivars of *G. hirsutum* [86], this type of material has so far been poorly characterized for its physiological and ecophysiological adaptive traits [87,88] and rarely exploited in breeding programs [89,90].

Lastly, a further understanding of the domestication process, through the comparison of the domesticated and wild pools of *G hirsutum*, for example at the transcriptome level [38], as well as for the identification of valuable phenotypic traits [91,92], can only benefit from an *ad hoc* categorization as attempted in the present study.

Conclusions

Ocean diffusion and ecological constraints, related to extreme aridity and low levels of competition, best explain the past and current distribution of truly wild populations of *G. hirsutum* restricted to littoral or related habitats, on the shores of the Caribbean Sea and the Gulf of Mexico from Venezuela to Florida, and even as far as Polynesian islands in the Pacific Ocean. The obvious niche conservatism expressed in the strong similarity of the natural habitats of all five allotetraploid species shows that their speciation was essentially driven by the geographic, rather than ecological, isolation of their highly scattered populations.

Our ecological and genetic data consistently support the hypothesis of Brubaker and Wendel [8], indicating that upland cotton domestication was very probably initiated in its largest native population, in northern Yucatán. Cultivated forms then diffused progressively to all the Mesoamerican cultural area, differentiating progressively into the five Mesoamerican races, following a process of geographical and cultural isolation [7]. The diffusion of both New World domesticated cottons, *G. hirsutum* and *G. barbadense*, would have allowed genetic introgression in southern Central America and/or northern South America, resulting in the development of race 'Marie-Galante'. The close genetic relatedness between 'Marie-Galante' and the Mesoamerican domesticated races shows that the introgression process was anterior to the diffusion of domesticated *G. hirsutum* to the Antilles.

Even where domesticated and TWC forms grow in close proximity, they hybridize only sporadically. As a result, the level of genetic divergence between them overwhelms differentiation among domesticated races and/or geographic regions.

Our understanding of plant evolution under domestication is more limited for perennial plants than for seed-propagated annual crops [93]. With their evolution from geographically limited wild populations and their concomitant diffusion and racial differentiation, allowing their establishment under warm temperate climates, the two cultivated tetraploid cottons present interesting parallels with the evolution and adaptation of maize in prehistoric and historic times. The persistence of truly wild populations of both species further increases their interest as unique models for understanding how the genomes of perennials respond to selection pressures operating on the relatively short time scale of the domestication process. The existence of three closely related wild species allows situating this process in the general context of the evolution of allotetraploid cottons from a unique hybridization event, 1–2 million years ago [94].

Wild forms of *G. hirsutum*, with seeds only sparsely covered with short fibers but with adaptation to extreme environmental conditions, contrast with cultivated cotton, with its highly valued long-fibered seeds but adaptation to less demanding ecologies. Owing to the advances of genomics and genome sequencing and the ability to scan the genomes of wild species for new and useful genes, we may now be in a position to unlock the genetic potential of the wild germplasm resources of crop plants [91,92], including cotton. The sequences of the two diploid species with genomes closest to the constitutive genomes of tetraploid cottons, the genome D of *G. raimondii* and the genome A of *G. arboreum*, have already been published [95,96]; and the sequencing of the AD genomes of *G. hirsutum* and *G. barbadense* is underway. It should be a relatively easy step now to systematically scan wild germplasm for useful genetic variants. However, this presupposes that *ex situ* collections are adequate, accessible and safe, and that *in situ* preservation efforts are effective in safeguarding material not yet in gene banks, and still evolving in the field. This work should facilitate the development not only of efficient strategies for exploiting cotton diversity for crop improvement, but also of strategies for its long-term conservation.

Supporting Information

Figure S1 Maps of the 11 locations/islands where truly wild cottons (red dots) were sampled for the SSR-based genetic analysis (Socorro Islands not shown). Purple dots represent sampling locations for feral populations in the vicinity. The comments attached to each map are extracted (and translated from French) from collecting reports as indicated.

Figure S2 Potential distribution of *G. hirsutum* during the Last Glacial Maximum (21,000 BP) in South America, extrapolated according to the MIROC climatic model.

Figure S3 Unrooted neighbor-joining tree based on dissimilarities between 110 perennial, and one cultivated, accessions of *Gossypium hirsutum* based for 26 SSR markers.

Figure S4 Prediction of the best value of K (ΔK method of Evanno et al., 2005), from K = 2 to 5 clusters, from the STRUCTURE analysis of 111 perennial cottons of *G. hirsutum*.

Table S1 Detailed records of the 111 accessions used in the SSR-based genetic analysis.

Table S2 List of 26 SSR markers, their chromosome localization on the consensus map (Blenda et al., 2012, *PlosONE* 0045739), and summary statistics (as calculated over 111 accessions): total number of alleles, *He* value as expected heterozygosity, numbers of alleles and unique alleles within groups (MG = race 'Marie-Galante', PU = race 'punctatum', TWC = truly wild cottons).

Table S3 Mean observed heterozygosity (in %) of 26 SSR markers as observed among 110 perennial accessions of *G. hirsutum* (cultivated variety FM966, as 111th accession was 100% homozygote).

Table S4 Mean dissimilarities within and between groups of 110 perennial accessions of *G. hirsutum*. MG, PU and TWC refer to 'Marie-Galante', 'punctatum' and truly wild cottons.

Acknowledgments

Authors are thankful to USDA and CIRAD seed banks curators, James Frelichowski and Dominique Dessauw, for providing seeds of accessions; to Manel Benhassen and Ronan Rivallan for their help in SSR genotyping; and to Luigi Guarino (Global Crop Diversity Trust) and Xavier Perrier (CIRAD) for their valuable comments and suggestions.

Author Contributions

Conceived and designed the experiments: GCE JML. Performed the experiments: GCE JML. Analyzed the data: GCE JML. Contributed to the writing of the manuscript: GCE JML.

References

1. Wendel JF, Brubaker CL, Seelenan T (2010) The origin and evolution of *Gossypium*. In: Stewart JM, Oosterhuis DM, Heitholt JJ, Mauney JR, editors. Physiology of cotton. Dordrecht: Spinger, Netherlands. pp. 1–18.
2. Brubaker CL, Bourland FM, Wendel JF (1999) The origin and domestication of cotton. In: Smith CW, Cothren JT, editors. Cotton, Origin, history, technology, and production: Wiley Series in crop science. pp. 3–31.
3. Stephens SG (1967) Evolution under domestication of the New world cottons (*Gossypium* spp.). Ciencia e cultura 19: 118–134.
4. Wendel JF, Brubaker CL, Percival AE (1992) Genetic diversity in *Gossypium hirsutum* and the origin of Upland cotton. Am J Bot 79: 1291–1310.
5. Wendel JF, Brubaker C, Alvarez I, Cronn RC, Stewart JM (2009) Evolution and natural history of the cotton genus. In: Paterson AH, editor. Plant genetics and genomics Crops and Models Volume 3 Genetics and genomics of cotton: Springer. pp. 3–22.
6. Hutchinson JB (1951) Intra-specific differentiation in *Gossypium hirsutum*. Heredity 5: 169–193.
7. Stephens SG (1973) Geographical distribution of cultivated cottons relative to probable centers of domestication in the New World. In: Adrian M, editor. Genes, Enzymes and Populations. pp. 239–254.
8. Brubaker CL, Wendel JF (1994) Reevaluating the origin of domesticated cotton (*Gossypium hirsutum*, Malvaceae) using nuclear restriction fragment length polymorphism (RFLP). Am J Bot 81: 1309–1326.
9. Stephens SG (1965) The effects of domestication on certain seed and fiber properties of perennial forms of cotton, *Gossypium hirsutum* L. The American Naturalist 94: 365–372.
10. Ulloa M, Stewart JM, Garcia EA, Godoy S, Gaytan A, et al. (2006) Cotton genetic resources in the Western states of Mexico: *in situ* conservation status and germplasm collection for *ex situ* preservation. Genet Resour Crop Evol 53: 653–668.
11. Sauer JD (1967) Geographic reconnaissance of seashore vegetation along the Mexican gulf coast; Mc Intire WG, editor. Baton Rouge. 59 p.
12. Sauer JD (1982) Cayman islands seashore vegetation. A study in comparative biogeography; Press UoC, editor. Berkeley. 161 p.
13. Hutchinson JB (1944) The cottons of Puerto-Rico. The Journal of Agriculture of the University of Puerto Rico 28: 35–42.
14. Stephens SG, Phillips LL (1966) Cotton collection in Colombia, Ecuador, N Peru and Surinam.
15. Stephens SG (1971) Some problems in interpreting transoceanic dispersal of the New world cotton. In: Riley CL, Kelley C, Pennington CW, Rands RL, editors. Man across the sea. University of Texas Press, Austin. pp. 401–415.
16. Hutchinson JB, Silow RA, Stephens SG (1947) The evolution of *Gossypium* and the differenciation of the cultivated cottons: Emprire Cotton Growing Corporation. 160 p.
17. Ano G, Fersing J, Lacape JM (1983) The cottons of Marie-Galante island. Cot Fib Trop 38: 206–208.
18. Ano G, Schwendiman J (1983) Multiphase collecting missions for cotton (I). Plant Genetic Resources Newsletter 54: 5p.
19. Schwendiman J, Percival AE, Belot J-L (1986) Cotton collecting on Carribean islands and South Florida. IBPGR: 4 p.
20. Fryxell PA, Moran R (1963) Neglected form of *Gossypium hirsutum* on Socorro island, Mexico. Empire Cotton Growing Review 40: 289–291.
21. Fosberg FR (1959) Vegetation and flora of Wake Island. Atoll Research Bulletin No67: 20p.
22. Stephens SG (1963) Polynesian cottons. Annals of the Missouri Botanical garden 50: 1–22.
23. Fryxell PA (1965) Stages in the evolution of *Gossypium*. Adv Fontiers Plant Sci 10: 31–56.
24. Fryxell PA (1979) The Natural History of the Cotton Tribe (Malvaceae, Tribe Gossypieae): Texas A&M University Press, College Station, TX. 245 p.
25. Mc Stewart J, Bertoni S (2008). The wild *Gossypium hirsutum* of Paraguay. Belwide Cotton Research Conferences. Nashville (Tennessee). pp. 833.
26. Lacape J-M, Dessauw D, Rajab M, Noyer JL, Hau B (2007) Microsatellite diversity in tetraploid *Gossypium* gene pool: assembling a highly informative genotyping set of cotton SSRs. Mol Breed 19: 45–58.
27. Ano G, Schwendiman J, Fersing J, Lacape J-M (1982) Les cotonniers primitifs de *G. hirsutum* race *yucatanense* de la Pointe des Châteaux en Guadeloupe et l'origine possible des cotonniers tétraploides du Nouveau Monde. Cot Fib Trop: 327–332.
28. Liu S, Cantrell RG, Mc Carty JCJ, Stewart JM (2000) Simple sequence repeat-based assessment of genetic diversity in cotton race stock accessions. Crop Sc 40: 1459–1469.
29. Rensch CR (1976) Comparative Otomanguean Phonology. Indiana University Publications (Language science monographs. Vol. 14). Bloomington.

30. Kaufman S (1990) Early Otomanguean homeland and cultures: Some premature hypotheses. University of Pittsburgh Working Papers in Linguistics 1. pp. 91–136.
31. Smith CE, Stephens SG (1971) Critical identification of Mexican archaeological cotton remains. Economic botany 25: 160–168.
32. Lubbers EL, Chee PW (2009) The worldwide gene pool of *G. hirsutum* and its improvement. In: Paterson AH, editor. Genetics and genomics of cotton: Springer. pp. 23–52.
33. Casas A, Otero-Arnaiz A, Perez-Negron E, Valiente-Banuet A (2007) In situ management and domestication of plants in Mesoamerica. Ann Bot 100: 1101–1115.
34. Ortiz F, Stoner KE, Perrez-Negron E, Casas A (2010) Pollination biology of *Myrtillocactus schenckii* (Cactaceae) in wild and managed populations of the Tehuacan Valley, Mexico. J Arid Env 74: 897–904.
35. Parra F, Casas A, Penaloza-Ramirez JM, Cortez-Palomec AC, Rocha-Ramirez V, et al. (2010) Evolution under domestication: ongoing artificial selection and divergence of wild and managed *Stenocereus pruinosus* (Cactaceae) populations in the Tehuacan Valley. Ann Bot 106: 483–496.
36. Guillen S, Terrazas T, la Barrera E, Casas A (2011) Germination differentiation patterns of wild and domesticated columnar cacti in a gradient of artificial selection intensity. Genet Resour Crop Evol 58: 409–423.
37. Butterworth KM, Adams DC, Horner HT, Wendel JF (2009) Initiation and early development of fiber in wild and cultivated cotton. Int J Plant Sci 170: 564–574.
38. Rapp RA, Haigler CH, Flagel L, Hovav RH, Udall JA, et al. (2010) Gene expression in developing fibers of Upland cotton (*Gossypium hirsutum* L.) was massively altered by domestication. BMC Biol 8: 139.
39. Wegier A, Pinero-Nelson A, Alarcon J, Galvez-Mariscal A, Alvarez-Buylla ER, et al. (2011) Recent long-distance transgene flow into wild populations conforms to historical patterns of gene flow in cotton (*Gossypium hirsutum*) at its centre of origin. Mol Ecol 20: 4182–4194.
40. Waltari E, Hijmans RJ, Peterson AT, Nyari AS, Perkins SL, et al. (2007) Locating pleistocene refugia: comparing phylogeographic and ecological niche model predictions. PLoS ONE 2: e563.
41. Martínez-Meyer E, Peterson AT (2006) Conservatism of ecological niche characteristics in North American plant species over the Pleistocene-to-Recent transition. Journal of Biogeography 33: 1779–1789.
42. Ano G, Schwendiman J (1981) Rapport de mission en Guyane française - Venezuela - Colombie sur la préservation des ressources génétiques du cotonnier. FAO-IBPGR. 44 p.
43. Ano G, Schwendiman J (1982) Rapport de mission au Mexique sur la préservation des ressources génétiques du cotonnier. FAO-IBPGR. 33 p.
44. Hijmans RJ, Cameron SE, Parra JL, Jones PG, Jarvis A (2005) Very high resolution interpolated climate surfaces for global land areas. Int J Climat 25: 1965–1978.
45. Phillips SJ, Anderson RP, Schapire RE (2006) Maximum entropy modeling of species geographic distributions. Ecol Modelling 190: 231–259.
46. Chan WL, Abe-Ouchi A, Oghaito R (2011) Simulating the mid-Pliocene climate with the MIROC general circulation model: experimental design and initial results. Geosci Model Dev 4.
47. Risterucci A-M, Grivet L, N'Goran JAK, Pieretti I, Flament MH, et al. (2000) A high density linkage map of *Theobroma cacao* L. Theor Appl Genet 101: 948–955.
48. Perrier X, Jaquemoud Collet JP (2006) DARwin software. Available: http://darwin.cirad.fr/darwin.
49. Saitou N, Nei M (1987) The neighbor-joining method: a new method for reconstructing phylogenetic trees. Mol Biol Evol 4: 406–425.
50. Pritchard JK, Stephens M, Donnelly P (2000) Inference of population structure using multilocus genotype data. Genetics 155: 945–959.
51. Evanno G, Regnau S, Goudet J (2005) Detecting the number of clusters of individuals using the software STRUCTURE: a simulation study. Mol Ecol 14: 2611–2620.
52. Earl DA, von Holdt BM (2012) STRUCTURE HARVESTER: a website and program for visualizing STRUCTURE output and implementing the Evanno method. Conservation Genet Resour 4: 359–361.
53. Metcalfe SE (2006) Late Quaternary Environments of the Northern Deserts and Central Transvolcanic Belt of Mexico Annals Missouri Botanical Garden 93: 258–273.
54. Percival AE, Kohel RJ (1990) Distribution, collection, and evaluation of *Gossypium*. Adv Agron 44: 225–256.
55. Hutchinson JB (1943) Notes on the native cottons of Trinidad. Tropical Agriculture 20: 235–238.
56. Krapovickas A, Seijo G (2008) *Gossypium ekmanianum* (Malvaceae), algodon silvestre de la Republica Dominicana. Bonplandia 17: 55–63.
57. Coville FV, Britton NL, Cook OF (1908) Wild Jamaica cotton. Science 24: 664–666.
58. Stoddart DR, Fosberg FR (1991) Plants of the Jamaican cays. Atoll Research Bulletin 352: 24 p.
59. Proctor GR (1980) Checklist of the plants of little Cayman. Atoll Research Bulletin 241: 71–80.
60. Stoddart DR (1980) Vegetation of little Cayman. Atoll Research Bulletin 241: 53–70.
61. Britton NL, Millspaugh CF (1920) The Bahama flora; Britton NL, editor. New York.
62. Nickrent DL, Eshbaugh WH, Wilson TK (2008) The vascular flora of Andros island, Bahamas.
63. Britton NL (1918) Flora of Bermuda. New York.
64. Stephens SG (1966) The potentiality for long range oceanic dispersal of cotton seeds. The American Naturalist 100: 199–210.
65. Pickersgill B, Barrett SCH (1975) Wild cotton in Northeast Brazil. Biotropica: 42–54.
66. Barroso PAV, Hoffmann LV, Freitas RBd, Araujo Batista CEd, Alves MF, et al. (2010) *In situ* conservation and genetic diversity of three populations of *Gossypium mustelinum* Miers ex Watt. Genet Resour Crop Evol 57: 343–349.
67. de Menezes IPP, Gaiotto FA, Hoffmann LV, Ciampi AY, Barroso PAV (2014) Genetic diversity and structure of natural populations of *Gossypium mustelinum*, a wild relative of cotton, in the basin of the De Contas River in Bahia, Brazil. Genetica 142: 99–108.
68. Wendel JF, Percy RG (1990) Allozyme diversity and introgression in the Galapagos Islands endemic *Gossypium darwinii* and its relationship to continental *G barbadense*. Bioch Syst Ecol: 517–528.
69. Fosberg FR, Sachet MH (1969) Wake island vegetation and flora. Atoll Research Bulletin 123.
70. Degener O, Degener I (1959) Canton island, South Pacific. Atoll Research Bulletin 64: 24 p.
71. Fosberg FR, Stoddart DR (1994) Flora of the Phoenix islands, Central Pacific. Atoll Research Bulletin 393: 61 p.
72. Bakineti T, Seluka S (1999) Country report for the Pacific sub-regional workshop on forest and tree genetic resources. 12–16 April 1999. In: SPRIG A, FAO, SPC, SPREP, editor. State of forest and tree genetic resources in the Pacific islands, and sub-regional action plan for their conservation and sustainable use. Apia, Samoa.
73. Fosberg FR, Sachet MH (1987) Flora of the Gilbert islands, Kiribati. Atoll Research Bulletin 295: 34 p.
74. Whistler A (2011) The rare plants of Samoa; Atherthon J, Duffy L, editors. Apia: Biodiversity Conservation. Lessons learned Technical Series. 212 p.
75. Peterson AT, Soberon J, Sanchez-Cordero V (1999) Conservatism of ecological niches in evolutionary time. Science 285: 1265–1267.
76. Fleming K, Johnston P, Zwartz D, Yokohama Y, Lambeck K, et al. (1998) Refining the eustatic sea-level curve since the Last Glacial Maximum using far- and intermediate-field sites. Earth and Planetary Science Letters 163: 327–342.
77. Grover C, Grupp KK, Wanzek RJ, Wendel JF (2012) Assessing the monophyly of polyploid *Gossypium* species. Plant Syst Evol 298: 1177–1183.
78. Wendel JF (2012) Genes, Jeans and Genomes: Exploring the Mysteries of Polyploidy in Cotton, in International Cotton Genome Initiative (ICGI) Research Conference, Raleigh (NC). Available: http://www.cottoninc.com/fiber/Agricultural-Research/Agricultural-Meetings-Conferences/ICGI-Presentations/p01-Wendel-Jonathan/01-Wendel-Jonathan.pdf. Accessed 2014 Jun 25.
79. Pulliam HR (2000) On the relationship between niche and distribution. Ecology Letters 3: 349–361.
80. Hutchinson JB (1954) New evidence on the origin of the old world cottons. Heredity 8: 225–241.
81. Miller AJ, Knouft JH (2006) GIS-based characterization of the geographic distributions of wild and cultivated populations of the Mesoamerican fruit tree *Spondias purpurea* (Anacardiaceae). Am J Bot 93: 1757–1767.
82. Morton J (1987) Purple Mombin. In: Morton JF, editor. Fruits of warm climates. Miami, FL. pp. 242–245.
83. Alves MF, Pereira FRA, de Andrade AM, de Menezes IPP, Hoffmann LV, et al. (2009) Molecular polymorphic markers between moco and herbaceous cotton. Rev Cienc Agron 40: 406–411.
84. Alves MF, Barroso PAV, Ciampi AY, Hoffmann LV, Azevedo VCR, et al. (2013) Diversity and genetic structure among subpopulations of *Gossypium mustelinum* (Mavaceae). Genetics and Molecular Research 12: 597–609.
85. Darwin-Foundation. *Gossypium darwinii*. Galapagos Species Checklist. Available: http://wwwdarwinfoundationorg/datazone/checklists/573/. Accessed 2014 Jun 25.
86. Stewart JM (1994). Potential for crop improvement with exotic germplasm and genetic engineering. In: Constable GA, Forrester NW, editors. Challenging the future, Proc World Cotton Res Conf -1. Brisbane (Aus). pp. 313–327.
87. Quisenberry JE, Jordan WR, Roark B, Fryrear DW (1981) Exotic cottons as genetic sources for drought resistance. Crop Sc 21: 889–896.
88. Wu T, Weaver DB, Locy RD, McElroy S, van Santen E (2014) Identification of vegetative heat-tolerant upland cotton (*Gossypium hirsutum* L.) germplasm utilizing chlorophyll fluorescence measurement during heat stress. Plant Breeding 133: 250–255.
89. Mc Carty JCJ, Jenkins JN, Wu J (2004) Primitive accession derived germplasm by cultivar crosses as sources for cotton improvement: II. Genetic effects and genotypic values. Crop Sc 44: 1231–1235.
90. Mc Carty JC, Wu J, Jenkins JN (2006) Genetic diversity for agronomic and fiber traits in day-neutral accessions derived from primitive cotton germplasm. Euphytica 148: 283–293.
91. Tanksley SD, Mc Couch S (1997) Seed banks and molecular maps: unlocking genetic potential from the wild. Science 277: 1063–1066.
92. McCouch S, Baute GJ, Bradeen J, Bramel P, Bretting PK, et al. (2013) Agriculture: Feeding the future. Nature 499: 23–24.
93. Miller AJ, Gross BL (2011) From forest to field: Perennial fruit crop domestication. Am J Bot 98: 1389–1414.

94. Wendel JF, Cronn RC (2003) Polyploidy and the evolutionary history of cotton. Adv Agron 78: 140–186.
95. Paterson AH, Wendel JF, Gundlach H, Guo H, Jenkins J, et al. (2012) Repeated polyploidization of *Gossypium* genomes and the evolution of spinnable cotton fibres. Nature 492: 423–427.
96. Li F, Fan G, Wang K, Sun F, Yuan Y, et al. (2014) Genome sequence of the cultivated cotton *Gossypium arboreum*. Nat Genet 46: 567–572.

4

Effects of Land Management Strategies on the Dispersal Pattern of a Beneficial Arthropod

Chiara Marchi[1]*, Liselotte Wesley Andersen[2], Volker Loeschcke[1]

1 Department of Bioscience, Aarhus University, Aarhus, Denmark, **2** Department of Bioscience, Aarhus University, Rønde, Denmark

Abstract

Several arthropods are known to be highly beneficial to agricultural production. Consequently it is of great relevance to study the importance of land management and land composition for the conservation of beneficial aphid-predator arthropod species in agricultural areas. Therefore our study focusing on the beneficial arthropod *Bembidion lampros* had two main purposes: I) identifying the physical barriers to the species' dispersal in the agricultural landscape, and II) assessing the effect of different land management strategies (i.e. use of pesticides and intensiveness) on the dispersal patterns. The study was conducted using genetic analysis (microsatellite markers) applied to samples from two agricultural areas (in Denmark) with different agricultural intensity. Land management effects on dispersal patterns were investigated with particular focus on: physical barriers, use of pesticide and intensity of cultivation. The results showed that *Bembidion lampros* disperse preferably through hedges rather than fields, which act as physical barriers to gene flow. Moreover the results support the hypothesis that organic fields act as reservoirs for the re-colonization of conventional fields, but only when cultivation intensity is low. These results show the importance of non-cultivated areas and of low intensity organic managed areas within the agricultural landscape as corridors for dispersal (also for a species typically found within fields). Hence, the hypothesis that pesticide use cannot be used as the sole predictor of agriculture's effect on wild species is supported as land structure and agricultural intensity can be just as important.

Editor: Anna-Liisa Laine, University of Helsinki, Finland

Funding: Funded by the Ministry of Food, Agriculture and Fisheries under the Finance and Appropriation Act, Sections 24.33.02.10. Part of the Ph.D. project is funded by AGSoS (Aarhus Graduate School of Science). The funders had no role in study design, data collection and analysis, decision to publish, or preparation of the manuscript.

Competing Interests: The authors have declared that no competing interests exist.

* E-mail: chiarapuspa@gmail.com

Introduction

A number of arthropods, typically living in agricultural areas, have been shown to be highly beneficial to agricultural production, because they act as predators on pest species [1]. These beneficial species are of major importance especially in relation to organic farming, and their abundance and dispersal pattern can be expected to be influenced by land management.

Bembidion lampros, a small polyphagus arthropod (of the Carabidae family) has been shown to be efficient as aphid pest controller in agricultural areas [1–3] and can also be used as indicator for insect abundance and diversity [4]. Effects of landscape structure and land use on *B. lampros* have been widely investigated, but focusing only on the species abundance rather than on the drivers of dispersal in the agricultural landscape.

The typical *B. lampros* life cycle is marked by an overwintering stage where individuals reside in hedges at the field margins and a dispersal stage in early spring where they move into the field [5]. The timing of dispersal is determined by air temperature and occurs in Denmark usually between early March and early April [5]. After reproducing in the field the individuals return in early autumn to the hedges to overwinter [6,7]. *B. lampros* is univoltine [6] and short-lived but it lives mostly more than a year. It is one of the early spring-moving species [5], thus it has probably a relatively larger effect on aphid pest control compared to later-moving species [3]. While its life cycle has been thoroughly studied, the dispersal routes within the agricultural land are largely unknown. In particular it is not known whether *B. lampros* cross hedgerows, tend to return to the same hedgerow after reproduction or easily cross ploughed fields. Such knowledge however is crucial for understanding the population dynamics of this highly beneficial species [2]. Since this information is difficult to get through mark-recapture studies (given the very high population number and the small size of this species), the use of molecular genetic markers may help to elucidate this part of its ecology. Knowing which factors act as barrier to dispersal can be used to predict the probability of an area to receive immigrants and consequently be re-colonized in case of a sudden drop in abundance (e.g. after pesticides spraying). This knowledge could allow the implementation of management strategies that prevent populations going extinct or suffer severe bottlenecks (which could then lead to local extinction).

Hedgerow types, soil type, presence of pesticides and number of mechanical soil treatments have a clear effect on the abundance of *B. lampros* [8–10]. In particular many commonly used insecticides are shown to be detrimental to this species, both under laboratory and field conditions [11,12]. Even pesticides legally used in organic fields in many European countries (but not in Denmark) have been proven to be harmful [9]. The use of pesticides in the field has led to a reduction in *B. lampros* abundance persisting up to four weeks after spraying; the following return to abundance levels preceding the spraying was ascribed to an effect of re-colonization from untreated nearby areas [13]. This might suggest that fields grown organically can act as genetic reservoirs for the re-colonization of conventional fields.

Several studies have shown the importance of cultivation intensity for the abundance and mobility of *B. lampros*, in particular of low yield cultivation with high weed abundance [14,15]. Therefore the effect of agricultural intensiveness was included in our study by comparing two areas with very different agricultural intensity. The organic fields in the two areas also differed in time since conversion from conventional fields.

The main objectives of our study were :

i) to analyze the population structure and gene flow pattern in the agricultural landscape and thereby identifying possibly dispersal barriers (with particular attention to fields, hedge-rows and roads).
ii) to test the hypothesis that organically grown fields act as genetic reservoirs for conventional grown fields and the hypothesis that areas with low intensity agricultural land use have a higher genetic diversity, effective population size and comparatively higher gene flow.

Materials and Methods

Sampling design

We selected two agricultural areas with different intensity of cultivation composed of conventional and organic fields separated by tree-covered hedgerows. Sampling was performed for two consecutive years in early autumn (when individuals return to the hedgerows to overwinter) at sampling sites along the hedgerows.

Individuals were sampled along the hedgerows from two agricultural areas in Denmark, Kalø 56.363° North 9.700° East, and Bjerringbro 56.290° Nord 10.500° East. The two areas represented an intensive (Bjerringbro) and an extensive (Kalø) type of management: Bjerringbro area had a higher average amount of cultivated areas (66%) compared to the national level (61%), while the opposite was true for the Kalø area where only 52% of land is used for agriculture (Beate Strandberg, pers. comm.). In Kalø, the fields have been organically managed for many years (>15 years), longer than in Bjerringbro, and the weed cover in the fields in Kalø was also much higher (Beate Strandberg pers. comm.). The sampling scheme was designed to obtain samples from hedgerows dividing two organic or two conventional fields as well as from hedgerows dividing organic and conventional fields (on both sides). Regarding the pesticide regime the organic fields were never treated with pesticides/herbicides while these were routinely applied to conventional fields. Therefore conventional vs. organic classification defines pesticide regime. Each hedgerow presented four sampling sites 50 meters apart, two on both sites just opposite each other with a width of one meter. A total of 1140 samples were hand-picked during the two seasons (see Table 1 and Fig. 1 for details). Before DNA-extraction all individuals were identified to species morphologically.

DNA extraction and PCR details

Samples were frozen dry upon collection and stored at −20°C. DNA was extracted using a standard CTAB buffer and proteinase-K procedure [16,17]. Two multiplex PCR runs were used to amplify 15 microsatellite markers developed specifically for this species (Marchi *et al.* unpublished) using Qiagen Multiplex enzyme following the manufactories instructions. First multiplex: BL8, BL11, BL19, BL22, BL28, BL28b, BL30, BL34; second multiplex: BL2, BL5, BL6, BL12, BL14, BL24, BL27). Thermal profile, for both multiplex runs, was: 95°C for 15 min; then 35 cycles of 30 s at 94°C, 45 s at 57°C and 60 s at 72°C, with a final extension at 60°C for 30 min. PCR products were analysed using an ABI 3730 automated sequencer and typed using GENEMAPPER version 4.1 (Applied Biosystems).

Genetic diversity and effective population size analysis

Genetic diversity for each sampling site and each year was estimated as allelic richness with FSTAT [18] and expected heterozygosity with GENALEX [19]. Tests for goodness of fit to Hardy-Weinberg expectations and linkage equilibrium were performed in FSTAT. Moreover effective population size (N_e) was estimated with NEESTIMATOR [20] for each sampling site for both years using "heterozygote excess method" [21] and Waples' [22] moments based approach. The differences in genetic diversity and N_e among conventional and organic sampling sites were tested using a Student t-test. The same tests were repeated, grouping the samples with regard to neighboring field's management as well (sampling site in conventional field with neighboring organic field versus sampling site in conventional field with neighboring conventional field and so on for all pairs).

Genetic structure analysis

Population structure was evaluated using a Bayesian based cluster analysis implemented in GENELAND R package [23]. The analyses were performed for Kalø and Bjerringbro data using geographic coordinates as prior and k(max) = 10 (based on five separate runs each with 20,000/100,000 burn-in/sampling iterations).

The population structure was also estimated using pairwise multilocus F_{ST} between sampling sites using ARLEQUIN v. 3.5 [24].

Gene flow analysis

The effect of environmental variables on the pattern of gene flow between sampling sites were assessed using BIMr [25]. BIMr estimates recent gene flow to find the best explanatory factors for the recovered pattern among the given environmental distance matrices using a linear model fit with the gene flow matrix acting as dependent variable (a null model, including none of the specified factors, is always included in the analysis). The variables included in the analysis were: distance, hedgerow of belonging, field of belonging (hedgerow or field nearest to the sampling site), management type (conventional or organic) and road side (only for Bjerringbro). The inclusion of distance between sampling sites was used to account for possible effects of spatial autocorrelation. BIMr analyses were run ten times for each area (using 1,000,000(burn in)/1,000,000(sampling) runs) and the gene flow values were obtained from the best run (which was selected on the basis of $D_{assignment}$ as suggested by Faubet et al. [26]; when more than one run presented the same $D_{assignment}$, posterior probability and the visual analyses of likelihood plot for each run were used to determine the best run). BIMr results were analyzed also on the basis of the alpha value recovered for each factor (or combination of factor): "the sign and the magnitude of the alpha's tell us about the direction and the strength of the environmental factors" [25].

Assignment test

Isolation of the clusters detected by GENELAND was checked using a series of assignment tests conducted in GENECLASS [27]. We performed four self-assignment tests using data from 2008 and 2009, separately, for each area. Then two assignment tests were performed assigning 2009 individuals using 2008 individuals as reference. For all six tests the individuals were divided according to

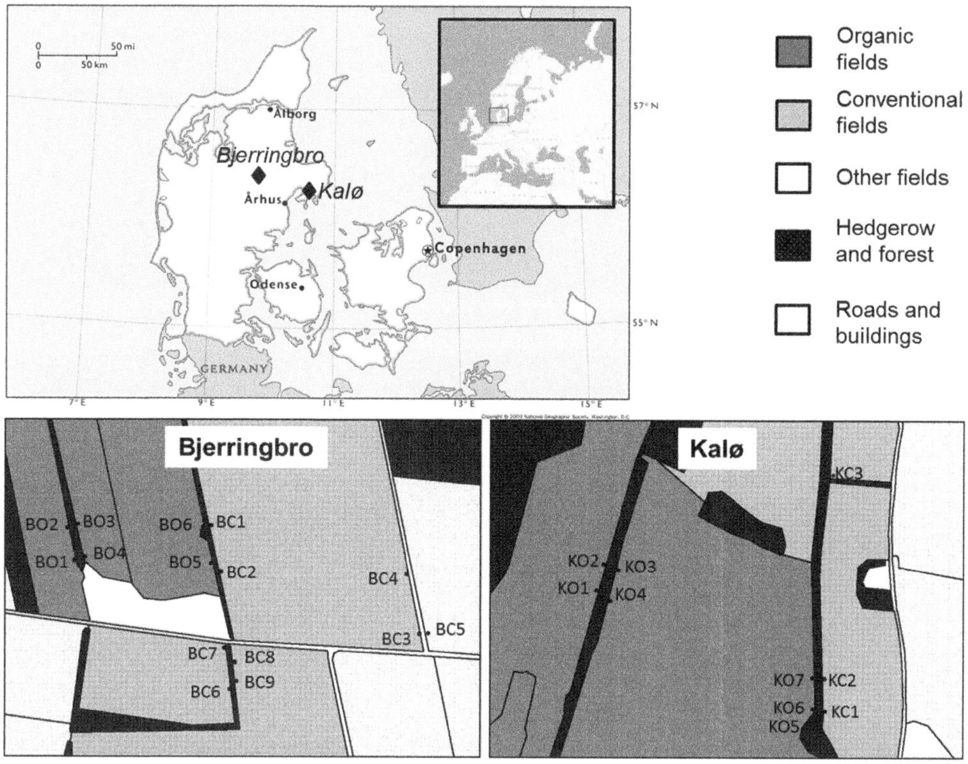

Figure 1. Map of sampling sites. The top map shows the location of the two areas used in this study. The maps at the bottom show the location and name of each sampling site for both areas (B = Bjerringbro, K = Kalø) and management types of the field (C for conventional or O for organic) and ordinal number.

the grouping recovered by the genetic structure analysis in GENELAND. All tests were performed using Rannala and Mountain [28] method; 10,000 simulations with Paetkau et al. [29] method and a type I error rate $P<0.01$.

Results

Genetic diversity and effective population size analysis

No significant deviation from Hardy-Weinberg- or linkage equilibrium was found. The genetic diversity indexes were very similar across sampling sites with different management and across years. The Kalø area showed a significantly higher allelic richness compared to Bjerringbro area (Table S1). Effective population size did not show any difference across area, year or management type (Tables S1 and S2).

Genetic structure analysis

GENELAND analysis performed on samples from both years detected five clusters in Kalø and eight in Bjerringbro. In both cases no migrants were identified and samples from 2008 clustered together with the corresponding ones from 2009. All the sampling sites that did not constitute a cluster of their own were grouped accordingly to the nearest hedgerow (Fig. 2).

The pairwise F_{ST} analyses showed higher F_{ST} values in Bjerringbro than in Kalø; despite the presence of significantly differentiated pairs, no isolated (group of) sampling sites were found (Fig. 3). However, in Kalø there were two more distinct groups: KO1-KO2-KO3-KO4 and KO7-KC1-KC2-KC3 (Fig. 3) which partly agreed with the results from GENELAND (Fig. 2) as the sampling sites in the two groups belonged to the same hedgerow (confirming the effect of fields as genetic barriers).

Gene flow analysis

BIMr analyses of Bjerringbro data converged in nine out of ten cases. "Hedgerow of belonging" was the most important factor in three out of nine runs. Further, in two single runs "road side" and "field of belonging", respectively, were the most important explanatory factors. In other three runs "distance" (two runs) and "distance plus road side" (one run) were most important. When two factors were combined, "road side" and "field of belonging" were the least important ones (lowest absolute value of alpha). One run found the best fit to be the null model. The highest posterior probability and highest alpha were given to "hedgerow of belonging" alone. For the best run, the most important factor was "hedgerow of belonging" and the gene flow was higher for pairs of sampling sites belonging to the same hedgerow (see alpha values in Table 2).

All ten runs regarding Kalø samples converged. In nine out of ten cases "management type" was the most important factor alone (four runs), together with "hedgerow of belonging" (4 runs) or together with" field of belonging" (1 run). In cases where two factors were present, "management type" was the most important one (highest absolute number of alpha). The best run found "management type" alone to be the most important factor (with a posterior probability of 0.582, see Table 3) and, as the correlation coefficient (alpha) was negative, the migration was lowest from conventional to organic sampling sites and highest from organic to conventional ones (see Materials and Methods for details).

In all cases, in both areas, belonging to the same hedgerow was found to increase gene flow (see alpha values for hedgerow in Tables 2 and 3).

Table 1. Sampling details.

	2008	2009	Total	Management
Bjerringbro				
BO1	28	0	28	Organic
BO2	28	0	28	Organic
BO3	24	17	41	Organic
BO4	29	27	56	Organic
BO5	34	28	62	Organic
BO6	20	22	42	Organic
BC1	30	29	59	Conventional
BC2	20	30	50	Conventional
BC3	23	9	32	Conventional
BC4	5	29	34	Conventional
BC5	0	25	25	Conventional
BC6	15	17	32	Conventional
BC7	9	7	16	Conventional
BC8	27	28	55	Conventional
BC9	26	26	52	Conventional
Total	318	294	612	
Kalø				
KO1	18	30	48	Organic
KO2	27	28	55	Organic
KO3	25	30	55	Organic
KO4	27	30	57	Organic
KO5	27	0	27	Organic
KO6	26	30	56	Organic
KO7	29	30	59	Organic
KC1	29	28	57	Conventional
KC2	30	30	60	Conventional
KC3	27	27	54	Conventional
Total	265	263	528	

Number of sampled individuals per sampling site, per year and in total (first to third column). Last column shows the type of management.

Assignment test

The self-assignment tests, performed on the samples from 2008 and 2009 for the two areas, were in good agreement with the population structure results as the percentage of mis-assigned individuals was generally low (Tables S3 and S4). The assignment test (2008 samples as reference for 2009) did not give clear cut results as most individuals could be assigned to more than one cluster. This was probably due to a lack of statistical power.

Discussion

The main objectives of this study were: I) to determine the gene flow and the population structure patterns of *B. lampros* in agricultural areas with respect to landscape features (particularly: fields, hedgerows and roads) and II) to determine the effect of different land management strategies (specifically, pesticide use and agricultural intensiveness) on such patterns.

Genetic diversity and effective population size analysis

Genetic diversity indexes and effective population sizes were similar across all sampling sites both within and across areas (with the exception of allelic richness which was much higher in Kalø). There was also no indication of an effect of management. The reason for this homogeneity might lie in the fact that the populations are not genetically isolated (as proven by the fact that the majority of pairs, in the F_{ST} analysis, did not differ significantly from each other).

Genetic structure analysis

The genetic structure analysis performed in GENELAND showed that fields act as barriers to gene flow. The GENELAND algorithm was chosen as it has been shown to be particularly good at finding genetic boundaries, especially when the population subdivision is recent [30]. Many sampling sites constituted a cluster of their own, while the ones that were grouped, always belonged to the same hedgerow. Management did not seem to affect the population structuring. It should also be noted that samples from the two separate years always clustered together showing that genetic structure is consistent through years.

The patterns recovered by the F_{ST} analysis were very different in the two areas (Fig. 4). The F_{ST} values were higher in Bjerringbro than in Kalø; however, there were very few significantly different pairs in Bjerringbro compared to Kalø. In Kalø the F_{ST} pattern is in good agreement with GENELAND results as sampling sites lying on different field's sides are more differentiated than the ones sharing the same hedgerow. This results suggest a higher gene flow (lower F_{ST} values), but an older isolation of sampling sites (more significantly different pairs) in Kalø compared to Bjerringbro. This difference might be due to a more consistent field structure in Kalø, where field management changed less over the years (Beate Strandberg, pers. comm.).

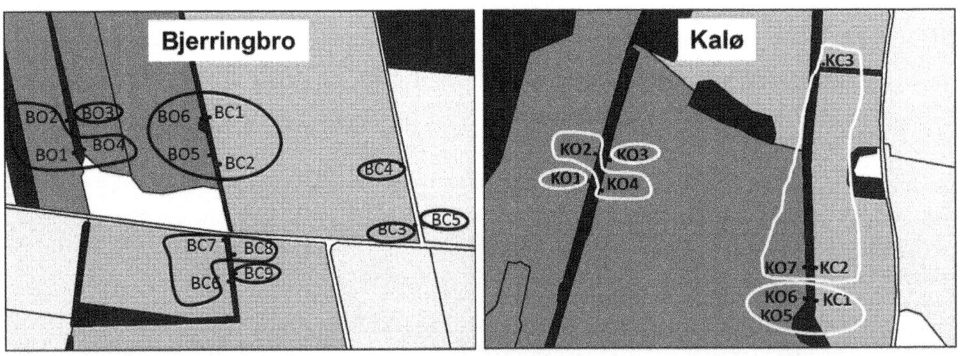

Figure 2. Genetic clusters recovered by GENELAND. The figure shows the genetic clusters for each area with each cluster being included in a solid line (black for Bjerringbro, white for Kalø).

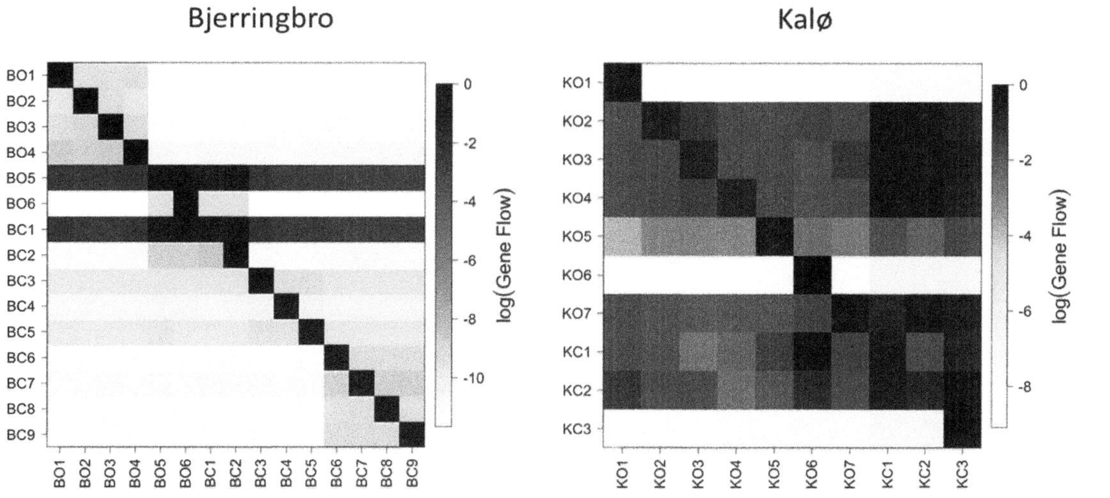

Figure 3. Pairwise F_{ST} values. The pairwise F_{ST} values for each pair of sampling sites are shown on a heatmap for both areas. Pairs that are significantly different, after sequential Bonferroni correction, are represented by a + sign on the heatmap.

Gene flow analysis

The analysis of gene flow confirmed the results of the F_{ST} analysis, showing a much higher gene flow in the Kalø area compared to Bjerringbro (Fig. 3). The barrier effect of fields was also confirmed and hedgerows were shown to be the main dispersal path in both areas (i.e. the highest gene flow occurred between sampling sites belonging to the same hedgerow, Fig. 4). A study by Eyre et al. [8] has shown the importance of vegetation cover and crop type on the activity of *B. lampros* and, despite not dealing directly with dispersal patterns, activity in the field was found not to be correlated with activity in the fields' margin. However, in the Kalø area there was a very clear effect of management on the strength and direction of gene flow. In particular, the highest gene flow went from organic towards conventional fields (Fig. 4 right, last three columns). The fact that this result was not found for Bjerringbro area suggests that the factor that makes a difference is the degree of intensity of the management and that only a low intensity of cultivation allows organic fields to act as genetic reservoirs. The positive effect of low-intensity cultivation has been shown for many wild species such as: corn buntings [31] and farmland birds in general [32], European hare [33] and for a large number of arthropods [34]. Together with the previously mentioned studies, our work confirms, from a genetic point of view, the importance of low-intensity cultivations for the conservation of biodiversity. Moreover we show that, while organic fields can act as reservoirs for the re-colonization of conventional fields, the absence of pesticide use is not enough for a field to become a fauna-source for more managed areas. Therefore, our study discloses that (at least for this species) pesticide use and agricultural intensity are interdependent variables that should not be evaluated separately. It is however not possible to infer from these results the long-term effect of pesticide use and agricultural intensity due to the limited time-scale of the study (two years).

Table 2. Results of gene flow analyses (Bjerringbro).

Most significant factors	Number of runs (total of 9 runs)	Posterior probability of the model	Alpha of the corresponding model
Hedgerow	3	*0.399*	*1.83*
		0.182	0.986
		0.352	1.51
Hedgerow + Road Side	1	0.242	Hedgerow:1.18
			Road Side:−0.9
Hedgerow + Field	1	0.191	Hedgerow: 0.996
			Field:−0.71
Distance	2	0.271	−1.58
		0.166	0.407
Distance + Road Side	1	0.103	Distance:−0.746
			Road Side:−0.477
Null model	1	0.155	1.35

For each of the 9 runs performed with BIMR the most significant factor(s), their posterior probability and the alpha value for that factor(s) are reported. The results from the best run (see Materials and methods for details) are shown in italics.

Table 3. Results of gene flow analyses (Kalø).

Most significant factors	Number of runs (total of 10 runs)	Posterior probability of the model	Alpha of the corresponding model	
Management	4	*0.582*	*−1.58*	
		0.441	−1.23	
		0.368	−2.36	
		0.497	−2.06	
Hedgerow + Management	4		Hedgerow	Management
		0.490	0.497	−2.44
		0.310	0.571	−1.58
		0.359	0.625	−1.07
		0.267	0.732	−1.9
Field + Management	1	0.603	Field: −0.686	
			Management: −1.89	
Null model	1	0.300	0.339	

For each of the 10 runs performed with BIMR the most significant factor(s), their posterior probability and the alpha value for that factor(s) are reported. The results from the best run (see Materials and methods for details) are shown in italics.

Assignment tests

The results of self-assignment tests showed a good agreement with the genetic structure results as the number of mis-assigned individuals was generally low for each cluster. The results from Bjerringbro gene flow analyses showed that BO5 and BC1 are the most important source of emigrants; this is in accordance with the assignment tests from 2008 in which a lot of individuals from other clusters were assigned to the cluster containing these two populations. The results from Kalø for 2009 are also in agreement with the gene flow results as a lot of samples from other clusters were assigned to the clusters containing the conventional fields.

Conclusions

All the presented results suggest that tree-covered hedgerows are the preferred dispersal route used by *B. lampros*. On the other hand, fields represent a barrier to gene flow, suggesting a tendency for individuals to return to the same hedgerow. However, low-intensity organic fields can be used for dispersal, probably due to a lower death rate caused by mechanical treatments and a higher concentration of aphids. This finding confirms the hypothesis of Huusela-Veistola [13] which suggested that the increase in *B. lampros* abundance following a sharp decrease after spraying was due to immigration from neighboring organic fields. Going beyond this, our results also confirm the importance of intensity of cultivation, and in particular of a low yield cultivation with high weed abundance [14,15]. Future longer term and geographically broader studies might be able to determine with more accuracy the extent of the effect of each single variable.

Given the importance of *B. lampros* as a pest control agent, the results of this study can be important for the development of management strategies that maximize its beneficial effects [2]. Moreover, the fact that tree-covered hedgerows and low-intensity cultivation were found to be fundamental for the gene flow in agricultural areas strengthen the importance of including land-

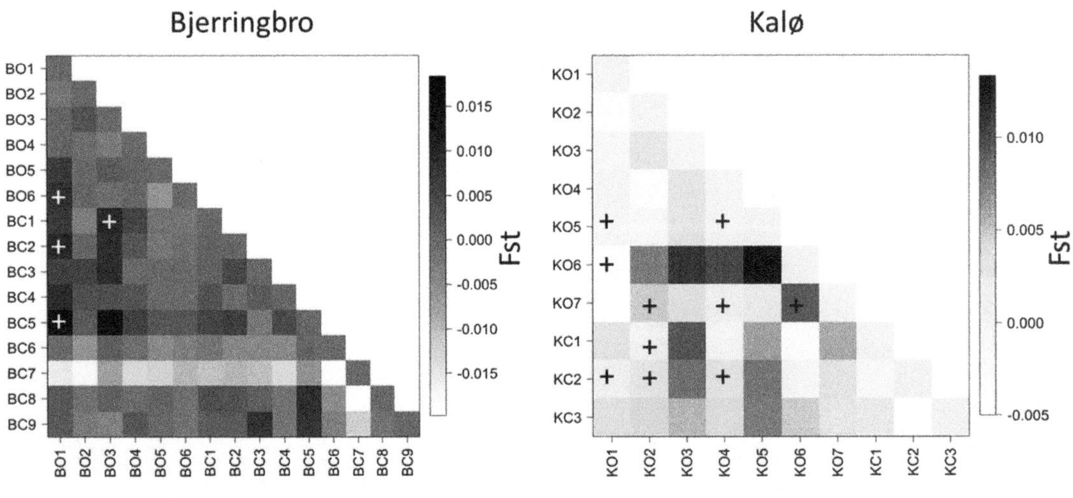

Figure 4. Gene flow values. The $\log_{10}$ of the gene flow values are shown on a heatmap for both areas. The populations of origin are shown on the vertical axis.

scape features and cultivation intensity as parameters for the evaluation of land management strategies.

Supporting Information

Table S1 Genetic diversity indexes for each sampling site and year. First column represent sampling site, (BO = Bjerringbro organic, BC = Bjerringbro conventional, KO = Kalø organic, KC = Kalø conventional) second column the year of sampling, third to fifth columns are the expected heterozigosity (H_e) allelic richness (Ar) and effective population size respectively (N_e) (see Materials and methods for details about their calculation).

Table S2 Effective population sizes for all sampling sites calculated with moment based method. BO = Bjerringbro organic, BC = Bjerringbro conventional, KO = Kalø organic, KC = Kalø conventional.

Table S3 Results of self-assignment tests for Bjerringbro. C1: BO1, BO2, BO4; C2: BO3; C3: BO5, BO6, BC1, BC2; C4: BC3; C5: BC4; C6: BC5; C7: BC6, BC7, BC8; C8: BC9.

Table S4 Results of self-assignment tests for Kalø. C1: KO; C2: KO2, BO4; C3: KO3; C4: KO5, KO6, KC1; C5: KO7, KC2, KC3.

Acknowledgments

We would like to thank all the people involved in the REFUGIA project for helpful discussions. Special thanks go to Zdenek Gavor, who performed the morphological identification of the beetle species.

Author Contributions

Conceived and designed the experiments: CM LWA VL. Performed the experiments: CM LWA. Analyzed the data: CM LWA VL. Contributed reagents/materials/analysis tools: CM LWA VL. Wrote the paper: CM LWA VL.

References

1. Edwards CA, Sunderland K, George K (1979) Studies on polyphagous predators of cereal aphids. Journal of Applied Ecology: 811–823.
2. Landis DA, Wratten SD, Gurr GM (2000) Habitat management to conserve natural enemies of arthropod pests in agriculture. Annual Review of Entomology 45: 175–201.
3. Chiverton PA (1986) Predator density manipulation and its effects on populations of Rhopalosiphumpadi (Horn.: Aphididae) in spring barley. Annals of Applied Biology 109: 49–60.
4. Cameron KH, Leather SR (2012) How good are carabid beetles (Coleoptera, Carabidae) as indicators of invertebrate abundance and order richness? Biodiversity and Conservation 21: 763–779.
5. Petersen MK (1999) The timing of dispersal of the predatory beetles *Bembidion lampros* and *Tachyporus hypnorum* from hibernating sites into arable fields. Entomologia Experimentalis et Applicata 90: 221–224.
6. Wallin H (1989) Habitat selection, reproduction and survival of two small carabid species on arable land: a comparison between *Trechus secalis* and *Bembidion lampros*. Ecography 12: 193–200.
7. Pederson M, Pedersen L, Abildgaard K (1990) Annual and diurnal activity of some Tachyporus species (Coleoptera: Staphylinidae) in two spring barley fields and a hedge. Pedobiologia 34: 367–378.
8. Eyre MD, Labanowska-Bury D, Avayanos JG, White R, Leifert C (2009) Ground beetles (Coleoptera, Carabidae) in an intensively managed vegetable crop landscape in eastern England. Agriculture, Ecosystems & Environment 131: 340–346.
9. Jansen JP, Defrance T, Warnier AM (2010) Effects of organic-farming-compatible insecticides on four aphid natural enemy species. Pest management science 66: 650–656.
10. Purvis G, Fadl A (2002) The influence of cropping rotations and soil cultivation practice on the population ecology of carabids (Coleoptera: Carabidae) in arable land. Pedobiologia 46: 452–474.
11. Hassan S, Bigler F, Bogenschütz H, Boller E, Brun J, et al. (1991) Results of the fifth joint pesticide testing programme carried out by the IOBC/WPRS-Working Group "Pesticides and Beneficial Organisms". BioControl 36: 55–67.
12. Hassan S, Bigler F, Bogenschütz H, Boller E, Brun J, et al. (1988) Results of the fourth joint pesticide testing programme carried out by the IOBC/WPRS-Working Group "Pesticides and Beneficial Organisms". Journal of Applied Entomology 105: 321–329.
13. Huusela-Veistola E (1996) Effects of pesticide use and cultivation techniques on ground beetles (Col., Carabidae) in cereal fields. Annales Zoologici Fennici 33: 197–206.
14. Honek A, Jarosik V (2000) The role of crop density, seed and aphid presence in diversification of field communities of Carabidae (Coleoptera). European Journal of Entomology 97: 517–526.
15. Purvis G, Curry J (1984) The influence of weeds and farmyard manure on the activity of Carabidae and other ground-dwelling arthropods in a sugar beet crop. Journal of Applied Ecology: 271–283.
16. Milligan B, Hoelzel A (1992) Plant DNA isolation. Molecular genetic analysis of populations: a practical approach: 59–88.
17. Andersen LW, Fog K, Damgaard C (2004) Habitat fragmentation causes bottlenecks and inbreeding in the European tree frog (*Hyla arborea*). Proceedings of the Royal Society of London, Series B: Biological Sciences 271: 1293–1302.
18. Goudet J (1995) FSTAT (version 1.2): a computer program to calculate F-statistics. Journal of Heredity 86: 485.
19. Peakall R, Smouse PE (2006) GENALEX 6: genetic analysis in Excel. Population genetic software for teaching and research. Molecular Ecology Notes 6: 288–295.
20. Ovenden JR, Peel D, Street R, Courtney AJ, Hoyle SD, et al. (2007) The genetic effective and adult census size of an Australian population of tiger prawns (*Penaeus esculentus*). Molecular Ecology 16: 127–138.
21. Pudovkin A, Zaykin D, Hedgecock D (1996) On the potential for estimating the effective number of breeders from heterozygote-excess in progeny. Genetics 144: 383–387.
22. Waples RS (1989) A generalized approach for estimating effective population size from temporal changes in allele frequency. Genetics 121: 379–391.
23. Guillot G, Mortier F, Estoup A (2005) GENELAND: a computer package for landscape genetics. Molecular Ecology Notes 5: 712–715.
24. Excoffier L, Lischer HEL (2010) Arlequin suite ver 3.5: a new series of programs to perform population genetics analyses under Linux and Windows. Molecular Ecology Resources 10: 564–567.
25. Faubet P, Gaggiotti OE (2008) A new Bayesian method to identify the environmental factors that influence recent migration. Genetics 178: 1491–1504.
26. Faubet P, Waples RS, Gaggiotti OE (2007) Evaluating the performance of a multilocus Bayesian method for the estimation of migration rates. Molecular Ecology 16: 1149–1166.
27. Piry S, Alapetite A, Cornuet JM, Paetkau D, Baudouin L, et al. (2004) GENECLASS2: a software for genetic assignment and first-generation migrant detection. Journal of Heredity 95: 536–539.
28. Rannala B, Mountain JL (1997) Detecting immigration by using multilocus genotypes. Proceedings of the National Academy of Sciences USA 94: 9197.
29. Paetkau D, Slade R, Burden M, Estoup A (2004) Genetic assignment methods for the direct, real-time estimation of migration rate: a simulation-based exploration of accuracy and power. Molecular Ecology 13: 55–65.
30. Safner T, Miller MP, McRae BH, Fortin MJ, Manel S (2011) Comparison of bayesian clustering and edge detection methods for inferring boundaries in landscape genetics. International Journal of Molecular Sciences 12: 865–889.
31. Brickle NW, Harper DGC, Aebischer NJ, Cockayne SH (2000) Effects of agricultural intensification on the breeding success of corn buntings *Miliaria calandra*. Journal of Applied Ecology 37: 742–755.
32. Smith HG, Danhardt J, Lindstrom A, Rundlof M (2010) Consequences of organic farming and landscape heterogeneity for species richness and abundance of farmland birds. Oecologia 162: 1071–1079.
33. Smith RK, Vaughan Jennings N, Harris S (2005) A quantitative analysis of the abundance and demography of European hares *Lepus europaeus* in relation to habitat type, intensity of agriculture and climate. Mammal Review 35: 1–24.
34. Hendrickx F, Maelfait J-P, Van Wingerden W, Schweiger O, Speelmans M, et al. (2007) How landscape structure, land-use intensity and habitat diversity affect components of total arthropod diversity in agricultural landscapes. Journal of Applied Ecology 44: 340–351.

5

The Bacterial Communities Associated with Honey Bee (*Apis mellifera*) Foragers

Vanessa Corby-Harris[1], Patrick Maes[2], Kirk E. Anderson[1,2*]

1 United States Department of Agriculture, Carl Hayden Bee Research Center, Tucson, Arizona, United States of America, **2** Center for Insect Science, University of Arizona, Tucson, Arizona, United States of America

Abstract

The honey bee is a key pollinator species in decline worldwide. As part of a commercial operation, bee colonies are exposed to a variety of agricultural ecosystems throughout the year and a multitude of environmental variables that may affect the microbial balance of individuals and the hive. While many recent studies support the idea of a core microbiota in guts of younger in-hive bees, it is unknown whether this core is present in forager bees or the pollen they carry back to the hive. Additionally, several studies hypothesize that the foregut (crop), a key interface between the pollination environment and hive food stores, contains a set of 13 lactic acid bacteria (LAB) that inoculate collected pollen and act in synergy to preserve pollen stores. Here, we used a combination of 454 based 16S rRNA gene sequencing of the microbial communities of forager guts, crops, and corbicular pollen and crop plate counts to show that (1) despite a very different diet, forager guts contain a core microbiota similar to that found in younger bees, (2) corbicular pollen contains a diverse community dominated by hive-specific, environmental or phyllosphere bacteria that are not prevalent in the gut or crop, and (3) the 13 LAB found in culture-based studies are not specific to the crop but are a small subset of midgut or hindgut specific bacteria identified in many recent 454 amplicon-based studies. The crop is dominated by *Lactobacillus kunkeei*, and Alpha 2.2 (Acetobacteraceae), highly osmotolerant and acid resistant bacteria found in stored pollen and honey. Crop taxa at low abundance include core hindgut bacteria in transit to their primary niche, and potential pathogens or food spoilage organisms seemingly vectored from the pollination environment. We conclude that the crop microbial environment is influenced by worker task, and may function in both decontamination and inoculation.

Editor: Nicole M. Gerardo, Emory University, United States of America

Funding: Funding was provided entirely by the USA tax-payer supported operating budget of the ARS-USDA Carl Hayden Bee Research Center, Tucson AZ. The funders had no role in study design, data collection and analysis, decision to publish, or preparation of the manuscript. There were no current external funding sources for this study.

Competing Interests: The authors have declared that no competing interests exist.

* E-mail: kirk.anderson@ars.usda.gov

Introduction

The honey bee, *Apis mellifera*, is critical for the pollination of many economically important crops. Continued colony losses have called for a deeper understanding of both symbiotic and pathogenic microbial interactions, particularly as they relate to food storage and the pollination environment. As part of a typical migratory beekeeping operation, the same population of honey bee colonies can be exposed to 5–10 agricultural ecosystems over the course of a year. At each site, foragers encounter a unique set of variables that includes climate, floral components, biocides, and water sources. Many biocides are later found in the wax and stored pollen, or "beebread", of commercial operations [1,2]. Evidence suggests that sub-lethal biocide exposure may alter the microbial balance of the individual, hive, or food stores resulting in long-term negative effects [3–5]. Recent results from comprehensive molecular surveys suggest the potential for microbial imbalance at many levels of organization [1,5–12], inviting a closer look at a variety of factors that influence the microbial constitution of the colony, particularly the origin and integrity of microbial communities in the gut, crop, and food stores.

The honey bee colony "superorganism" consists of individual, group, and hive components, complete with a large repertoire of socially interactive and homeostatic behaviors [13]. Anatomically, the foregut (hereafter crop) is the honey bee's social/nutritional interface. This portion of the alimentary tract is essentially an inflatable storage bag used to transport nectar from the flower to the hive, share liquid nutrition with sibling nestmates, and selectively pass pollen into the midgut [14]. More generally, the crop represents the microbial intersection of food sharing, food storage and the pollination environment. At the flower, pollen foragers use their forelegs and tongue to gain access to nectar and pollen. Pollen accumulated on the head and body hairs is consolidated using the forelegs, mixed with liquid sugars from the crop, and this sticky mixture is packed into hindleg pollen baskets called corbiculae. Once returned to the hive, corbicular pollen pellets are packed tightly into wax cells and become beebread, a nutrient dense mixture of pollen, honey and various microbes. Beebread has been likened to a silage environment wherein the protoplasmic nutrients locked within the durable pollen coat are fermented, or essentially "pickled" for future consumption [15–17]. Pollen storage is critical because stored food provides the honey bee with essential nutrients during the winter or periods of pollen dearth.

The present understanding of crop bacteria is culture-dependent, and maintains that "13 different beneficial bacteria reside inside the honey crop of bees, are placed on pollen at the flower, and work synergistically to protect beebread from degradation" [18–24]. These conclusions are provisional however, as they include culture bias and highly inconsistent detection of the 13 proposed strains, all labeled "lactic acid bacteria" (LAB) although four of the strains belong to the genus *Bifidobacterium*. An alternative hypothesis states that these 13 "crop-specific" LAB are not specific to the crop but are instead part of a larger subset of bacteria occurring preferentially in other gut compartments or hive microenvironments [17]. Newly emerged bees are seemingly free of bacteria, and appear to acquire their core hindgut bacteria from the hive environment, food stores, or older individuals [17,25]. Combined with the fact that honey bees are constantly swapping food via oral trophallaxis [26], bacteria evolved to occupy the midgut, hindgut or food storage niche would likely occur with some frequency in the crop of most bees.

Recent non-culture based investigations of the honey bee microbiome have focused on bees captured from within the hive, sampling either the entire alimentary tract, or only the midgut and hindgut [25,27–35]. Less attention has been paid to the microbial diversity of foraging bees, food stores and the pollination environment. Independent findings from solitary and social pollinators suggest that both potentially pathogenic and beneficial microbes are regularly vectored from the pollination environment or floral sources [7,36–39]. Some sources of floral nectar contain bacteria similar to those found in the stored food or hive materials of honey bees, suggesting that the crop acts as a "semi-permeable filter", a selective environment wherein particular bacteria can survive to be vectored between floral and hive environments [17,40–50].

Here we investigate the crop and gut microbial diversity of returning pollen foragers and the corbicular pollen they collected using 454 amplicon assays of the 16S rRNA gene. To determine the origin of the bacterial communities on inbound pollen, we compare forager gut and corbicular pollen communities for two seasonal time points in the same geographic location. We also test the hypothesis that bacteria associated with the honey bee crop and placed on corbicular pollen are composed of the 13 putative core crop bacteria. We compare our results for known pollen foragers with past studies of gut microbial communities of typically younger in-hive bees.

Methods

Ethics Statement

All colonies were sampled from apiaries located at the USDA Carl Hayden Bee Research Center in Tucson AZ. Our field collections did not involve endangered or protected species and no specific permissions were required because the study was conducted by USDA employees.

Bee samples and DNA preparation

Pollen foragers were collected in the fall (3 December 2012) and spring (20 March 2013) from two colonies housed side by side in the same apiary at the USDA Carl Hayden Bee Research Center in Tucson, Arizona. In December, the availability of pollen sources had decreased and the colonies had ceased brood rearing and were storing pollen for the winter. In contrast, colonies were actively brood rearing in March, collecting pollen from a variety of different plant sources and quickly converting this pollen into new brood. For each colony collection, the hive entrance was blocked and returning pollen foragers were collected in less than 5 minutes per hive. In both the spring and fall, fourteen individual pollen foragers per colony were captured with sterile soft forceps, placed in sterile Falcon tubes, and chilled on ice. Pollen loads were removed from both corbiculae (i.e., pollen baskets) before dissecting each bee's GI tract, containing the entire length of the alimentary tract from the crop through to the rectum. To compare crop and corbicular pollen diversity, additional pollen foragers were sampled similarly in the spring from 14 different colonies in the apiary. Initial samples composed of individual crops produced negligible template DNA and inconsistent PCR products. Thus, the crops and corbicular pollen of ten pollen foragers per colony were pooled, yielding 14 libraries composed of 10 crops each, and 14 libraries composed of the associated corbicular pollen from those same 10 foragers (10 foragers×2 legs = 20 corbicular loads per library).

Total genomic DNA was extracted from individual GI tract samples, pooled crop samples, and pooled and individual corbicular pollen samples. Prior to DNA extraction, gut and crop samples were processed similarly by bead beating the tissue for 30 sec in a sterile 2 mL centrifuge tube containing 350 µl of 0.5 mm silica beads and 1.5 ml TE/Triton × lysis buffer (20 mM Tris-HCl, 2 mM EDTA, 1.2% Triton X-100, pH 8.0). The supernatant was transferred to a new 1.5 ml centrifuge tube and centrifuged for 30 minutes. The supernatant was removed and 180 µl TE/Triton × lysis buffer (20 mM Tris-HCl, 2 mM EDTA, 1.2% Triton X-100, pH 8.0; 20 mg/ml lysozyme added immediately before use) was added. Upon removal from the bee, corbicular pollen was added to 1.5 ml of TE/Triton × buffer (20 mM Tris-HCl, 2 mM EDTA, 1.2% Triton X-100, pH 8.0) and vortexed for 5 minutes. Each sample was briefly centrifuged at low speed and the supernatant was transferred to a new 2 ml bead beating centrifuge tube where it was centrifuged for 10 minutes on high to pellet the bacterial cells. This wash cycle was repeated 4 times to maximize the amount of bacterial detachment from the pollen grains. After the final wash cycle 350 µl of 0.5 mm silica beads were added to the 2 ml tube and bead beaten for 30 sec. The supernatant was then transferred to a new 1.5 ml centrifuge tube and centrifuged for 30 minutes. The supernatant was removed and 180 µl TE/Triton × lysis buffer (20 mM Tris-HCl, 2 mM EDTA, 1.2% Triton X-100, pH 8.0; 20 mg/ml lysozyme added immediately before use) was added. Samples in lysis buffer were then subjected to genomic DNA extraction using the GeneJet Genomic DNA Purification Kit (Fermentas) following the protocol for gram-positive bacteria.

PCR and pyrosequencing

The V1-V2 region of the 16S rRNA gene of the samples was PCR amplified using universal 16S rRNA primers fitted with 454 FLX Titanium adapter sequences (27F 5′-CCATCTCATCCCTGCGTGTCTCCGACTCAG-NNNNNNNNNN-agagtttgatcctggctcag -3′; 338R: 5′- CCTATCCCCTGTGTGCCTTGGCAGTCTCAG-tgctgcctcccgtaggagt -3′; uppercase letters denote the adapter sequences, N's indicate library-specific barcodes, lowercase letters indicate universal 16S rRNA primers). Amplicons were sequenced using Roche 454 GS FLX Titanium sequencing.

Pyrotagged sequence analysis

Sequences were processed using mothur v.1.26.0 [51]. Sequences in the.sff files were quality filtered using the trim.flows command (minflows = 360, maxflows = 720) and all sequences less than 150 base pairs (bp) with more than 2 base mismatches to the 27F primer sequence or 1 mismatch to the 10 bp pyrotag after trimming were eliminated using the trim.seqs command. Pyrotags were removed and the sequences were aligned to Silva SSURef database (v102) using the align.seqs command. Sequences that started after the 27F sequence or that were shorter than the 98%

of the sequences were eliminated using the screen.seqs command. Sequences that were within 1% similarity were clustered together using the pre.culster command. Chimeras were removed using UCHIME [52] and any sequences that were of mitochondrial, chloroplast, Archaeal, Eukaryote, or unknown origin were removed. The sequence libraries were then concatenated and aligned as described above. A distance matrix was constructed for the aligned sequences using the dist.seqs command and the default parameters. Sequences were then binned into operational taxonomic units (OTUs) based on 97% sequence similarity. It should be noted that our investigation did not warrant a single experiment-wide cutoff for delineating taxonomy. Honey bee gut taxonomy has been refined by many recent papers [28–35], but there is far less taxonomic information concerning bacterial communities associated with the hive or general pollination environment. Representative sequences from each 97% OTU were characterized in two ways. First, these sequences were classified with the RDP Naïve Bayesian Classifier using a manually constructed training set that contained sequences from the greengenes 16S rRNA database (version gg_13_5_99 accessed May 2013), the RDP version 9 training set, and all full length honey bee associated gut microbiota listed in NCBI (accessed July 2013) trimmed to the V1-V2 region of the 16S rRNA gene. Next, representative OTU's classified by RDP were then subject to a BLAST query using the NCBI nt database, taking the hit with the lowest e-value less than or equal to 1×10^{-10}. Any remaining sequences that were of chloroplast or mitochondrial origin were removed as well as sequences classified with less than 100% confidence at the Phylum level using the RDP Naïve Bayesian Classifier [53]. Rarefaction curves were generated for each of the libraries using the rarefaction.single command. For sample types that were completely or near completely sampled (according to the rarefaction curves generated), individual libraries were pooled by sample type (i.e., fall guts or spring guts) and Chao estimates of species richness and Good's estimate of coverage were determined using the summary.shared command.

Data accessibility

Sequences can be found in the NCBI Sequence Read Archive (SRA) under accession number SRP035369. Table S1 links the pyrosequencing barcodes to each library consisting of gut, crop or corbicular pollen samples from individual bees or bees pooled by colony.

Analyses of crop- and gut-associated "lactic acid bacteria" (LAB)

The present paradigm claims that the crop contains 13 strains of "Lactic Acid Bacteria" (LAB; 9 *Lactobacillus* spp., 4 *Bifidobacterium* spp.) that are core to the crop, that the bee deposits on corbicular pollen at the flower, and that act in synergy to preserve beebread [18–24]. An alternate hypothesis is that most of LAB found in the crop are simply in transit to the hindgut [17]. We tested these conflicting hypotheses in two ways. First, we conducted a BLAST search using full-length sequences of the 16S rRNA gene from these 13 LAB as query sequences and the sequences found in the fall gut, spring gut, and spring crop libraries. The BLASTn algorithm was used (-task = blastn) and successful hits were those with an e-value$\leq 1 \times 10^{-10}$ and 100% sequence identity for the length of the alignment. Tallies were taken to determine the number of 100% hits to the 13 LAB to determine whether these bacterial strains could be present in our samples and whether they were more prevalent in the crop than the guts. The majority of sequences that were significant hits to these 13 LAB matched the *Lactobacillus* spp. but the number of *Bifidobacterium* spp. hits was negligible (see Results). We next constructed a phylogeny to examine the similarity of the *Lactobacillus* sp. isolated from the pooled crops and individual guts of spring or fall bees amongst themselves and to known sequences of *Lactobacillus* spp. A sequence database was constructed using 16 *Lactobacillus* OTUs from the present study, 9 putative crop-specific *Lactobacillus* sequences, and 27 published full-length *Lactobacillus* sp. 16S rRNA sequences as a reference. These published full-length sequences, and the sequences identified as *Lactobacillus* spp. in the present study were aligned using Muscle v3.8.31 [54] and manually edited for quality using BioEdit [55]. The alignment was then cropped to include only the V1/V2 region, and gaps were eliminated, leaving 272 positions in the final alignment. A Neighbor-Joining phylogeny was built using MEGA version 5 [56] and 500 bootstrap replicates. The analysis included 52 nucleotide sequences and all codon positions were included. Rate variation among sites was modeled with a gamma distribution. We used this phylogeny to illustrate whether the crop-associated bacteria were distinct from the gut-associated bacteria found in this and other studies.

Crop plate counts

We determined by direct plate count the number of bacteria in the crops of bees performing different tasks because (1) there are stark differences concerning the predicted number of bacteria in the crop as defined in the literature [18,22,25], and (2) we experienced consistent difficulties extracting sufficient quantities of bacterial DNA and/or performing PCR amplification of crop bacteria from individual crops. Combined with the reliable amplification of the positive PCR control (previously verified *Apis mellifera* gut sample), this indicated relatively low numbers of bacterial cells in crop samples. In-hive nectar processors, nurse bees, general foragers, and pollen foragers were collected in the spring of 2012 from a three frame observation hive separate from the colonies used to gather the pyrosequencing data. In-hive nectar processors were defined as bees that made 5 consecutive visits to stored honey in the hive. Nurses were bees that made 5 consecutive visits to developing larvae. General foragers were bees returning from flights that did not have pollen in their pollen baskets, and pollen foragers had pollen loads in their corbiculae. Individuals were collected into sterile Falcon tubes on wet ice and brought to the lab for immediate dissection. The crops of 12 nectar processors and 12 nurse bees were dissected and their contents were recorded (full, empty, or half full). Most of the nectar processors' crops were empty and most of the nurse crops were full so for consistency only these samples were used in the analysis (nectar processors $N = 9$, nurses $N = 10$). The crops of 12 general foragers and 12 pollen foragers were dissected and weighed. Each crop was homogenized in 1 ml of physiological saline and 100 μl was plated onto deMan-Rosaga-Sharp (MRS) and Tryptic-Soy agar (TSA) plates and placed in aerobic or microaerophilic (5% CO_2) conditions at 35°C for 24 h. Counts of the number of colony forming units (CFUs) per plate were made and, where appropriate, the log number of CFUs per gram of crop was calculated to compare to other studies [47,57]. Comparisons were made between CFUs growing in aerobic versus microaerophilic conditions for each media type or between the CFUs growing on MRS versus TSA plates incubated in similar environmental conditions.

Results

Reads analyzed and taxa generated

A total of 1,616,883 reads were generated across the six sample types and 1,452,224 reads remained after the initial quality trimming (Table 1). On average, 1.4% of these reads were chimeric and were removed. There were large differences among sample types in the number of reads that were chloroplast in origin (cpDNA). Of the gut and crop samples, an average of 2.6% of the remaining non-chimeric reads were cpDNA, while 87% of the corbicular pollen reads were cpDNA. This cpDNA contamination resulted in drastically different average read numbers for the crop and gut libraries compared to the corbicular pollen libraries. A total of 812 operational taxonomic units (OTUs) were resolved across the six sample types at the 97% level of similarity (Table S2). BLAST results from the NCBI nucleotide database were in complete agreement with the RDP classifier result, which suggested that the RDP classification was adequate for grouping the taxa as gut microbiota or otherwise as presented in the tables and figures. Rarefaction curves generated for each sample type and library show that the diversity of the crop and gut samples was completely or near completely sampled, the diversity of the spring corbicular pollen samples was not completely sampled, and the diversity of the fall corbicular pollen samples was near completely sampled (Table S2, Fig. S1). Chao estimates of species richness and coverage were calculated for the four sample types that were completely or near completely sampled (i.e., the crop, fall guts, spring guts, and fall corbicular pollen). Coverage was above 99% for all four sample types and the richness of the fall corbicular pollen was significantly greater than the fall guts, spring guts, and spring crops, which all had comparable species richness (Table 2).

Venn diagrams comparing the fall gut, spring gut, and spring crop libraries showed that 30 OTUs were shared among all sample types and 70 were specific to the crop (Fig. 1). Moreover, the shared sequences comprised 99.6%, 99.9%, and 96.1% of the fall gut, spring gut, and spring crop sample types, respectively (Fig. 1). The majority of these shared sequences are considered core gut bacteria. Crop-specific taxa were comprised mostly of singletons or OTUs with few sequences (4.16±9.42 s.d. sequences per crop-specific OTU; Table S2). The most abundant OTU found only in the crop contained just 72 sequences and was closely related to *Peptoniphilus* sp., a gram-positive anaerobic coccus that has been found in both floral nectar and on the surface of honey bees [58].

The majority of the fall gut (94%) and spring gut (99%) sequences were comprised of core gut bacteria (Figs. 2 and 3). The entire core gut microbiota (Acetobacteraceae (Alpha 2.1), *Lactobacillus* sp. (Firm 4), *Lactobacillus* sp. (Firm 5), *Frischella perrara* (Gamma 2), *Gilliamella apicola* (Gamma 1), *Snodgrassella alvi* (Beta), and a honey bee associated *Bifidobacterium* sp. [30,59,60]) was represented to some degree in at least one of the sample groups. Compared to previous studies of in-hive bee guts employing a similar approach [25,32,33], we observed more Alpha 2.1 (Table 2; fall guts: 8.5%, spring guts: 2.0%), and more *S. alvi* sequences (Table 3; fall guts: 14.5%, spring guts: 13.8%). Only two *F. perrara* sequences were observed across all sample types (Table 3; fall guts: 0.0005%, spring guts: 0.0006%). Twenty percent of the sequences found in the crop were previously defined as core gut microbiota (Figs. 2 and 3). There was some variation among individuals' guts in the degree that the core was represented (Fig. 3). Most guts were comprised almost entirely of core gut taxa, but this core was less abundant in a few individuals. Where the core taxa did not dominate the whole gut, the remainder of the library was comprised of sequences belonging to the Enterobacteriaceae, Pseudomonadales, or Alpha 1, a relative of *Bartonella* spp. identified previously from bees (Fig. 3).

The crop harbored a microbiota highly distinct from the guts (Figs. 2 and 3). Crop bacteria were dominated by Alpha 2.2 (42%) and *L. kunkeei* (25%; Fig. 2). There was considerable variation in the distribution and composition of the microbial communities among libraries, and crops often contained high levels of Enterobacteriaceae (Fig. 2). The most numerous core gut sequences found in the crop were *Lactobacillus* sp. Firm 5 (8% of sequences), *G. apicola* (6% of sequences), and *S. alvi* (5% of sequences; Fig. 2 and Table S2).

Corbicular pollen contained a diverse microbiota dominated by non-gut and non-crop bacteria including Actinomycetales, Alphaproteobacteria, Enterobacteriaceae, Pseudomonadales, Firmicutes, and Xanthomonadaceae, (Figs. 2 and 3). Much like the crop libraries, there was considerable variation in taxon richness and abundance among corbicular pollen carried by individual foragers (Fig. 3). Libraries were not consistently dominated by one taxon, but Actinomycetales was abundant in the majority of corbicular pollen samples (Fig. 3 and Table S2).

Similarity of reads to the 13 putative crop-specific "lactic acid bacteria" (LAB)

We examined two alternate hypotheses: (1) the crop harbors 13 specific strains of LAB (9 *Lactobacillus* and 4 *Bifidobacterium* species) that act in synergy to preserve hive food stores [18–24], and (2) bacteria specific to the hindgut are frequently sampled from the crop due to high rates of trophallaxis among individuals and the incidental movement of bacteria from the hive environment through the crop on their way to the hindgut [17,61]. To determine the number of sequences from the crop and gut that are

Table 1. Summary of pyrotagged sequence processing.

season	tissue or substrate	Total libraries	N[A] initial reads	N after quality trimming	% chimeras	N after cpDNA[B] removed	% cpDNA	N after manual culling[E]
fall	corbicular pollen	14[C]	285,365	257,832	0.23	58,218	77	33,284
fall	guts	14[C]	251,856	229,503	0.24	228,563	0.17	191,037
spring	corbicular pollen	14[C]	288,351	259,928	0.12	30,680	88	5,428
spring	guts	14[C]	298,257	267,324	6.4	249,893	0.11	163,468
spring	corbicular pollen	14[D]	245,430	216,426	0.10	9,175	96	4,879
spring	crop	14[D]	247,624	221,211	1.5	201,122	7.5	195,264

[A]Number of reads produced by the 454-instrument.
[B]Chloroplast DNA (cpDNA).
[C]Each library contained material from one individual forager.
[D]Contained 10 pooled individuals per library. Each library represents a single colony.
[E]Sequences that were not cpDNA or mitochondrial DNA in origin, and classified at a confidence level of 100% at the Phylum level according to the RDP classifier.

Table 2. Estimates of species richness and coverage for adequately sampled libraries.

Sample type	OTUs	Good's Coverage	Chao estimate of species richness	LCI[A]	HCI[A]
Fall corbicular pollen	630	0.994	813.82	753.99	902.52
Fall guts	82	0.999	130.33	100.91	205.49
Spring guts	85	0.999	99.44	90.15	125.46
Spring crops	116	0.999	159.05	135.87	209.26

[A]Low (LCI) and high (HCI) 95% confidence intervals for the Chao estimate of species richness. Libraries were randomly sampled in equal numbers for 1000 replicates.

a significant match to the putative 13 crop-specific strains, we performed BLAST searches between the fall gut, spring gut, and spring crop samples and a reference database containing these 13 bacterial strains (Table 4). Two of the *Lactobacillus* strains were not detected in any sample type. Sequences corresponding to the four *Bifidobacterium* strains were uniformly rare, on average representing<0.01% of all alimentary tract sequences combined. Among the nine queried *Lactobacillus* spp., only the sequence corresponding to strain Hma11 (within OTU050) occurred with noticeable frequency in both the gut and crop samples (Table 4). Sequences matching *L. kunkeei* strain Fhon2 were abundant in the crop, but virtually absent from the gut samples (Table 4). Of all sequences classified within the broader taxonomy of *Lactobacillus* Firm 4 or Firm 5, only 1.7% were detected in the crop and matched the putative crop-specific strains. From this same group however, 12.5% of sequences matched the putative crop-specific strains and were found in the gut. Similarly, of all sequences classified as *Bifidobacterium*, only 1.3% matched the putative crop-specific strains and were found in the crop, while 20.8% of all *Bifidobacterium* sequences matched the putative crop-specific strains and were found in the gut.

A Neighbor-Joining phylogeny of the *Lactobacillus* species identified in our samples revealed 4 distinct groups (Fig. 4). The largest group was closely related to *Lactobacillus* sp. Firm 5. *Lactobacillus* sp. Firm 4 resolved into 2 well-supported clades, while *L. kunkeei* formed its own group (Fig. 4). Again only one sequence matching the putative core crop bacteria [18–24], *L. kunkeei* strain Fhon2, was prevalent in the crop (Table 4). After super-imposing these groups onto the sequence tallies, the spring crops were comprised primarily of sequences related to *L. kunkeei*, while the spring and fall guts were populated mostly by *Lactobacillus* Firm4 and Firm5 (Fig. 4).

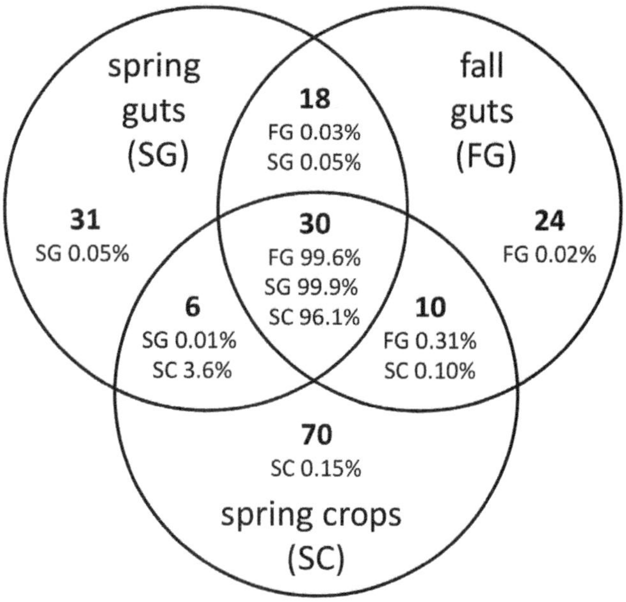

Figure 1. Bacterial taxa shared among the pyrosequencing libraries from the gut and crop samples. Libraries were pooled by sample type (fall gut, spring gut, spring crop) and the number of OTU's (in bold) was defined based on ≥97% sequence similarity. The number of sequences in each of the gut and crop sample sets was calculated for each OTU and the percentage of sequences in each sample type that was either shared among sample sets or unique to a particular sample set was calculated. For example, the 70 OTU's unique to the spring crop samples account for only 0.15% of the total spring crop sequences, while the 30 OTU's shared among all sample sets includes 96.1% of all spring crop sequences.

Crop counts

Crops of honey bees performing different behaviors were plated as a dilution series on acidic (MRS) and neutral (TSA) media in aerobic or microaerophilic conditions. Nearly all colonies that grew on the plates were indistinguishable according to size, color and colony morphologies. Overall, foragers contained a crop microbiota that flourished in both ambient and microaerophilic atmosphere on both MRS and TSA media (Table 5). In general, bees performing tasks in the hive had substantially less crop bacteria than presumably older foragers. Under ambient atmosphere, the acidic MRS media harbored significantly more CFUs than the neutral TSA media for all honey bee types. Under microaerophilic conditions, more bacterial growth was supported on the acidic media compared to the neutral media, but the difference was non-significant. After adjusting the CFUs for the amount of crop plated (1/10 of each crop plated, see methods; Table 5), the number of CFUs in the crop was many orders of magnitude less than counts reported for the midgut (>10 log CFUs per gram) and rectum (>10 log CFUs per gram) of summer and winter bees [47,57] and less than the log 6 CFUs in the honey crop reported in [22]. However, the number of CFUs observed here was comparable to the number of CFUs in honey bee crops reported in [18].

Discussion

Food storage is a critical fitness component of perennial social insects [62]. As a step towards understanding the microbial succession that occurs during pollen storage in the honey bee, we sought to differentiate between bee-mediated microbial inoculations occurring at the flower and microbes introduced as a consequence of environmental exposure. We captured inbound pollen foragers and sequenced the bacterial DNA from individual guts and pooled foreguts (crops), and from the associated corbicular pollen on their hind legs using high throughput

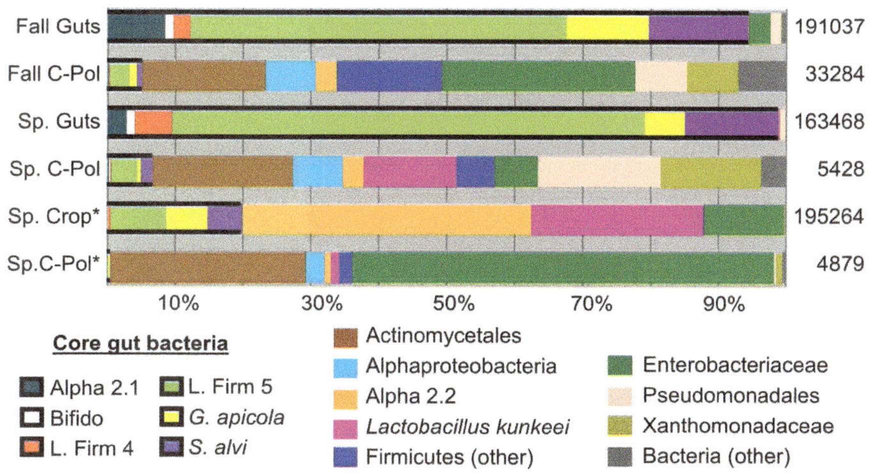

Figure 2. Relative abundances of bacterial groups from each sample type. Libraries (individuals or colonies) were pooled by sample type and the number of sequences belonging to each taxon relative to the total number of sequences in that sample type was determined. Black boxes around portions of each bar denote core gut bacteria. Asterisks (*) denote pooled colony samples, and total reads post filtering are displayed to the right of each sample type.

sequencing. Our results do not support the presence of 13 core crop "LAB" bacteria and associated beebread preservation proposed by previous culture dependent studies [18–24]. Similar to acidic and sugar rich environments of fermented food and silage, the crops of foragers were dominated by *Lactobacillus* and Alpha 2.2 (Acetobacteraceae), but also contained a small number of sporadically abundant Enterobacteriaceae that likely have their origins in the pollination environment. We found that bacteria considered core to the midgut and hindgut can be found in the crops of pollen foragers, but on average, less than 4% of the bacteria identified from corbicular pollen can be considered core gut bacteria. Similarly, 12 of 13 putative crop-specific LAB strains [18–24] are simply a subset of the core gut bacteria, occur at low frequency in the crop and gut, and are not placed on corbicular pollen at the flower. The majority of bacteria found in corbicular pollen does not originate with the crop or gut, but appears to originate from the pollination environment.

We used a combination of 16S rRNA gene sequences and culture-based methods to determine the diversity and abundance of bacteria found in the crop, entire alimentary tract and associated corbicular pollen. Sampled microenvironments differed for both the richness and abundance of bacterial taxa (Fig. 2). Data from the entire alimentary tract (guts) was similar across the two sampled time points, and confirms previous definitions of the core gut microbiota based on in-hive bees [25,29,30,32,33,35], but reveals differences in relative taxon abundance which may reflect a lack of pollen in the forager diet. Despite far less sequence coverage for corbicular pollen libraries, they were much more diverse than libraries of forager guts and crops (Table 2). Although we used different primers, tissues and methods (i.e. exclusion of the crop, isolation of DNA/RNA, in-hive bees vs. foragers) compared to past studies employing 454 amplicons [32,33,35], we still found very similar gut communities when sampling returning pollen foragers. The gut communities of forgers in this study were derived from complete alimentary tracts (guts) of individual bees including the crop, midgut, and hindgut. Tissue specific plate counts for the crop, midgut, and hindgut indicate that the crop bacterial community will not be represented when sequencing the entire alimentary tract using 454-amplicon sequencing (Table 5) [47,57]. This is because bacteria present at 10^5–10^6 (crop) comprise less than 0.01% of the total bacteria in the alimentary tract. For the crop tissues, it was therefore necessary to pool the contents of many foragers per hive to attain sufficient template DNA to overcome potential chance effects that can occur during PCR. Our results agree with [25] that the majority of core gut bacteria are localized in the midgut and hindgut. However, it is clear from this and other studies that core gut bacteria or those typically found in the food stores or pollination environment are also found in the crop with varying frequency.

Bacteria in the crops of pollen foragers

Our results provide no support for the hypothesis that 13 different beneficial bacteria reside inside the honey crop of bees, are placed on pollen at the flower, and work synergistically to protect beebread from degradation [18–24]. Given relatively equal coverage and read number from the crop and gut, we might expect (1) the occurrence of the 13 bacteria at equal or greater frequency in the crop than in the gut, and (2) some level of OTU or sequence read correspondence between the pooled crop and corbicular pollen libraries. A BLAST search of our results reveals that only one of the 13 putative crop bacteria fits both of the above criteria: *Lactobacillus kunkeei*, a species found worldwide in flowers, beehives, many types of bees and their food stores [17,38,39,44,63]. The 12 remaining "core crop" bacteria occur at much greater frequency in the hindgut than the crop, a pattern consistent with their inclusion in the broad phylogenetic groupings described previously by many independent research groups; *Lactobacillus* sp. Firm 4, *Lactobacillus* sp. Firm 5, and *Bifidobacterium* (Table 4, Fig. 4).

As in [18,22], our bees were sampled from intentionally untreated, essentially wild colonies. Unfortunately, we cannot provide a direct comparison of our crop abundance measures with published estimates because the methods used to associate plate counts with the abundance of different taxonomic groups in [18] is not evident. The various LAB identified here and in a previous study [17] were morphologically indistinguishable from one another, thus requiring a significant genotyping effort to estimate taxon abundance. As a crude estimate we can use our two crop-specific data sets to estimate the fraction of total crop reads for a given phylotype (Table S2), and express this value as a proportion

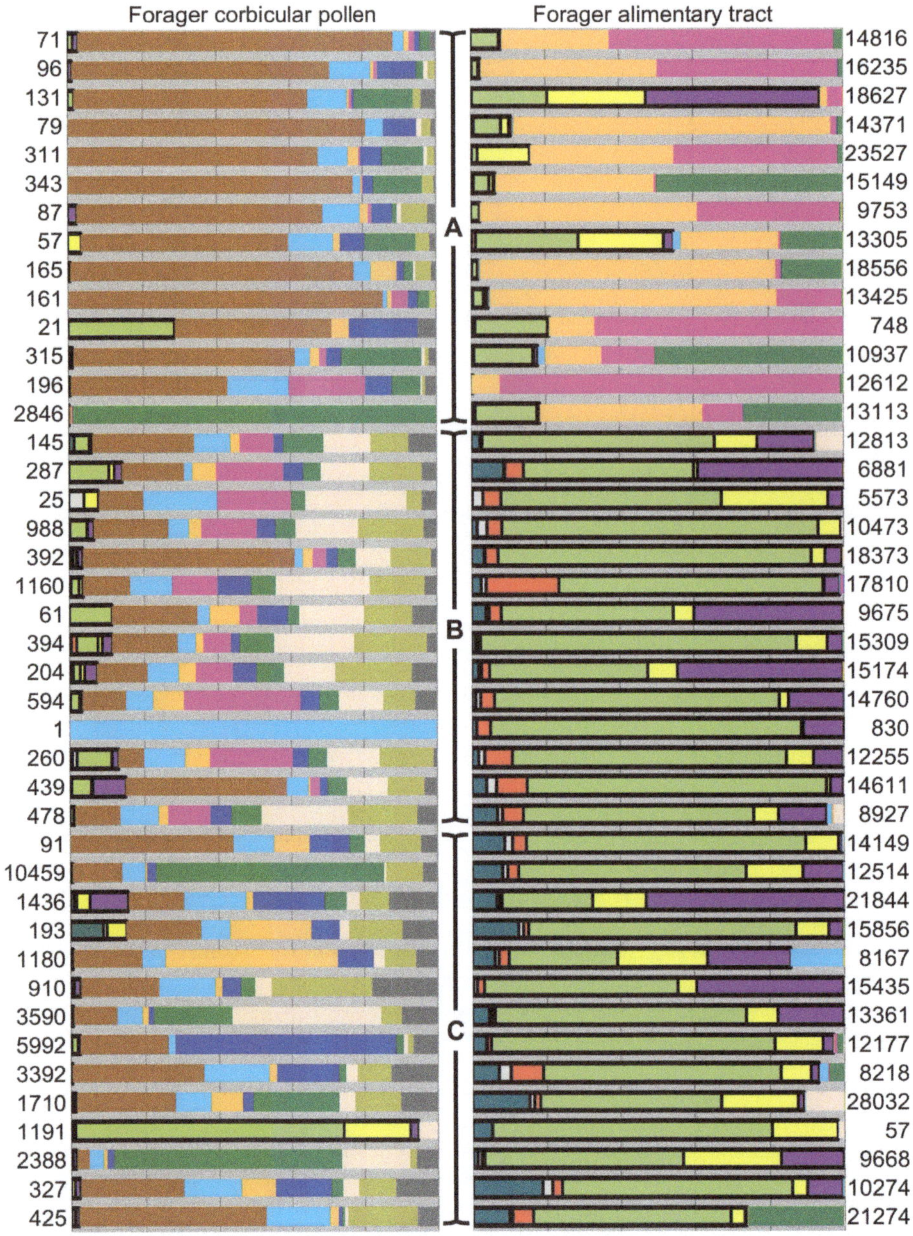

Figure 3. Relative abundances of bacterial groups within each individual library by sample type. Color key is the same as figure 2. In the left column, corbicular pollen libraries were derived from the same bee(s) depicted in the adjacent alimentary tract column. Bars indicated by "A" each denote 10 pooled spring crops and associated corbicular pollen from 14 pooled colony samples; bars within "B" are 14 individual forager guts and their associated corbicular pollen sampled from a single colony in spring, and "C" is 14 individual forager guts and their associated pollen sampled from a single colony in fall. Black boxes around portions of each bar denote the core gut bacteria, and each bar represents a library flanked by its total read number.

of the mean plate count for an individual pollen forager crop (7×10^5 CFU's). This estimate suggests that an average individual forager crop contains 5.6×10^4 CFU's of *Lactobacillus* Firm 5, 1.8×10^3 CFU's of *Lactobacillus* Firm 4, and 1.7×10^2 CFU's of *Bifidobacterium*. As a culture independent comparison, extrapolation from qRT-PCR values based on universal bacterial primers indicates only 10^4 bacterial 16S rDNA gene copies in the crop [25]. It therefore appears that previous estimates of crop abundance based on plate counts [18,22] are overestimated by at least an order of magnitude. Based on the sum of available results we conclude that 12 of the 13 putative core crop bacteria are not only inconsistent with the hypotheses detailed earlier [18–24], but don't actually occur in the crop or gut with any appreciable frequency or abundance (Table 4, Fig. 4).

In general, the crop contained a microbial community very different from that found in guts (Fig. 3). The crop was dominated by Alpha 2.2 (Acetobacteraceae) and *Lactobacillus kunkeei*, a finding consistent with past results culturing the crops of in-hive bees and honey, as well as culture-dependent and culture-independent assessments of beebread [17,25]. These two taxa are not considered part of the core gut bacteria, and thrive in sugar-rich, acidic environments such as the crop, beebread and honey

Table 3. Relative abundance[A] of the core microbiota[B] from honey bee guts and crops.

Study	Site[C]	Sample	Life stage	Total n sequences	A2	BT	G1	G2	F4	F5	BF	Other[D]
Martinson *et al.* 2011	AZ	whole bee	in hive	271	1.1	11.1	11.8	0.0	10	63.8	0.7	1.5
Martinson *et al.* 2011	AZ	pooled guts	in hive	267	0.0	3.7	9.7	0.0	10.5	60.7	13.1	2.2
Martinson *et al.* 2012	AZ	dissected guts	9d old nurses (N = 3)	78,595	0.0	20.0	12.0	17.0	0.2	49.0	0.8	0.1
Martinson *et al.* 2012	AZ	dissected guts	30d old forager (N = 1)	17,910	0.0	19.8	0.11	57.0	0.32	22.0	0.6	0.0
Sabree *et al.* 2012	MA	dissected guts	12 days old	106,344	1.2[E]	6.7	49.1	1.1	11.1	21.0	5.4	5.2
Moran *et al.* 2012	AZ & MD	dissected guts	in hive, outer frames	329,550	1.0	9.1	11.9	2.0	23.2[F]	45.4[F]	5.4	1.9
present	AZ	dissected guts	pollen forager (fall)	191,037	8.5	14.5	12.1	0.0[G]	2.5	55.4	1.2	5.8
present	AZ	dissected guts	pollen forager (spring)	163,468	2.8	13.8	5.8	0.0[G]	5.5	69.6	1.2	1.3
present	AZ	dissected crops	pollen forager (spring)	195,264	0.0	5.1	6.4	0.0	0.3	8.6	0.1	79.5

[A]Values rounded to the nearest tenth of a percent.
[B]A2 is Alpha 2.1; Acetobacteraceae, BT is *Snodgrassella alvi* [60], G1 is *Gilliamella apicola* [60], G2 is *Frischella. perrara* [59], F4 and F5 are *Lactobacillus* species Firm4 and Firm5, and BF is *Bifidobacterium*.
[C]Site corresponds to U.S. state; AZ = Arizona, MA = Massachusetts, MD = Maryland.
[D]Bacteria not found consistently across published studies when sampling the entire alimentary tract or the mid/hindgut (see text).
[E]1.2% of the library was found to contain Alpha 2.1 when smaller amplicon lengths were allowed.
[F]Values for *Lactobacillus* sp. Firm 4 and Firm 5 [33] were switched in [32] and are corrected in this table.
[G]One sequence was found in each sample type.

[17,44]. They may be considered core hive bacteria, as they are associated with nurse workers and developing larvae [17,50]. Alternatively, *L. kunkeei* has been detected worldwide in many different flowers and the pollen provisions of both solitary and social bees suggesting it harbors considerable genetic diversity, and can be readily acquired via the pollination environment [17,38,39,44,64]. That extremophilic bacteria dominate the crop suggests that it is not an optimal niche for microbial growth, but likely acts as a selective sieve allowing relatively few bacteria to flourish [12,17]. This conclusion is supported by the consistent differences in bacterial abundance seen among the crops of worker bees performing different tasks (Table 5). In-hive bees contained much lower microbial loads than presumably older bees exposed to the foraging environment. Although crop temperature is fairly constant, and simple sugars likely plentiful, conditions favoring microbial growth in the crop may fluctuate widely according to worker task including pH, osmotic conditions, and the availability protein and other micronutrients.

Almost all crop samples contained the core gut microbiota to some degree (Fig. 3). In most cases, the crop libraries had fewer than 20% of the core gut phylotypes. However, in two of the 14 libraries (colonies 7 and 12), core gut phylotypes contributed 55% and 95% of the crop sequences, respectively. On average, the gut-specific taxa in greatest abundance in the forager crop corresponded to *Lactobacillus* (Firm 5), *Gilliamella apicola* (Gamma1) and *Snodgrassella alvi* (Beta). This contrasts with past culturing results from the crops of newly emerged bees and nurse bees [17]. From that assay, *G. apicola* and *S. alvi* were absent from the crop and grew primarily on pH neutral media, suggesting exclusion from acidic niches and associated media bias. However, the culturing of crop bacteria *L. kunkeei* and Alpha 2.2 were seemingly unaffected by a broad range of pH values. Collectively, this may suggest that the forager crop experiences a reduction in pH, but this remains an open question as foragers are known to differ from younger bees (nurses) for a number of crop and head gland specific components [65–67]. Another bacteria considered core to the hindgut, a 97% OTU corresponding to *G. apicola* (OTU 211), was found in the majority of crop libraries with sporadic abundance. It also occurred with much greater abundance in the crop than in the guts, but was not detected with any frequency in corbicular pollen (Table S1). This is an intriguing result as *G. apicola* has been detected at low levels in the crop, and some of the strains can utilize pectin found in the pollen wall [25,34]. However, neither culturing, cloning, nor qRT-PCR with species-specific primers have detected this taxonomic group in beebread [17,25].

Bacteria in the guts of pollen foragers

Most of the recent studies concerning the gut microbiota of honey bees document the bacterial phylotypes in bees that have not yet transitioned to foraging. The forager diet differs from that of a nurse bee. Young (nurse) bees consume large amounts of pollen, producing nutrient rich glandular secretions to feed worker larvae, queens, and older bees [68,69]. The oldest bees do not ingest pollen, but forage for resources using stored honey and nectar to fuel their flight muscles. Foragers utilize mostly simple sugars, but also receive protein-rich worker jelly at unknown frequency via trophallaxis [61]. Gut microbial communities are strongly affected by host diet [70,71], and so the results shown in this study add to the existing literature to address how microbial communities change as bees transition to different life stages and diets. Despite diet differences between younger in-hive bees and older foragers, we find that forager guts are, like younger bees,

Table 4. Number of sequences from the guts and crops of pollen foragers showing 100% sequence similarity to the 13 putative crop-specific bacterial strains.

Strain [A]	Group [B]	Accession number	Sample type[C]					
			Fall guts		Spring guts		Spring crops	
***Lactobacillus* spp.**								
Hma8	A (Firm 5)	EF187243.1	227	0.1	7	0.004	15	0.008
Bma5	A (Firm 5)	EF187242.1	2,297	1.2	1,877	1.2	489	0.2
Biut2	A (Firm 5)	EF187241.1	3,688	1.9	4,282	2.6	116	0.06
Hma2	A (Firm 5)	EF187240.1	0	0.0	0	0.0	0	0.0
Hma11	A (Firm 5)	EU753689.1	11,535	6.0	3,259	2.0	3,471	1.8
Bin4	B (Firm4)	EF187245.1	703	0.4	2,301	1.4	201	0.1
Fhon2	C (*L. kunkeei*)	EF187239.1	88	0.05	118	0.07	37,001	18.9
Fhon13	C (*L. kunkeei*)	HM534758.1	0	0.0	0	0.0	0	0.0
Hon2	D (Firm 4)	EF187244.1	0	0.0	1,319[D]	0.8	0	0.0
***Bifidobacterium* spp.**								
Hma3	n/a	EF187236.1	24	0.01	55	0.03	0	0.0
Bma6	n/a	EF187237.1	118	0.06	274	0.2	36	0.02
Bin7	n/a	EF187234.1	121	0.06	185	0.1	14	0.007
Bin2	n/a	EF187231.1	40	0.02	104	0.06	8	0.004

[A]Strain names according to [18–23].
[B]Refers to the 4 clades depicted in Figure 4.
[C]The first column under each sample type represents the total number of reads from all 14 libraries having 100% sequence similarity to the queried strain. The second column shows the percent of the total sequence reads from that sample type matching the queried LAB strain sequence at 100%.
[D]Value attained when comparing all trimmed sequences passing quality filtering, but classified at a confidence level <100% at the Phylum level according to the RDP classifier.

comprised of only a few distinct core phylotypes. After omitting the single deficient gut library for the fall individual 12 we found that every individual forager gut contained *Lactobacillus* (Firm 5), *Snodgrassella alvi* (Beta), *Gilliamella apicola* (Gamma 1), Acetobacteraceae (Alpha 2.1), and *Lactobacillus* (Firm 4). *Bifidobacterium* was detected in 21 of 23 guts. In forager guts collected in the spring and fall, these core bacteria represented 99% and 94% of sequences, respectively. One of three major OTUs corresponding to *Lactobacillus* Firm 5 (050, Table S1) accounted for 12% of Firm 5 reads, is a 100% match to a *Lactobacillus* sp. Firm 5 found in bumble bees [72], and has also been sampled from the apple phyllosphere [73] suggesting some degree of horizontal transmission.

As shown for in-hive bees [33], the relative frequencies of core phylotypes varied considerably among individuals in the same hive and in many cases bacteria identified as core to the gut were not consistently found in foragers. In stark contrast to recent amplicon based work [25], *Frischella perrara* (Gamma 2) was represented as only one read from one forager gut from each sampled time period (Table S1). Further investigation revealed that the primers used in this and other studies [32,33] were a 100% match to the *F. perrara* 16S rDNA sequence. Thus the finding that foragers contain little to no *F. perrara* relative to younger bees appears authentic. We identified no sequences corresponding to Gamma 4, a Gammaproteobacteria occurring sporadically in previous samples of in-hive bees from this same Arizona location [33].

Many of the individuals with low levels of the core gut microbiota possessed what might be considered environmental bacteria occurring at sporadic abundance consistent with the hypothesis of a disease state. As an example, the gut of fall individual 1 contained the least amount of the core microbiota, but exhibits a high level of non-core Enterobacteriaceae sequences. The putative *Bartonella*-like species Alpha1 (OTU021) and a *Pseudomonas* species (OTU084) also showed a similar pattern in the guts of fall individual 10 and spring individual 14, respectively. That these same bacteria did not occur with any frequency in corbicular pollen suggests their potential as opportunistic gut pathogens, as opposed to recently vectored plant bacteria. In a contrasting pattern, OTU159 classified as *Arsenophonus*, was detected at high frequency from both the crop and corbicular pollen of one pooled colony sample (see [8,74]).

Corbicular pollen has a rich microbial community distinct from the gut and crop

Corbicular pollen harbored significantly more taxa than microbial communities of the gut or crop (Table S2). The core gut bacteria were rarely found in corbicular pollen and, on average, comprised less than 3.5% of the corbicular pollen libraries (Fig. 3). In a couple of libraries, the strong presence of core gut bacteria (i.e. 90% in fall individual 4) may have been caused by inadvertently pinching the forager's abdomen with forceps during collection, resulting in contamination of the corbicular pollen pellet with bacteria-rich feces. In general, however, most of the corbicular pollen samples were dominated by bacterial groups often found in wind-blown sediment, soil, flowers, the rhizosphere and phyllosphere such as Gammaproteobacteria, Actinomycetales, and Enterobacteriaceae [42,75,76]. Diversity measures indicate that the sampling of corbicular pollen was nearly complete in the fall, whereas spring libraries of corbicular pollen are underestimated. Additionally, it is difficult to determine whether the crop contents simply reflect taxa passing through the crop on their way to the gut and/or whether the crop serves as a source of inoculum for newly collected pollen. Despite

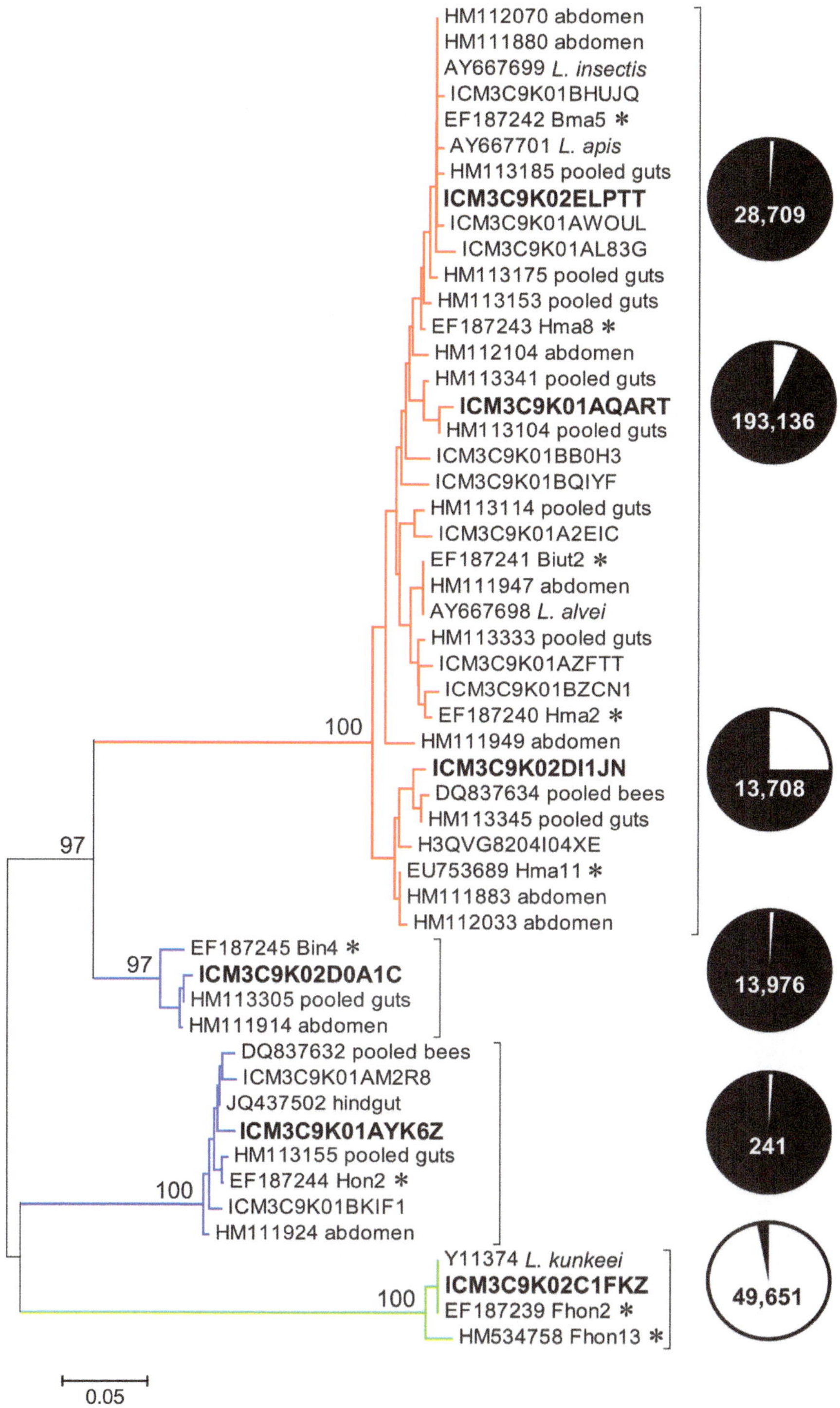

Figure 4. Phylogenetic tree of *Lactobacillus* spp. identified from the guts and crops of foragers. A Neighbor-Joining phylogeny was constructed using the V1/V2 region of the 16S rRNA gene of *Lactobacillus* spp. from alimentary tracts in this study. Published full-length *Lactobacillus* spp. 16S rRNA sequences from gut samples, and putative crop specific sequences (marked with an asterisk) were used for reference. Numbers at the branch nodes indicate the level of bootstrap support for 500 replicates. Representative sequence titles for each OTU (97% sequence similarity) identified in the present study are shown in bold font. Values within pie charts are the number of sequences comprising each of the six *Lactobacillus* OTUs containing more than 100 reads. Pie charts represent the proportion of reads in each OTU occurring in the gut (black) vs. the crop (white). Colored clades correspond to the four major groups identified in Table 4. Please note that the two Firm 4 clades typically form a single clade in phylogenies constructed from full-length 16S rRNA sequences.

Table 5. Bacterial counts (CFUs)[A] for honey bee crops[B] cultured under different conditions.

Growth conditions		Workers sampled inside the hive		Workers sampled outside the hive	
Media[C]	Atm[D]	Nectar[E] (9)	Nurses[F] (10)	General (12)	Pollen (12)
MRS	O_2[A]	0.69±0.85	3.60±1.31	11.93±6.40	7.18±5.54
MRS	CO_2	1.97±1.06	3.99±1.49	8.75±4.90	6.78±4.39
TSA	O_2	0.013±0.012	0.026±0.009	3.26±1.45	2.11±1.37
TSA	CO_2	0.83±0.52	2.21±0.96	3.41±1.21	5.63±3.56

[A]Mean ± SE colony forming units per crop; multiply listed values by 10^5,
[B]Crop is the foregut or social stomach. Sample size in parentheses.
[C]MRS: de Man, Rogosa and Sharpe media; pH 5.7, TSA: Tryptic soy agar; pH 7.3
[D]Atmosphere; O_2 = Ambient conditions, CO_2 = Ambient conditions containing 5% CO_2.
[E]Workers with empty crops processing nectar.
[F]Workers with full crops attending to brood.

these difficulties, informative patterns were evident when comparing OTU frequency and occurrence across microenvironments.

We might infer that OTUs highly abundant in corbicular pollen and either absent or found at much lower frequency in the gut and crop represent bacteria acquired from the pollination environment (Table S2). Microbes abound in the phyllosphere, flowers and nectar of plants, and are highly diverse, often showing strong taxonomic affiliations with their host plant [77]. It is tempting to hypothesize that some of the variation among corbicular pollen samples reflects each individual bees' foraging flight. In a single foraging trip, a worker can collect both nectar and pollen, but foraging typically favors only one of these resources [78]. Individual foragers show strong constancy in the plant species they pollinate and the pollen on a returning forager is frequently derived from a single plant species [79]. In addition, pollen can be inoculated with flower bacteria, and airborne bacteria that stick to the corbicular mass as the bee returns to the hive. Therefore, the bacterial community in corbicular pollen might not only reflect microbes found on the floral resource, but the path traveled by foragers. Abundant corbicular OTUs belonged to environmental bacteria Xanthomonadaceae, Pseudomonadaceae, Enterobacteriacae, Acinetobacter, and many different families of Actinomycetales.

Many of the Actinobacteria uniformly present in individual corbicular pollen samples have also been found at appreciable levels in beebread [17], suggesting they can survive for extended periods in the food stores. Actinobacteria found in beebread were represented by at least 12 families and included a variety of *Streptomyces* [17]. Many Actinobacteria are apparently mutualists, found in the food stores, hive materials, or cuticle of a variety of Hymenoptera including solitary and social bees [49,80–85]. Actinobacteria are renowned for their vast arsenal of metabolic weaponry used to inhibit fungal growth, a common cause of pollen spoilage [37,86]. Actinobacteria can grow slowly, form spore-like structures, and survive on minimal media [87], such that transmission between hives, environments and life stages may occur on the bodies of individual honey bees. Actinobacteria in corbicular pollen, beebread and brood casings is particularly interesting given the hypothesis that both solitary and social Hymenoptera, developing in close association with resources prone to fungal infection, have strong selective pressure to evolve protective symbioses [88].

Bacteria most abundant in the gut were considered gut-specific, and were consistently found at much lower frequency in the crop and corbicular pollen. These core bacteria may therefore be transmitted to new offspring via trophollaxis, or "leaked" from the crop, as opposed to direct or indirect coprophagy [25]. Their presence in the corbicular pollen could suggest that they function in pollen preservation or nutrition. However, recent cultures, clones, and qRT-PCR with core gut-specific 16S rDNA primers suggest that core gut bacteria do not survive well in beebread. Only two crop inhabiting bacteria, Alpha 2.2 and *Lactobacillus kunkeei*, show a pattern suggestive of corbicular pollen inoculation followed by long term beebread survival [17,25]. As detailed previously, these two bacteria were much more abundant in the crop than the gut, and also found with considerable frequency in corbicular pollen. Interestingly however, *Lactobacillus kunkeei* was found only in spring corbicular pollen, perhaps due to an environmental or hive-specific effect. In contrast, Alpha 2.2 was consistently abundant in corbicular pollen across both sampling periods. Combined with previous findings, this suggests that Alpha 2.2 is a "core hive" bacterium capable of survival in royal jelly and honey, the most extreme of hive environments [17,50].

Conclusion

We catalogued the diversity of bacteria associated with the guts, crops, and corbicular pollen of honey bee foragers. Our results do not support the existing hypothesis of a core crop microbiota comprised of 13 LAB strains [18–24]. Rather, when sequences corresponding to these 13 bacterial strains were detected, they were at very low relative frequencies from both the foregut (crop) and entire alimentary tract, with the single exception of *L. kunkeei*, abundant in forager crops. We find additional evidence for a core gut microbiota and add to the existing literature focused on younger in-hive bees [25,32,33]. Lastly, we find that corbicular pollen is a microbially diverse environment and likely the source of several plant associated bacteria commonly found in hive food stores.

Acknowledgments

Disclaimer: The USDA is an equal opportunity employer and provider. The authors acknowledge two anonymous reviewers for comments on a previous version of the manuscript and the members of the Anderson lab for technical assistance. The corresponding author thanks Belynda Starr, Ariel Calypso and Isaak Edward for their valuable input. We thank the Bio5 Institute, at the University of Arizona Genomics Core for the generation of 454 amplicons.

Author Contributions

Conceived and designed the experiments: KEA PM. Performed the experiments: PM. Analyzed the data: KEA VC-H. Contributed reagents/materials/analysis tools: KEA VC-H. Wrote the paper: KEA VC-H.

References

1. VanEngelsdorp D, Evans JD, Donovall L, Mullin C, Frazier M, et al. (2009) "Entombed Pollen": A new condition in honey bee colonies associated with increased risk of colony mortality. J Invertebr Pathol 101: 147–149.
2. Mullin CA, Frazier M, Frazier JL, Ashcraft S, Simonds R, et al. (2010) High levels of miticides and agrochemicals in North American apiaries: implications for honey bee health. PLoS One 5: e9754. Available: http://www.pubmedcentral.nih.gov/articlerender.fcgi?artid = 2841636&tool = pmcentrez&rendertype = abstract.
3. Gilliam M, Valentine DK (1974) Enterobacteriaceae isolated from foraging worker honey bees, *Apis mellifera*. J Invertebr Pathol 23: 38–41. Available: http://dx.doi.org/10.1016/0022-2011(74)90069-X.
4. Pettis JS, VanEngelsdorp D, Johnson J, Dively G (2012) Pesticide exposure in honey bees results in increased levels of the gut pathogen *Nosema*. Naturwissenschaften 99: 153–158.
5. Pettis JS, Lichtenberg EM, Andree M, Stitzinger J, Rose R, et al. (2013) Crop pollination exposes honey bees to pesticides which alters their susceptibility to the gut pathogen *Nosema ceranae*. PLoS One 8: e70182. Available: http://www.pubmedcentral.nih.gov/articlerender.fcgi?artid = 3722151&tool = pmcentrez&rendertype = abstract. Accessed 2013 Sep 20.
6. Runckel C, Flenniken ML, Engel JC, Ruby JG, Ganem D, et al. (2011) Temporal analysis of the honey bee microbiome reveals four novel viruses and seasonal prevalence of known viruses, *Nosema*, and *Crithidia*. PLoS One 6: e20656. Available: http://www.pubmedcentral.nih.gov/articlerender.fcgi?artid = 3110205&tool = pmcentrez&rendertype = abstract. Accessed 2013 Feb 28.
7. Singh R, Levitt AL, Rajotte EG, Holmes EC, Ostiguy N, et al. (2010) RNA Viruses in Hymenopteran Pollinators: Evidence of Inter-Taxa Virus Transmission via Pollen and Potential Impact on Non-*Apis* Hymenopteran Species. PLoS One 5: e14357. Available: http://www.pubmedcentral.nih.gov/articlerender.fcgi?artid = 3008715&tool = pmcentrez&rendertype = abstract.
8. Cornman RS, Tarpy DR, Chen Y, Jeffreys L, Lopez D, et al. (2012) Pathogen webs in collapsing honey bee colonies. PLoS One 7: e43562. Available: http://www.pubmedcentral.nih.gov/articlerender.fcgi?artid = 3424165&tool = pmcentrez&rendertype = abstract. Accessed 2013 Mar 8.
9. Ravoet J, Maharramov J, Meeus I, De Smet L, Wenseleers T, et al. (2013) Comprehensive Bee Pathogen Screening in Belgium Reveals *Crithidia mellificae* as a New Contributory Factor to Winter Mortality. PLoS One 8: e72443. Available: http://www.pubmedcentral.nih.gov/articlerender.fcgi?artid = 3753275&tool = pmcentrez&rendertype = abstract. Accessed 2013 Sep 19.
10. Dainat B, Evans JD, Chen YP, Gauthier L, Neumann P (2012) Predictive markers of honey bee colony collapse. PLoS One 7: e32151. Available: http://www.pubmedcentral.nih.gov/articlerender.fcgi?artid = 3285648&tool = pmcentrez&rendertype = abstract. Accessed 2013 Mar 15.
11. Hamdi C, Balloi A, Essanaa J, Crotti E, Gonella E, et al. (2011) Gut microbiome dysbiosis and honeybee health. J Appl Entomol 135: 524–533.
12. Anderson KE, Sheehan TH, Eckholm BJ, Mott BM, DeGrandi-Hoffman G (2011) An emerging paradigm of colony health: microbial balance of the honey bee and hive (*Apis mellifera*). Insectes Soc 58: 431–444.
13. Hölldobler B, Wilson EO (2008) The Superorganism: The Beauty, Elegance, and Strangeness of Insect Societies. W. W. Norton & Company. 522 p.
14. Blatt J, Roces F (2002) The control of the proventriculus in the honeybee (*Apis mellifera carnica* L.) I. A dynamic process influenced by food quality and quantity? J Insect Physiol 48: 643–654.
15. Pain J, Maugenet J (1966) Recherches biochimiques et physiologiques sur le pollen emmagsiné par les abeilles. Ann Abeille 9: 209–236.
16. Loper GM, Standifer LN, Thompson MJ, Gilliam M (1980) Biochemistry and microbiology of bee-collected almond (*Prunus dulcis*) pollen and beebread. I- Fatty Acids, Sterols, Vitamins and Minerals. Apidologie 11: 63–73.
17. Anderson KE, Sheehan TH, Mott BM, Maes P, Snyder L, et al. (2013) Microbial Ecology of the Hive and Pollination Landscape: Bacterial Associates from Floral Nectar, the Alimentary Tract and Stored Food of Honey Bees (*Apis mellifera*). PLoS One 8: e83125. Available: http://www.pubmedcentral.nih.gov/articlerender.fcgi?artid = 3866269&tool = pmcentrez&rendertype = abstract. Accessed 2014 Feb 3.
18. Olofsson TC, Vásquez A (2008) Detection and identification of a novel lactic acid bacterial flora within the honey stomach of the honeybee *Apis mellifera*.
19. Vásquez A, Olofsson TC (2009) The lactic acid bacteria involved in the production of bee pollen and bee bread. J Apic Res 48: 189–195.
20. Olofsson TC, Vásquez A (2009) Phylogenetic comparison of bacteria isolated from the honey stomachs of honey bees *Apis mellifera* and bumble bees *Bombus* spp. J Apic Res 48: 233–237.
21. Olofsson TC, Vásquez A, Sammataro D, Macharia J (2011) A scientific note on the lactic acid bacterial flora within the honeybee subspecies *Apis mellifera* (Buckfast), *A. m. scutellata*, *A. m. mellifera*, and *A. m. monticola*. Apidologie 42: 696–699.
22. Vásquez A, Forsgren E, Fries I, Paxton RJ, Flaberg E, et al. (2012) Symbionts as Major Modulators of Insect Health: Lactic Acid Bacteria and Honeybees. PLoS One 7: e33188. Available: http://dx.plos.org/10.1371/journal.pone.0033188.
23. Vásquez A, Olofsson TC (2011) The honey crop – the holy Grail when antibiotics fail. Microbiol Today 38: 226–229.
24. Butler E, Alsterfjord M, Olofsson TC, Karlsson C, Malmstrom J, et al. (2013) Proteins of novel lactic acid bacteria from *Apis mellifera mellifera*: an insight into the production of known extra-cellular proteins during microbial stress. doi:10.1186/1471-2180-13-235.
25. Martinson VG, Moy J, Moran NA (2012) Establishment of characteristic gut bacteria during development of the honey bee worker. Appl Environ Microbiol 78: 2830–2840. Available: http://www.ncbi.nlm.nih.gov/pubmed/22307297.
26. Brodschneider R, Crailsheim K (2010) Nutrition and health in honey bees. Apidologie 41: 278–294.
27. Mohr KI, Tebbe CC (2006) Diversity and phylotype consistency of bacteria in the guts of three bee species (Apoidea) at an oilseed rape field. Environ Microbiol 8: 258–272.
28. Babendreier D, Joller D, Romeis J, Bigler F, Widmer F (2007) Bacterial community structures in honeybee intestines and their response to two insecticidal proteins. FEMS Microbiol Ecol 59: 600–610.
29. Cox-foster DL, Conlan S, Holmes EC, Palacios G, Evans JD, et al. (2007) A metagenomic survey of microbes in honey bee colony collapse disorder. Science 318: 283–287.
30. Martinson VG, Danforth BN, Minckley RL, Rueppell O, Tingek S, et al. (2011) A simple and distinctive microbiota associated with honey bees and bumble bees. Mol Ecol 20: 619–628.
31. Disayathanoowat T, Young JPW, Helgason T, Chantawannakul P (2011) T-RFLP analysis of bacterial communities in the midguts of *Apis mellifera* and *Apis cerana* honey bees in Thailand. Fems Micriobiology Ecol 79: 273–281.
32. Sabree ZL, Hansen AK, Moran NA (2012) Independent Studies Using Deep Sequencing Resolve the Same Set of Core Bacterial Species Dominating Gut Communities of Honey Bees. PLoS One 7: e41250. Available: http://dx.plos.org/10.1371/journal.pone.0041250.
33. Moran NA, Hansen AK, Powell JE, Sabree ZL (2012) Distinctive Gut Microbiota of Honey Bees Assessed Using Deep Sampling from Individual Worker Bees. PLoS One 7: e36393. Available: http://dx.plos.org/10.1371/journal.pone.0036393.
34. Engel P, Martinson VG, Moran NA (2012) Functional diversity within the simple gut microbiota of the honey bee. Proc Natl Acad Sci USA 109: 1–6.
35. Ahn J-H, Hong I-P, Bok J-I, Kim B-Y, Song J, et al. (2012) Pyrosequencing analysis of the bacterial communities in the guts of honey bees *Apis cerana* and *Apis mellifera* in Korea. J Microbiol 50: 735–745.
36. Gilliam M, Prest DB, Lorenz BJ (1989) Microbiology of pollen and bee bread: taxonomy and enzymology of molds. Apidologie 20: 53–68.
37. Batra LR, Batra SWT, Bohart GE (1973) The mycoflora of domesticated and wild bees (Apoidea). Mycopathol Mycol Appl 49: 13–44.
38. McFrederick QS, Wcislo WT, Taylor DR, Ishak HD, Dowd SE, et al. (2012) Environment or kin: whence do bees obtain acidophilic bacteria? Mol Ecol 21: 1754–1768.
39. McFrederick QS, Cannone JJ, Gutell RR, Kellner K, Plowes RM, et al. (2013) Specificity between Lactobacilli and Hymenopteran Hosts Is the Exception Rather than the Rule. Appl Environ Microbiol 79: 1803–1812.
40. Piccini C, Antunez K, Zunino P (2004) An approach to the characterization of the honey bee hive bacterial flora. J Apic Res 43: 101–104.
41. Loncaric I, Heigl H, Licek E, Moosbeckhofer R, Busse H-J, et al. (2009) Typing of Pantoea agglomerans isolated from colonies of honey bees (*Apis mellifera*) and culturability of selected strains from honey. Apidologie 40: 40–54.
42. Fridman S, Izhaki I, Gerchman Y, Halpern M (2012) Bacterial communities in floral nectar. Environ Microbiol Rep 4: 97–104.
43. Alvarez-Pérez S, Lievens B, Jacquemyn H, Herrera CM (2013) *Acinetobacter nectaris* sp. nov. and *Acinetobacter boissieri* sp. nov., isolated from floral nectar of wild Mediterranean insect-pollinated plants. Int J Syst Evol Microbiol 63: 1532–1539.
44. Neveling DP, Endo A, Dicks LMT (2012) Fructophilic *Lactobacillus kunkeei* and *Lactobacillus brevis* Isolated from Fresh Flowers, Bees and Bee-hives. Curr Microbiol. Available: http://www.ncbi.nlm.nih.gov/pubmed/22797888.

45. Endo A, Futagawa-Endo Y, Dicks LMT (2009) Isolation and characterization of fructophilic lactic acid bacteria from fructose-rich niches. Syst Appl Microbiol 32: 593–600.
46. Ruiz-Argueso T, Rodriguez-Navarro A (1975) Microbiology of ripening honey. Appl Microbiol 30: 893–896.
47. Rada V, Máchová M, Huk J, Marounek M, Dušková D (1997) Microflora in the honeybee digestive tract: counts, characteristics and sensitivity to veterinary drugs. Apidologie 28: 357–365.
48. Gilliam M (1997) Identification and roles of non-pathogenic microflora associated with honey bees. FEMS Microbiol Lett 155: 1–10.
49. Promnuan Y, Kudo T, Chantawannakul P (2009) Actinomycetes isolated from beehives in Thailand. World J Microbiol Biotechnol 25: 1685–1689.
50. Vojvodic S, Rehan SM, Anderson KE (2013) Microbial Gut Diversity of Africanized and European Honey Bee Larval Instars. PLoS One 8: e72106. Available: http://dx.plos.org/10.1371/journal.pone.0072106. Accessed 26 August 2013.
51. Schloss PD, Westcott SL, Ryabin T, Hall JR, Hartmann M, et al. (2009) Introducing mothur: open-source, platform-independent, community-supported software for describing and comparing microbial communities. Appl Environ Microbiol 75: 7537–7541.
52. Edgar RC, Haas BJ, Clemente JC, Quince C, Knight R (2011) UCHIME improves sensitivity and speed of chimera detection. Bioinformatics 27: 2194–2200.
53. Wang Q, Garrity GM, Tiedje JM, Cole JR (2007) Naive Bayesian classifier for rapid assignment of rRNA sequences into the new bacterial taxonomy. Appl Environ Microbiol 73: 5261–5267.
54. Edgar RC (2004) MUSCLE: multiple sequence alignment with high accuracy and high throughput. Nucleic Acids Res 32: 1792–1797.
55. Hall TA (1999) BioEdit: a user friendly biological sequence alignment editor and analysis program for Windows 95/98/NT. Nucleic Acids Symp Ser 41: 95–98.
56. Tamura K, Peterson D, Peterson N, Stecher G, Nei M, et al. (2011) MEGA5: molecular evolutionary genetics analysis using maximum likelihood, evolutionary distance, and maximum parsimony methods. Mol Biol Evol 28: 2731–2739.
57. Kacániová M, Chlebo R, Kopernický M, Trakovická A (2004) Microflora of the honeybee gastrointestinal tract. Folia Microbiol 49: 169–171.
58. Aizenberg-Gershtein Y, Izhaki I, Halpern M (2013) Do honeybees shape the bacterial community composition in floral nectar? PLoS One 8: e67556. Available: http://www.pubmedcentral.nih.gov/articlerender.fcgi?artid = 3701072&tool = pmcentrez&rendertype = abstract. Accessed 23 September 2013.
59. Engel P, Kwong WK, Moran NA (2013) *Frischella perrara* gen. nov., sp. nov., a gammaproteobacterium isolated from the gut of the honey bee, *Apis mellifera*. Int J Syst Evol Microbiol 63: 3646–3651.
60. Kwong WK, Moran NA (2012) Cultivation and characterization of the gut symbionts of honey bees and bumble bees: *Snodgrassella alvi* gen. nov., sp. nov., a member of the Neisseriaceae family of the Betaproteobacteria; and *Gilliamella apicola* gen. nov., sp. nov., a member of Orbaceae. Int J Syst Evol Microbiol. Available: http://www.ncbi.nlm.nih.gov/pubmed/23041637. Accessed 9 October 2012.
61. Crailsheim K (1992) The flow of jelly within a honeybee colony. J Comp Physiol B 162: 681–689.
62. Judd TM (2011) The role of food storage and communication in the evolution of perennial social Hymenopteran colonies. In: Stewart EM, editor. Social insects: Structure, Function and Behavior. Nova. pp 1–24.
63. Endo A, Irisawa T, Futagawa-Endo Y, Takano K, du Toit M, et al. (2012) Characterization and emended description of *Lactobacillus kunkeei* as a fructophilic lactic acid bacterium. Int J Syst Evol Microbiol 62: 500–504.
64. Endo A, Futagawa-Endo Y, Dicks LMT (2009) Isolation and characterization of fructophilic lactic acid bacteria from fructose-rich niches. Syst Appl Microbiol 32: 593–600.
65. Ohashi K, Natori S, Kubo T (1999) Expression of amylase and glucose oxidase in the hypopharyngeal gland with an age-dependent role change of the worker honeybee (*Apis mellifera* L.). Eur J Biochem 265: 127–133.
66. Feng M, Fang Y, Li J (2009) Proteomic analysis of honeybee worker (*Apis mellifera*) hypopharyngeal gland development. BMC Genomics doi:10.1186/1471-2164-10-645.
67. Leoncini I, Le Conte Y, Costagliola G, Plettner E, Toth AL, et al. (2004) Regulation of behavioral maturation by a primer pheromone produced by adult worker honey bees. Proc Natl Acad Sci USA 101: 17559–17564.
68. Toth AL, Robinson GE (2005) Worker nutrition and division of labour in honeybees. Anim Behav 69: 427–435.
69. Toth AL, Kantarovich S, Meisel AF, Robinson GE (2005) Nutritional status influences socially regulated foraging ontogeny in honey bees. J Exp Biol 208: 4641–4649.
70. Tang X, Freitak D, Vogel H, Ping L, Shao Y, et al. (2012) Complexity and variability of gut commensal microbiota in polyphagous lepidopteran larvae. PLoS One 7: e36978. Available: http://www.pubmedcentral.nih.gov/articlerender.fcgi?artid = 3398904&tool = pmcentrez&rendertype = abstract. Accessed 17 March 2013.
71. Dillon RJ, Dillon VM (2004) The gut bacteria of insects: nonpathogenic interactions. Annu Rev Entomol 49: 71–92.
72. Koch H, Abrol DP, Li J, Schmid-Hempel P (2013) Diversity and evolutionary patterns of bacterial gut associates of corbiculate bees. Mol Ecol: 2028–2044.
73. Yashiro E, Spear RN, McManus PS (2011) Culture-dependent and culture-independent assessment of bacteria in the apple phyllosphere. J Appl Microbiol 110: 1284–1296.
74. Mcfrederick QS, Mueller UG, James RR (2014) Interactions between fungi and bacteria influence microbial community structure in the *Megachile rotundata* larval gut. Proc. R. Soc. B 281: 20132653. http://dx.doi.org/10.1098/rspb.2013.2653
75. Gardner T, Acosta-Martinez V, Calderón FJ, Zobeck TM, Baddock M, et al. (2012) Pyrosequencing reveals bacteria carried in different wind-eroded sediments. J Environ Qual 41: 744–753.
76. Junker RR, Loewel C, Gross R, Dötterl S, Keller A, et al. (2011) Composition of epiphytic bacterial communities differs on petals and leaves. Plant Biol http://doi.wiley.com/10.1111/j.1438-8677.2011.00454.x.
77. Jackson CR, Denney WC (2011) Annual and seasonal variation in the phyllosphere bacterial community associated with leaves of the southern *Magnolia* (*Magnolia grandiflora*). Microb Ecol 61: 113–122.
78. Dreller C, Page Jr RE, Fondrk MK (1999) Regulation of pollen foraging in honeybee colonies: effects of young brood, stored pollen, and empty space. Behav Ecol Sociobiol 45: 227–233.
79. Grüter C, Moore H, Firmin N, Helanterä H, Ratnieks FLW (2011) Flower constancy in honey bee workers (*Apis mellifera*) depends on ecologically realistic rewards. J Exp Biol 214: 1397–1402.
80. Inglis GD, Sigler L, Goette MS (1993) Aerobic microorganisms associated with alfalfa leafcutter bees (*Megachile rotundata*). Microb Ecol 26: 125–143.
81. Barke J, Seipke RF, Grüschow S, Heavens D, Drou N, et al. (2010) A mixed community of actinomycetes produce multiple antibiotics for the fungus farming ant *Acromyrmex octospinosus*. BMC Biol http://www.biomedcentral.com/1741-7007/8/109
82. Currie CR, Scott JA (1999) Fungus-growing ants use antibiotic-producing bacteria to control garden parasites. Nature 398: 701–705.
83. Seipke RF, Barke J, Brearley C, Hill L, Yu DW, et al. (2011) A Single Streptomyces Symbiont Makes Multiple Antifungals to Support the Fungus Farming Ant *Acromyrmex octospinosus*. PLoS One 6: 8. Available: http://dx.plos.org/10.1371/journal.pone.0022028.
84. Heuer H, Krsek M, Baker P, Smalla K, Wellington EM (1997) Analysis of actinomycete communities by specific amplification of genes encoding 16S rRNA and gel-electrophoretic separation in denaturing gradients. Appl Environ Microbiol 63: 3233–3241.
85. Kaltenpoth M, Göttler W, Herzner G, Strohm E, Go W (2005) Symbiotic Bacteria Protect Wasp Larvae from Fungal Infestation. Curr Biol 15: 475–479.
86. Gilliam M, Roubik DW, Lorenz BJ (1990) Microorganisms associated with pollen, honey, and brood provisions in the nest of a stingless bee, *Melipona fasciata*. Apidologie 21: 89–97.
87. Kaltenpoth M, Goettler W, Koehler S, Strohm E (2009) Life cycle and population dynamics of a protective insect symbiont reveal severe bottlenecks during vertical transmission. Evol Ecol 24: 463–477.
88. Kaltenpoth M (2009) Actinobacteria as mutualists: general healthcare for insects? Trends Microbiol 17: 529–535.

6

Global Ex-Situ Crop Diversity Conservation and the Svalbard Global Seed Vault: Assessing the Current Status

Ola T. Westengen[1,2]*, Simon Jeppson[1], Luigi Guarino[3]

1 Nordic Genetic Resource Center (NordGen), Alnarp, Sweden, 2 Centre for Development and the Environment, University of Oslo, Oslo, Norway, 3 The Global Crop Diversity Trust, Bonn, Germany

Abstract

Ex-situ conservation of crop diversity is a global concern, and the development of an efficient and sustainable conservation system is a historic priority recognized in international law and policy. We assess the completeness of the safety duplication collection in the Svalbard Global Seed Vault with respect to data on the world's *ex-situ* collections as reported by the Food and Agriculture Organization of the United Nations. Currently, 774,601 samples are deposited at Svalbard by 53 genebanks. We estimate that more than one third of the globally *distinct* accessions of 156 crop genera stored in genebanks as orthodox seeds are conserved in the Seed Vault. The numbers of safety duplicates of *Triticum* (wheat), *Sorghum* (sorghum), *Pennisetum* (pearl millet), *Eleusine* (finger millet), *Cicer* (chickpea) and *Lens* (lentil) exceed 50% of the estimated numbers of distinct accessions in global *ex-situ* collections. The number of accessions conserved globally generally reflects importance for food production, but there are significant gaps in the safety collection at Svalbard in some genera of high importance for food security in tropical countries, such as *Amaranthus* (amaranth), *Chenopodium* (quinoa), *Eragrostis* (teff) and *Abelmoschus* (okra). In the 29 food-crop genera with the largest number of accessions stored globally, an average of 5.5 out of the ten largest collections is already represented in the Seed Vault collection or is covered by existing deposit agreements. The high coverage of ITPGRFA Annex 1 crops and of those crops for which there is a CGIAR mandate in the current Seed Vault collection indicates that existence of international policies and institutions are important determinants for accessions to be safety duplicated at Svalbard. As a back-up site for the global conservation system, the Seed Vault plays not only a practical but also a symbolic role for enhanced integration and cooperation for conservation of crop diversity.

Editor: Wengui Yan, National Rice Research Center, United States of America

Funding: The authors have no support or funding to report.

Competing Interests: OTW and SJ are employees of Nordic Genetic Resource Centre (NordGen). LG is an employee of The Global Crop Diversity Trust. Both NordGen and The Global Crop Diversity Trust are nonprofit organizations involved in the administration of the Svalbard Global Seed Vault. This does not alter the authors' adherence to all the PLOS ONE policies on sharing data and materials.

* E-mail: ola.westengen@nordgen.org

Introduction

The use of genetic diversity to adapt crops to human needs is as old as the Neolithic revolution [1]. In the 1920s the Russian geneticist and botanist Nicolai Vavilov started systematically collecting and conserving genetic diversity as *a resource* for crop breeding, making *ex-situ* (off-site) conservation part of the agricultural R&D system [2]. Crops producing seeds that can be conserved at low relative humidity and low temperature (orthodox seeds) are now commonly conserved *ex-situ* in genebanks. The two-fold rationale for genebanks is, on the one hand, to conserve diversity that is threatened *in-situ* (in farmers' fields or in the wild) and, on the other hand, to make genetic resources accessible to users [3]. In the 1930s, the barley breeder Harry V. Harlan was among the first to sound the alarm on *genetic erosion* of crop genetic resources [4], and, in 1967, a conference in the UN Food and Agriculture Organization (FAO) initiated what has become the genetic resources movement [5]. In the early 1970s, other hallmark conferences laid out practical action plans for the FAO and the Consultative Group on Agricultural Research (CGIAR) to establish an international network of conservation activities and genebanks [6,7]. While the initial focus was on establishing a small number of genebanks with a global mandate, the FAO currently reports that there are 1750 genebanks around the world [8]. Precarious funding, in combination with less than perfect collaboration and coordination among genebanks, has called into question the ability of many of these facilities to ensure long-term conservation, and genetic erosion *inside* genebanks has become a major concern [8,9]. The need for proper safety duplication of the world's unique crop genetic resources is therefore an important international priority [8,10,11].

The Svalbard Global Seed Vault was established with the "*objective to provide a safety net for the international conservation system of plant genetic resources, and to contribute to the securing of the maximum amount of plant genetic diversity of importance to humanity for the long term in accordance with the latest scientific knowledge and most appropriate techniques*" [12].The Seed Vault is managed in partnership by the Government of Norway, the Nordic Genetic Resource Center (NordGen) and the Global Crop Diversity Trust (the Trust). NordGen is a public regional institute supported by the governments of the Nordic countries, and the Trust an independent international organization based in Bonn, Germany. The Norwegian Ministry of Agriculture and Food is the legally responsible authority for the Seed Vault, and its operation is overseen by an International Advisory Council consisting of international technical and policy experts representing, among others, the FAO, national genebanks, the CGIAR and the Governing Body of the International Treaty on Plant Genetic Resources for Food and Agriculture (ITPGRFA). The Seed Vault provides free-of-charge, long-term storage of duplicates from genebanks around the world and works as an insurance policy against incremental or catastrophic loss of the original collections (Fig. 1). The international community has called for an effective, efficient and sustainable global system to conserve Plant Genetic Resources for Food and Agriculture (PGRFA) in the Global Plan of Action for Plant Genetic Resources for Food and Agriculture (GPA) [13,14] and the ITPGRFA [15]. The Seed Vault has, in its five years of operation, become a cornerstone in the global system emerging from within this international policy and legal framework, and good progress has been made towards the target of duplicating all the distinct accessions of PGRFA conserved as orthodox seeds around the world. At its fifth anniversary in February 2013, the collection stood at 774,601 seed samples, originating from 95% of the 193 UN member states. All seed samples are safety duplicates of accessions already stored in conventional genebanks, with 53 genebanks having deposited material so far (Fig. 2).

While the idea of establishing an international collection in the permafrost at Svalbard dates back to the early 1980 s, it took almost 20 years before the technical, legal and political context allowed the idea to be realized [16]. The most important factors were the long, and at times politically polarized, international processes and negotiations regarding the conservation and sustainable use of PGRFA [17]. Of particular importance is the legally binding ITPGRFA [15], which, in effect, was a prerequisite for the establishment of the Seed Vault [16,18]. In exercising their sovereign rights over genetic resources, the Contracting Parties to the ITPGRFA agreed to make their genetic diversity and related information about the crops stored in their genebanks available through a Multilateral System (MLS) for Access and Benefit Sharing (ABS) [15]. It is in this spirit of global collaboration to conserve a resource for which all countries are interdependent that the Seed Vault was established. The management of the Seed Vault has, in its first five years, prioritized storage of material conserved under the terms of the Treaty. But while the Treaty is the international legal and policy framework, the Seed Vault is not a direct instrument of the Treaty. Since the Seed Vault's purpose is to secure as much as possible of all the distinct PGRFA in the world, political and property-rights issues are addressed by ensuring that there is no transfer of legal ownership by depositing seeds in the Seed Vault in its Standard Deposit Agreement (SDA) [19]. Accessions can only be returned to the genebank that deposited them, and distribution to other users is entirely the responsibility of that genebank. This is a standard procedure for safety duplication of PGRFA [11,20], and the Seed Vault is no exception.

Genebank operation can be grouped into three service categories: genetic resource conservation, distribution, and information assembly and management [21]. To properly fulfill the conservation function, the accessions of genetic resources must be safety duplicated. The new edition of FAO's Genebank Standards, which are still in draft form at the time of writing, devotes a section to guidelines for safety duplication [11]. Because of the costs involved, developing a rational approach to safety duplication is a priority in international collaboration in global *ex-situ* conservation efforts [20]. The new FAO draft standards state that "*a safety duplicate sample for every original accession is stored in a geographically distant area, under the same or better conditions than those in the original genebank*" and, furthermore, that "*the location is chosen to minimize possible risks and provides the best possible storage facilities. To minimize risks that can arise in any individual country safety duplication will be ideally undertaken outside that country.*" [11]. The draft Genebank Standards explicitly state that Svalbard Global Seed Vault now offers a safety-duplication facility which complies with its recommendations.

This is in line with conclusions reached in 2004 by the international committee that undertook the feasibility study for establishing the Seed Vault. That committee formulated the need for the Seed Vault based on the following threats to the *ex-situ* system: "*Individual genebanks are vulnerable to a host of problems that can endanger their collections, including poor management, lack of adequate funding, equipment failures, and natural catastrophes, civil strife, war, and acts of terrorism.*" [22]. The committee concluded that Svalbard is a unique and appropriate location for an international safety duplication facility due to several security, physical and political factors: 1) The permafrost offers natural freezing and provides unparalleled insulation properties for the seeds, even in the unlikely case of breakdown in the cooling equipment; 2) Svalbard offers a unique combination of remoteness and accessibility, providing security from human-related dangers while at the same time allowing transportation of seeds in and out; 3) Military activity is prohibited under the terms of the International Treaty of Svalbard; 4) The political situation is stable, and the highly

Figure 1. The Svalbard Global Seed Vault. The exterior and interior of the Svalbard Global Seed Vault. Printed under a CC BY license, with permission from Photographer Mari Tefre.

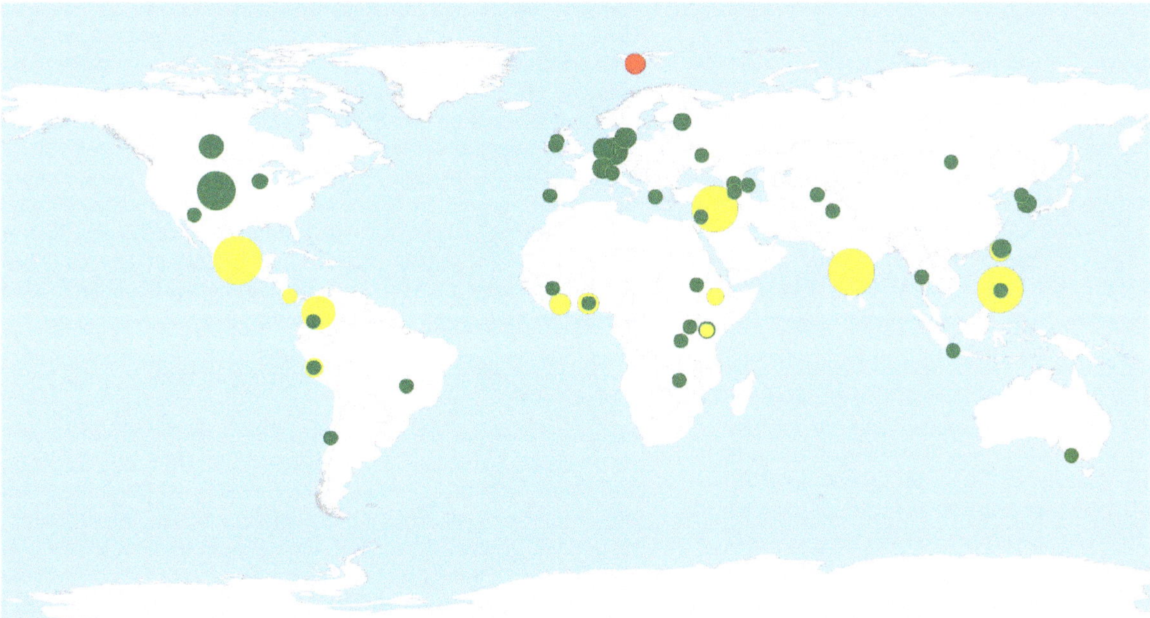

Figure 2. Genebanks with safety deposits in the Svalbard Global Seed Vault. The radius of the circles is relative to the number of samples deposited, and the circle size reflects the size of the deposits according to 25 size classes. Yellow circles are International Agricultural Research Centers, and green circles are regional, national or subnational genebanks. The radius of the red SGSV circle is not relative to the holdings.

competent local government is supportive; 5) The location inside a mountain increases security, and the area is geologically stable with low radiation levels; and 6) Norway has the necessary trust in the international efforts on PGRFA conservation [22]. On the basis of this feasibility study the Norwegian government decided to put into practice these plans; in June 2006, the five Nordic prime ministers participated in the cornerstone-laying ceremony at Svalbard, and in February 2008 the Seed Vault opened officially.

In this paper, we assess the completeness of the Seed Vault collection in relation to global conservation efforts and discuss the role of the Seed Vault as back-up storage site in a global *ex-situ* conservation system for PGRFA. We assess available data on the world's collections of PGRFA and compare these results to statistics on what is currently backed-up in the Seed Vault. We establish five different benchmarks for the assessment: 1) The total number of collections and accessions reported in the FAO database World Information and Early Warning System on PGRFA (WIEWS) [23]; 2) the number of *distinct* accessions, as estimated in the Second Report on the State of the World's Plant Genetic Resources for Food and Agriculture (SoW2) from the FAO [8]; 3) the total number of accessions in the GENESYS accession level database [24]; 4) the number of accessions covered by the ITPGRFA, and; 5) the number of accessions held in the ten institutions with the largest number of accessions for a selected array of important food crops. Based on the different framings of the global *ex-situ* collections represented by these five benchmarks, we discuss crop and institutional coverage of the current safety duplication collection in the Seed Vault.

Methods

We analyzed the current status of the Svalbard Global Seed Vault collection in relation to two different databases on global holdings of PGRFA: 1) the genebank-level data in the World Information and Early Warning System on PGRFA (WIEWS) [23] maintained by the FAO and 2) the accession-level data in GENESYS [24]. As of February 2013, data in the WIEWS database covered more than 190 countries and 1750 genebanks, and GENESYS covered 2,348,549 accessions from 365 participating genebanks in Europe, the USA and the international CGIAR centers. Database downloads of the WIEWS and the GENESYS databases were provided by database managers at the FAO and Bioversity International, respectively. The Svalbard Global Seed Vault database is managed by NordGen, and we also used a download of the accession-level descriptors from that database. Although all these datasets are freely available on the internet, it is not possible to carry out the analyses we required online. We used a PostgreSQL [25] database with data from all three databases (updated as of 01.01.2013) to make taxonomic and institutional queries.

We first categorized the WIEWS and GENESYS data taxonomically and selected relevant genera for our analysis. We included only genera represented with more than 1000 accessions conserved *ex-situ* globally according to WIEWS and those with taxa suitable for *ex-situ* conservation as seeds, following Kew's Seed Information Database (SID) v.7.1 [26] to classify genera according to the predominance of recalcitrant, intermediate or orthodox seed types. We eliminated those that are not currently represented in the Seed Vault, thereby assuming that the genera in the current Seed Vault collection are representative of the genera suitable for conservation of this type. We further classified genera into crop groups (e.g., Cereals, Food Legumes, etc.) by using the classification scheme from FAO's SoW2 [8]. We estimated the number of distinct accessions by using the same method as FAO in SoW2 and counted: a) Accessions in national collections with origins reported from the country of the genebank; b) The international collections held in trust for the world community by the International Agricultural Research Centers (IARCs).

Furthermore, we assessed the number of accessions covered by the global legal framework of the ITPGRFA. We estimated the number of accessions that are covered by the Treaty by extracting data on all taxa covered in the Treaty's Annex 1 from all institutions located in countries that are parties to the Treaty. In addition, we included the collections which are part of the multilateral system under Article 15 agreements with IARCs. Information on participation in the Treaty and the Article 15 collections is available from the ITPGRFA website [27]. We excluded the IARCs with Article 15 agreements in the first query to avoid counting them twice. To determine which taxa to include, we used the Annex 1 list elaborated for an analysis of the taxonomic composition of the collections in the European genebank network [28]. Because of exceptions excluding and including certain species and genera within some crops in the Annex 1 crop list, the Treaty analysis was conducted at the species level for the collections not covered by Article 15.

To assess institutional coverage in the Seed Vault's holdings, we identified the world's ten largest collections (according to WIEWS) for a selection of important food crops. We compared this list of institutions and the number of accessions they hold with the current representation of these collections in the Seed Vault.

Results and Discussion

On the fifth anniversary of the Svalbard Global Seed Vault on February 26, 2013, the total safety back-up collection stood at 774,601 seed samples deposited by 53 institutions from around the world (Fig. 2). Twelve of these are international, and the rest are regional, national, or subnational, including national agricultural research systems, universities and NGOs. The total number of accessions in the recent FAO WIEWS database is 7,205,007.

From the list of genera with more than 1000 accessions stored in genebanks, we removed names that did not refer to actual genera, combined synonymous genera when warranted, and removed six non-PGRFA genera: *Arabidopsis, Picea, Pinus, Populus, Pseudotsuga* and *Rhododendron*. This resulted in a list of 206 genera. Based on the assumption that the current Seed Vault collection is representative in terms of taxa currently commonly stored *ex-situ* as seeds, we excluded all genera for which no accessions are stored among the current 774,601 in the Seed Vault (*Amygdalis, Anacardium, Ananas, Annona, Bohemeria, Carya, Colocasia, Dioscorea, Ficus, Gladiolus, Ilex, Jathropha, Juglans, Malus, Manihot, Morus, Opuntia, Oxalis, Phoenix, Pistacia, Prunus, Punica, Tulipa, Ullucus, Vitis, Zingiber,*), including genera with recalcitrant and intermediate seeds *(Bactris, Camellia, Carica, Castanea, Cinnamonum, Citrus, Cocos, Coffea, Corylus, Curcuma, Durio, Elaeis, Garcinia, Hevea, Mangifera, Musa, Nephelium, Persea, Piper, Poncirus, Pouteria, Quercus, Salix, Theobroma*), ending up with a final list of 156 genera. The total number of accessions for these 156 genera was 5,979,663 in WIEWS. By using the FAO's method for estimating the number of distinct samples, we identified 2,185,452 distinct accessions in national and IARC collections. And, by using the information on participation in the ITPGRFA and the Annex 1 crop list from [28], we identified 2,625,646 Annex 1 accessions. Together with the 693,752 accessions held by Article 15 institutions according to the Treaty webpage, the total number of accessions covered by the ITPGRFA is about 3.3 million accessions. This ITPGRFA coverage estimate is for the total number of accessions, not only distinct accessions. Our estimates of the size of global *ex-situ* collections of selected genera and crop groups are presented in Table 1, and a list of the 156 crops with more than 1000 accessions in genebanks worldwide is presented in supplementary Table S1.

In our assessment of institutional coverage, we included only food crops and excluded fiber, sugar, oil and forage crop genera (*Aegilops, Carthamus, Dactylis, Festuca, Gossypium, Linum, Medicago, Panicum, Saccharum, Trifolium*) from the list of 39 genera with most accessions in WIEWS, resulting in a list of 29 genera for which we harvested data on the holdings of the ten largest institutional collections (Table 2, Table S2). For these 29 crop genera, on average 3.8 of the ten largest collections of each genus are currently backed-up at Svalbard. Considering the legal and institutional arrangements of all the individual collections among the ten largest holders, we find that, on average, 5.5 are either already represented in the Seed Vault or covered by existing SDAs. Genebanks in 26 countries with the ten largest collections in the 29 crop genera have not yet safety duplicated in the Seed Vault. All of the CGIAR centers among the top-ten holders have duplicated all or part of their collections.

Crop coverage in the Seed Vault collection

Out of the total number of accessions of genetic resources conserved *ex-situ* in the world according to WIEWS, about 10% are currently backed up in the Seed Vault. However, this is an underestimate of the coverage of the current Seed Vault collection with regard to its objective to back-up the *maximum amount of genetic diversity*, which is not the same as the *maximum number of accessions*. It is important to distinguish between numbers and diversity because according to estimates presented in FAO's SoW2 only about 25–30% of the 7.4 million accessions conserved *ex-situ* are distinct, while the remaining are duplicates held within the same institutions or by different ones [8]. The Seed Vault will as far as possible only conserve safety duplicates of *distinct* accessions and avoid internal duplication [19]. Furthermore, the Seed Vault can only back-up accession of taxa suitable for conservation as seeds, whereas WIEWS includes data on a range of taxa not suitable for conservation in this form.

We found that the relative representation of crop groups in the current Seed Vault collection is dominated by crop groups containing crop genera with large global production and orthodox seeds. Cereals and food legumes together constitute 87% of the accessions in the Seed Vault, whereas the figure is 65% for the same groups in the WIEWS database (Fig. 3). Among the 206 genera with more than 1000 accessions in WIEWS, 50 are not being stored in the Seed Vault. Twenty-three of these can be explained by having recalcitrant or intermediate seeds, and among the remaining 27 genera not stored at Svalbard are many genera that produce orthodox seeds, but are not normally conserved *ex-situ* because the seeds are not true to type (e.g. *Coffea, Dioscorea, Malus, Manihot, Prunus, Pyrus, Vitis*). In theory, all orthodox seed species can be conserved as seeds, but in practice genebanks predominantly store seeds from crops whose seeds are "true to type". Most root and tuber crops, for example, are normally asexually (clonally) propagated, as sexual reproduction results in plants that are markedly different from their parents. *Manihot* is an example of a genus with orthodox seeds, but whose propagation is usually done clonally, and, consequently, whose *ex-situ* conservation focuses on *in-vitro*, cryo-conservation or tuber storage in combination with field genebanks. However, there are also examples of tubers where conservation as true seeds is now common, such as potato (*Solanum tuberosum*) [29]. In the case of potato and sweet potato (*Ipomoea batatas*), CIP uses true seed conservation in addition to other methods and that center is currently safety duplicating these seed collections in the Seed Vault [30]. Currently, however, no cassava, taro (*Colocasia*) or yam (*Dioscorea*) accessions are conserved this way in the Seed Vault. We stress that the list of genera in Table 1 and Table S1 is not meant to be comprehensive in terms of PGRFA genera that *can* be stored *ex-situ* as seeds, but rather reflects *the current practice* of most of the world's genebanks.

Table 1. Selected crop genera and their status in different databases.

Crop	WIEWS total	GENESYS total	WIEWS distinct	WIEWS ITPGRFA	SGSV (total)
Triticum (Wheat)	855,639	367,994	270,237	599,876	145,698
Oryza (Rice)	773,948	192,983	440,313	553,235	145,540
Hordeum (Barley)	469,590	171,603	138,722	339,448	61,390
Zea (Maize)	323,802	104,518	134,185	145,921	32,822
Sorghum (Sorghum)	235,690	88,801	75,355	167,769	40,695
Avena (Oat)	131,332	56,489	24,619	79,349	11,302
Pennisetum (Pearl millet)	65,447	24,910	37,024	89,688	20,444
Setaria (Foxtail millet)	46,606	3,193	5,623	1,675*	2,519
Aegilops (Goatgrass)	42,268	16,238	19,453	5,173*	3,525
Triticale (Wheat x Rye)	40,470	2,833	27,587	31,042	21,455
Eleusine (Finger millet)	35,382	7,516	14,602	38,105	7,636
Amaranthus (Amaranth)	28,313	5,620	14,290	316*	1,300
Secale (Rye)	21,452	15,592	6,091	13,633	935
Chenopodium (Quinoa)	16,263	1,648	9,028	34*	111
Eragrostis (Teff)	8,820	1,760	6,126	152*	38
Cereals	**3,095,022**	**1,061,698**	**1,223,255**	**2,065,416**	**495,410**
Phaseolus (Common bean)	262,491	96,433	98,890	246,606	35,230
Glycine (Soybean)	230,091	54,797	74,890	2,102*	17,778
Vigna (Cowpea)	149,590	48,993	69,302	91,731	26,129
Arachis (Groundnut)	128,435	28,489	31,191	22,857*	14,462
Cicer (Chickpea)	98,319	48,347	48,432	115,802	28,872
Pisum (Pea)	95,290	37,258	20,783	69,193	9,670
Vicia (Fababean)	81,470	36,343	32,632	75,492	11,573
Lens (Lentil)	58,430	19,557	20,589	56,685	11,946
Cajanus (Pigeon pea)	40,820	13,707	31,702	53,326	10,076
Lupinus (Lupin)	38,053	13,567	7,503	183*	591
Psophocarpus (Winged bean)	4,217	454	1,582	88*	12
Food Legumes	1,187,206	397,945	437,496	734,065	166,339
Medicago (Alfalfa)	92,019	27,946	28,413	9,408	10,912
Trifolium (Clover)	74,707	34,019	24,062	6,057	3,944
Panicum (Switchgrass)	48,850	20,448	12,342	2,273*	2,195
Forage Crops	**389,225**	**161,796**	**139,298**	**47,995**	**35,103**
Solanum (Potato, Eggplant, Tomato)	147,566	58,781	52,180	48,241	15,733
Ipomoea (Sweet potato)	35,478	7,927	16,369	22,864	19,36
Roots and Tubers	**183,044**	**66,708**	**68,549**	**71,105**	**17,669**
Brassica (Cabbage)	101,353	27,889	26,025	63,624	7,121
Capsicum (Peppers)	73,520	21,831	23,945	1,163*	2,163
Lycopersicon (Tomato old name)	59,039	23,863	10,801	475*	2,866
Cucumis (Cantaloupe)	44,402	16,674	8,000	1*	2,428
Cucurbita (Squash)	39,599	14,795	14,411	2,612*	780
Allium (Onion)	30,064	12,800	9,816	0*	630
Vegetables	**397,069**	**137,084**	**107,672**	**73,209**	**17,988**
Sesamum (Sesame)	50,462	3,204	17,946	1*	1,727
Helianthus (Sunflower)	39,850	10,921	5,575	20,509	1,811
Carthamus (Safflower)	29,195	3,494	6,079	0*	511
Oil Crops	**140,196**	**21,030**	**39,729**	**20,511**	**4,068**
Major crops total (taxa with more than 1000 accessions in *ex-situ*)	**5,979,663**	**2,040,606**	**2,185,452**	**3,117,472**	**766,292**
All crops total (all taxa in WIEWS)	**7,205,007**	**2,334,747**	**2,498,098**	**3,319,398**	**774,601**

Table 1. Cont.

Crop	WIEWS total	GENESYS total	WIEWS distinct	WIEWS ITPGRFA	SGSV (total)
Collections	**1,366**	**365**	**na**	**na**	**53**

Column headings: *WIEWS total* reports the total number of accessions in the World Information and Early Warning System of Plant Genetic Resources for Food and Agriculture database; *GENESYS total* reports the total number of accessions in the GENESSYS database; *WIEWS distinct* reports the estimates of the number of distinct accessions in the WIEWS database; *WIEWS ITPGRFA* reports the estimate for accessions covered by the International Treaty on Plant Genetic Resources for Food and Agriculture, including Annex 1 crop representation in genebanks located in Contracting Parties and the collections in international genebanks under Article 15 agreements (Exclusion of species and lumping of genera under common crop names are done according to the Annex 1crop list for collections not under Article 15); SGSV total reports the current total holding in the Seed Vault. The numbers reported for crop groups are for 156 crop genera with more than 1000 accessions in genebanks worldwide according to WIEWS (See table S1 for full list). *Genera not represented in the Annex 1 list of the ITPGRFA.

In terms of total number of accessions, the largest genera in WIEWS are those of wheat (*Triticum*), rice (*Oryza*), barley (*Hordeum*), maize (*Zea*), common bean (*Phaseolus*), sorghum (*Sorghum*), soybean (*Glycine*), cowpea (*Vigna*), potato, tomato and eggplant (*Solanum*), oat (*Avena*), groundnut (*Arachis*), cotton (*Gossypium*) and, cabbage (*Brassica*), all with more than 100,000 accessions stored in genebanks globally. As shown by Kilian and Graner [31], there is a correlation between the aggregate size of *ex-situ* collections of the ten largest crops worldwide and their global acreages in agriculture. We found that the correspondence between number of accessions of crops stored *ex-situ* and agricultural importance goes beyond the big staple crops; comparing the list of species found to "feed the world" in Prescott-Allen and Prescott-Allen (1990) [32] with the WIEWS data showed that only eight of the 103 species providing more than 90% of plant commodities worldwide are from genera represented by less than 1000 accessions conserved in *ex-situ* collections worldwide.

The GENESYS database contains data on 2 million accessions in the 156 genera included in this assessment, corresponding to about one third of the number in the WIEWS database. This proportion also holds for the major crop groups (cereals, food legumes, roots and tubers and vegetables), whereas there is a smaller relative representation of other crop groups, such as oil and fiber crops in GENESYS compared to WIEWS. The GENESYS accession-level database serves a different and broader user-group than does the WIEWS database, most notably breeders and other plant scientists that needs accession-level information. WIEWS does not provide data on descriptors for every accession like genebank databases do, it merely provides data on the number of accessions institutions hold for different taxa. WIEWS is based on a range of sources updated at different times, such as surveys of institutional holdings, reporting on GPA implementation and SoW country reports. GENESYS, on the other hand, taps directly into genebank databases and should therefore be more up to date. The 365 genebanks that have provided data to GENESYS hold a relatively larger share of the 156 genera than we found for the average holding per genebank in WIEWS. A comparative analysis revealed that for some genera there are more accessions in the Seed Vault than the total numbers in GENESYS (e.g., *Eleusine* and *Triticale*) while the median value across the 156 genera is as low as 14% for the same comparison. Not all of the genebanks with safety duplicates in the Seed Vault are data providers to GENESYS, and comparing numbers is not the same as assessing how much the GENESYS providers have duplicated at Svalbard. The reason why we found it interesting to include GENESYS as a benchmark is that GENESYS itself is emerging as the unifying database for the global *ex-situ* conservation system called for in the GPA and the ITPGRFA. Starting with the accession level data from the CGIAR centers, GENESYS now reports if accessions are backed up in the Seed Vault, thereby connecting these two elements of the emerging global system.

The ratio of accessions stored in the Seed Vault vs. the estimated numbers of distinct accessions in WIEWS is about one third. There is good correspondence between the size of the WIEWS total world holdings and the current collection at Svalbard; the ten largest crop genera in terms of distinct accessions are all among the 15 largest genera in the Seed Vault. The cereals *Triticum*, *Sorghum*, *Pennisetum*, *Triticale* and *Eleusine* are represented with numbers corresponding to >50% of the estimate for distinct world holdings. The same is true for *Lens* and *Cicer* among the food legumes. The least well represented cereal genera (in the broad sense used by FAO) are *Amaranthus*, *Chenopodium* and *Eragrostis*, and the least well represented food legume genera are *Lupinus* and *Psophocarpus*. In terms of importance for global agriculture and food security, we found that the genera that include the top four cereals in terms of world production, *Triticum*, *Sorghum*, *Oryza*, and *Zea*, are represented by 54%, 54%, 33% and 24% of the estimates of distinct accessions in WIEWS, respectively. We found a relative under-representation of genera of some crops that are of great importance for food security in parts of the tropics, such as the genera of *Abelmoschus*, *Amaranthus*, *Chenopodium*, *Eragrostis*, *Ipomoea* and *Lupinus*.

In our assessment of what is distinct in the global conservation system, we have assumed that indigenous accessions are likely to be distinct, while accessions with origins outside the country of the genebank are likely to be duplicates of those stored in the country of origin. We use this method to be consistent with the figures reported in the SoW2. This method is supported by two characteristics of the global system; the international collecting by the IARCs was reduced in the 1990–2000 period and collecting is increasingly done by institutions based in the country where the collecting takes place [8], and, for IARC collecting, it has been a common practice for expeditions to deposit samples in local genebanks. However, this method does not account for accessions collected in the era when PGRFA was largely considered a common heritage, and some international collection missions collected without depositing seeds in local genebanks, in many cases because there were no local genebanks. Neither does the method account for cases where the collected accessions are lost in the local genebanks. Furthermore, it does not account for accessions that are duplicated within a country, which is the case when countries have reported both working collections and base collections in WIEWS, and also in cases where national genebanks are conserving duplicates of IARC accessions. We therefore think that the estimate of distinct accessions is at the higher end of the likely range and consequently that the proportion of distinct accessions conserved in the Seed Vault reported in this paper are conservative. Collection-specific analyses are needed to obtain more accurate estimates of what is distinct in the different

Table 2. Representation in the Seed Vault of the world's 10 largest collections of selected crop genera.

Genus (major crops)	IARCs and Countries of location of 10 largest collections	No of collections in SGSV	No of collections under SDA conditions
Oryza (Rice)	IRRI (107), India (0), China (0,0,0), Japan (0), South Korea (4), USA (20), WARDA (57)	4	5
Triticum (Wheat)	CIMMYT (69), USA (17), China (0), India (0), ICARDA (95), Japan (0), Russia (3), Italy (0), Germany (5), Australia (0)	5	7
Hordeum (Barley)	Canada (22), USA (23), Brazil (0), ICARDA (97), Japan (0), Germany (11), China (0), South Korea (19), Russia (6), Ethiopia (0)	6	7
Sorghum (Sorghum)	ICRISAT (83), USA (13, 62), China (0), India (0), Ethiopia (0), Brazil (0), Kenya (10), Japan (0), Australia (0)	4	7
Zea (Maize)	CIMMYT (86), Portugal (0), USA (4), China (0), Mexico (0,0,0,0), Russia (10), India (0)	3	5
Glycine (Soybean)	China (0,0,0), USA (45), South Korea (11), AVRDC (0), Brazil (0,0), Japan (0), Russia (0)	4	6
Phaseolus (Common bean)	CIAT (84), USA (5), Brazil (0,0,0), Mexico (0,0), Germany (19), China (0), Russia (3)	5	6
Vigna (Cowpea)	IITA (74), China (0), USA (1), India (0,0,0), AVRDC (76), Philippines (8), Japan (0), Brazil (0)	4	7
Avena (Oat)	Canada (13,0), USA (26), Russia (2), Germany (7), Kenya (0), Australia (0), China (0), UK (0), Poland (0)	5	7
Arachis (Groundnut)	ICRISAT (58), India (0,0,0), USA (3), Argentina (0,0), Niger (0), China (0,0)	2	3
Cicer (Chickpea)	ICRISAT (85), India (0,0), ICARDA (79), Australia (0), USA (0), Iran (0), Pakistan (19), Russia (3), Turkey (0)	6	7
Solanum (Potato, Eggplant, Tomato)	Russia (3), AVRDC (18), France (0), CIP (59), USA (98,0,0), Germany (15), India (0), Japan (0)	5	6
Pennisetum (Pearl millet)	ICRISAT (88), Brazil (0), India (0,0), France (0), Canada (0), India (0), Niger (0), Uganda (0), USA (13)	3	4
Brassica (Cabbage)	India (0,0,0,0), China (0,0), Australia (0), Japan (0), Russia (3), UK (0)	2	3
Vicia (Fababean)	ICARDA (63), Russia (3), Australia (0), Germany (14), China (0), Italy (0), Spain (0), USA (11), Turkey (0), Bulgaria (0)	5	5
Pisum (Pea)	Australia (4), Russia (1), ICARDA (62), Germany (16), USA (28), Italy (0), China (0), UK (0,0), India (0)	5	6
Setaria (Foxtail millet)	China (0), India (0), France (0,0), Japan (0), ICRISAT (92), USA (71), Kenya (0), UK (0)	4	5
Lens (Lentil)	ICARDA (99), India (0,0), Australia (0), Iran (0), USA (0), Russia (6), Chile (0), Canada (32), Hungary (0)	5	7
Cajanus (Pigeon pea)	ICRISAT (73), India (0,0,0,0), Kenya (4), Philippines (3), Australia (0), Brazil (0), Nepal (0)	3	7
Capsicum (Peppers)	AVRDC (4), USA (5), Mexico (0,0,0), India (0), Brazil (0,0), Japan (0), Philippines (0)	3	5
Sesamum (Sesame)	India (0,0,0), China (0,0), Israel (0), Kenya (0), Brazil (0), Japan (0), Mexico (0)	1	5
Eleusine (Finger millet)	India (0,0,0), ICRISAT (98) Kenya (19), Ethiopia (0), Uganda (0), Zambia (3), Nepal (0), USA (88)	4	5
Triticale (Wheat x Rye)	CIMMYT (114), Russia (0), USA (5), Canada (5), Ukraine (0), Poland (0), Germany (7), Bulgaria (0), Slovakia (0), Uzbekistan (0)	5	6
Cucumis (Cucumbers, True Melons)	USA (23), Japan (0), Russia (0), China (0,0), Brazil (0), Kazakhstan (0), France (0), Germany (22), India (0)	3	5
Helianthus (Sunflower)	Serbia (0), USA (37), China (0), France (0), Brazil (0,0), Russia (0), Australia (0), India (0), Morocco (0)	2	5
Lupinus (Lupin)	Australia (0,0), Germany (6), Russia (1), France (0), Peru (0,0,0), Spain (0), UK (0)	3	4
Ipomoea (Sweet potato)	CIP (30), Japan (0,0,0), USA (0), Papua New Guinea (0), Brazil (0), China (0), Taiwan ROC (0)	3	3
Allium (Onion)	India (0), Russia (0), Japan (0), USA (0,0), Germany (12), AVRDC (0), UK (0,0), Hungary (0)	3	5

Table 2. Cont.

Genus (major crops)	IARCs and Countries of location of 10 largest collections	No of collections in SGSV	No of collections under SDA conditions
Amaranthus (Amaranth)	India (0,0,0), USA (29), Brazil (0), Peru (0), China (0), Hungary (0), AVRDC (33), Argentina (0)	2	6
Average		3,8	5,5

The table lists the genera with the largest number of accessions conserved *ex-situ*. The list excludes fiber, sugar, oil and forage crop genera (*Gossypium, Medicago, Trifolium, Panicum, Linum, Saccharum, Aegilops, Festuca, Dactylis, Carthamus*). The percentages of the individual collections backed up in the Seed Vault compared to the total number of accessions reported in WIEWS are given in brackets. The collections/countries are sorted by decreasing number of accessions. If a country holds more than one collection the percentage share of all collections are given in the bracket after the first appearance of the country on the list.

collections. Such analyses can to a certain extent be done by comparing passport and descriptor data from different collections [33], but since the necessary information is not always available, expert knowledge on the different collections is often the most useful way to identify the most important collections in terms of unique diversity conserved [34].

Our estimate of the number of genebank accessions covered by the ITPGRFA's Multilateral System is 3.3 million in total and 3.1 million when considering only the 156 genera. Our assessment of representation of genera in the Seed Vault in relation to the ITPGRFA world collection revealed a variety of situations. Some genera are absent or nearly so from the Multilateral System, although they are fairly well represented in the Seed Vault collection (e.g., *Cucumis* and *Linum*). Others are represented by only a small share in the Seed Vault collection compared to the number covered by the ITPGRFA (e.g., *Ipomoea* and *Helianthus*). Since Annex 1 only covers 35 food crops and 29 forages, it represents only a subset of global *ex-situ* conservation holdings, and this is reflected in our assessment. But while the Seed Vault does hold a large number of non-Annex 1 crops, some genera have conspicuously low representation. The genera *Amaranthus, Allium, Capsicum, Chenopodium, Cucurbita, Eragrostis, Glycine, Lupinus* and *Psophocarpus* are examples of important non-Annex 1 food crops with representation in the Seed Vault corresponding to <10% of the distinct accessions in WIEWS. A notable exception to this low representation of non-Annex 1 crops is *Arachis*, where the share of accessions in the Seed Vault vs. distinct accessions in WIEWS is nearly 50%, chiefly because of the global collections maintained by ICRISAT under Article 15.

Global participation

The four largest national collections in the world are located in the USA, Russia, India and China. The National Plant Germplasm System (NPGS) in the USA and the N.I. Vavilov All-Russian Scientific Research Institute of Plant Industry (VIR) in Russia are both in the process of backing-up their collections at Svalbard, while the National Bureau of Plant Genetic Resources (NBPGR) in India has signed the SDA, but has not yet started safety duplication. The Institute of Crop Germplasm Resources (ICGR-CAAS) in China is not participating so far. Out of the five other national institutions with more than 100,000 accessions in storage, the national genebank in Japan is not currently a depositor, whereas the national genebanks in Brazil, Canada, Germany and the Republic of Korea have all deposited seeds at Svalbard. The most significant origin country gap is India, which appears 40 times in Table 2. However, safety duplication of the Indian collections is expected in the future since most Indian collections in Table 2 are covered by an SDA between the NBPGR and the Seed Vault. In Brazil, USA and Australia, the base collections have safety duplicated at Svalbard while working

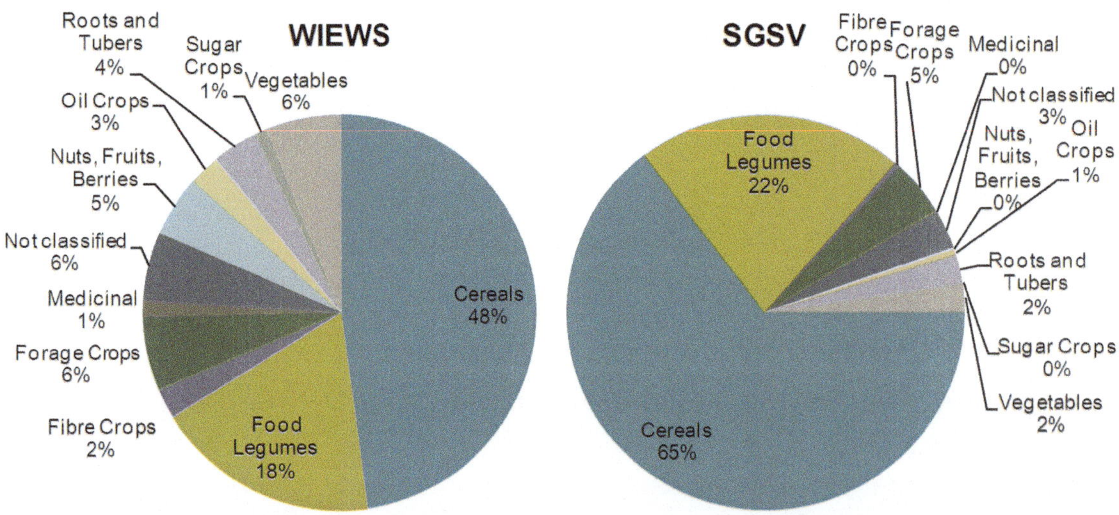

Figure 3. Crop group representation. All genera with more than 1000 accessions reported in the WIEWS database of the FAO and in the safety back-up collection in the Svalbard Global Seed Vault.

collections located in other parts of the same countries also appear in Table 2. There is therefore considerable duplication in the WIEWS data from these countries, illustrating that the genus-level estimate for distinct accessions in the previous section is conservative.

Considering the institutional coverage at the genus level, we found that eight genera have SDA coverage for seven of the ten largest global holders. All genera are represented with accessions from at least one of the largest collections. More than 25% of the aggregate holdings of the ten largest collections of the genera *Cajanus, Cicer, Lens, Oryza, Pennisetum, Sorghum, Triticale* and *Triticum* are backed up in the Seed Vault today. For another group of genera, less than 1% is safety duplicated at Svalbard: *Sesamum, Allium, Capsicum, Lupinus* and *Brassica*. These differences in institutional coverage among genera can again to a large extent be explained by the high number of safety duplicates from the CGIAR. The genera with relatively high coverage are those for which a CGIAR center has a global conservation and research mandate, and the genera with relatively low coverage are those for which no such mandate exists. That all of the CGIAR centers on the top tenlist have safety duplicated all or parts of the collections at Svalbard shows that the Seed Vault is now an integral part of the CGIAR's genetic resource management. The CGIAR centers are strategically located in important areas of diversity of their mandate crops, and, for many of the genera in Table 2, the location of the centers and national collection holders match well with areas of high importance of conservation: e.g., IRRI, WARDA and the South Korean national genebank for *Oryza*; ICARDA for *Triticum, Hordeum, Cicer, Vicia, Pisum* and *Lens*; ICRISAT for *Sorghum, Pennisetum, Setaria*, and *Cajanus*; CIMMYT for *Zea*; CIAT for *Phaseolus*; IITA for *Vigna*; CIP for *Solanum*; ICRISAT and the Kenyan national genebank for *Eleusine*; and the USA national genebank for *Helianthus*.

National and subnational genebanks typically conserve a higher number of accessions collected in their local region than from other regions [8] and for some genera in Table 2 certain institutions currently not safety duplicating in the Seed Vault are located in areas of important diversity: e.g., *Oryza* from the Indian and Chinese genebanks; *Sorghum, Pennisetum* and *Hordeum* from the Ethiopian genebank; *Zea, Phaseolus* and *Capsicum* from the Mexican genebank; *Glycine* from the Chinese genebank; *Cicer* and *Lens* from the Iranian and Indian genebanks; *Pennisetum* from the Nigerian genebank; *Cajanus* from the Indian genebank; and *Amaranthus* from the Indian, Peruvian and Argentinian genebanks. Accessions of these genera and genebanks possibly represent important gaps in the current safety collection at Svalbard, but further analyses are needed to see if they are distinct from those already stored in the Seed Vault.

Conclusion

The high coverage of ITPGRFA Annex 1 crops and of those crops for which there is a CGIAR mandate in the current Seed Vault collection indicates that international policies and institutions are important determinants for accessions to be safety duplicated at Svalbard. Agriculture in all countries in the world has, at least since the beginning of global trade, been dependent on genetic resources originating elsewhere. While some regions have been richer sources of crops than others [2,35,36], there are hardly any regions that today do not grow crops which originated in distant lands. Since crops are subject to the same evolutionary imperative as all other organisms (adapt, move or perish), the long-term success of crop cultivation depends on the availability of genetic diversity. Therefore, at varying levels, all countries are interdependent with regard to access to PGRFA. This interdependence has inspired a sense of global unity with respect to the issue of conserving PGRFA. The need for this has been repeatedly articulated throughout the history of the field, where there have been numerous calls for the creation of a global conservation system [37]. The first mandate to address global crop conservation was given to the International Board for Plant Genetic Resources (now Bioversity International), established in 1974, and in 1983 the important ethical and political aspects of the issue led to the creation of an intergovernmental forum at FAO, the Commission on Genetic Resources for Food and Agriculture (CGRFA) [17]. The CGRFA negotiated and developed the GPA and the ITPGRFA as policy instruments for the development of an efficient and sustainable system of *ex-situ* conservation [14,15]. The ITPGRFA provides an internationally agreed, partial solution to the politically polarized ABS issue that has arguably resulted in institutional proliferation and duplication of conservation efforts. In principle, the ITPGRFA fosters a global system where each distinct accession only needs to be conserved and made available from one (or at most a small number of) long-term storage site with appropriate safety duplication arrangements. The Global Crop Diversity Trust manages an endowment with the objective of "*providing a permanent source of funds to support the long-term conservation of the ex situ germplasm collections on which the world depends for food security*" [38]. The Svalbard Global Seed vault is one of the recipients of such long-term grants, and, by 2012, the Trust had also financed incoming shipment of 75% of the safety deposits stored at Svalbard [38]. In addition to the Trust, the Seed Vault is also supported by a range of other critical institutions and instruments in the emerging global *ex-situ* conservation system. At its establishment in 2007, the Seed Vault was welcomed by the 172 national members, plus the EU, of the CGRFA [39]. The importance of the Seed Vault has been highlighted in SoW2 [8] and in the second GPA [14], and its role in safety duplication will be highlighted in the new international genebank standards from FAO [11]. Most importantly, 53 of the world's genebanks have confirmed the need and importance of the Seed Vault by depositing a substantial part of their collections at Svalbard within its first five years of operation. Thus, the Seed Vault is today well established within the legal and institutional framework of the international genetic resources movement.

A symbol for the larger cause

The Seed Vault has gained considerable international media attention and even fame [40]. We believe that media reports, though occasionally inaccurate, have contributed to increased public awareness about the importance of crop diversity. While conservation of genetic resources has been part of the environmental movement since the seminal UN conference on the human environment held in Stockholm in 1972, it has often been overshadowed by other issues. The Seed Vault has contributed towards raising the profile of this issue on the broader environmental and food security agenda. In the words of the late Nobel Laureate, Wangari Maathai: "*The significant public interest in the seed vault project indicates that collectively we are changing the way we think about environmental conservation. We now understand that along with international movements to save endangered species and the rainforests of the world, it is just as important for us to conserve the diversity of the world's crops for future generations.*" The Seed vault is not a panacea for securing the future's food supply, but it is an important element in safely conserving the genetic resources necessary for agricultural development. As the UN Secretary General Ban Ki-moon said on the occasion of his visit in the Seed Vault in 2009: "*Sustainable food production may not begin in this cold arctic environment, but it does begin by conserving crop diversity.*"

In light of the huge media attention directed towards the Seed Vault, it is important to stress that it only makes sense as a part of a global conservation system. The conventional genebanks spread around the world are doing the essential job of conserving, regenerating, multiplying and distributing seeds to those that use them for applied and basic research for agricultural development and increased food security. The Seed Vault is, on the one hand, a high-profile environment and development project and, on the other, a low-tech practical solution increasingly serving a basic global need for the safety duplication of seeds held in conventional genebanks, as documented in our analyses. There are important synergies between these two aspects, and the Seed Vault plays an important symbolic role for enhanced integration and cooperation in the global *ex-situ* conservation efforts.

Supporting Information

Table S1 Representation of 156 crop genera in WIEWS with more than 1000 accessions stored *ex-situ* globally. Only genera commonly stored as seeds are included.

Table S2 The ten largest collections worldwide of the 29 largest food crop genera in WIEWS. Genebanks are identified by WIEWS coding [23].

Acknowledgments

We thank the genebank managers of the institutions with safety duplicates in the Seed Vault for providing accession-level data as well as information about legal and institutional arrangements at their institutions. Thanks to Jessica Raneri of Bioversity International and Stefano Diulgheroff of the FAO for providing data downloads of the GENESYS and the WIEWS databases, respectively. We are grateful to Theo van Hintum of the Centre for Genetic Resources, the Netherlands and to colleagues in the Global Crop Diversity Trust and the Nordic Genetic Resource Center for valuable advice and assistance. The manuscript was improved by critical and insightful comments made by three reviewers.

Author Contributions

Conceived and designed the experiments: OW. Performed the experiments: OW SJ. Analyzed the data: OW SJ LG. Contributed reagents/materials/analysis tools: OW SJ. Wrote the paper: OW SJ LG.

References

1. Darwin C (2007) The variation of animals and plants under domestication Vol. 1. Middlesex, UK: The Echo Library. 362 p.
2. Harlan JR (1995) The living fields. Our agricultural heritage. Cambridge, UK, New York, USA, Melbourne, Australia: Cambridge University Press. 273 p.
3. Fowler C, Hodgkin T (2004) Plant genetic resources for food and agriculture: Assessing global availability. Annual Review of Environment and Resources 29: 143–179.
4. Gepts P (2006) Plant genetic resources conservation and utilization: The accomplishments and future of a societal insurance policy. Crop Science 46: 2278–2292.
5. Frankel O, Bennett E, editors (1970) Genetic resources in plants – their exploration and conservation. Oxford, United Kingdom: Blackwell. 554 p.
6. Frankel O, Hawkes J, editors (1975) Crop genetic resources for today and tomorrow. Cambridge University Press Cambridge, England & New York, USA: 492 p.
7. Pistorius R (1997) Scientists, plants and politics – a history of the plant genetic resources movement. Rome, Italy: IPGRI. 134 p.
8. FAO (2010) The Second Report on the State of the World's Plant Genetic Resources. Rome, Italy: FAO. 370 p.
9. FAO (1998) The State of the World's Plant Genetic Resources for Food and Agriculture. Rome, Italy: FAO. 510 p.
10. Frankel O (1987) Genetic resources: The founding years. Part IV: After twenty years. Diversity 3: 25–27.
11. FAO (2012) Draft revised genebank standards for the conservation of orthodox seeds. Available from: http://www.fao.org/agriculture/crops/core-themes/theme/seeds-pgr/conservation/gbs/en/Accessed 15 February 2013.
12. NMAF GCDT, NordGen (2007) Agreement between the Royal Norwegian Minsitry of Agricultue and Food, The Global Crop Diversity Trust and the Nordic Gene bank providing for the funding, management and operation of the Svalbard Global Seed Vault. http://www.regjeringen.no/en/dep/lmd/campain/svalbard-global-seed-vault.html?id = 462220 Accessed 15 February 2013.
13. FAO (1996) The Global Plan of Action for the Conservation and Sustainable Utilization of Plant Genetic Resources for Food and Agriculture. Rome, Italy: FAO. 63 p.
14. FAO (2011) Second Global Plan of Action for Plant Genetic Resources for Food and Agriculture. Rome, Italy: FAO. 91 p.
15. FAO (2009) The International Treaty on Plant Genetic Resources for Food and Agriculture. Rome, Italy: FAO. 60 p.
16. Qvenild M (2008) Svalbard Global Seed Vault: a 'Noah's Ark'for the world's seeds. Development in Practice 18: 110–116.
17. Esquinas-Alcazar J (2005) Protecting crop genetic diversity for food security: political, ethical and technical challenges. Nature Reviews Genetics 6: 946–953.
18. Fowler C (2008) The Svalbard Seed Vault and crop security. Bioscience 58: 190–191.
19. NMFA NordGen (2012) Standard Agreement between the Depositor and the Norwegian Minsitry of Agriculture and Food.
20. Engels J, Visser B, editors (2003) A guide to effective management of germplasm collections. IPGRI Handbooks for Genebanks, No. 6. Rome, Italy: IPGRI. 172 p.
21. Koo B, Pardey PG, Wright BD, editors Saving Seeds: the Economics of Conserving Crop Genetic Resources Ex Situ in the Future Harvest Centres of the CGIAR. Wallingford UK and Cambridge USA: Cabi publishing. 219 p.
22. Fowler C, George W, Shands H, Skovmand B (2004) Study to Assess the Feasibility of Establishing a Svalbard Arctic Seed Depository for the International Community. Prepared for the Ministry of Foreign Affairs (Norway). Ås, Norway: Center for International Environment and Development Studies. Agricultural University of Norway and the Nordic Gene Bank. 21 p.
23. FAO (2013) World Information and Early Warning System on PGRFA (WIEWS).
24. Bioversity (2013) GeneSys. http://www.genesys-pgr.org/. Accessed 15 February 2013.
25. PostgreSQL (2013) PostgreSQL. PostgreSQL. http://www.postgresql.org/. Accessed 15 February 2013.
26. Royal Botanic Gardens Kew. (2008) Seed Information Database (SID). Version 7.1. Available from: http://data.kew.org/sid/Accessed 15 February 2013.
27. ITPGRFA (2013). http://www.planttreaty.org/.Accessed 15 February 2013.
28. van Hintum T, Knupffer H (2010) Current taxonomic composition of European genebank material documented in EURISCO. Plant Genetic Resources-Characterization and Utilization 8: 182–188.
29. Simmonds NW (1997) A review of potato propagation by means of seed, as distinct from clonal propagation by tubers. Potato Research 40: 191–214.
30. CIP (2013). http://cipotato.org/genebank/conservation-methods.Accessed 15 February 2013.
31. Kilian B, Graner A (2012) NGS technologies for analyzing germplasm diversity in genebanks. Briefings in Functional Genomics 11: 38–50.
32. Prescott-Allen R, Prescott-Allen C (1990) How many plants feed the world? Conservation Biology 4: 365–374.
33. van Hintum TJ, Visser DL (1995) Duplication within and between germplasm collections. Genetic resources and crop evolution 42: 135–145.
34. Khoury C, Laliberte B, Guarino L (2010) Trends in ex situ conservation of plant genetic resources: a review of global crop and regional conservation strategies. Genetic Resources and Crop Evolution 57: 625–639.
35. Vavilov NI (1992) Origin and geography of cultivated plants. Cambridge University Press, Cambridge, UK: 505 p.
36. Myers N (1997) Biodiversity's genetic library. Nature's services: Societal dependence on natural ecosystems. In: Daily GC (ed.). 255–273. Washington, DC: Island Press. 255–273.
37. Frison E, Mitteau M, Sharrock S, Visser B (1994) Sharing responsibilities. In: Engels J, Visser B, editors. A guide to effective management of germplasm collections IPGRI Handbooks for Genebanks, No 6. Rome, Italy: IPGRI. 107–121.
38. CGRFA (2012) CGRFA/WG-PGR-6/12/Inf.6: Report from the Global Crop Diversity Trust to the Intergovernmental Technical Working Group on Plant Genetic Resources for Food and Agriculture. Available from http://www.fao.org/agriculture/crops/core-themes/theme/seeds-pgr/itwg/6th/en/Accessed 15 February 2013.
39. CGRFA (2007) CGRFA-11/07/Report: Eleventh Regular Session of the Commission on Genetic Resources for Food and Agriculture Rome, Italy, 11 – 15 June 2007. Available from http://www.fao.org/agriculture/crops/core-themes/theme/seeds-pgr/itwg/6th/en/Accessed 15 February 2013.
40. GCDT (2013) Svalbard Global Seed Vault – Fame. http://www.croptrust.org/content/fame Accessed 15 February 2013.

7

Fitness of Crop-Wild Hybrid Sunflower under Competitive Conditions: Implications for Crop-to-Wild Introgression

Kristin L. Mercer[1]*, D. Jason Emry[2,3], Allison A. Snow[4], Matthew A. Kost[1], Brian A. Pace[1], Helen M. Alexander[2]

1 Ohio State University, Department of Horticulture and Crop Science, Columbus, Ohio, United States of America, **2** University of Kansas, Department of Ecology and Evolutionary Biology, Lawrence, Kansas, United States of America, **3** Washburn University, Topeka, Kansas, United States of America, **4** Ohio State University, Department of Evolution, Ecology, and Organismal Biology, Columbus, Ohio, United States of America

Abstract

Understanding the likelihood and extent of introgression of novel alleles in hybrid zones requires comparison of lifetime fitness of parents and hybrid progeny. However, fitness differences among cross types can vary depending on biotic conditions, thereby influencing introgression patterns. Based on past work, we predicted that increased competition would enhance introgression between cultivated and wild sunflower (*Helianthus annuus*) by reducing fitness advantages of wild plants. To test this prediction, we established a factorial field experiment in Kansas, USA where we monitored the fitness of four cross types (Wild, F_1, F_2, and BC_w hybrids) under different levels of interspecific and intraspecific competition. Intraspecific manipulations consisted both of density of competitors and of frequency of crop-wild hybrids. We recorded emergence of overwintered seeds, survival to reproduction, and numbers of seeds produced per reproductive plant. We also calculated two compound fitness measures: seeds produced per emerged seedling and seeds produced per planted seed. Cross type and intraspecific competition affected emergence and survival to reproduction, respectively. Further, cross type interacted with competitive treatments to influence all other fitness traits. More intense competition treatments, especially related to density of intraspecific competitors, repeatedly reduced the fitness advantage of wild plants when considering seeds produced per reproductive plant and per emerged seedling, and F_2 plants often became indistinguishable from the wilds. Wild fitness remained superior when seedling emergence was also considered as part of fitness, but the fitness of F_2 hybrids relative to wild plants more than quadrupled with the addition of interspecific competitors and high densities of intraspecific competitors. Meanwhile, contrary to prediction, lower hybrid frequency reduced wild fitness advantage. These results emphasize the importance of taking a full life cycle perspective. Additionally, due to effects of exogenous selection, a given hybrid generation may be especially well-suited to hastening introgression under particular environmental conditions.

Editor: Norman Johnson, University of Massachusetts, United States of America

Funding: The study was conducted at and partially supported by the University of Kansas Field Station, a research unit of the Kansas Biological Survey and the University of Kansas. It was also supported by Biotechnology Risk Assessment Grant Program competitive grant no. 2006-39454-17438 to AAS, KLM, and HMA from the United States Department of Agriculture, National Institute of Food and Agriculture; www.nifa.usda.gov. Salaries and research support also provided by State and Federal funds appropriated to the Ohio Agricultural Research and Development Center, Ohio State University: manuscript no. HCS-14-07. The funders had no role in study design, data collection and analysis, decision to publish, or preparation of the manuscript.

Competing Interests: The authors have declared that no competing interests exist.

* Email: mercer.97@osu.edu

Introduction

Hybridization among differentiated plant taxa can introduce novel variation on which selection can then act (e.g., [1]). Thus, in any hybrid zone, questions arise about how hybridization and introgression (i.e., stable incorporation) may affect the evolutionary trajectories of the parent taxa and their hybrid progeny. Will there be introgression of introduced alleles into a given population or species? Further, which conditions promote or retard introgression? Addressing these questions requires an ecological genetics perspective, and much can be learned by measuring fitness components in common garden field experiments. Selection within hybrid zones can be environmentally independent and/or dependent, referred to as endogenous and exogenous respectively (reviewed in [2]). With endogenous selection, hybrid generations may have consistent, inherent fitness advantages or disadvantages relative to their parents and each other, i.e., due to the expression of genetic incompatibilities (e.g., [3]). By contrast, exogenous selection infers that selection operating on hybrids in hybrid zones may differ based on the environmental conditions. Such a situation implies the existence of genotype-by-environment (G×E) interactions, i.e., fitness differences among hybrids and their parents are influenced by local conditions (e.g., [4]). This environmental-dependence of the relative fitness of various hybrid generations can be crucial to evolutionary dynamics in hybrid zones [2,5].

In this study, we explored the effect of the biotic environment on hybrid and wild parental fitness and specifically how inter- and intraspecific competition can affect rates of introgression. Interspecific competitors can differentially influence plant fitness and, ultimately, a genotype's presence/dominance in that community [6–8]. Population density can be a major driver of the outcome of competition between species or among individuals within a species [9,10] thereby influencing plant fitness. With regard to intraspecific competition, in addition to density, the genetic composition of intraspecific competitors can influence plant fitness. In hybrid zones, the relative frequencies of hybrids *vs.* non-hybrid genotypes are expected to vary depending on the distances between hybridizing individuals, the relative sizes of each population, pollinator behavior (if relevant), and other ecological factors [11,12]. When the frequency of particular genotypes or cross types within a population affects fitness, frequency-dependent selection also has the capacity to influence introgression dynamics [13–15]. Knowledge of the differential ability of hybrids and parental types to complete their entire life cycle within the context of a range of realistic competitive environments should allow us to better predict the conditions under which we may or may not see introgression. However, studies that provide these data are rare, as described in [36].

A common and well-studied case of hybridization between differentiated populations is gene flow between crops and their wild relatives [11]. The advent of genetically modified crops raised concerns about the potential for introgression of novel alleles into wild populations [16,17] possibly increasing invasiveness or altering the wild population's genetic structure (see [11] for review). The possibility of introgression of crop alleles, and the rate at which it occurs, depends on the fitness of various hybrid generations relative to their wild counterpart [18]. Yet the relative fitness of wild and crop-wild hybrid generations can be strongly influenced by the environment in which they are compared (e.g., [19]). Biotic and abiotic factors that can affect hybrid fitness include pathogen or herbivore species [20–23], unspecified differences between locations [24,25], and competitive conditions [19,26,27], among others. Many fitness studies are performed in highly controlled conditions (e.g., in a greenhouse: [20]) or in field conditions mimicking farm fields (e.g., [19]) even though crop-wild hybrid zones often extend into non-cultivated areas. To clarify the context dependence of crop allele introgression on the unmanaged landscape, we need field studies that rigorously address how fitness of multiple hybrid generations is influenced by a range of relevant biotic conditions [28].

Our current research focuses on *Helianthus annuus* (common sunflower), which exhibits a particularly high gene flow rate in the USA, with as many as 66% of cultivated fields surveyed overlapping in flowering time with an adjacent, conspecific wild population [29]. Up to 25% of seeds produced in wild sunflower populations alongside crop fields have been shown to be crop-wild hybrids; such hybrids have also been found in populations up to 1 km away from crop fields [30]. Moreover, crop alleles can remain in wild populations for >5 years and crop-to-wild introgression can be relatively common [31,32]. In past studies, we found that F_1 crop-wild hybrid sunflowers produced fewer seeds per plant relative to wild sunflowers, but this disadvantage of hybrids diminished when compared under more competitive conditions [19,33], indicating that rates of introgression might increase accordingly under such conditions. Increased competition reduced branching in wild genotypes, while the faster seedling growth in the hybrids may have increased their performance under a denser canopy [34]. However, this past work only considered F_1 hybrids, and thus did not include many of the hybrid generations found in hybridizing wild sunflower populations. Further, it did not assess fitness throughout the life cycle under a range of natural competitive environments.

Here we report findings from a large, manipulative competition experiment in Kansas, USA. We measured the survival and fecundity of wild and three crop-wild hybrid sunflower cross types grown for their entire life cycle under a range of competitive conditions in the field. First, we hypothesized that the presence of interspecific competitors and increased density of intraspecific competition would reduce fitness advantages of the wild plants relative to the hybrids [19]. This was confirmed, although wild plants still retained superior fitness over hybrids, especially once we accounted for their better overwintering seed survival and seedling emergence. Second, we predicted that a higher frequency of the faster growing F_1 seedlings should lead to a more competitive environment and a reduction in the wild advantage over hybrids. This prediction was not upheld and we found that the effects of frequency were minor relative to the other treatments. We conclude that knowledge of the differential ability of hybrids and wild types to survive and reproduce across this range of competitive environments should allow us to better predict the biotic conditions that affect introgression.

Methods

Seed sources

Seed sources and crossing design for the hybrid generations are explained in Weiss et al. [35] and Alexander et al. [36]. In brief, we collected achenes (hereafter, seeds) from common sunflower (*Helianthus annuus*) populations in and around Lawrence, Kansas, in the fall of 2006 from five habitats where common sunflower is often found: roadside, construction zone, agricultural field, abandoned field, and wetland. Crop sunflowers are uncommon in northeastern Kansas. By collecting from these environments, we attempted to include all possible adaptive diversity from the local gene pool. Seeds from these original populations were pooled and sown for use in hand-pollinations in Columbus, OH, in 2007 to produce an F_1 hybrid cross type between the wild and crop sunflower. A crop sunflower inbred line known as HA 89 was used as the pollen parent for F_1 crosses. Crop-wild F_1 hybrid progeny produced on 20 wild maternal parents in 2007 were then used in 2009, along with the original wild seeds and HA 89, to produce four cross types: a new set of wild (wild×wild) and F_1 hybrid (wild×crop) progeny, as well as F_2 hybrid (F_1×F_1) and BC_w hybrid (wild×F_1) generations. (Maternal parent is noted first in the parenthetical crosses). As discussed more fully in Alexander et al. [36], F_1 plants from these populations appeared to be self-incompatible (KLM, personal observation). Wild, F_1, and BC_w cross types were produced on the same 18 wild maternal parents, while F_2 seeds were produced on 18 F_1 maternal parents. Wild crosses on a given maternal parent were sired by approximately five wild pollen parents, while F_1 crosses were sired by up to five pollen parents of the genetically uniform crop parent. BC_W and F_2 crosses were sired with pollen from two F_1 pollen parents or with bulked F_1 pollen from multiple pollen parents. Seeds from up to five inflorescences (hereafter, heads) were used to make up a given maternal family. With this procedure, all seeds for the four cross types were of similar age and were produced on bagged inflorescences of field-grown plants.

Establishment of the field experiment

Seeds of these four sunflower cross types were then used to establish a large field experiment in the fall of 2009 at the University of Kansas Field Station (Jefferson County, Kansas, USA). We planted seeds in the fall because natural seed dispersal occurs from October - December; this planting time allowed for fitness to include the ability of seeds to overwinter. Overall, seeds used in this study represented three categories: focal, matrix, and buffer seeds. The focal seeds represented all four cross types and we followed their fate from emergence through seed production. Matrix seeds consisted of mixtures of wild and F_1 crop-wild hybrid seeds and were used to create the different intraspecific competition treatments (density and frequency of hybrids). Buffer seeds were wild seeds collected from our crossing blocks that were spread in the outer 15 cm of each plot to reduce edge effects.

In overview, the experiment consisted of six blocks that were established in an old field environment dominated by brome grass (*Bromus* spp.) in November 2009. (Blocks had been rototilled in the spring of 2009 to allow for emergence of weeds and to confirm that wild sunflower was not in the seed bank). Within each block, we established two 14.3 m×1.35 m strips with a wide aisle between them, which together contained a total of twelve 1.35 m×1.8 m plots. As described in detail below, these 12 plots consisted of factorial combinations of levels of the three competitive factors. We refer to these as environmental treatments. Within each plot, we followed the fate of 72 focal seeds (four cross types×18 families per cross type); these will be referred to as genetic treatments. From a statistical perspective, this is a split plot design with randomized locations of the environmental treatments (=main plots) within each block and randomized locations of genetic treatments within each main plot (=subplots).

The 12 environmental treatments were factorial combinations of the various levels of an interspecific competition factor and two intraspecific competition factors (density of seed rain (hereafter, seed density) and crop-wild hybrid frequency). We chose to manipulate competition around the focal plants because wild sunflower populations establish within various vegetation contexts–from monospecific stands to highly diverse communities (H.M.A, Pers. Obs.) and from low to high densities [37,38]. To create two levels of interspecific competition, we either weeded all plants other than matrix sunflower (often ragweed) or allowed these species and sunflowers to coexist. We created three levels of intraspecific seed density (100, 255, and 495 seeds m^{-2}) by altering the amount of matrix seeds. We manipulated a second intraspecific competitive factor, the frequency of hybrids, by altering the frequency of F_1 seeds in the matrix seeds to mimic wild populations experiencing substantial gene flow from cultivated sunflower fields. We chose two levels, 15% and 40%. Arias and Rieseberg [30] found that crop-wild hybrid sunflower seeds can be produced at a frequency of as high as 0.60 on plants just three meters from a crop field (average of 0.27), while that frequency drops to 0.10 by 300 m and to closer to 0.01 by 1000 m. See Text S1 for analyses to confirm that our manipulations were successful in altering these three factors.

Into each plot experiencing a given set of environmental treatments, we planted our genetic treatments (72 focal seeds = 4 cross types×18 families per cross type). Due to incomplete emergence, families could not be included as a factor in our analyses of fitness and will not be discussed further. Focal seeds were affixed to labeled plastic cocktail stirrers (hereafter, swizzle sticks) with Gorilla Glue (Gorilla Glue Company, Cincinnati, Ohio) to allow us to maintain the identity of emerging focal seedlings in the following spring, as in Mercer et al. [39]. We planted focal seeds just below the soil surface at 10 cm spacing. We then scattered matrix seeds over the whole plot while protecting each focal seed with a small cup to maintain a small, cleared zone around it. After removing the cups, we covered the area with a 1 cm layer of sieved field soil. Focal seeds were planted in the center portion of each plot and excluded from a 15 cm buffer zone to reduce edge effects on fitness measures. We lowered sample sizes of hybrid cross types in the low density, low hybrid frequency plots to maintain the correct hybrid frequency. In total, we planted 4824 focal seeds in November 2009. The focal seeds that successfully overwintered and emerged in the spring were marked as focal plants and were observed for the rest of the season. See Table S1 for the sample sizes of focal seeds that emerged from each treatment combination.

Data collection

We focused on components of lifetime fitness of focal seeds, namely emergence, survival to reproduction, number of mature heads per plant, and seeds produced per head. Focal seedlings began to emerge March 22, 2010, and emergence was monitored every two to three days until May 27, when emergence became rare (i.e., only one seedling had emerged in eight days). At the end of the season, we counted the number of heads per plant. Sunflower plants typically have only one primary head (derived from their apical meristem) and can have numerous secondary heads (here defined as heads that are produced on ends of branches or on branches off branches). However, damage to the apical meristem can lead to loss of the primary head; such plants have increased branching and thus more secondary heads. We categorized developing primary and secondary heads as high or low quality. Low quality heads were those covered in larval frass or either too hard or too soft in the bud stage–all harbingers of poor or absent seed development (see Alexander et al. [36] for further discussion). Heads that did not have time to mature before killing frosts were not counted.

To estimate seed production, we collected the primary and first secondary head produced on each plant and counted their seeds by hand. All other secondary heads were randomly subsampled in such a way that any given head had an 80% chance of being selected for collection and counting. When seeds were counted in a given head, we noted whether each seed was in good condition or whether it had a hole in the pointed or blunt end or a bite removed from the pointed end. Holes towards the seed's pointed end were likely made by *Isophrictis similiella* (Lepidoptera); holes towards the seed's blunt end were likely caused by weevils (*Smicronyx fulvus* and *S. sordidus*, Coleoptera); and the insects that made the larger bites in the seeds are unknown [40] (D. Pilson, personal communication). Only good quality seed were included here in fitness estimates. Low quality heads covered in larval frass (noted above) had been infested with *Homeosoma electellum* larva, which resulted in total seed loss in most affected heads, thereby reducing overall fitness. The use of bridal veil to cover maturing seed heads may have deterred some seed predators, including other insects and birds, although we do not expect this to be a major factor in estimating seed production. The apical meristems of some plants at our site were attacked by disease, stem borers, or gall producing insects, resulting in unusually high levels of branching and head production. We collected data on the presence of this meristem damage during the season in order to be able to take it into account in our analysis.

Data analysis

We used Glimmix in SAS (version 9.3) to run restricted maximum likelihood ANOVAs to test the effects of our environmental and genetic treatments on our fitness response variables. Given the split-plot design, we tested for the effects of our main plot factors – seed density, hybrid frequency, and interspecific competition (and their interactions) – using the interaction of the block, seed density, hybrid frequency, and interspecific competition factors as an error term. Analysis of our genetic treatment was restricted to cross type since family was poorly replicated, as noted above; thus, cross type was applied at the subplot level and its effect and all of the interactions between cross type and our three main plot factors were tested with the pooled error term composed of the interaction of the block, seed density, hybrid frequency, interspecific competition, and cross type factors. Blocks were considered random, as were any interactions with block.

Fitness estimates depended on a) probabilities of reaching life cycle stages (such as becoming reproductive) and b) seed production per reproductive plants. For the former, we first analyzed the probability that focal seeds planted in the fall emerged and produced a mature head (i.e., survived to reproduce). Least squares means of these probabilities were predicted using the binomial distribution option in Proc Glimmix accounting for all treatment effects (as above). Probabilities for individuals surviving to reproduce were conditional on emergence. (See Table S1 for sample sizes for number of focal seeds that emerged and survived to reproduce). For seed produced per reproductive plant, we used estimated numbers of seeds per head and known counts of numbers of heads. Estimates of number of seeds per head for a particular plant were based on knowledge of whether the head was primary or secondary (and high or low quality) and the cross type and environmental treatments of the plant (Text S2, Table S2).

We estimated seed production per reproductive plant as:

(presence of high quality primary head×num. viable seeds per high quality primary head) + (presence of low quality primary head×num. viable seeds per low quality primary head) +

(num. high quality secondary heads×num. viable seeds per high quality secondary head) +

(num. low quality secondary heads×num. viable seeds per low quality secondary head).

While this estimated seed per reproductive plant, we ultimately produced average estimates for each cross type in each plot for subsequent analyses. Similarly, we created two increasingly integrative fitness measures. First, also by cross type and plot, we estimated the number of seeds produced per emerged seedling. This was defined as the probability of survival to reproduction× number of seeds produced per reproductive plant. Second, we defined the number of seeds produced per planted seed as the probability of emerging×probability of survival to reproduction× number of seeds produced per reproductive plant. These analyses of seed numbers included a binary variable for meristem damage to account for effects on seeds via enhanced head production.

Subsequent calculations of the fitness of a given hybrid cross type, i, relative to the wild (relative fitness, w_i) were simply calculations of wild fitness/hybrid fitness in the same set of treatment combinations and employed back-transformed least squares means from the analyses above. Calculations of the % change in w_i as interspecific competition was added was calculated as (w_i with interspecific competition – w_i without interspecific competition)/w_i without interspecific competition for each level of intraspecific density. Similarly, the % change in w_i as density of intraspecific competitors increases = (w_i under high density – w_i under low density)/w_i under low density for with and without interspecific competition. Finally, the % change in relative fitness going from low density, without interspecific competitors to high density, with interspecific competitors = (w_i high density, with interspecific competition – w_i low density, without interspecific competition)/w_i low density, without interspecific competition.

Results

Effects of competitive treatments and cross type identity on emergence and survival to reproduction

Competitive environment did not differentially affect the emergence of cross types and only cross type itself had a significant effect on seedling emergence in the spring (Table 1). Wild seed emerged at the highest proportion (0.68, s.e. = 0.017), followed by BC_w (0.61, s.e. = 0.019), F_1 (0.50, s.e. = 0.019), and F_2 seeds (0.40, s.e. = 0.018) (LS means; all significantly different with a Tukey-Kramer test). By contrast, only density of intraspecific competitors affected the probability of survival to reproduction once a seedling had emerged (Table 1). Plants in low and medium density plots had high probabilities of reproducing (low density: 0.92 (s.e. = 0.013); medium density: 0.87 (s.e. = 0.015)), while the probability was 0.74 (s.e. = 0.023) for high density plots (low and medium significantly different from high, Tukey-Kramer test).

Differential effects of competition on seed production of the cross types

Given our original hypotheses regarding how the fitness of crop-wild hybrids relative to their wild counterparts would increase under more competitive conditions, we were most interested in discerning how interactions between competitive factors and cross type influenced seed production (i.e., G×E interactions). If found, the presence of such higher order interactions makes discussion of main effects irrelevant (Table 1).

The interaction between cross type, the density of intraspecific competitors, and the presence of interspecific competitors had the greatest effect on changing the magnitude of fitness measures, which largely supported our expectations (Table 1, Figure 1). Under the least competitive conditions (low density, no interspecific competition), wild plants were far more fit than BC_w, which were more fit than F_1 or F_2 cross types (Figure 1). However, fitness differences among cross types were much reduced in magnitude under the more competitive high density treatments or when interspecific competition was applied (Figure 1). (Fitness differences would have appeared eliminated had we only assayed heads per plant (Figure S1)). In fact, for the number of seeds produced per reproductive plant, the difference between the wild and F_2 cross types was 22 times as great under low density, without interspecific competition (Figure 1A) as under high density, with interspecific competition (Figure 1B). Thus, F_2 hybrids could not be distinguished from wild plants under medium density, without interspecific competition (Figure 1A) or at high densities, with interspecific competition (Figure 1B). Moreover, once survival to reproduction was also taken into account, there were even more cases where more competitive conditions eliminated the fitness differences between wilds and particular hybrid generations (in this case, both F_2 and BC_w hybrids; Figure 1C, D, Table 1). However, when emergence of seeds in the spring was included in the compound fitness measure (i.e., for seed production per planted seed), none of the hybrids were equivalent to wild cross type, even under high density (Figure 1E, F). Still, the magnitude of difference between wild and F_2 cross types remain 19 times greater under low density, without interspecific competition (Figure 1E) as under high density, with interspecific competition (Figure 1F). Therefore, for seed production per planted seed, the wild cross type ultimately maintained a slight fitness advantage,

Table 1. ANOVA on fitness measures for crop-wild hybrid sunflowers grown under competitive conditions in Kansas, USA.

			Prob. Emerged[1]		Prob. Surv. Reprod.[1]		Seeds per Repr. Plt[2]		Seeds per Emgd Sdling[2]		Seeds per Seed Plted[2]	
Source	**ndf[3]**	**ddf[3]**	**F**	**P-value**	**F**	**P-value**	**F**	**P-value**	**F**	**P-value**	**F**	**P-value**
Frequency of Hybrids	1	55	0.23	0.6371	0.14	0.7102	6.19	**0.0159**	7.57	**0.008**	5.46	**0.0231**
Interspec Competitors	1	55	0	0.9707	0.74	0.3947	61.06	**<.0001**	50.19	**<.0001**	36.72	**<.0001**
Density of Seeding	2	55	0.06	0.9456	33.41	**<.0001**	292.8	**<.0001**	315.6	**<.0001**	215.53	**<.0001**
Freq*Interspec	1	55	0.04	0.8514	0.26	0.6138	0.52	0.4742	0.23	0.63	0	0.9872
Freq*Dens	2	55	1.07	0.3499	0.08	0.9254	1.81	0.1736	1.85	0.167	2.56	0.0864
Interspec*Dens	2	55	0.05	0.9498	0.18	0.8333	2.9	0.0636	3.14	0.0513	3.14	0.0511
Freq*Interspec*Dens	2	55	0.47	0.6251	0.92	0.4063	0.08	0.9229	0.01	0.9907	0.36	0.6993
Cross Type	3	179–186	65.09	**<.0001**	1.45	0.2305	197.9	**<.0001**	159.5	**<.0001**	168.17	**<.0001**
Freq*Cross	3	179–186	0.24	0.8667	1.55	0.2036	1.66	0.1778	4.09	**0.0078**	2.81	**0.0412**
Interspec*Cross	3	179–186	1.18	0.3178	0.75	0.5227	1.06	0.3679	1.8	0.1492	1.88	0.1341
Dens*Cross	6	179–186	0.59	0.7406	1.48	0.1875	14.68	**<.0001**	13.39	**<.0001**	11.52	**<.0001**
Freq*Interspec*Cross	3	179–186	0.29	0.8338	0.68	0.5654	0.93	0.4292	0.39	0.7576	0.46	0.7125
Freq*Dens*Cross	6	179–186	0.59	0.7414	0.49	0.8174	0.69	0.6548	0.48	0.8259	0.63	0.7057
Interspec*Dens*Cross	6	179–186	0.31	0.9306	0.33	0.9185	2.35	**0.033**	2.19	**0.046**	1.4	0.2167
Freq*Interspec*Dens*Cross[4]	6	179–186	0.94	0.4668	–	–	0.12	0.9945	0.22	0.9687	0.55	0.7677
Meristem Damage	3	1941	–	–	–	–	3.95	**0.0485**	0.11	0.7377	2.53	0.1138

Notes: These analyses explore the effects of experimental treatments on the probability that focal seeds emerged and survived to reproduce, the number of seeds produced by reproductive plants, the number of seeds produced by emerged seedlings, and the number of seeds produced by each seed planted the previous fall (see text for details). P-values>0.05 are bolded to emphasize significance.

[1]Binary life cycle data (probability of emergence and survival to reproduction) analyzed using residual pseudo-likelihood and the logit link function in SAS Glimmix.

[2]Seeds per reproductive plant, Seeds per emerged seedling, and Seeds per planted seed were all natural log transformed (ln(y+1)).

[3]ndf and ddf, numerator and denominator degrees of freedom for each trait, respectively.

[4]This 4-way interaction was left out of the probability of survival to reproduction analysis because the model could not otherwise converge.

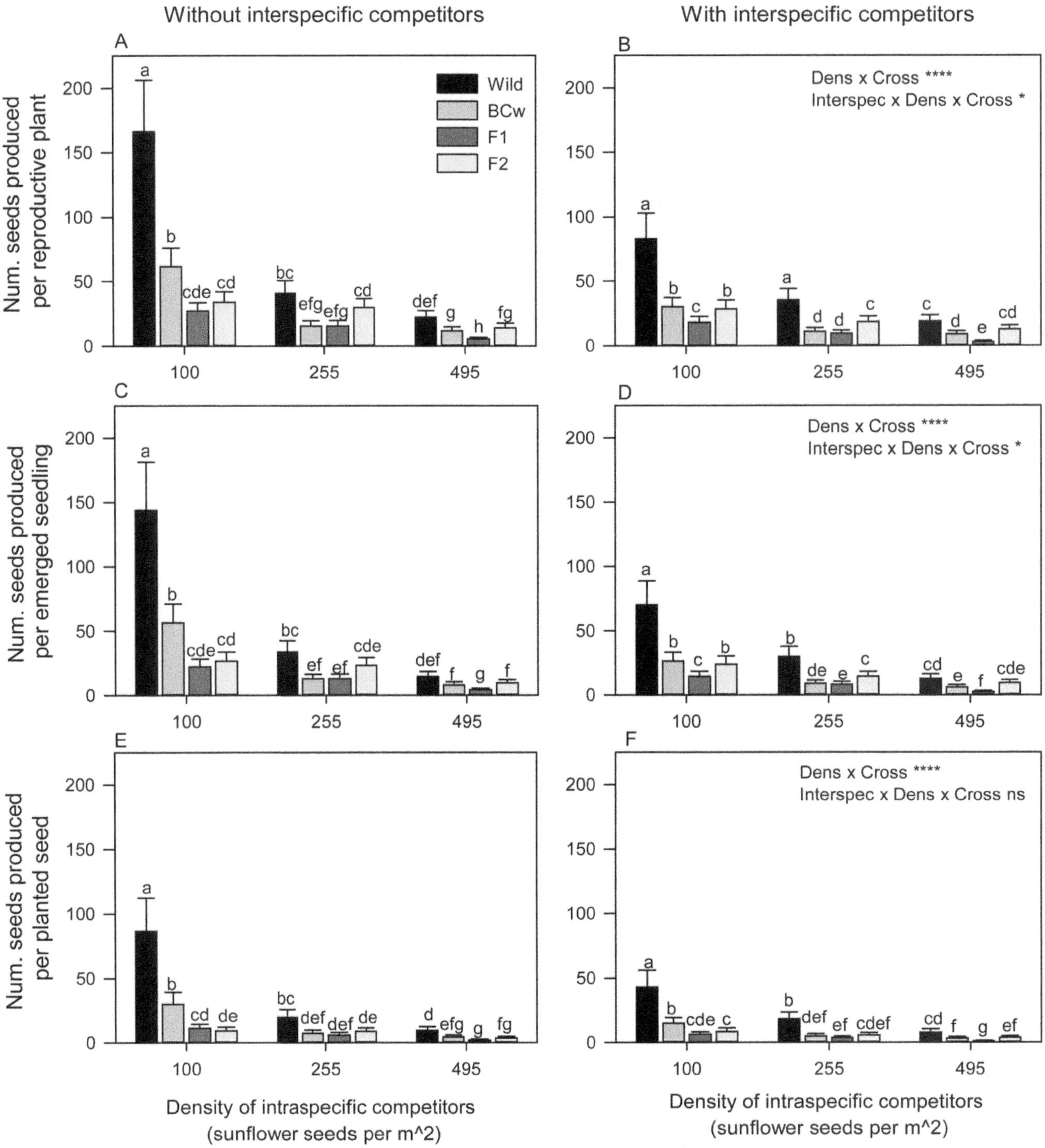

Figure 1. Combined effects of interspecific competitors, density of intraspecific competitors, and crop-wild hybrid cross type on three fitness measures in sunflower. Three fitness measures integrate increasing proportions of the life cycle: number of seeds produced per reproductive plant (A, B), number of seeds produced per emerged seedling (C, D), and number of seed produced per planted seed (E, F). Values are back-transformed least squares means with 95% confidence intervals (only the larger upper portion of asymmetrical interval is shown). Values sharing the same letter within a panel are not significantly different based on Tukey-Kramer multiple comparison tests. ANOVA effects for reference correspond to Table 1. Dens = Density of intraspecific competitors; Cross = Cross type; Interspec = Interspecific competition; *$P<0.05$, ****$P<0.0001$, ns $P\geq0.05$.

despite responding more negatively to the competitive conditions than certain hybrids.

Assessing this same G×E interaction using relative fitness values rather than absolute values can further illustrate selection pressures. Using the most complete fitness measure, numbers of seeds produced per planted seed, we calculated values of the fitness of each hybrid cross type relative to the wild under factorial combinations of interspecific competition (with and without) and under the highest and lowest densities of intraspecific competition. These relative fitness values ranged from 0.34–0.46 for BC_w, 0.13–0.20 for F_1, and 0.11–0.48 for F_2 cross types (Table 2). None of these values is greater than one and as we compare treatment combinations (i.e., from low to high density, with interspecific competitors), we see no large rearrangements in ordering or fitness rankings. However, relative values do change in response to treatments. All of the lowest relative fitness values are found under low density and all the highest are found under high density, although F_1 values remain generally low throughout. Increasing density, without interspecific competition, had the greatest positive effect on relative fitness for BC_w and F_1 cross types, with changes of 32% and 56%, respectively (Table 2). By contrast, adding interspecific competition and increasing density increased relative fitness for F_2 cross types by 353% (Table 2). On the low end, relative fitness values for BC_w and F_1 cross types declined with interspecific competition under high density by 12 and 33%, respectively (Table 2). Under those same conditions, F_2 relative fitness increased by 26%, which was by far the smallest change seen in F_2 relative fitness with increased competition. As a caveat, without error terms on these estimates, it is hard to know which differ significantly from zero.

The frequency of hybrids also differentially affected the fitness of cross types (Table 1, Figure 2). For two cross types (F_1 and BC_w), seed production was equivalent across the treatments, and for two others (Wild and F_2), seed production decreased (or trended to) as frequency of hybrids declined (Figure 2). Importantly, the magnitude of differences among cross types declined with decreases in hybrid frequency due to non-significant, but substantial reductions in wild seed production (Figure 2). Declines in hybrid frequency also tended to align the seed production of the F_2 and BC_w cross types (Figure 2A, B). However, once emergence was accounted for, the F_2 and BC_W cross types were equivalent no matter the hybrid frequency (Figure 2C). Ultimately, we did identify differences in the fitness of the hybrid cross types relative to the wild in seeds produced per planted seed as hybrid frequency declined: both the BC_W and F_1 cross types increased in relative fitness (37% and 21%, respectively); the relative fitness of the F_2 declined slightly by 9%. Regardless, changes in hybrid frequency did not have a strong enough effect to eliminate the wild fitness advantage over the various hybrids, no matter the measure.

Discussion

Effects of environmental treatments on hybrid vs. wild fitness

The likelihood of crop alleles introgressing into wild populations depends on the overall fitness of early-generation hybrids, plus the fitness effects of particular alleles and any linked loci [41]. We found evidence that fitness of crop-wild hybrids relative to wild sunflowers was not constant, but depended on the competitive context. We can thus begin to infer the ecological conditions consistent with higher rates of introgression and the cross types that are more likely to contribute to that introgression.

Specifically, we had predicted that competitive environments created with the presence of interspecific competitors, high intraspecific density, and higher hybrid frequency might enhance introgression by increasing the fitness of crop-wild hybrids relative to wild genotypes. Our data supported some of these predictions, but not all. We did find that higher density and the presence of interspecific competitors generally increased the fitness of hybrids relative to the wild, especially for the F_2. High competition reduced the magnitude of differences between cross types, such that wild genotypes became indistinguishable from some hybrids

Table 2. Relative fitness and percent changes in relative fitness for crop-wild hybrid sunflowers grown under competitive conditions in Kansas, USA.

		Density of intraspecific competitors								
		Low (100 seeds per m²)			High (495 seeds per m²)			% Change as density increases		
		Cross type			Cross type			Cross type		
		BC_W	F_1	F_2	BC_W	F_1	F_2	BC_W	F_1	F_2
Interspecific Competition	Without	0.35	0.13	0.11	0.46	0.20	0.38	32%	56%	260%
	With	0.34	0.14	0.20	0.41	0.13	0.48	19%	−6%	146%
% Change as interspecific competition added		−2%	11%	84%	−12%	−33%	26%	16%[1]	5%[1]	353%[1]

Note: Relative fitness, change in relative fitness, and cross types are as defined in the text (Materials and Methods section).
[1]These cells represent changes in relative fitness from low density, without interspecific competition to high density, with interspecific competition.

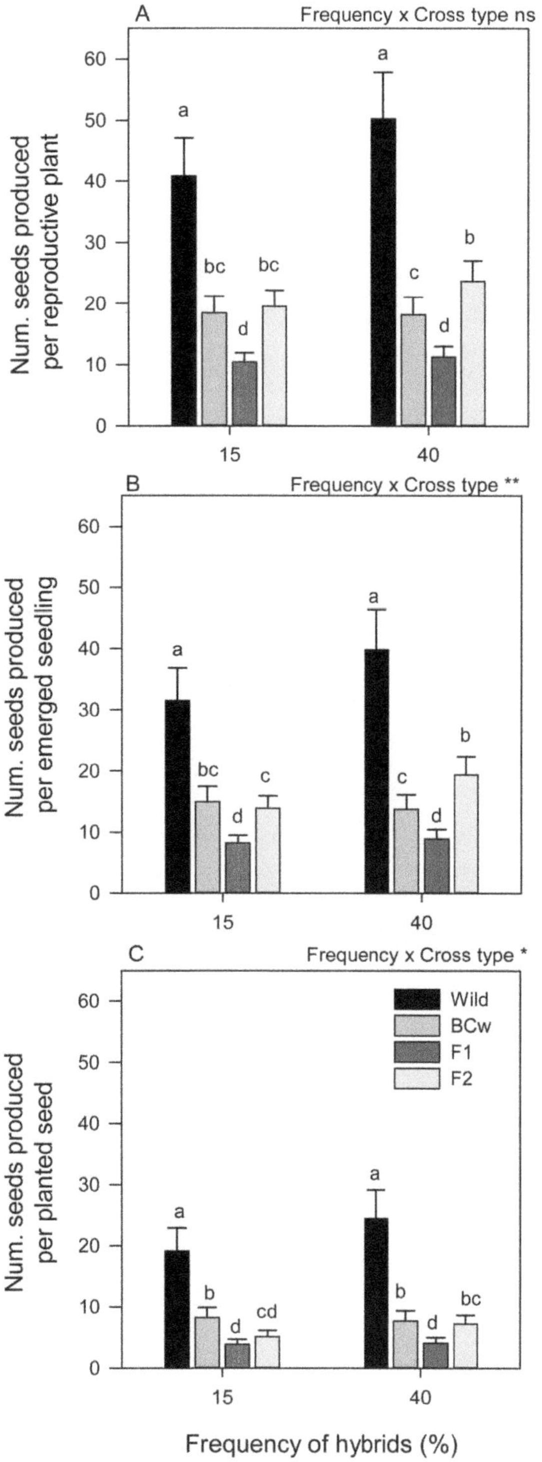

Figure 2. The combined effects of the frequency of crop-wild hybrids and cross type on three fitness measures in sunflower. Three fitness measures integrate increasing proportions of the life cycle: number of seeds produced per reproductive plant (A), number of seeds produced per emerged seedling (B), and number of seed produced per planted seed (B). Values are back-transformed least squares means with 95% confidence intervals (only upper portion of asymmetrical interval is shown). Values sharing the same letter within a panel are not significantly different based on Tukey-Kramer multiple comparison tests. ANOVA effects for reference correspond to Table 1. $^{*}P<0.05$, $^{**}P<0.01$, ns $P\geq0.05$.

(F_2 and occasionally BC_W) in terms of seeds per emerged seedling (Figure 1C, D). Yet using our most complete fitness measure, seeds per planted seed, hybrid and wild fitness was never equivalent (Figure 1E, F). Hybrid frequency also affected fitness, although low frequency of hybrids proved to be the set of conditions which reduced fitness differences between cross types (Fig. 2). This latter observation was contrary to our initial prediction, but previous work confirms that increasing hybrid frequency can reduce the relative fitness of hybrids [26] and that frequency-dependent responses can be quite profound, even causing reversals of rankings among hybrid cross types [27]. Thus, habitats with high intraspecific density, with the presence of interspecific competitors, or with low hybrid frequencies may ultimately increase the likelihood of crop allele introgression.

Exogenous selection by intra- and interspecific competition

Our study is consistent with the phenomenon of exogenous selection acting within crop-wild hybrid zones. For some studies of exogenous selection, such as where there is local adaptation across an environmental cline [42], one might expect reversals of rankings of genetic classes across that cline, so relative fitness values would shift dramatically [2,5]. However, for investigations of G×E interactions that could promote introgression under particular sets of conditions, reversals of fitness rankings may not be apparent or necessary. In our experiment, competition greatly reduced the magnitude of differences in fitness among cross types, but we did not see any radical shifts in their rankings. Wild plants maintained an actual, if not statistical, advantage over most hybrid classes across most components of fitness (i.e., from emergence through seeds per seed planted). Yet the relative fitness of all hybrids tended to increase with competition (Table 2). In other crop-wild hybrid systems, competition has also been shown to increase the relative fitness of hybrids, though not always ([26] and references therein). Our work adds to the few studies that have dissected the effects of different forms of competition on hybrid fitness (e.g., [26,27]). To our knowledge, ours is the first study of its type that has followed plants from seed to seed under field conditions.

It is important to emphasize that, despite the lack of radical shifts in ranking, the effects of competition on hybrid fitness were not equivalent and relative fitness values did change across competition treatments. F_1 and BC_W relative fitness values were mildly responsive to density; they both changed only an average of 25% as density increased (Table 2). By contrast, the exogenous selection affecting the F_2 cross type was stronger. Increasing density of intraspecific competitors more than doubled or tripled the fitness of the F_2 cross type relative to its wild counterpart and increasing density while also adding interspecific competitors quadrupled it (Table 2). This magnitude of change in relative fitness should be sufficient to alter evolutionary processes and enhance introgression of crop alleles under these biotic conditions. Thus, the F_2 cross type will be much more likely to ferry crop alleles to the subsequent generation under competitive conditions than when competition is low.

Factors contributing to competitive resilience in F_2

While the BC_W cross type did best among the hybrids under lower competition, the F_2 cross type equaled or surpassed it as competition increased (Figure 1). Several factors could account for improved competitive ability in F_2 progeny. The first relates to differences between F_2 progeny and their fellow hybrids with 50% crop contribution–the F_1 cross type. By having different maternal parents (F_1 seeds were produced on a wild maternal plant and F_2

seeds were produced on an F_1 maternal plant), maternal genetic effects could produce cross types differences due to any maternally inherited seed coverings, organelles, or organellar genomes [43]. The seeds and seedlings of the F_2 cross type were larger than those of the F_1; this could have enhanced competitive ability as seedling size was associated with a greater chance of surviving to flowering in this same experiment (MAK, personal observation). Others have seen similar fitness benefits of seed or seedling size under competitive conditions [44,45]. It should be noted, however, that in a sunflower study with lower competition, maternal genetic effects in sunflower were not observed to affect fitness late in the life cycle [36]. Large seed size can also have negative implications in this species, such as higher herbivory [38].

Increased F_2 competitive ability could also be the result of selection earlier in the season. Because F_2 genomes constitute recombining wild and crop genomes, particular combinations of homozygous wild loci may lead to individuals that have wild phenotypes and surprisingly high fitness. Likewise, individuals that are crop-like at these loci could have experienced premature germination or overwintering mortality [36], as evidenced by the F_2 cross type having the lowest emergence. Thus, strong selection during the overwintering phase may have selected for a more fit or competitive subgroup of the F_2 progeny, if alleles on which selection was operating were physically linked to, or had pleiotropic effects on, traits with effects during later life stages.

Life cycle stages and fitness estimates

It is rare for researchers to label individual seeds of known genetic background and follow their performance to the next generation while creating a realistic competitive environment. By taking this approach, we were able to include a seed's overwintering survival and successful germination as components of fitness. Yet there still remain gaps in our understanding of seed-related fitness components. For instance, premature germination of hybrid seeds prior to the spring [35,36] may be what reduced the most comprehensive fitness measure, seeds produced per planted seed, and kept wilds and hybrids from parity. On the other hand, the comprehensive estimates of F_2 fitness may actually be conservative here because an adjacent study registered higher emergence for the F_2 than for wild seedlings [36]. Thus, wild and hybrid equivalence might be possible for all fitness metrics under competitive conditions. Nevertheless, we could not account for seed dormancy in this study. Including dormancy would have likely further enhanced the fitness of cross types produced on the wild maternal parent–especially the wilds themselves. Ungerminated wild seeds are more likely to overwinter safely, remain dormant in the soil seed bank, and may emerge another year, while ungerminated hybrid seeds with more crop-like maternal parents and greater percentages of crop ancestry are more likely to die as seeds or prematurely germinate [46]. Clearly, accounting for early stages of the life cycle facilitates a better understanding of fitness and relative fitness differences among hybrid cross types. We are aware, however, that our results may depend on the unique abiotic and biotic conditions present in the year of our study.

Overall, our results showed that the fitness deficits experienced by crop-wild hybrids compared to wilds were diminished when various forms of inter- and intraspecific competition were applied, indicating the potential for exogenous selection within naturally occurring crop-wild hybrid zones. However, using more complete information from across the life cycle also clarified that early traits (seed overwintering ability, emergence, seedling size) can reduce hybrid fitness, but also play a role in enhancing the competitive ability of some cross types (e.g., F_2). These early traits are all controlled to some extent by maternal genetic effects, yet the potential role of maternal genetic effects in altering rates of introgression does not appear to be well-studied, especially beyond the F_1 generation (but see [47]). Further, maternal effects are notoriously difficult to account for in evolutionary processes more generally [48]. Crop-wild hybrid zones and other (animal or plant) hybrid zones where taxa are differentiated for traits controlled by maternal genetics may prove to be excellent laboratories for such study. In conclusion, full life cycle assessments of fitness differences among hybrid generations are useful for assessments of opportunities for crop allele introgression in the field (e.g., [28,49]). Also, while more competitive conditions appear to facilitate the introgression of crop alleles into wild populations, some cross types (e.g., the F_2) may particularly benefit due to their unique characteristics. We encourage further research into the complexities of factors influencing introgression of novel alleles into wild populations across the landscape.

Supporting Information

Figure S1 Combined effects of density of wild sunflower, frequency of hybrid sunflower, and cross type on number of mature heads.

Table S1 Sample sizes for number of focal seeds that emerged (before slash) or that survived to reproduce (after the slash).

Table S2 ANOVA on number of viable seeds per head on crop-wild hybrid sunflowers grown in the field in Kansas, USA.

Text S1 Did our treatments alter the biotic environment?

Text S2 Estimates of seed per head and counts of number of heads per plant.

Acknowledgments

We thank numerous undergraduates and staff in Kansas and Ohio who assisted in the field, as well as J. Cardina for helpful comments on the manuscript.

Author Contributions

Conceived and designed the experiments: KLM HMA AAS DJE. Performed the experiments: HMA DJE KLM MAK BAP. Analyzed the data: KLM. Contributed reagents/materials/analysis tools: KLM HMA AAS. Contributed to the writing of the manuscript: KLM HMA AAS.

References

1. Lenormand T (2002) Gene flow and the limits to natural selection. Trends in Ecology & Evolution 17: 183–189.
2. Campbell DR, Waser NM (2007) Evolutionary dynamics of an *Ipomopsis* hybrid zone: confronting models with lifetime fitness data. The American Naturalist 169: 298–310.

3. Barton NH, Hewitt GM (1985) Analysis of hybrid zones. Annual Review of Ecology and Systematics 16: 113–148.
4. Arnold ML (1997) Natural hybridization and evolution. New York: Oxford University Press.
5. Kimball S, Campbell DR, Lessin C (2008) Differential performance of reciprocal hybrids in multiple environments. Journal of Ecology 96: 1306–1318.
6. Goldberg DE, Barton AM (1992) Patterns and consequences of interspecific competition in natural communities: a review of field experiments with plants. The American Naturalist 139: 771–801.
7. Mangla S, Sheley R, James J, Radosevich S (2011) Intra and interspecific competition among invasive and native species during early stages of plant growth. Plant Ecology 212: 531–542.
8. Tilman D (1988) Plant strategies and the dynamics and structure of plant communities: Princeton University Press.
9. Stanton-Geddes J, Tiffin P, Shaw RG (2012) Role of climate and competitors in limiting fitness across range edges of an annual plant. Ecology 93: 1604–1613.
10. Weiner J (1990) Asymmetric competition in plant populations. Trends in Ecology & Evolution 5: 360–364.
11. Ellstrand NC (2003) Dangerous liaisons?: when cultivated plants mate with their wild relatives. Baltimore: Johns Hopkins University Press. 244 p.
12. Nagy ES (1997) Selection for native characters in hybrids between two locally adapted plant subspecies. Evolution 51: 1469–1480.
13. Ayala FJ, Campbell CA (1974) Frequency-dependent selection. Annual Review of Ecology and Systematics 5: 115–138.
14. Heino M, Metz JAJ, Kaitala V (1998) The enigma of frequency-dependent selection. Trends in Ecology & Evolution 13: 367–370.
15. Xiao Z, Gao X, Steele MA, Zhang Z (2010) Frequency-dependent selection by tree squirrels: adaptive escape of nondormant white oaks. Behavioral Ecology 21: 169–175.
16. Andow DA, Zwahlen C (2006) Assessing environmental risks of transgenic plants. Ecology Letters 9: 196–214.
17. Snow AA, Palma PM (1997) Commercialization of transgenic plants: potential ecological risks. Bioscience 47: 86–96.
18. Ellstrand NC, Meirmans P, Rong J, Bartsch D, Ghosh A, et al. (2013) Introgression of crop alleles into wild or weedy populations. Annual Review of Ecology, Evolution, and Systematics 44: 325–345.
19. Mercer KL, Wyse DL, Shaw RG (2006) Effects of competition on the fitness of wild and crop-wild hybrid sunflowers from a diversity of wild populations and crop lines. Evolution 60: 2044–2055.
20. Vacher C, Weis AE, Hermann D, Kossler T, Young C, et al. (2004) Impact of ecological factors on the initial invasion of Bt trangenes into wild populations of birdseed rape (*Brassica rapa*). Theoretical and Applied Genetics 109: 806–814.
21. Laughlin KD, Power AG, Snow AA, Spencer LJ (2009) Risk assessment of genetically engineered crops: fitness effects of virus-resistance transgenes in wild *Cucurbita pepo*. Ecological applications 19: 1091–1101.
22. Dechaine JM, Burger JC, Chapman MA, Seiler GJ, Brunick R, et al. (2009) Fitness effects and genetic architecture of plant–herbivore interactions in sunflower crop–wild hybrids. New Phytologist 184: 828–841.
23. Yang X, Xia H, Wang W, Wang F, Su J, et al. (2011) Transgenes for insect resistance reduce herbivory and enhance fecundity in advanced generations of crop–weed hybrids of rice. Evolutionary Applications 4: 672–684.
24. Campbell LG, Snow AA, Ridley CE (2006) Weed evolution after crop gene introgression: greater survival and fecundity of hybrids in a new environment. Ecology Letters 9: 1198–1209.
25. Hovick SM, Campbell LG, Allison AS, Whitney KD (2012) Hybridization alters early life-history traits and increases plant colonization success in a novel region. The American Naturalist 179: 192–203.
26. Campbell LG, Snow AA (2007) Competition alters life history and increases the relative fecundity of crop–wild radish hybrids (*Raphanus* spp.). New Phytologist 173: 648–660.
27. Hauser TP, Damgaard C, Jorgensen RB (2003) Frequency-dependent fitness of hybrids between oilseed rape (*Brassica napus*) and weedy B. rapa (Brassicaceae). American Journal of Botany 90: 571–578.
28. Hails RS, Morley K (2005) Genes invading new populations: a risk assessment perspective. Trends in Ecology & Evolution 20: 245–252.
29. Burke JM, Gardner KA, Rieseberg LH (2002) The potential for gene flow between cultivated and wild sunflower (*Helianthus annuus*) in the United States. American Journal of Botany 89: 1550–1552.
30. Arias DM, Rieseberg LH (1994) Gene flow between cultivated and wild sunflowers. Theoretical and Applied Genetics 89: 655–660.
31. Linder CR, Taha I, Seiler GJ, Snow AA, Rieseberg LH (1998) Long-term introgression of crop genes into wild sunflower populations. Theoretical and Applied Genetics 96: 339–347.
32. Whitton J, Wolf DE, Arias DM, Snow AA, Rieseberg LH (1997) The persistence of cultivar alleles in wild populations of sunflowers five generations after hybridization. Theoretical and Applied Genetics 95: 33–40.
33. Snow AA, Moran-Palma P, Rieseberg LH, Wszelaki A, Seiler GJ (1998) Fecundity, phenology, and seed dormancy of F1 wild-crop hybrids in sunflower (*Helianthus annuus*, Asteraceae). American Journal of Botany 85: 794–801.
34. Mercer KL, Andow DA, Wyse DL, Shaw RG (2007) Stress and domestication traits increase the relative fitness of crop–wild hybrids in sunflower. Ecology Letters 10: 383–393.
35. Weiss AN, Primer SB, Pace BA, Mercer KL (2013) Maternal effects and embryo genetics: germination and dormancy of crop–wild sunflower hybrids. Seed Science Research 23: 241–255.
36. Alexander HM, Emry DJ, Pace BA, Kost MA, Sparks KA, et al. (2014) Roles of maternal effects and nuclear genetic composition change across the life cycle of crop-wild hybrids. American Journal of Botany 101: 1176–1188.
37. Alexander HM, Foster BL, Ballantyne F, Collins CD, Antonovics J, et al. (2012) Metapopulations and metacommunities: combining spatial and temporal perspectives in plant ecology. Journal of Ecology 100: 88–103.
38. Cummings CL, Alexander HM (2002) Population ecology of wild sunflowers: effects of seed density and post -dispersal vertebrate seed predators. Oecologia 130: 274–280.
39. Mercer KL, Alexander HM, Snow AA (2011) Selection on seedling emergence timing and size in an annual plant, *Helianthus annuus* (common sunflower, Asteraceae). American Journal of Botany 98: 975–985.
40. Snow AA, Pilson D, Rieseberg LH, Paulsen MJ, Pleskac N, et al. (2003) A Bt transgene reduces herbivory and enhances fecundity in wild sunflowers. Ecological Applications 13: 279–286.
41. Rieseberg LH, Burke JM (2001) The biological reality of species: gene flow, selection, and collective evolution. Taxon 50: 47–67.
42. Kawecki TJ, Ebert D (2004) Conceptual issues in local adaptation. Ecology Letters 7: 1225–1241.
43. Roach DA, Wulff RD (1987) Maternal effects in plants. Annual Review of Ecology and Systematics 18: 209–235.
44. Stanton ML (1984) Seed variation in wild radis: effect of seed size on components of seedling and adult fitness. Ecology 65: 1105–1112.
45. Wulff RD (1986) Seed size variation in *Desmodium paniculatum*. III. Effects on reproductive yield and competitve ability. Journal of Ecology 74: 115–121.
46. Pace BA (2012) Maternal effects on multiple generations of *Helianthus annuus* crop-wild hybrid seed: overwinter germination, dormancy and survival. M.Sc. Thesis, Columbus, Ohio: Ohio State University. 85 p. Available: http://rave.ohiolink.edu/etdc/view?acc_num=osu1354696610. Accessed 11 June 11 2014.
47. Allainguillaume J, Harwood T, Ford CS, Cuccato G, Norris C, et al. (2009) Rapeseed cytoplasm gives advantage in wild relatives and complicates genetically modified crop biocontainment. New Phytologist 183: 1201–1211.
48. Räsänen K, Kruuk L (2007) Maternal effects and evolution at ecological time-scales. Functional Ecology 21: 408–421.
49. Hooftman DAP, Oostermeijer JGB, Jacobs MMJ, Den Nijs HCM (2005) Demographic vital rates determine the performance advantage of crop-wild hybrids in lettuce. Journal of Applied Ecology 42: 1086–1095.

8

Development and Characterization of Polymorphic EST-SSR and Genomic SSR Markers for Tibetan Annual Wild Barley

Mian Zhang[1], Weihua Mao[2], Guoping Zhang[1], Feibo Wu[1]*

1 Department of Agronomy, College of Agriculture and Biotechnology, Zijingang Campus, Zhejiang University, Hangzhou, P.R. China, **2** Center of Analysis and Measurement, Zhejiang University, Hangzhou, P.R. China

Abstract

Tibetan annual wild barley is rich in genetic variation. This study was aimed at the exploitation of new SSRs for the genetic diversity and phylogenetic analysis of wild barley by data mining. We developed 49 novel EST-SSRs and confirmed 20 genomic SSRs for 80 Tibetan annual wild barley and 16 cultivated barley accessions. A total of 213 alleles were generated from 69 loci with an average of 3.14 alleles per locus. The trimeric repeats were the most abundant motifs (40.82%) among the EST-SSRs, while the majority of the genomic SSRs were di-nuleotide repeats. The polymorphic information content (PIC) ranged from 0.08 to 0.75 with a mean of 0.46. Besides this, the expected heterozygosity (He) ranged from 0.0854 to 0.7842 with an average of 0.5279. Overall, the polymorphism of genomic SSRs was higher than that of EST-SSRs. Furthermore, the number of alleles and the PIC of wild barley were both higher than that of cultivated barley, being 3.12 *vs* 2.59 and 0.44 *vs* 0.37. Indicating more polymorphism existed in the Tibetan wild barley than in cultivated barley. The 96 accessions were divided into eight subpopulations based on 69 SSR markers, and the cultivated genotypes can be clearly separated from wild barleys. A total of 47 SSR-containing EST unigenes showed significant similarities to the known genes. These EST-SSR markers have potential for application in germplasm appraisal, genetic diversity and population structure analysis, facilitating marker-assisted breeding and crop improvement in barley.

Editor: Xianlong Zhang, National Key Laboratory of Crop Genetic Improvement, China

Funding: The project was supported by National Natural Science Foundation of China (31171488), the National 863 Program (2012AA101105), and the Key Research Foundation of Science and Technology Department of Zhejiang Province of China (2012C12902-2). The funders had no role in study design, data collection and analysis, decision to publish, or preparation of the manuscript.

Competing Interests: The authors have declared that no competing interests exist.

* E-mail: wufeibo@zju.edu.cn

Introduction

Barley (*Hordeum vulgare* L.) is the fourth important cereal crop worldwide. With the rapid development of beer and feed industry, the demand for barley keeps increasing. However, during the long-term domestication of the cultivated barley, especially after the modern breeding and intensive cultivation, the genetic variation degraded significantly, resulting in missing lots of genes, including some rare alleles [1]. The monotonous genetic background of cultivated barley has become the bottleneck of the effectiveness of breeding, while the abundant diversity of wild barley can provide a pool of alleles for barley breeding and improvement [2,3]. Morphological, archaeological cytogenetic and isozyme data revealed that wild barley on the Qinghai-Tibet Plateau is different from the Fertile Crescent wild barley [4]. Researches so far have shown even rich genetic diversity in Tibetan wild barley than in Ethiopian barley [5]. Novel germplasm has been identified from the Tibetan wild barley tolerant to drought, salinity and aluminum toxicity [6–8].

Increasing efficient molecular markers would be valuable in diversity analyses, resource conservation and beneficial alleles exploitation for wild barley. Comprehensive sets of expressed sequence tags (ESTs) sequences have been generated in many plants (http://www.ncbi.nlm.nih.gov/dbEST). The availability of increasing sequence databases enables the identification of functional genes with similar sequences in related species [9]. EST-based SSR markers (EST-SSRs) have been widely employed as powerful molecular genetic tools in a large number of cereal crop species due to their high level of transferability, close association to genes with known function, codominant inheritance, and low cost for development with available development from public databases [10–12]. Jaikishan et al. [13] used 25 EST-SSRs and 25 genomic SSRs to predict grain yield heterosis; multiple EST-SSRs were generated for wheat (*Triticum aestivum* L.) and these markers showed high transferability between wheat and the other crops, such as barley, maize, rice, and sorghum [14–16]. Up to date, polymorphic EST-SSRs were identified to establish *Hordeum chilense* evolutional relationships [17] and new EST-SSRs and genomic SSRs were complemented to the published Australian barley genetic maps [18]. However, to our knowledge, little work has been performed to develop EST-SSRs and apply them for population structure in Tibetan wild barley.

In the present study, with the objective of exploiting new SSRs from EST databases and confirming the published genomic SSRs in the Tibetan wild and cultivated barley accessions, 49 EST-SSRs and 20 genomic SSRs were developed and characterized. These markers can be utilized to evaluate the genetic variation and phylogenetic relationships of 96 barley genotypes. Furthermore,

Table 1. Characterization of 49 polymorphic EST-SSR makers in barley (*Hordeum vulgare*L.).

Primer	SSR motif	Primer sequence (5′-3′) F/R	Expected size(bp)	Na	Ne	Ho	He	PIC
P181	(GAGAG)4	GTCGTCTCCCTCCCTTCA	227	5	3.23	0.1979	0.6944	0.6379
		CATTGCCAGCACTGTTTC						
P129	(GCC)7	CGAGGAGTTCGAGGTGGA	260	4	3.26	0.6042	0.697	0.635
		ACTCTGCGTCCCAGTTCTT						
P184	(TGC)9	CCTACCAAACAACGGAATA	276	4	3.04	0.1053	0.6746	0.6232
		CAGCCAGAAGGTCTACGA						
P50	(AATC)5	ACAAGCAGATCACCGACG	215	3	2.95	0.2727	0.6649	0.5868
		AACCCGACTGAACAAATAAT						
P91	(TC)13	CGAGGCTCCTCATCTCCT	211	3	2.89	0.3542	0.657	0.5796
		CCAGCATCGTCGCAAACT						
P8	(AG)15	TCGTTGATCCGAACTTTACC	197	3	2.82	0.2188	0.6484	0.5735
		CACCGCAGACGCTGAGTA						
P29	(ATAC)13	CTGCTTAGTTCTAGGAGGCT	140	3	2.57	0.1935	0.6141	0.5391
		CTCGGTTCGATTGTTCAT						
P103	(CTG)9	CATTTGGCATTGGTTGAT	100	3	2.53	0.0104	0.6073	0.5362
		AGTTCTTCTTCGCTGGAA						
P32	(GATG)6	GCAGAATGGCAGAAACAG	233	3	2.6	0.0957	0.6187	0.5352
		CAAGAATGAGCGAAAGGT						
P168	(TTC)7	TTCCTCCAGTCCTTCTCC	169	4	2.49	0.3511	0.6013	0.5344
		CTGCTGCTACCGTTCTTAT						
P99	(ATC)7	GATGTGATCTGATGCCATTT	273	3	2.46	0.25	0.5966	0.5266
		TTTCTTCGGTGTTCTTTCC						
P152	(CT)11	ACCAAGCCCACGAGTAGCA	251	3	2.4	0.0521	0.5867	0.5186
		CGACCCGAGGACGACAGAT						
P144	(CT)11	CTTCGTTCCCTCCTCACC	134	5	2.19	0.4545	0.5473	0.5127
		TCCGCTTCCACGATTGAC						
P121	(TACAT)4	CCCAGGAATAAGAACAGACAC	287	4	2.25	0.3684	0.5587	0.4971
		CACCGCCTAATAGCAACAA						
P34	(CTTC)6	GGCGAGGAACTGTTGTTG	252	3	2.33	0.2083	0.5738	0.4898
		GATCGGCTTCATCGTCTACT						
P101	(ATC)12	CCCCGTATAAACCACCCA	245	3	2.18	0.2556	0.5439	0.4827
		GGCAGAACTTCAGCACCC						
P149	(AGC)9	CTTGGCACGCTTTGTTTG	259	3	2.17	0.191	0.5431	0.478
		ACTTTCCCACGGCATCAG						
P150	(GAGC)5	TAAGTAGGTTTGAGGAAGGGAA	265	3	2.22	0.1149	0.5533	0.4693
		CAACATAGACAAGGTGCTGGA						
P83	(AAGAA)4	CTCGGCAAACAGAGGACA	278	4	2.21	0.2083	0.5506	0.468
		TTGTAGCAGCGGATGGTC						
P30	(ATGT)12	ACTGCCACTCCATTTAGG	241	3	2.16	0.1684	0.5407	0.4626
		CTGTCGTAGGCTTGCTTT						
P63	(AGC)9	GGCTTGGCACGCTTTGTT	259	3	2.07	0.1146	0.5206	0.4573
		TTTCCCACGGCATCAGTC						
P90	(GAT)7	CGCAAGCCACAGAGCACA	177	3	2.12	0.1146	0.5301	0.4507
		TCCGTCCGTTCGTCCATC						
P9	(AC)11	ATCACAAACAGCCACTGTCCTA	111	4	2.01	0.2812	0.5047	0.4388
		GTGGTGAACCTTGCCCTTG						
P3	(GA)10	GCGAGGATGATGTATAAACCG	132	4	1.95	0.3077	0.4887	0.4256
		TGCATTCTGTGCCCTAACTAA						

Table 1. Cont.

Primer	SSR motif	Primer sequence (5′-3′) F/R	Expected size(bp)	Na	Ne	Ho	He	PIC
P45	(GGTT)5	CCCACAACACCAACAAAC GCCCGTAGAATGAACAAGTA	229	3	2.08	0.1771	0.5219	0.413
P55	(CTG)9	TTGATGGAGAAGGAGCAT ACATAGTAGGATAGATAGACCC	264	3	1.78	0.0319	0.4419	0.3926
P105	(CCTCG)4	GCGACTACCAGGACGACAA CACCGACCGATACAGACAGA	297	3	1.78	0.0632	0.4415	0.381
P56	(CTG)7	AGTGATCTGAGGCGGTAT CGTACGTCCAATGTTGTC	176	2	1.99	0.1875	0.5007	0.374
P66	(CTCTT)4	CAAATGTGCCAGTAGAAA GGATGAGTTGCAGGTGAT	293	2	1.99	0.234	0.5006	0.374
P67	(TTG)12	AGAAACAAACAGACAGACCCAT ATTCCACCACCGTCACCA	284	2	1.97	0.5729	0.496	0.3717
P180	(CAG)8	ATTCTCGCCGCCAACAACT CCACGTAGAAAGGGAGGGTCA	217	2	1.97	0.2	0.4946	0.371
P80	(GGTTG)4	ACTCCTGCTGCTGCTGAC CGGTATTAGGCGACTCTTC	149	2	1.95	0.3229	0.4903	0.3688
P57	(AATA)5	ATAACAGCCGTTGATGAG GATCCGTTCCACAAACAT	260	2	1.94	0.2604	0.4869	0.367
P54	(ATC)7	CAGCACCACTACTAATCAAGAA GCCACCAACAAGACCTCC	245	2	1.93	0	0.4849	0.366
P137	(GAAGA)4	AGAGGACAAGCCAAGGAAG CACGGAAACGGAACAAAA	161	2	1.91	0.1739	0.479	0.3629
P106	(CTG)8	CGAGCCGTTGCTTAGGTC TCTACTGCCAGGGCGTGA	206	2	1.85	0.1383	0.4612	0.3535
P139	(GCAT)5	ACTCACATAGTAATCGAAGGG GGGCAAGAACGAATCTCC	287	2	1.83	0.4896	0.4568	0.3512
P186	(CTGA)5	GGTAGTTCCGCCATCAGA CCTCCTGTGGACGAAGAT	177	2	1.72	0.2604	0.4197	0.3303
P187	(GCACA)4	CTCGGACGACCATTTATT TTCAAAGTTCAAGGGTGC	209	2	1.7	0.1875	0.4154	0.3278
P53	(CCAA)5	AGGGAAAGAAATCCTAAC TTGACTTGCTTATACACCT	224	2	1.63	0.0968	0.3902	0.3128
P13	(AT)19	CACATGCGTTAGTGTCCC GCGATTATCTTCGTCCAG	298	2	1.63	0	0.3899	0.3126
P16	(TG)11	CGAGCAGGCATAGCCATAT GACGCTGAGTACGTTGAGGT	256	3	1.44	0.069	0.3097	0.2853
P61	(GCA)8	CAAATGGAGCCAAGCAAC CCATCCTTGACGCACATC	235	2	1.47	0.1828	0.3204	0.2679
P81	(CTG)8	GCAGGATAGGCGACACTC GAGACGGAGAAGGAGCAG	141	2	1.38	0.1333	0.2793	0.2392
P185	(CGG)8	AAACGGCTTTCACATCTCCC CGCCCAAACAAGTCCTCC	201	2	1.38	0.0625	0.2792	0.2392
P120	(AGC)7	GAAATACTCCCAGGACAGC AGCAAGTGCCAGTTCTACC	249	2	1.33	0.0106	0.2473	0.2157
P100	(CACG)6	CACATAAACAACCGAACCAA CGACATACGCAGGGAGTG	245	2	1.23	0.0208	0.1876	0.1693
P21	(GAC)7	AACCTATGCCGCCTACTT CCACCCGTCCACTCTTTT	241	2	1.11	0.0417	0.0993	0.0939

Table 1. Cont.

Primer	SSR motif	Primer sequence (5′-3′) F/R	Expected size(bp)	Na	Ne	Ho	He	PIC
P44	(GCAA)5	AGTCCCGTAAACCTACCTGAG	165	2	1.09	0	0.0854	0.0813
		TGCCGGAGAATGTAATCG						

Note: Na, number of alleles; Ne, number of effective alleles; Ho, observed heterozygosity; He, expected heterozygosity; PIC, polymorphic information content.

polymorphism, and genetic diversity in the Tibetan wild barley accessions were evaluated which would be particularly useful for identification of novel genes with traits of interest, and marker-assisted breeding in barley.

Materials and Methods

Plant materials

A total of 96 barley accessions were used in this study including 80 Tibetan annual wild barley from Qinghai-Tibet Plateau provided by Huazhong Agricultural University barley germplasm collection, and 16 cultivars from China which were stored at the Institute of Crop Science, Zhejiang University, Hangzhou, China (Table S1). These accessions were collected on public land. And no specific permits were required for the collection. Seeds were surface sterilized with 3% H_2O_2 for 30 minutes and thoroughly rinsed with distilled water, followed by germination in nutrient rich soil in an incubator (22/18°C, day/night) for 10 days. Total genomic DNA was extracted from barley leaves using the Plant Genomic DNA Kit (TianGen, Beijing, China).

Sequence screening and primer designing

A total of 525999 barley ESTs were acquired from the EST database of GenBank (up to September 2012) (http://www.ncbi.nlm.nih.gov/Genbank/). Redundant sequences were removed from these ESTs using CD-HIT-EST (http://cd-hit.org) with the identity parameter of 95%. The presence of SSRs was screened using Simple Sequence Repeat Identification Tool (SSRIT) software (http://www.gramene.org/gramene/searches/ssrtool). The criteria for di-, tri-, tetra-, and penta-nucleotides were 10, 7, 5, and 4 repeat units, respectively. A total of 188 EST-SSRs were randomly selected and primers were designed using Primer5.0 with a length ranging from 18–22 bp, and product sizes of 100 to 300 bp. The reverse primers were marked with 6-FAM or HEX fluorescent dye at 5′ side for each pair. Based on the previous study of barley, 41 genomic SSR markers were selected and SSR primers were designed with the same criteria as mentioned above.

PCR amplification and sequencing

PCR amplification was performed in a total of 20 μL reaction mixture that contained 1 μL of genomic DNA, 1 U ExTaq DNA polymerase (Takara Inc.), 2 μL of 10×Ex Taq Buffer (Mg^{2+} Plus), 0.2 mM dNTPs mix, 0.05 μM forward primers, 0.1 μM reverse primers and fluorescent primers (FAM or HEX).

The PCR protocol used was as follows: initial denaturation for 5 min at 94°C, followed by 5 cycles of denaturation for 30 s at 94°C, annealing for 30 s at 50°C, and extension for 30 s at 72°C, subsequently followed by 32 cycles of denaturation for 30 s at 94°C, annealing for 30 s at 55°C, extension for 30 s at 72°C, with a final extension for 10 min at 72°C and a 4°C holding temperature. PCR products were diluted and tested on a MegaBACE 1000 DNA analysis system (Amersham Biosciences, Piscataway, NJ) at the Center of Analysis and Measurement in Zhejiang University. The lengths of PCR fragments were calculated using the ET550-R size standard and Genetic Profiler version 2.2.

Calculation of polymorphism

The polymorphism of EST- and genomic SSR alleles were scored for the presence (1) and absence (0) for 96 accessions. Alleles with frequency less than 5% (rare alleles) in the population were removed and considered as missing data for the polymorphism calculation and population structure analysis [19]. The genetic diversity was evaluated by the number of alleles (Na), the effective number of alleles (Ne), observed heterozygosity (Ho), and expected heterozygosity (He) using POPGENE v.1.31 [20]. Polymorphism information content (PIC) was calculated by applying software PIC_CALC version 0.6.

Population structure

Population structure was assessed using the STRUCTURE software v2.3.3 based on the admixture model [21]. Models were tested for clusters (k) from 1 to 15, each with ten independent runs and 100,000 MCMC (Markov Chain Monte Carlo) iterations. The most likely number of clusters (k) was indicated by Δk, the change rate of the estimated log probability of the data (LnP[D]) [22].

Gene function blast

EST-SSRs associated unigene sequences were blasted against the GenBank non-redundant (nr) protein database using BLASTX (http://www.ncbi.nlm.nih.gov/BLAST) with an expected value (E-value) of 10^{-10} for the function of polymorphic EST-SSRs.

Results

Characterization of polymorphic SSRs

In total, 69 SSR primer pairs, including 49 (26% out of 188) EST-SSRs and 20 (49% out of 41) genomic SSRs (Tables 1 and 2), showed polymorphism among 96 accessions. A total of 213 alleles were generated from 69 loci with an average of 3.14 alleles per locus. The ratio of the EST-SSR repeat motifs was not equally distributed. The di-, tri-, tetra-, and penta-nucleotides accounted for 16.32%, 40.82%, 26.53%, and 16.32%, respectively. Whilst most of the genomic SSRs selected were composed of dinucleotide repeats. According to the results of POPGENE for the 69 SSRs, the observed number of alleles per locus (Na) ranged from 2 to 6 (mean = 3.14) and the effective number of alleles per locus (Ne) varied from 1.09 to 4.54 (mean = 2.30). The average Na was 3.12 and 2.59 for wild and cultivated barley, respectively (Table 3). Besides this, the polymorphic information content (PIC) ranged

Table 2. Characterization of 20 genomic-SSR makers in barley.

Primer	SSR motif	Primer sequence (5′-3′) F/R	Expected size (bp)	Na	Ne	Ho	He	PIC
S40	(AT)29	ACACCTTCCCAGGACAATCC	182	6	4.54	0.022	0.7842	0.748
		CAGAGCACCGAAAAAGTCTGTA						
S22	(GT)13,(AG)19	AAGCTCTTTCTTGTATTCGTG	158	5	4.09	0.0526	0.7595	0.7162
		GTCCATACTCTTTAACATCCG						
S18	(CT)28	CTGGGATTGGATCACTCTAA	107	5	3.9	0.0211	0.7474	0.7016
		AAAACAAGTACTGAAAATAGGAGA						
S7	(AC)20	ATAGATCACCAAGTGAACCAC	177	5	3.49	0.0833	0.7175	0.6776
		GGTTATCACTGAGGCAAATAC						
S37	(CT)18	CCGACAACATGCTATGAAGC	131	5	3.35	0.0521	0.7049	0.6596
		CTGCAGCAAATACCCATGTG						
S2	(AC)7T(CA)15 (AT)9	CCATCAAAGTCCGGCTAG	215	4	3.32	0.0326	0.703	0.6504
		GTCGGGCCTCATACTGAC						
S11	(AG)15	TCCATGATGATGTGTGCATAGA	173	5	3.01	0.0909	0.672	0.6121
		CGGATCCCAACAAACACAC						
S4	(AT)6(AC)16	GCTATGGCGTACTATGTATGGTTG	173	4	3.04	0.0549	0.6749	0.6106
		TCACGATGAGGTATGATCAAAGA						
S41	(TG)8	AGTATGGGGAATTTATTTGG	136	4	2.79	0.0312	0.6455	0.5864
		GCTGCAAAGTATGACAATATG						
S25	(CT)24	TTTGTGACATCTCAAGAACAC	158	4	2.77	0.1889	0.6428	0.5845
		TGACAAACAAATAATCACAGG						
S38	(GA)17(GA)7	CTATCACACGACGCAACATG	169	5	2.73	0.5106	0.6376	0.5828
		CCTGAGAAAGAAAGCGCAAC						
S30	(GC)5GGG (GT)16	CAAATCAATCAAGAGGCC	153	3	2.74	0	0.6384	0.5615
		TTTGAAGTGAGACATTTCCA						
S21	(AG)7C(AG)30-(AG)6	GGGAACTTGCTAATGAAGAG	150	3	2.67	0	0.6284	0.5546
		AATGTAAGGGAGTGTCCATAG						
S19	(AG)19	CCCTAGCCTTCCTTGAAG	135	3	2.46	0.0316	0.5973	0.5292
		TTACTCAGCAATGGCACTAG						
S29	(GT)16	AGAATCAAGATCGACCAAAC	124	4	2.19	0.0233	0.5464	0.5027
		AAAAACATGAACCGATGAA						
S15	(CT)16	ATTCATCGATCTTGTATTAGTCC	174	3	2.16	0.0319	0.5391	0.4749
		ACATCATGTCGATCAAAGC						
S31	(CT)21	CTATTTTCTAATGCTTGGACC	149	3	2.18	0.0947	0.5437	0.4647
		TGTCTAGTTCATCATCATTGC						
S36	(CA)9	GGATTTTCTCAAGAACACTT	239	3	2.13	0.766	0.5324	0.4597
		GCGTGAGTGCATAACATT						
S1	(AC)11	GTCCTTTACGCATGAACCGT	138	3	2.1	0.0316	0.5256	0.4547
		ACATACGCCAGACTCGTGTG						
S8	(AC)13(AT)9	GCTCTCTCTCAGAAAAATGAA	177	3	1.63	0.0444	0.3899	0.3492
		GAATTATTCTAGGGCTGTGAA						

from 0.08 to 0.75 with a mean of 0.46, and the PIC of wild barley was higher than that of cultivars with 0.44 *vs* 0.37. The expected heterozygosity (He) ranged from 0.0854 to 0.7842 with an average of 0.5279, while the observed heterozygosity (Ho) ranged from 0 to 0.766 with an average of 0.1677. As an indicator of genetic diversity, the average He was 0.5098 in wild barley accessions and 0.4333 in cultivated accessions.

Gene functions of the 49 unigene sequences containing polymorphic EST-SSRs

Functions of the 49 polymorphic EST-SSRs were determined and 47 unigenes showed significant similarities to the known genes (Table 4), for instance, zinc finger protein MAGPIE, transcription factor LAF1, photosystem II reaction center PSB28 protein, xyloglucan endotransglycosylase (XET), and protein kinase

Table 3. Polymorphism of SSR makers in Tibetan wild and cultivated barley.

	No. of alleles		PIC		He			No. of alleles		PIC		He	
Marker	Wild	Cultivated	Wild	Cultivated	Wild	Cultivated	Marker	Wild	Cultivated	Wild	Cultivated	Wild	Cultivated
P3	4	2	0.450	0.156	0.5259	0.1754	P129	4	4	0.581	0.658	0.6544	0.7359
P8	3	3	0.582	0.482	0.6597	0.5565	P137	2	1	0.374	0	0.5019	0
P9	4	2	0.467	0.110	0.5454	0.1210	P139	2	2	0.332	0.366	0.4226	0.4980
P13	2	2	0.271	0.371	0.3251	0.5081	P144	5	2	0.523	0.346	0.5553	0.4598
P16	3	3	0.256	0.375	0.2757	0.4456	P149	3	2	0.495	0.332	0.5609	0.4345
P21	2	1	0.110	0	0.1179	0	P150	3	3	0.460	0.456	0.5357	0.5701
P29	3	3	0.484	0.520	0.5637	0.6048	P152	3	3	0.442	0.450	0.5048	0.5222
P30	3	3	0.431	0.398	0.4908	0.4758	P168	4	2	0.571	0.258	0.6388	0.3145
P32	3	2	0.542	0.315	0.6201	0.4046	P180	2	1	0.374	0	0.5019	0
P34	3	2	0.507	0.366	0.5878	0.4980	P181	4	3	0.618	0.478	0.6731	0.5544
P44	2	2	0.075	0.110	0.0783	0.1210	P184	4	3	0.567	0.468	0.6177	0.5484
P45	3	3	0.365	0.294	0.4623	0.3306	P185	2	2	0.229	0.283	0.2653	0.3528
P50	3	2	0.545	0.305	0.6285	0.3871	P186	2	2	0.345	0.195	0.4458	0.2258
P53	2	2	0.288	0.374	0.3503	0.5149	P187	2	2	0.347	0.110	0.4500	0.1210
P54	2	2	0.372	0.258	0.4981	0.3145	S1	3	2	0.488	0.195	0.5651	0.2258
P55	3	2	0.382	0.359	0.4237	0.4839	S2	4	3	0.655	0.440	0.7106	0.5425
P56	2	2	0.371	0.359	0.4953	0.4839	S4	4	3	0.615	0.561	0.6740	0.6587
P57	2	2	0.347	0.283	0.4500	0.3528	S7	4	2	0.583	0.359	0.6318	0.4839
P61	2	2	0.280	0.195	0.3389	0.2258	S8	3	3	0.275	0.528	0.2988	0.6323
P63	3	2	0.475	0.323	0.5386	0.4173	S11	5	3	0.640	0.327	0.6996	0.3730
P66	2	2	0.364	0.258	0.4821	0.3145	S15	3	3	0.428	0.563	0.4927	0.6621
P67	2	2	0.375	0.305	0.5024	0.3871	S18	5	4	0.704	0.592	0.7511	0.6694
P80	2	2	0.372	0.323	0.4981	0.4173	S19	3	3	0.457	0.412	0.5197	0.4966
P81	2	2	0.248	0.195	0.2916	0.2258	S21	3	3	0.521	0.460	0.5902	0.5652
P83	4	3	0.473	0.438	0.5550	0.5423	S22	5	5	0.683	0.687	0.7344	0.7581
P90	3	3	0.378	0.544	0.4479	0.6371	S25	4	4	0.560	0.515	0.6307	0.6000
P91	3	3	0.587	0.327	0.6650	0.3730	S29	4	4	0.429	0.607	0.4713	0.7059
P99	3	2	0.549	0.305	0.6223	0.3871	S30	3	3	0.528	0.354	0.5985	0.4113
P100	2	2	0.110	0.337	0.1179	0.4435	S31	3	2	0.499	0.110	0.5846	0.1210
P101	3	2	0.499	0.315	0.5637	0.4046	S36	3	3	0.423	0.548	0.5021	0.6414
P103	3	2	0.470	0.371	0.5252	0.5081	S37	5	3	0.584	0.555	0.6318	0.6452
P105	3	2	0.380	0.195	0.4634	0.2258	S38	5	4	0.510	0.626	0.5732	0.7011
P106	2	2	0.329	0.349	0.4185	0.4657	S40	5	3	0.694	0.363	0.7422	0.4203
P120	2	2	0.159	0.359	0.1749	0.4839	S41	4	4	0.530	0.483	0.5809	0.5565
P121	4	3	0.485	0.367	0.5584	0.4529							
							Average	**3.12**	**2.59**	**0.441**	**0.373**	**0.5098**	**0.4333**

APK1B. In addition, the results revealed that the most annotated proteins were from *Triticum urartu* (17, 36.2%), and the species *Hordeum vulgare* and *Aegilops tauschii* accounted for the same percentage (11, 23.4%).

Population structure and genetic distance

To detect the population structure in the 96 barley genotypes, we performed STRUCTURE program for Bayesian clustering analysis using 69 SSR markers, assuming that the number of populations (K) ranged from 1 to 15. The highest log likelihood score (Δk) was at K = 8 (Figure 1A), indicating that the most suitable number of subpopulations was eight. The frequency of each accession assigned to a subpopulation was shown in Table S1. If the threshold of frequency was set at 0.5, only six accessions were defined as admixed. However, about 80% of the accessions can be derived from the subpopulations when the threshold was at 0.7. The output of structure analysis demonstrated that wild and cultivated barleys were assigned to different subpopulations (Figure 1B). Most of the cultivated barleys were classified into the subpopulation 4, except for A74, Tadmor, B1342 and B1031. Fifty percent of the wild barley accessions studied were assigned to subpopulation 1.

According to the values of genetic distance of the eight subpopulations, we get the dendrogram showing the genetic

Table 4. The putative proteins identified by BLASTX of 49 unigene sequences containing polymorphic EST-SSRs.

Primer	Accession No.	Putative protein	Organism	E-value
P181	CA032876.1	Hypothetical protein TRIUR3_30088	*Triticum urartu*	4.00E-51
P129	CV063130.1	Putative SKP1 protein	*T.aestivum*	1.00E-77
P184	CB858539.1	Hypothetical protein TRIUR3_19075	*T.urartu*	1.00E-46
P50	DN178534.1	UCW116, putative lipase	*H. vulgare* subsp. *vulgare*	3.00E-125
P91	FD524685.1	Putative syntaxin-131	*Aegilops tauschii*	1.00E-93
P8	AL506646.1	Zinc finger protein MAGPIE	*T.urartu*	4.00E-41
P29	AV943994.1	RNA polymerase sigma factor rpoD	*T.urartu*	7.00E-116
P103	CA009356.1	GID1-like gibberellin receptor	*H. vulgare* subsp. *vulgare*	4.00E-04
P32	EX593207.1	Disease resistance protein RGA2	*Aegilops tauschii*	8.70E-02
P168	BU997138.1	Hypothetical protein TRIUR3_09517	*T.urartu*	1.00E-04
P99	GH218162.1	Two-component response regulator ARR9	*T.urartu*	2.00E-64
P152	AV938130.1	Predicted protein	*H. vulgare* subsp. *vulgare*	1.10E-01
P144	EX598444.1	No hit	-	-
P121	CK569829.1	ACC oxidase	*H. vulgare*	9.00E-74
P34	DN186304.1	Predicted: UDP-glucose 6-dehydrogenase-like	*Brachypodium distachyon*	5.00E-65
P101	GH223749.1	FT-like protein	*H. vulgare* subsp. *vulgare*	1.00E-45
P149	EX583185.1	Condensin-2 complex subunit G2	*T.urartu*	5.00E-54
P150	FD519288.1	Curcuminoid synthase	*T.urartu*	5.00E-59
P83	FD527549.1	Putative pectinesterase 53	*Aegilops tauschii*	1.00E-76
P30	DN177250.1	Hypothetical protein F775_31773	*Aegilops tauschii*	1.00E-05
P63	EX577085.1	Condensin-2 complex subunit G2	*T.urartu*	6.00E-69
P90	FD528427.1	Photosystem II reaction center PSB28 protein	*T.urartu*	2.00E-83
P9	AL505258.1	Hypothetical protein f775_27232	*Aegilops tauschii*	6.00E-113
P3	BJ547928.1	Hypothetical protein TRIUR3_27885	*T.urartu*	1.00E-113
P45	FD523777.1	Hypothetical protein OsI_14737	*Oryza sativa* Indica Group	3.00E-50
P55	AL505545.1	No hit	-	-
P105	CA014373.1	Eukaryotic translation initiation factor 1A	*Leymus chinensis*	5.00E-72
P56	EX584572.1	Hypothetical protein F775_08651	*Aegilops tauschii*	2.00E-37
P66	FD518055.1	Predicted: protein LOC100843116	*B.distachyon*	5.00E-51
P67	FD520223.1	Hypothetical protein TRIUR3_27901	*T.urartu*	8.00E-36
P180	CA030489.1	Hypothetical protein TRIUR3_23016	*T.urartu*	4.00E-73
P80	FD523499.1	Casein kinase I-2-like protein	*A.tauschii*	1.00E-75
P57	EX599270.1	Hypothetical protein ZEAMMB73_419738	*Zea mays*	7.00E-56
P54	AL500476.1	PM2	*H. vulgare* subsp. *vulgare*	5.00E-67
P137	DN180922.1	PREDICTED: protein LOC100846358	*B.distachyon*	2.00E-02
P106	CA031374.1	OSJNBa0074L08.11	*Oryza sativa* Japonica Group	1.00E-46
P139	AL501810.1	GDSL esterase/lipase	*A.tauschii*	3.00E-40
P186	CB864664.1	Protein kinase APK1B, chloroplastic	*A.tauschii*	4.00E-50
P187	CB864737.1	Inactive ubiquitin carboxyl-terminal hydrolase 54	*T.urartu*	1.00E-17
P53	EH090859.1	TBC1 domain family member 15	*A.tauschii*	2.00E-57
P13	CK569261.1	Hypothetical protein TRIUR3_25268	*T.urartu*	3.50E-01
P16	CB873886.1	Phospholipid transfer protein precursor	*H. vulgare* subsp. *vulgare*	2.00E-43
P61	EX573461.1	Predicted protein	*H. vulgare* subsp. *vulgare*	6.00E-60
P81	FD521065.1	Predicted protein	*H. vulgare* subsp. *vulgare*	1.00E-81
P185	CB860073.1	Peptide transporter PTR2	*A.tauschii*	5.00E-60
P120	CK569159.1	Xyloglucan endotransglycosylase (XET)	*H. vulgare* subsp. *vulgare*	5.00E-69
P100	GH216950.1	Rho GDP-dissociation inhibitor 1	*T.urartu*	7.00E-69
P21	CK122115.1	Predicted protein	*H. vulgare* subsp. *vulgare*	5.00E-116
P44	CV063055.1	Transcription factor LAF1	*T.urartu*	3.00E-70

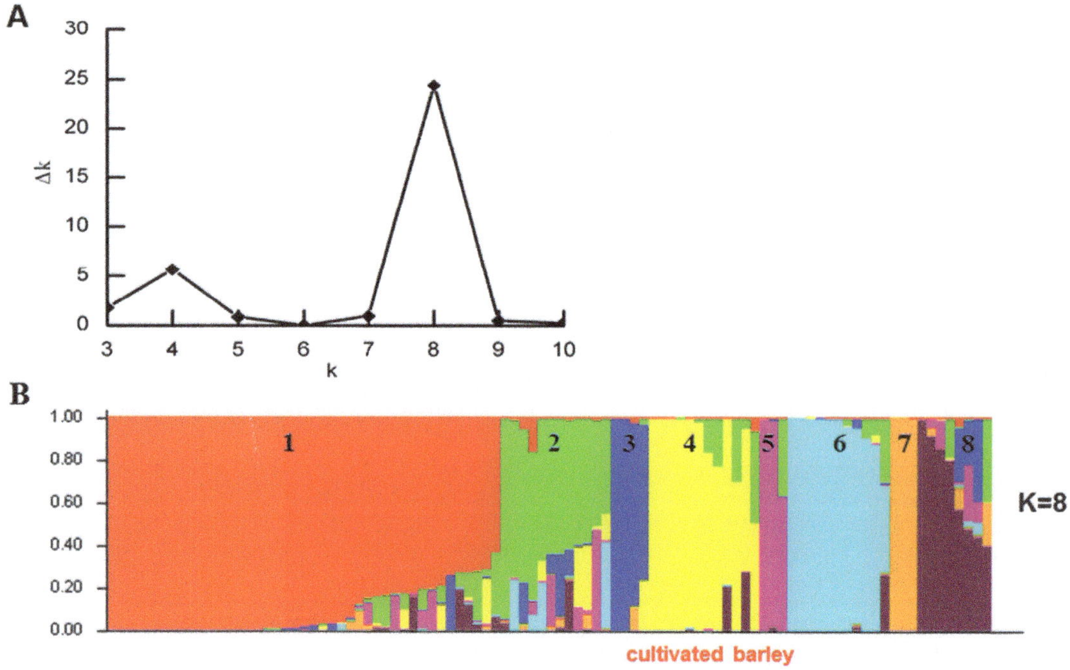

Figure 1. Δk and population structure. Estimation of the likelihood of clusters (k) for the most appropriate subpopulations (Δk) (A), and the population structure of 96 barley accessions in k = 8 clusters (B).

relationship of the subpopulations via UPGMA clustering analysis (Figure 2). The dendrogram showed that the subpopulation 3 was most close to the cultivated barleys (subpopulation 4) with the genetic distance of 132.188. The subpopulation 7 had the largest genetic distance (165.167) with the cultivated subpopulation.

Discussion

In recent years, different kinds of molecular markers have been used widely, including marker-assisted breeding, study of genetic relationships between populations, and screening candidate genes associated with the target traits [23]. The simple sequence repeats (SSRs) are increasingly important due to their high polymorphism and convenient techniques. However, EST-SSRs are superior to genomic SSRs for their transcriptional sequence and suitable application in cross-species [24]. In the present study, we developed 49 EST-SSR and 20 genomic SSR markers for wild barley. These novel EST-derived markers will be a valuable resource for tagging and mapping of genes related to agronomic and stress-resistant traits of interest. In addition, these markers are advantageous for identifying functional diversity of unique adaptive germplasm because of their genic function.

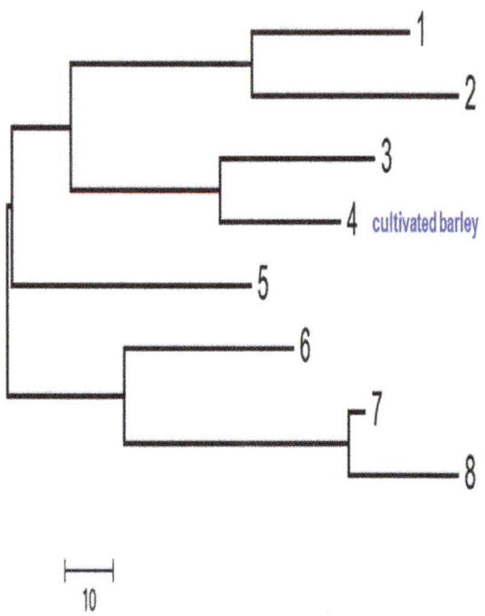

Figure 2. The dendrogram of the eight subpopulations according to the genetic distance using UPGMA clustering analysis.

In many plants, the di- and tri-nucleotides repeat motifs were the major types, but the predominant motifs were different in various species [25,26]. In our research, the tri-meric repeats were the most abundant motifs (40.82%), followed by the tetra-meric repeats accounted for 26.53%, and the di-meric and penta-meric repeat motifs were at the same frequency (16.32%).The polymorphism of SSRs can be divided into three degrees: high (PIC>0.5), medium (0.5>PIC>0.25) or low (PIC<0.25) [27]. In our study, the genetic diversity of genomic SSRs was higher than the EST-SSRs, with the mean PIC value of 0.57 (high) and 0.41 (medium), respectively, resulting in the general medium polymorphism (mean = 0.46). This finding was in line with previous results, and the lower level of polymorphism of EST-SSRs might be due to the selection against the variation in the conserved regions of the EST-SSRs [28]. Moreover, the expected levels of heterozygosity at EST-SSRs were also not as high as that of genomic SSRs, ranging from 0.0854 to 0.697 *vs* 0.3899 to 0.7842. Pompanon et al. [29] contributed the deficiency of heterozygosity to the primer problems, the deletion of alleles and appearance of invalid alleles at the annealing points.

Studies of the genetic variation in barley suggested that Tibetan wild barley showed higher polymorphism than cultivated barley [30–32]. The results of our study were consistent with the previous studies. The number of alleles and the PIC of wild barley were both higher than that of cultivated barley, being 3.12 *vs* 2.59 and 0.44 *vs* 0.37. The expected heterozygosity (He) showed the same trend, with 0.5098 and 0.4333 for wild and cultivated barley, respectively. The richness of genetic diversity in Tibetan wild barley may be the source of novel genes contributing to the tolerance of biotic and abiotic stresses, which is important in the barley breeding.

BLASTX analysis indicated that 47 (96%) of the 49 unigenes containing EST-SSRs can be matched to at least one important proteins in the NCBI nr protein database. For futher study, we can search the candidate genes of interest via association analysis referring to the function of markers in the metabolism pathways. Furthermore, these EST-SSR markers can be utilized as affirmative markers for comparative studies in the related species, for example, *Triticum urartu* and *Aegilops tauschii*.

In the present investigation, the findings of population structure analysis demonstrated that the developed EST-SSRs and genomic SSRs could distinguish between the cultivated and wild barley genotypes clearly. The 96 genotypes were divided to eight subpopulations. The subpopulation 3 (XZ161, XZ163, XZ165, XZ168) was most closely related to the cultivated barley (subpopulation 4), and the subpopulation 7 (XZ120, XZ151, XZ153) and the cultivated barleys were two most genetically distant populations. The genetic relation of the subpopulations suggested that the subpopulation 3 contained the most domesticated genotypes among the studied wild barley. Futhermore, the other subpopulations of wild barley, especially subpopulation 7, may be the important germplasm resource for the improvement of cultivars tolerant of abiotic and biotic stresses. These results were consistent with recent clustering studies in the Tibetan wild barley genotype using DArT markers and SNPs[3]. This indicates that the cluster analysis using EST-SSR and SSR markers is an effective way to determine the structure of populations and can constitute a solid foundation for the genetic variation study.

Conclusion

The 49 novel EST-SSRs and 20 genomic SSR markers developed from 96 barley genotypes were highly polymorphic and could be employed to examine genetic diversity, evolution, linkage mapping, comparative genomics, and population structure. The Tibetan wild barley showed higher genetic variation than cultivated barley, and the cultivated subpopulation could be separated from the wild barley clearly. For further studies, these developed markers could be useful in identifying trait-marker association of interest in the marker-assisted breeding programs in barley.

Author Contributions

Conceived and designed the experiments: FBW MZ. Performed the experiments: MZ. Analyzed the data: MZ WHM. Contributed reagents/materials/analysis tools: FBW GPZ WHM. Wrote the paper: MZ FBW GPZ WHM.

References

1. Russell J, Booth A, Fuller J, Harrower B, Hedley P, et al. (2004) A comparison of sequence-based polymorphism and haplotype content in transcribed and anonymous regions of the barley genome. Genome 47:389–398.
2. Nevo E, Apelbaum-Elkaher I, Garty J, Beiles A (1997) Natural selection causes microscale allozyme diversity in wild barley and lichen at 'Evolution Canyon' Mt Carmel Israel. Heredity 78: 373–382.
3. Dai F, Nevo E, Wu DZ, Comadran J, Zhou MX, et al. (2012) Tibet is one of the centers of domestication of cultivated barley. Proc Natl Acad Sci USA 109: 16969–16973.
4. Ren XF, Nevo E, Sun DF, Sun GL (2013) Tibet as a Potential Domestication Center of Cultivated Barley of China. PloS One 8: e62700.
5. Zhang QF, Dai XK (1992) Comparative assessment of genetic variation at 6 isozyme loci in barley from two centers of diversity: Ethiopia and Tibet. Acta Genet Sin 19: 236–243.
6. Zhao J, Sun HY, Dai HX, Zhang GP, Wu FB (2010) Difference in response to drought stress among Tibet wild barley genotypes. Euphytica 172: 395–403.
7. Dai HX, Shan WN, Zhao J, Zhang GP, Li CD, et al. (2011) Difference in response to aluminum stress among Tibetan wild barley genotypes. J Plant Nutr Soil Sc 174: 952–960.
8. Wu DZ, Qiu L, Xu LL, Ye LZ, Chen MX, et al. (2011) Genetic variation of HvCBF genes and their association with salinity tolerance in Tibetan annual wild barley. PloS One 6: e22938.
9. Michalek W, Weschke W, Pleissner KP, Graner A (2002) EST analysis in barley defines a unigene set comprising 4,000 genes. Theor Appl Genet 104:97–103.
10. Varshney RK, Graner A, Sorrells ME (2005) Genic microsatellite markers in plants: features and applications. Trends Biotechnol 23: 48–55.
11. Zeng SH, Xiao G, Guo J, Fei ZJ, Xu YQ, et al. (2010) Development of a EST dataset and characterization of EST-SSRs in a traditional Chinese medicinal plant, *Epimedium sagittatum* (Sieb. Et Zucc.) maxim. BMC Genomics 11: 94–104.
12. Li M, Zhu L, Zhou CY, Lin L, Fan YJ, et al. (2012) Development and characterization of EST-SSR markers from *Scapharca broughtonii* and their transferability in *Scapharca subcrenata* and *Tegillarca granosa*. Molecules 17: 10716–10723.
13. Jaikishan I, Rajendrakumar P, Ramesha MS, Viraktamath BC, Balachandran SM, et al. (2010) Prediction of heterosis for grain yield in rice using 'key' informative EST-SSR markers. Plant Breeding 129: 108–111.
14. Mohan A, Goyal A, Singh R, Balyan HS, Gupta PK (2006) Physical mapping of wheat and rye expressed sequence tag-simple sequence repeats on wheat chromosomes. Crop Sci 47(S_1): S3–S13.
15. Tang J, Gao L, Cao Y, Jia J (2006) Homologous analysis of SSR-ESTs and transferability of wheat SSR-EST markers across barley, rice and maize. Euphytica 151: 87–93.
16. Li L, Wang J, Guo Y, Jiang F, Xu Y (2008) Development of SSR markers from ESTs of gramineous species and their chromosome location on wheat. Prog Nat Sci 18: 1485–1490.
17. Castillo A, Budak H, Varshney RK, Dorado G, Graner A (2008) Transferability and polymorphism of barley EST-SSR markers used for phylogenetic analysis in Hordeum chilense. BMC Plant Biol 8: 97.
18. Willsmore KL, Eckermann P, Varshney RK, Graner A, Langridge P (2006) New eSSR and gSSR markers added to Australian barley maps. Crop Pasture Sci 57: 953–959.
19. Breseghello F, Sorrells ME (2006) Association mapping of kernel size and milling in wheat (*Triticum aestivum* L.) cultivars. Genetics 172:1165–1177.
20. Yeh FC, Yand RC, Boyle T (1999) POPGENE (Version 1.31): Microsoft Window-bases freeware for population genetic analysis, University of Alberta and the Centre for International Forestry Research.
21. Pritchard JK, Stephens M, Donnelly P (2000) Inference of population structure using multilocus genotype data. Genetics 155: 945–959.
22. Evanno G, Regnaut S, Goudet J (2005) Detecting the number of clusters of individuals using the software STRUCTURE: a simulation study. Mol Ecol 14: 2611–2620.
23. Gupta PK, Rustgi S (2004) Molecular markers from the transcribed/expressed region of the genome in higher plants. Funct Integr Genomics 4: 139–162.
24. Mian MA, Saha MC, Hopskins AA, Wang ZY (2005) Use of tall fescue EST-SSR markers in phylogenetic analysis of cool-season for age grasses. Genome 48: 637–647.
25. Varshney RK, Thiel T, Stein N, Langridge P, Graner A (2002) In silico analysis on frequency and distribution of microsatellites in ESTs of some cereal species. Cell Mol Biol Lett 7:537–546.
26. Kumpatla SP, Mukhopadhyay S (2005) Mining and survey of simple sequence repeats in expressed sequence tags of dicotyledonous species. Genome 48:985–998.
27. Xie WG, Zhang XQ, Cai HW, Liu W, Peng Y (2010) Genetic diversity analysis and transferability of cereal EST-SSR markers to orchardgrass (*Dactylis glomerata* L.). Biochem Syst Ecol 38: 740–749.
28. Scott KD, Eggler P, Seaton G, Rossetto M, Ablett EM, et al. (2000) Analysis of SSRs derived from grape ESTs. Theor Appl Genet 100:723–726.
29. Pompanon F, Bonin A, Bellemain E, Taberlet P (2005) Genotyping errors: causes, consequences and solutions. Nat Rev Genet 6:847–859.

30. Ellis RP, Forster BP, Robinson D, Handley LL, Gordon DC, et al. (2000) Wild barley: a source of genes for crop improvement in the 21st century. J Exp Bot 51: 9–17.
31. Jin XL, Cai SG, Han Y, Wang J, Wei K, et al. (2011) Genetic variants of *HvGlb1* in Tibetan annual wild barley and cultivated barley and their correlation with malt quality. J Cereal Sci 53: 59–64.
32. Sun DF, Ren WB, Sun GL, Peng JH (2011) Molecular diversity and association mapping of quantitative traits in Tibetan wild and worldwide originated barley (*Hordeum vulgare* L.) germplasm. Euphytica 178: 31–43.

9

Expression of Cry1Ab and Cry2Ab by a Polycistronic Transgene with a Self-Cleavage Peptide in Rice

Qichao Zhao[1]ⓐ, Minghong Liu[1]ⓐ, Miaomiao Tan[1], Jianhua Gao[2], Zhicheng Shen[1]*

1 State Key Laboratory of Rice Biology, Institute of Insect Sciences, Zhejiang University, Hangzhou, China, **2** College of Life Science, Shanxi Agricultural University, Taigu, China

Abstract

Insect resistance to *Bacillus thuringiensis* (Bt) crystal protein is a major threat to the long-term use of transgenic Bt crops. Gene stacking is a readily deployable strategy to delay the development of insect resistance while it may also broaden insecticidal spectrum. Here, we report the creation of transgenic rice expressing discrete Cry1Ab and Cry2Ab simultaneously from a single expression cassette using 2A self-cleaving peptides, which are autonomous elements from virus guiding the polycistronic viral gene expression in eukaryotes. The synthetic coding sequences of Cry1Ab and Cry2Ab, linked by the coding sequence of a 2A peptide from either foot and mouth disease virus or porcine teschovirus-1, regardless of order, were all expressed as discrete Cry1Ab and Cry2Ab at high levels in the transgenic rice. Insect bioassays demonstrated that the transgenic plants were highly resistant to lepidopteran pests. This study suggested that 2A peptide can be utilized to express multiple Bt genes at high levels in transgenic crops.

Editor: Mario Soberón, Instituto de Biotecnología, Universidad Nacional Autónoma de México, Mexico

Funding: This research was supported by The National Natural Science Foundation of China (31021003, http://www.nsfc.gov.cn/) received by Zc Shen. The funders had no role in study design,data collection and analysis, decision to publish, or preparation of the manuscript.

Competing Interests: The authors have declared that no competing interests exist.

* Email: zcshen@zju.edu.cn

ⓐ These authors contributed equally to this work.

Introduction

Since first demonstrated in transgenic tobacco in 1987, Bt crystal toxin genes have been widely utilized in transgenic crops for pest management [1,2]. However, due to the widespread application of Bt toxins, several major insect pests, including fall armyworm (*Spodoptera frugiperda*), dimondback moth (*Plutella xylostella*), maize stalk borer (*Busseola fusca*), cotton bollworm (*Helicoverpa armigera*), western corn rootworm (*Diabrotica virgifera*) and cabbage looper (*Trichoplusia ni*) have developed resistance to Bt toxins in field or greenhouse, which threatens the long-term utilization of Bt crops in the future [3–5].

Several strategies have been proposed and/or deployed to cope with the development of insect resistance to Bt toxins, including high-dose/refuge strategy, discovery of novel insecticidal genes with novel modes of actions, modification of used Bt genes and Bt gene stacking [6–8]. Bt gene stacking strategy introduces more than one Bt genes into plant and has been demonstrated to be an effective way to delay the development of insect resistance to Bt toxins [9–11]. Expression of multiple genes using conventional approaches has several potential limitations, most notably imbalanced expression among different genes and a large T-DNA size required to include multiple genes [12]. Efficient and stoichiometric expression of discrete proteins may be achieved by a polycistronic system involving self-cleaving peptides such as the 2A peptide from foot and mouth disease virus (F2A) or porcine teschovirus-1 (P2A) [12,13]. When expressed in eukaryotic system, nascent 2A peptide can interact with the ribosome exit tunnel to dictate a stop-codon-independent termination at the final proline codon of 2A peptide [14]. Subsequently translation is reinitiated on the same proline codon. When linked by a 2A peptide coding sequence, different genes are co-expressed from a single open reading frame.

2A self-cleaving peptides have been extensively studied previously [15]. In this study, the DNA encoding F2A or P2A was used to link two potent insecticidal Bt genes, the truncated *Cry1Ab* encoding the N-terminal 648 amino acids of active Cry1Ab endotoxin (Genbank:AAG16877.1) and the full-length *Cry2Ab* encoding Cry2Ab endoxin (Genbank:AAA22342.1) with 634 amino acid residues, to generate a polycistronic gene for co-expression in transgenic rice. Analysis of the obtained transgenic rice lines revealed that discrete Cry1Ab and Cry2Ab were indeed co-expressed at a level approximately comparable to transgenic plants expressing traditional monocistronic Bt genes. Insect bioassays demonstrated that the transgenic rice generated was highly resistant to its target insects.

Materials and Methods

Rice cultivar

Elite rice (*Oryza sativa spp. Japonica*) cultivar Xiushui 134 originated from the Jiaxing Academy of Agricultural Science in Zhejiang Province was used for *Agrobacterium*-mediated transformation. The homozygous transgenic rice line KMD1, containing a synthetic truncated *Cry1Ab* gene under control of maize ubiquitin promoter [16,17], was kindly provided by Dr. Gongyin Ye from Institute of Insect Sciences, ZheJiang University.

Construction of binary vector for rice transformation

pCambia1300 (Cambia, Canberra, Australia) was used for construction of binary vectors for plant transformation. This vector was first modified by substituting the hygromycin-resistant gene expression cassette with a glyphosate-tolerant *5-enolpyruvylshikimate-3-phosphate synthase* (*EPSPS*) gene expression cassette. The modified vector was named as pCambia1300–GLY, and was further used to clone an expression cassette for a polycistronic gene encoding Cry1Ab and Cry2Ab linked by a 2A peptide (Fig. 1).

Four polycistronic genes of *Cry1Ab* and *Cry2Ab* linked by the coding sequence of F2A or P2A were generated. They were named as *Cry1Ab-F2A-Cry2Ab* (*Cry1-F-Cry2*), *Cry1Ab-P2A-Cry2Ab* (*Cry1-P-Cry2*), *Cry2Ab-F2A-Cry1Ab* (*Cry2-F-Cry1*), and *Cry2Ab-P2A-Cry1Ab* (*Cry2-P-Cry1*), respectively (Fig. 1). *Cry1-F-Cry2* is a fusion gene, in the order from 5' to 3', of coding sequences of Cry1Ab, F2A, and Cry2Ab (GenBank: KJ716232); *Cry1-P-Cry2* is a fusion gene of Cry1Ab, P2A, and Cry2Ab (GenBank: KJ716233); *Cry2-F-Cry1* is a fusion gene of Cry2Ab, F2A, and Cry1Ab (GenBank: KJ716234); and *Cry2-P-Cry1* is a fusion gene of Cry2Ab, P2A, and Cry1Ab (GenBank: KJ716235).

Maize ubiquitin-1 promoter (pUbi) was obtained by PCR from maize genome with primers pUbi-F (5' attaagcttagcttgcatgcctacagtg 3', with *Hin*dIII restriction site underlined) and pUbi-R (5' taaggatccctctagagtcgacctgca 3', with *Bam*HI restriction site underlined). The pUbi fragment digested with *Hin*dIII and *Bam*HI, the polycistronic gene fragment digested with *Bam*HI and *Kpn*I and the vector 1300-GLY predigested with *Hin*dIII and *Kpn*I were ligated to generate plasmid p1300-Cry1-F-Cry2, p1300-Cry1-P-Cry2, p1300-Cry2-F-Cry1 and p1300-Cry2-P-Cry1, respectively (Fig. 1).

Agrobacterium-mediated transformation

The T-DNA plasmids were separately transformed into *Agrobacterium tumefaciens* LBA4404 by electroporation. The *Agrobacterium*-mediated transformation of rice was carried out according to Hiei Y. *et al.*, except that 2 mM glyphosate (Sigma-Aldrich, St. Louis, MO, USA) was used for selection [18].

RT-PCR analysis of transgenic rice

Total RNA was extracted from rice leaf with Trizol reagent (Invitrogen, Carlsbad, CA,USA). Concentration of RNA was determined by a spectrophotometer (Thermo Scientific, DE, USA). One μg RNA was treated with Dnase I and used as template to synthesize the first strand of cDNA using cDNA synthesis Kit (Thermo Scientific, DE, USA). Finally the synthesized cDNA was used as PCR template. The primers P1-2-F (5' CAGCGGCAACGAGGTGTACA 3') and P1-2-R (5' TAGGCGTCGCAGATGGTGGT 3') were used to amplify the joint parts in Cry1-F-Cry2 and Cry1-P-Cry2 (Fig. 2**a**). The PCR products were expected to be 321 bp. The primers P2-1-F (5' GCCGCTCGACATCAACGTGA 3') and P2-1-R (5' TCAGGCTGATGTCGATGGGG 3') were used to amplify the joint parts in Cry2-F-Cry1 and Cry2-P-Cry1 (Fig. 2**a**). The PCR products were expected to be 314 bp. The procedure for both PCRs was pre-denaturation at 95°C for 3 min; then 30 cycles of denaturation at 95°C for 40 s, annealing at 50°C for 30 s, and extension at 72°C for 30 s; finally followed with extension at 72°C for 5 min. The PCR products were analyzed by electrophoresis in 1% (w/v) agarose gel.

Western blot analysis

Leaf sample of 0.01 g collected from 1-month-old rice was frozen in liquid nitrogen for 45 s and disrupted with TissueLyserII (Qiagen, Hilden, Germany) at 40 Hz for 45 s. The sample was then homogenized in 200 μL 1×PBS Buffer by shaking vigorously for 30 s and centrifuged at 12000 rpm for 15 min at 4°C. The supernatants were collected as protein samples for western blot analysis. These samples were separated by electrophoresis on 8% polyacrylamide gel and transferred to PVDF membrane (Pall, Ann Arbor, MI, USA). The blotted membrane was blocked in 5% (w/v) skim milk in Tris Buffer Saline with 0.1% Tween-20 (TBST) for 1 h at room temperature. The membrane was then incubated with primary antibody against either Cry1Ab or Cry2Ab (diluted 1:1000 in 1% skim milk/TBST) for 1 h at room temperature with gentle shaking. The polyclonal antisera against Cry1Ab and Cry2Ab were prepared from New Zealand white rabbits immunized with purified recombinant Cry1Ab and Cry2Ab from *Escherichia coli*, respectively. After three times of 5 minute wash in TBST, the membrane was incubated with HRP-conjugated secondary antibody (Promega, Madison, WI, USA) diluted at

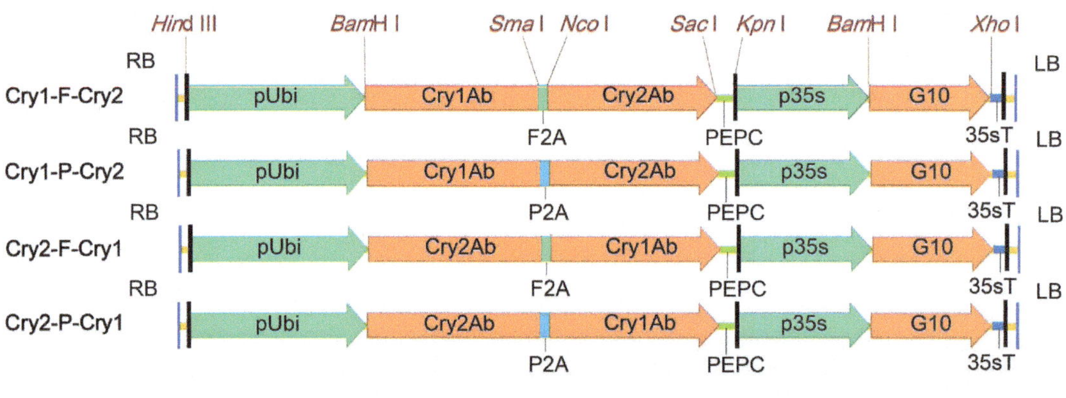

Figure 1. Schematic diagram of T-DNAs for transgenic expression of 2A linked polycistronic genes. Each T-DNA contained two expression cassettes, one for insecticidal polycistronic gene under control of maize ubiquitin-1 promoter (pUbi), the other for glyphosate-tolerant *5-enolpyruvylshikimate-3-phosphate synthase (EPSPS)* gene (G10) under control of Cauliflower mosaic virus 35s promoter (p35s). Cry1Ab and Cry2Ab,synthetic Bt inseciticidal gene; F2A and P2A, synthetic DNAs encoding Foot and Mouth Disease Virus 2A and Porcine teschovirus-1 2A, respectively; RB and LB, right border and left border of T-DNA; PEPC, maize *phosphoenolpyruvate carboxylase* gene terminator; 35sT, Cauliflower mosaic virus 35S gene terminator.

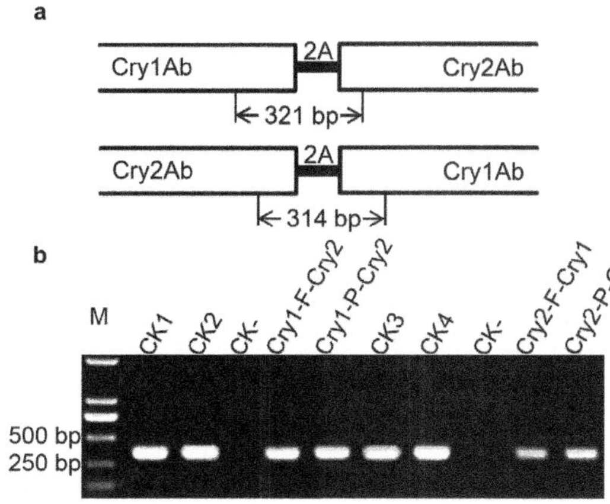

Figure 2. RT-PCR detection of mRNA integrity of the polycistronic genes. a, schematic diagram of the fragments amplified by RT-PCR; **b**, electrophoresis analysis of RT-PCR products. CK1, plasmid containing *Cry1-F-Cry2*; CK2, plasmid containing *Cry1-P-Cry2*; CK3, plasmid containing *Cry2-F-Cry1*; CK4, plasmid containing *Cry2-P-Cry1*; CK-, cDNA prepared from non-transgenic rice; M, DNA marker.

1:5000 with 1% skim milk/TBST for 1 h at room temperature with gentle shaking. After another three times of 5 minute wash in TBST, signals were visualized with DAB substrate (Sigma-Aldrich).

Protein quantification of Cry1Ab and Cry2Ab

Concentrations of Cry1Ab and Cry2Ab in the leaf of transgenic rice at tillering stage were determined by enzyme-linked immunosorbent assay (ELISA) using Envirologix kit AP003 and AP005 (Envirologix, Portland, ME, USA), respectively, according to the manufacturer's recommendation. For each construct, 3 transgenic lines were selected, and the samples prepared with the same method from non-transgenic rice were used to eliminate the basal absorption at 450 nm. After quantification of Cry1Ab and Cry2Ab, the molar ratio of the upstream protein to downstream protein of each transgenic line selected was calculated and analyzed by One-Way ANOVA [19] in SPSS.

Insect bioassay

Insect bioassays were conducted with cotton bollworm (Bt-susceptible and Cry1Ac-resistant), beet armyworm (*Spodoptera exigua*), striped stem borer (*Chilo suppressalis*) and rice leaf roller (*Cnaphalocrocis medinalis*). For beet armyworm and striped stem borer, detached leaf bioassay was carried out [20]. Two leaf blades from the same line were collected at 2-3 cm length at seedling stage and placed in a 70-mm-diameter petri dish lined with a pre-moistened filter paper. The leaf samples in the petri dish were infested with 10 newly hatched neonates. The petri dishes were then sealed with parafilm membrane and placed in dark at 28°C. The result was recorded after 3 days incubation. For cotton bollworm, detached leaf bioassay was carried out except that one leaf blade was placed in a petri dish and infested with one neonate to avoid cannibalism of cotton bollworm. Each line was infested with a total of 10 neonates. Eggs of Bt-susceptible cotton bollworm (CB-S), beet armyworm and striped stem borer were obtained from Genralpest Biotech (Genralpest Biotech, Bejing, China). Eggs of Cry1Ac-resistant cotton bollworm (CB-RR) were kindly provided by Dr. Kongming Wu from Institute of Plant Protection, Chinese Academy of Agricultural Sciences. For rice leaf roller, whole plant assay was carried out. Tillering-stage rice grown in greenhouse was infested with 10 newly hatched neonates. The result was recorded 2 weeks after infestation. Eggs of rice leaf roller used in the assays were obtained from caged moths collected from rice field. Each assay was repeated for 3 times.

Results

Transgene transcribed as a long intact mRNA

About 30 independent transgenic lines were generated via *Agrobacterium*-mediated transformation for each of the four constructs. To investigate whether the transcript of each polycistronic transgene is a long intact mRNA with sequences encoding both Cry1Ab and Cry2Ab, RT-PCR was used to detect the sequences in the area connecting the two Bt genes, which include sequences from both *Cry1Ab* and *Cry2Ab*, and the 2A peptide coding sequence in between (Fig. 2**a**). The RT-PCR products with expected sizes were detected clearly in the transgenic lines from all of the four constructs (Fig. 2**b**), suggesting that each of the polycistronic genes was transcribed into a long intact mRNA as expected.

Co-expression of discrete Cry1Ab and Cry2Ab from polycistronic transgenes

Transgenic lines from each construct were analyzed by western blot analysis using antisera against Cry1Ab and Cry2Ab,respectively. When detected with antiserum against Cry1Ab, a band of approximately 72 kD was detected among different transgenic lines, indicating that the Cry1Ab protein was expressed as a discrete protein rather than a fusion protein (Fig. 3 Upper panel). When detected by antiserum against Cry2Ab, a band of about 68~70 kD was detected among all the transgenic lines (Fig. 3 Lower panel), indicating a discrete Cry2Ab protein was expressed. A possible fusion protein of "Cry-2A-Cry" would be at a size of approximately 147 kD. However, no significant signal of proteins at such size was detected by antiserum against either Cry1Ab or Cry2Ab, suggesting that there was no significant amount of fusion protein expressed in the transgenic rice. The upstream Cry2Ab detected in the Cry2-F-Cry1 and the Cry2-P-Cry1 transgenic lines was slightly larger than the downstream Cry2Ab in the Cry1-F-Cry2 and the Cry1-P-Cry2. This increased size was likely due to the 2A peptide residues attached to the upstream protein during translation,which has been demonstrated in other 2A polycistronic transgenes [21].

Determination of gene expression level and 2A cleavage efficiency

To determine the expression levels of Bt genes with 2A polycistronic transgene and evaluate the cleavage efficiency of 2A peptide in each construct, the amounts of Cry1Ab and Cry2Ab expressed in the transgenic rice leaves were determined by ELISA. Three transgenic lines at tillering stage from each construct were selected. The expression levels of Cry1Ab and Cry2Ab were different among different lines as expected. The concentrations of the soluble Cry1Ab and Cry2Ab in the leaf were in the range of 0.67 to 1.82 μg/g and 0.69 to 2.31 μg/g of leaf fresh weight (LFW), respectively (Fig. 4**a**).

The molar ratio (R) of the upstream protein to the downstream protein of each construct was calculated to estimate the cleavage efficiency. The R value for polycistronic gene using F2A as 2A cleavage peptide was lower than that using P2A (Fig. 4**b**), indicating that the cleavage efficiency of F2A peptide was higher than P2A. The R value of polycistronic gene with Cry1Ab at the upstream was lower than that with Cry2Ab at the upstream (Fig. 4**b**), suggesting that the Cry1Ab at the upstream was likely

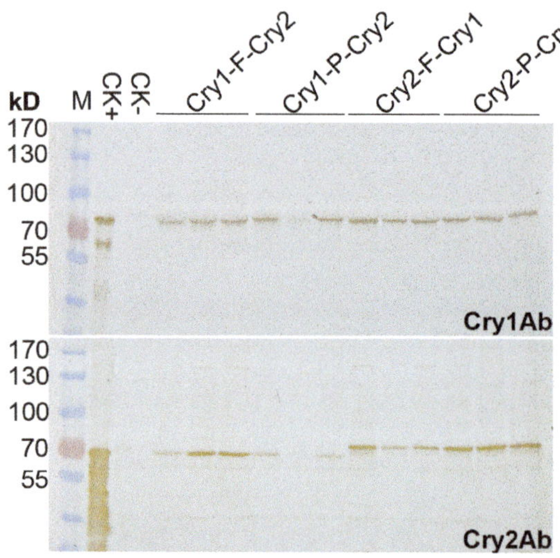

Figure 3. Western bolt analysis of Cry1Ab and Cry2Ab in transgenic rice. Three transgenic lines from each construct were selected for the analysis. Each sample was detected with antiserum against Cry1Ab (upper) and Cry2Ab (bottom) respectively. Cry1Ab or Cry2Ab protein expressed by *E.coli* was used as positive control (CK+). Sample prepared from non-transgenic rice was used as negative control (CK-). M, prestained protein ladder.

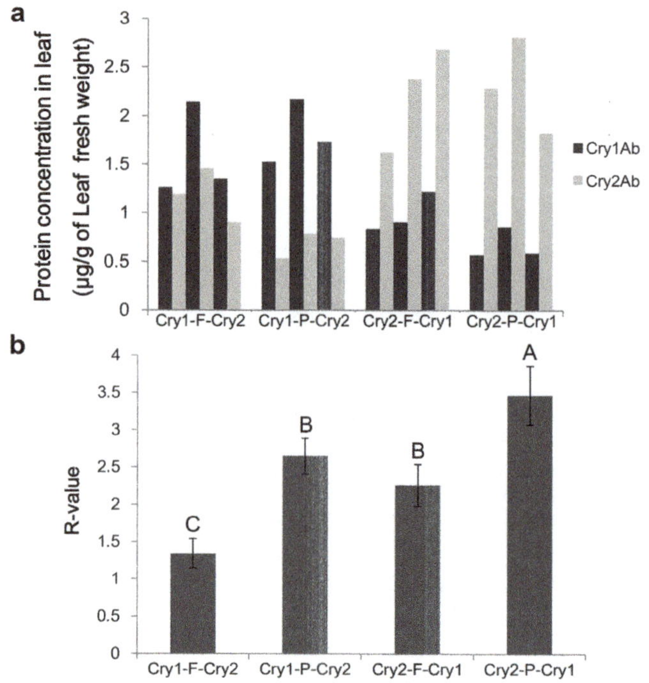

Figure 4. Cry1Ab and Cry2Ab concentrations in different transgenic rice lines and comparison of 2A cleavage efficiency. **a**, concentrations of Cry1Ab and Cry2Ab in leaves of 3 selected transgenic lines from each construct; **b**, R values of different 2A constructs. R value is the molar ratio of the upstream protein to the downstream protein. Different capital letters on each bar indicate extreme significant differences ($p<0.01$, Fisher's least-significant difference and Duncan's multiple range test).

more efficient in cleavage than the Cry2Ab at the upstream, regardless of whether F2A or P2A was the cleavage peptide in the synthetic construct. The transgenic lines from the construct of Cry1-F-Cry2 had the R value close to 1, indicating that the two Bt proteins were expressed almost equally (Fig. 4**a**).

Insect-resistant activity of transgenic rice

The transgenic rice lines were assayed for their insecticidal activity using neonates of CB-S, striped stem borer and rice leaf roller. Three transgenic lines from each construct were selected and age-matched non-transgenic rice plants were used as control. Mortalities of all the 3 insect species feeding on transgenic rice were 100% while those on the non-transgenic rice were much lower (Fig. 5). In the assays with CB-S and striped stem borer, the transgenic leaves were only slightly bitten by the insects while non-transgenic controls were severely consumed (Fig. 6**a**). In the assays with rice leaf roller, no rolled leaf was observed in the transgenic plant while several leaves were rolled in the non-transgenic plant (Fig. 6**b**). The bioassays demonstrated that transgenic rice plants generated from the 2A constructs were highly insect-resistant.

The transgenic rice lines were also assayed for their activity against beet armyworm and CB-RR. The KMD1 transgenic rice at the same growing stage was used as control. The mortalities of beet armyworm feeding on the 2A transgenic rice lines were obvious higher than that feeding on KMD1 ($p<0.01$) (Figure 7). The KMD1 showed a rather low activity to CB-RR, while the 2A transgenic rice showed much higher activity ($p<0.01$) (Figure 7).

Discussion

In this study, Cry1Ab and Cry2Ab, two potent Bt crystal proteins with different receptors in the insect midgut [22], were expressed by the polycistronic transgene using 2A peptide. To our best knowledge, this is the first application of 2A peptide for expressing two Bt genes in a transgenic crop. Utilization of two or more Bt genes simultaneously is desirable for management of insect resistance to Bt toxins as well as for broadening insecticidal spectrum of Bt crops. Self-cleaving 2A peptides and 2A-like sequences have also been found in insect virus, and some of them have high self-cleavage activity [23]. While there is no scientific base for any safety concern for using 2A peptide from mammalian virus, it may be better for public perception to use 2A peptides from insect virus for the development of transgenic rice.

Introduction of multiple genes into plant permits complex and sophisticated manipulation of traits in transgenic crops [2]. The number of genes being introduced into crops for genetic engineering is increasing steadily. For instance, the recently released commercial transgenic corn "Genuity SmartStax" (Monsanto, St Louis, MS, USA) has a total of 8 transgenes for insect-resistance and herbicide-tolerance. With more traits under research and development for transgenic improvement, more genes are expected to be utilized for future transgenic crops. Clearly polycistronic strategy using 2A peptide can simplify the process of gene stacking significantly, as it enables us to introduce multiple genes into plants using a single T-DNA with a single promoter [2]. Moreover, polycistronic strategy alleviates the concern of gene silencing induced by the insertion of homologous promoters or multiple T-DNA sequences into recipient genome [2,24]. Additionally, equal expressions of different genes could be achieved by the 2A polycistronic strategy [12,24].

The expression levels of Cry1Ab and Cry2Ab by 2A polycistronic transgene in this study were comparable to traditional monocistronic transgene. The Cry1Ab concentration of transgenic rice KMD1, which is highly resistant to striped stem borer and rice leaf roller [16], was 2.95 µg/g of LFW [25]. The Cry1C concentration in homozygous rice line T1c-19 harboring a

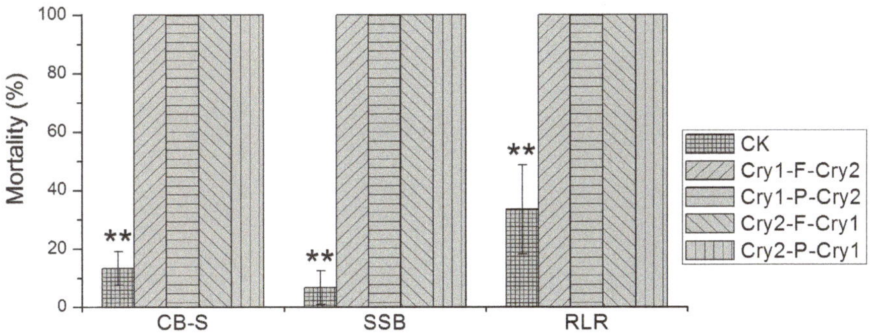

Figure 5. Mortality of Bt-susceptible cotton bollworm, striped stem borer and rice leaf roller feeding on transgenic rice plants. Non-transgenic rice at the same growing stage was used as control (CK). ** on the bar indicate extreme significant differences ($p<0.01$, Kruskal Wallis test). CB-S, Bt-susceptible cotton bollworm; SSB, striped stem borer; RLR, rice leaf roller.

synthetic Cry1C gene under control of maize ubiquitin promoter was 1.38 μg/g of LFW at the heading stage [26]. The Cry2A protein concentrations of the homozygous transgenic rice lines harboring a synthetic Cry2 gene driven by maize ubiquitin promoter was reported at the range from 9.65 to 12.11 μg/g of LFW [27]. In the transgenic *indica* rice line harboring both *Cry1Ac* and *Cry2Ab* gene, the Cry2Ab concentration was reported to be around 1 μg/g LFW [28]. Thus, the expression levels of the Cry1Ab and Cry2Ab in the transgenic rice lines obtained by this study were well within the range of the expression levels of the monocistronic transgene reported previously.

Although cotton bollworm was not a pest of rice, it has been long a good target insect for evaluating Bt toxicity. Moreover, cotton bollworm is a good model insect for the study of insect resistance to Bt toxins because of the availabilities of both sensitive and resistant lines to Cry1Ac. The 2A transgenic lines obtained in this study showed high insect-resistant activity against cotton bollworm, either sensitive or resistant to Cry1Ac, while the KMD1 transgenic rice had little activity toward Cry1Ac-resistant cotton bollworm. This suggested that little or no cross resistance existed between Cry1Ac and Cry2Ab and the 2A peptide based transgenic Bt rice will be useful for the management of Bt resistance developed by insect pests.

The 2A transgenic lines obtained in this study conferred significantly higher activity toward beet armyworm than the KMD1 transgenic rice containing Cry1Ab only [16], suggesting that the Cry2Ab expressed in the 2A transgenic lines indeed contributed to the enhancement of insecticidal activity. A similar phenomenon was also observed in transgenic cotton BollgardII. While Bollgard cotton only had moderate toxic activity to beet armyworm, BollgardII cotton showed greatly improved activity due to the addition of Cry2Ab [29].

The upstream proteins expressed with 2A polycistronic strategy are usually attached with the 2A peptide residues at the C-terminus [30]. The western blot analysis in this study appeared to agree with it. However, due to the limitation of the western blot analysis in determination of protein size, C-terminal amino acid sequencing and mass spectrum are required to confirm. While no adverse impact was observed for the utilization of 2A peptide in gene stacking [12], the Bt toxins attached with a 2A peptide will be considered as different proteins according to the regulation of transgenic crops, and thus additional safety studies will be required for commercialization of transgenic crops using 2A cleaving peptide.

Figure 6. Insect bioassay for transgenic rice with a polycistronic Bt transgene. a, bioassay with Bt-susceptible cotton bollworm (CB-S) and striped stem borer (SSB). **b,** bioassay with rice leaf roller. For each assay, age-matched non-transgenic rice was used as control (CK).

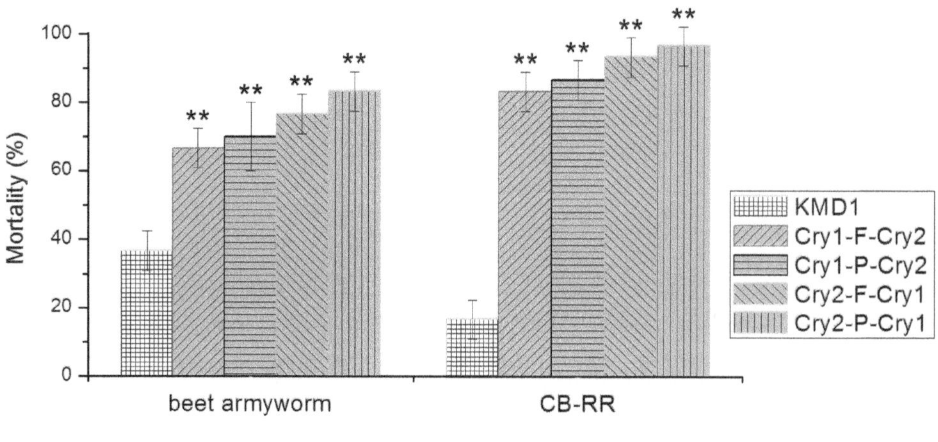

Figure 7. Mortality of beet armyworm and Cry1Ac-resistant contton bollworm feeding on KMD1 and various 2A transgenic rice lines. The transgenic rice KMD1 was used as control. ** on the bar indicate extreme significant difference ($p<0.01$, Fisher's least-significant difference and Duncan's multiple range test). CB-RR, Cry1Ac-resistant cotton bollworm.

Acknowledgments

We thank Kongming Wu and Gemei Liang for the supply of Cry1Ac-resistant cotton bollworm. We thank Gongyin Ye for the supply of KMD1 transgenic rice.

Author Contributions

Conceived and designed the experiments: QcZ MhL JHG ZcS. Performed the experiments: QcZ MhL MmT. Analyzed the data: QcZ MhL MmT. Contributed reagents/materials/analysis tools: QcZ MhL JHG. Wrote the paper: QcZ MhL JHG ZcS.

References

1. Romeis J, Meissle M, Bigler F (2006) Transgenic crops expressing Bacillus thuringiensis toxins and biological control. Nature Biotechnology 24: 63–71.
2. Halpin C (2005) Gene stacking in transgenic plants – the challenge for 21st century plant biotechnology. Plant Biotechnology Journal 3: 141–155.
3. Gassmann AJ, Petzold-Maxwell JL, Keweshan RS, Dunbar MW (2011) Field-Evolved Resistance to Bt Maize by Western Corn Rootworm. PLoS ONE 6.
4. Tabashnik BE, Van Rensburg JBJ, Carriere Y (2009) Field-Evolved Insect Resistance to Bt Crops: Definition, Theory, and Data. Journal of Economic Entomology 102: 2011–2025.
5. Wang P, Zhao J-Z, Rodrigo-Simon A, Kain W, Janmaat AF, et al. (2006) Mechanism of resistance to Bacillus thuringiensis toxin Cry1Ac in a greenhouse population of cabbage looper, Trichoplusia ni. Applied and Environmental Microbiology: 01834–01806.
6. Bates SL, Zhao JZ, Roush RT, Shelton AM (2005) Insect resistance management in GM crops: past, present and future. Nature Biotechnology 23: 57–62.
7. Shelton AM, Tang JD, Roush RT, Metz TD, Earle ED (2000) Field tests on managing resistance to Bt-engineered plants. Nature Biotechnology 18: 339–342.
8. Tabashnik BE, Carriere Y, Dennehy TJ, Morin S, Sisterson MS, et al. (2003) Insect resistance to transgenic Bt crops: Lessons from the laboratory and field. Journal of Economic Entomology 96: 1031–1038.
9. Cao J, Zhao JZ, Tang JD, Shelton AM, Earle ED (2002) Broccoli plants with pyramided cry1Ac and cry1C Bt genes control diamondback moths resistant to Cry1A and Cry1C proteins. Theoretical and Applied Genetics 105: 258–264.
10. Yang Z, Chen H, Tang W, Hua HX, Lin YJ (2011) Development and characterisation of transgenic rice expressing two Bacillus thuringiensis genes. Pest Management Science 67: 414–422.
11. Zhao JZ, Cao J, Li YX, Collins HL, Roush RT, et al. (2003) Transgenic plants expressing two Bacillus thuringiensis toxins delay insect resistance evolution. Nature Biotechnology 21: 1493–1497.
12. Szymczak-Workman AL, Vignali KM, Vignali DAA (2012) Design and Construction of 2A Peptide-Linked Multicistronic Vectors. Cold Spring Harbor Protocols 2012: pdb.ip067876.
13. Ha S-H, Liang YS, Jung H, Ahn M-J, Suh S-C, et al. (2010) Application of two bicistronic systems involving 2A and IRES sequences to the biosynthesis of carotenoids in rice endosperm. Plant Biotechnology Journal 8: 928–938.
14. Brown JD, Ryan MD (2010) Ribosome "Skipping": "Stop-Carry On" or "StopGo" Translation Recoding: Expansion of Decoding Rules Enriches Gene Expression. InAtkinsJFGestelandRFeditors: New York:Springerpp101–121.
15. Luke G, Escuin H, De Felipe P, Ryan M (2010) 2A to the fore - research, technology and applications. Biotechnology and Genetic Engineering Reviews 26: 223–260.
16. Shu QY, Ye GY, Cui HR, Cheng XY, Xiang YB, et al. (2000) Transgenic rice plants with a synthetic cry1Ab gene from Bacillus thuringiensis were highly resistant to eight lepidopteran rice pest species. Molecular Breeding 6: 433–439.
17. Ye GY, Shu QY, Yao HW, Cui HR, Cheng XY, et al. (2001) Field evaluation of resistance of transgenic rice containing a synthetic cry1Ab gene from Bacillus thuringiensis Berliner to two stem borers. Journal of Economic Entomology 94: 271–276.
18. Hiei Y, Komari T, Kubo T (1997) Transformation of rice mediated by Agrobacterium tumefaciens. Plant Molecular Biology 35: 205–218.
19. Heiberger R, Neuwirth E (2009) One-Way ANOVA. R Through Excel: New YorkSpringerpp165–191.
20. Chen Y, Tian J-C, Shen Z-C, Peng Y-F, Hu C, et al. (2010) Transgenic Rice Plants Expressing a Fused Protein of Cry1Ab/Vip3H Has Resistance to Rice Stem Borers Under Laboratory and Field Conditions. Journal of Economic Entomology 103: 1444–1453.
21. Halpin C, Cooke SE, Barakate A, Amrani AE, Ryan MD (1999) Self-processing 2A-polyproteins – a system for co-ordinate expression of multiple proteins in transgenic plants. The Plant Journal 17: 453–459.
22. Hernandez-Rodriguez CS, Hernandez-Martinez P, Van Rie J, Escriche B, Ferre J (2013) Shared Midgut Binding Sites for Cry1A. 105, Cry1Aa, Cry1Ab, Cry1Ac and Cry1Fa Proteins from Bacillus thuringiensis in Two Important Corn Pests, Ostrinia nubilalis and Spodoptera frugiperda. Plos One 8.
23. Ryan M (2014) 2A and 2A like sequences. Available: http://www.st-andrews.ac.uk/ryanlab/2A_2Alike.pdf/. Accessed 4 September 2014.
24. Sainsbury F, Benchabane M, Goulet M-C, Michaud D (2012) Multimodal Protein Constructs for Herbivore Insect Control. Toxins 4: 455–475.
25. Tian J-C, Chen Y, Li Z-L, Li K, Chen M, et al. (2012) Transgenic Cry1Ab Rice Does Not Impact Ecological Fitness and Predation of a Generalist Spider. PLoS ONE 7: e35164.
26. Tang W, Chen H, Xu C, Li X, Lin Y, et al. (2006) Development of insect-resistant transgenic indica rice with a synthetic cry1C* gene. Molecular Breeding 18: 1–10.
27. Chen H, Tang W, Xu C, Li X, Lin Y, et al. (2005) Transgenic indica rice plants harboring a synthetic cry2A* gene of Bacillus thuringiensis exhibit enhanced resistance against lepidopteran rice pests. Theoretical and Applied Genetics 111: 1330–1337.
28. Bashir K, Husnain T, Fatima T, Latif Z, Mehdi SA, et al. (2004) Field evaluation and risk assessment of transgenic indica basmati rice. Molecular Breeding 13: 301–312.
29. Perlak FJ, Oppenhuizen M, Gustafson K, Voth R, Sivasupramaniam S, et al. (2001) Development and commercial use of Bollgard cotton in the USA – early promises versus today's reality. The Plant Journal 27: 489–501.
30. Szymczak-Workman AL, Vignali KM, Vignali DAA (2012) Verification of 2A Peptide Cleavage. Cold Spring Harbor Protocols 2012: pdb.prot067892.

10

Global Wild Annual *Lens* Collection: A Potential Resource for Lentil Genetic Base Broadening and Yield Enhancement

Mohar Singh[1]*, Ishwari Singh Bisht[1], Sandeep Kumar[1], Manoranjan Dutta[1], Kailash Chander Bansal[1], Moreshwar Karale[1], Ashutosh Sarker[2], Ahmad Amri[3], Shiv Kumar[4], Swapan Kumar Datta[5]

1 National Bureau of Plant Genetic Resources, Pusa, New Delhi, India, **2** International Centre for Agricultural Research in Dry Areas, South Asia and China Regional Programme, NASC Complex, DPS Marg, Pusa, New Delhi, India, **3** International Centre for Agricultural Research in Dry Areas, Tunisia, Tunis, **4** International Centre for Agricultural Research in Dry Areas, Rabat, Morocco, **5** Division of Crop Science, Indian Council of Agricultural Research, Krishi Bhawan, New Delhi, India

Abstract

Crop wild relatives (CWRs) are invaluable gene sources for various traits of interest, yet these potential resources are themselves increasingly threatened by the impact of climate change as well as other anthropogenic and socio-economic factors. The prime goal of our research was to cover all aspects of wild *Lens* genetic resource management like species characterization, agro-morphological evaluation, diversity assessment, and development of representative sets for its enhanced utilization in lentil base broadening and yield improvement initiatives. We characterized and evaluated extensively, the global wild annual *Lens* taxa, originating from twenty seven counties under two agro-climatic conditions of India consecutively for three cropping seasons. Results on various qualitative and quantitative characters including two foliar diseases showed wide variations for almost all yield attributing traits including multiple disease resistance in the wild species, *L. nigricans* and *L. ervoides* accessions. The core set developed from the entire *Lens* taxa had maximum representation from Turkey and Syria, indicating rich diversity in accessions originating from these regions. Diversity analysis also indicated wide geographical variations across genepool as was reflected in the core set. Potential use of core set, as an initial starting material, for genetic base broadening of cultivated lentil was also suggested.

Editor: Qin Zhang, China Agricultrual University, China

Funding: This research was funded by the DAC-ICAR-ICARDA collaborative research programme (Department of Agriculture and Cooperation Govt of India-Indian Council of Agricultural Research and International Centre for Agricultural Research in Dry Areas, Aleppo, Syria). The funders had no role in study design, data collection and analysis, decision to publish, or preparation of the manuscript.

Competing Interests: The authors have declared that no competing interests exist.

* Email: singhmohar_2003@yahoo.com

Introduction

Crop wild relatives (CWRs) are the reservoir of useful genes and alleles that can be used in breeding new and better adapted varieties resistant to biotic and abiotic stresses, and more importantly to adapt to adverse effect of climate change. [1,2]. Despite their potential value, many CWRs are, however, not adequately collected and conserved in gene banks across the world. Increasing threats to natural habitats and farming systems makes it imperative to collect, conserve and characterize CWRs in order to make them available for use in mitigating the impact of major biotic and abiotic stresses caused by climate change [1] and other factors. The genus *Lens* Miller is part of the family *Fabaceae* (*Leguminosae*), and is placed in either subfamily *Faboideae* tribe *Fabeae* [3] or in subfamily *Papilionaceae* tribe *Vicieae* [4]. Lentil is an annual self-pollinating true diploid ($2n = 2x = 14$) species with an estimated genome size of 4063 Mbp/C [5]. The cultivated lentil (*Lens culinaris* Medikus ssp. *culinaris*) encompasses two groups on the basis of agro-morphological traits, the small-seeded (*microsperma*), large-seeded (*macrosperma*) and *L. culinaris* ssp. *orientalis* (Boiss.) Ponert, is considered its immediate wild progenitor. The other sub-species and species are *L. culinaris* ssp. *tomentosus* (Ladiz.), *L. culinaris* ssp. *odemensis* (Ladz.), *L. ervoides* (Brign.), *L. nigricans* (Bieb.) and *L. lamottei* Czefr. [6]. Globally lentil productivity has increased from an average yield of 560 kg ha^{-1} in 1961–63 to 950 kg ha^{-1} during 2010–2011. Despite this increase, the current yields are much lower as compared to many other grain legume species, because of the limited yield potential of modern lentil cultivars. Major yield limiting factors include, poor seedling vigour, slow leaf area development, high rate of flower drop, low pod setting, poor dry matter, low harvest index, lack of lodging resistance, and exposure to several biotic and abiotic stresses [7]. Further, lentil breeding programme in the past has been primarily depended on landraces of cultivated lentil. Recently developed lentil varieties, however, have superiority over traditional cultivars in terms of their yield potential and disease resistance. For the development of these modern varieties, a small number of improved local landraces have been utilized mainly through pure-line selection following hybridization between lines adapted to specific environmental conditions. Substantial improvement in the lentil productivity could not be achieved due to the loss of invaluable alleles for higher productivity including resistance/tolerance to major biotic and abiotic stresses and low extent of genetic variation in lentil cultivars [8–11]. Wild relatives of lentil are good sources of disease resistance as well as many useful agronomic characters like, high number of pods plant^{-1}, high number of pods cluster^{-1}, high number of branches plant^{-1}, and short internodes [12,13]. Furthermore, substantial genetic diversity is currently not available in *ex-situ* conserved collections. The treasure of lentil genetic resource is held up in populations poorly or incompletely sampled or completely unsampled wild *Lens* taxa. The gap not only limits its use, it also renders precious genetic diversity inaccessible and vulnerable to extinction [14]. Maxted et al. (2012a) [15] suggested a comprehensive strategy for gap analysis on CWRs including lentil for germplasm conservation and how existing georeferenced passport data associated with the collections of *Lens* species can be utilized in identifying gaps in the current *ex-situ* holdings and develop reliable *in-situ* conservation strategy.

The efficient utilization of desired diversity of crop wild relatives will be the solution to achieve substantial gains in productivity of cultivated gene pool. Maintaining and evaluating a large set of accessions of wild *Lens* species is often difficult due to various reasons like inadequate seed germination, high pod shattering and photoperiod sensitivity and certain wild *Lens* species do not flower under normal winter cropping season. These photoperiod sensitive accessions would require artificially lengthened days in the greenhouse or growing them under long days in summer season.

Establishing a core collection of the existing holding is a favoured approach for efficient utilization of novel variations in wild gene pool with a manageable number. The core collection is a limited set of accessions representing, with minimum repetitiveness, the genetic diversity of a crop and its wild relatives [16–19]. Genebank curators have the responsibility for conservation, regeneration, safety duplication, documentation, characterization and evaluation as well as facilitating utilization of the genetic resources to the breeder for crop improvement. The core collection, being smaller in size in comparison to the whole collection, can be used very effectively as a starting point for research involving screening of the germplasm collection for sources of desirable traits. Since, the inception of the idea of core collections, many approaches for selecting core collections have been proposed and used e.g. M-Strat [20], Genetic Distance Sampling [21], PowerCore [22] and Core Hunter [23]. Similarly, core has been developed using several kind of data ranging from genealogical data in the Czech spring wheat [24], agronomic data in groundnut [25,26] and molecular data or integration of data, in bread wheat [27] and in rice [28,29]. The core collection of wild *Lens* gene pool could serve as a working collection for the users, which then needs to be extensively evaluated and used in lentil improvement. Focused Identification of Germplasm Strategy (FIGS approach), based on co-evolution between the plant and the environment conditions, uses trait and environmental data (climate data and phenological data) to develop *a priori* information based on the quantification of the trait-environment relationship. This relationship when detected can be used to develop subsets of accessions with a high probability of containing the sought after traits. This approach was used efficiently in wheat, barley and faba bean to select novel sources of resistance and tolerance to diseases, insect, drought and salinity [30,31].

In view of the above, the objectives of the present study were (1) to systematically characterize and evaluate global wild annual *Lens* collection for different morphological and agronomic traits including major foliar biotic stresses, (2) to develop core set of wild *Lens* taxa for enhanced utilization in lentil genetic base broadening and yield enhancement, (3) to assess the diversity available in global wild annual *Lens* collection, and compare it with that of core set and (4) to analyse the geographical distribution of wild *Lens* taxa representing maximum diversity, and suggest potential areas for future collecting and conserving diversity *in-situ* on-farm.

Materials and Methods

Plant material

A total of 405 diverse global wild annual *Lens* accessions comprising 4 of cultivated *L. culinaris* ssp. *culinaris*, 171 of *L. culinaris* ssp. *orientalis*, 42 of *L. culinaris* ssp. *odemensis*, 20 of *L. culinaris* ssp. *tomentosus*, 35 of *L. nigricans*, 124 of *L. ervoides* and 9 of *L. lamottei* (Table 1) were studied encompassing various aspects of genetic resource management. These include species characterization and agro-morphological evaluation, development of core sets for its enhanced utilization in lentil genetic base broadening, diversity assessment, and geographical distribution of *Lens* taxa representing maximum diversity. The wild *Lens* accessions were obtained from the Genebank of the International Centre for Agricultural Research in the Dry Areas (ICARDA) Aleppo, Syria. The original identity of accessions in the present

Table 1. List of wild annual *Lens* accessions introduced from ICARDA, Aleppo, Syria.

S. No.	Species	Accessions	Origin
1	*Lens culinaris* ssp. *culinaris*	04	India, Israel
2	*Lens culinaris* ssp. *orientalis*	171	Turkey, Iran, Cyprus, Israel, Syria, Jordan, Uzbekistan, Tajikistan, Turkmenistan, Lebanon, Armenia, Azerbaijan, Russia, Czech Republic
3	*Lens culinaris* ssp. *odemensis*	42	Syria, Palestine, Turkey, Israel
4	*Lens culinaris* ssp. *tomentosus*	20	Turkey, Syria
5	*Lens nigricans*	35	Syria, Italy, France, Spain, Croatia, Bahrain, Turkey, Ukraine, Montenegro, Croatia
6	*Lens ervoides*	124	Ukraine, Turkey, Croatia, Italy, Former Soviet Union, Bahrain, Israel, Palestinian Territories, Syrian, Jordan, Tajikistan, Lebanon, Former Yugoslavia, Montenegro, Armenia, Azerbaijan
7	*Lens lamottei*	09	Spain, Turkey, France

Table 2. Frequency distribution of qualitative plant characteristics in different annual *Lens* species.

Character	Class (descriptor state)	Frequency						
		L. culinaris ssp. *culinaris*	*L. culinaris* ssp. *orientalis*	*L. culinaris* ssp. *odemensis*	*L. culinaris* ssp. *tomentosus*	*L. nigricans*	*L. ervoides*	*L. lamottei*
Seedling stem pigmentation	0: Absent	1.00	0.09	0.03	0.11	0.54	0.15	0.43
	1: Present	0.00	0.91	0.97	0.89	0.46	0.85	0.57
Leaf pubescence	0: Absent	0.00	0.03	0.00	0.00	0.14	0.30	0.29
	3: Slight	0.75	0.50	0.35	0.21	0.64	0.66	0.71
	7: Dense	0.25	0.47	0.65	0.79	0.22	0.04	0.00
Leaflet size	3: Small	0.00	0.53	0.30	0.16	0.82	0.37	0.29
	5: Medium	0.50	0.46	0.68	0.79	0.14	0.49	0.71
	7: Large	0.50	0.01	0.02	0.05	0.04	0.14	0.00
Tendril length	1: Rudimentary	0.17	0.29	0.14	0.11	0.25	0.33	0.71
	2: Prominent	0.83	0.71	0.86	0.89	0.75	0.67	0.29
Pod shedding	0: None	1.00	0.00	0.00	0.00	0.00	0.00	0.00
	3: Low	0.00	0.11	0.20	0.00	0.23	0.11	0.50
	5: Medium	0.00	0.28	0.60	0.36	0.54	0.26	0.33
	7: High	0.00	0.61	0.20	0.64	0.23	0.63	0.17
Pod dehiscence	0: None	1.00	0.00	0.00	0.00	0.00	0.00	0.00
	3: Low	0.00	0.00	0.20	0.00	0.18	0.11	0.33
	5: Medium	0.00	0.90	0.80	0.64	0.64	0.56	0.50
	7: High	0.00	0.10	0.00	0.36	0.18	0.33	0.17
Flower colour	1: White	0.17	0.09	0.00	0.00	0.22	0.00	0.29
	2: Yellow	0.00	0.00	0.00	0.00	0.00	0.00	0.00
	3: Red	0.00	0.00	0.00	0.00	0.00	0.00	0.00
	4: Purple	0.83	0.91	1.00	1.00	0.78	1.00	0.71
Ground colour of testa	1: Green	0.00	0.00	0.00	0.00	0.00	0.00	0.00
	2: Grey	0.00	0.07	0.35	0.06	0.37	0.26	0.57
	3: Brown	1.00	0.93	0.65	0.94	0.63	0.74	0.43
	4: Black	0.00	0.00	0.00	0.00	0.00	0.00	0.00
	5: Pink	0.00	0.00	0.00	0.00	0.00	0.00	0.00
Pattern of testa	0: Absent	0.50	0.01	0.00	0.06	0.00	0.04	0.00
	1: Dotted	0.50	0.17	0.05	0.00	0.11	0.08	0.00
	2: Spotted	0.00	0.12	0.11	0.06	0.15	0.08	0.14
	3: Marbled	0.00	0.58	0.16	0.76	0.59	0.74	0.57
	4: Complex	0.00	0.12	0.68	0.12	0.15	0.06	0.29
Cotyledon colour	1: Yellow	0.25	0.04	0.78	0.00	0.19	0.10	0.71
	2: Orange	0.75	0.96	0.22	1.00	0.81	0.90	0.29

Table 2. Cont.

Character	Class (descriptor state)	Frequency						
		L. culinaris ssp. *culinaris*	*L. culinaris* ssp. *orientalis*	*L. culinaris* ssp. *odemensis*	*L. culinaris* ssp. *tomentosus*	*L. nigricans*	*L. ervoides*	*L. lamottei*
	3: Olive-green	0.00	0.00	0.00	0.00	0.00	0.00	0.00

Table 3. Range of variations for different quantitative agro-morphological traits in annual *Lens* species.

Character	*Lens culinaris* ssp. *culinaris*			*Lens culinaris* ssp. *orientalis*			*Lens culinaris* ssp. *odemensis*		
	Range	Mean ± SE	CV (%)	Range	Mean ± SE	CV (%)	Range	Mean ± SE	CV (%)
Days to 50% flowering	34.50–43.00	39.75±1.89	9.53	27.00–135.00	93.76±2.35	32.81	88.00–143.00	126.42±1.81	9.30
Days to maturity	72.00–86.00	79.50±2.93	7.38	76.00–168.00	130.15±1.97	19.80	124.00–174.00	155.02±1.50	6.29
Plant height (cm)	17.00–24.00	20.50±1.55	15.12	12.33–95.50	26.58±0.68	33.48	20.33–59.94	33.38±1.59	30.94
No. of branches plant^{-1}	6.50–9.50	8.12±0.80	19.70	4.00–35.00	13.47±0.68	66.07	3.50–22.00	14.69±0.32	29.06
No. of pods plant^{-1}	18.50–208.00	92.50±44.50	96.22	3.00–250.00	59.58±3.68	80.89	5.50–185.50	74.75±6.95	60.29
No. of seeds pod^{-1}	1.51–1.83	1.63±0.07	8.58	0.74–2.33	1.41±0.02	23.40	1.00–1.96	1.78±0.03	12.35
No. of seeds plant^{-1}	28.50–316.50	147.37±67.62	91.76	4.00–409.00	82.53±5.1	80.84	6.00–357.50	136.26±13.83	65.77
100-seed weight (g)	0.75–2.50	1.62±0.38	47.53	0.15–4.00	0.91±0.03	56.04	0.40–1.10	0.73±0.02	24.65
Seed yield plant^{-1} (g)	0.25–4.25	1.85±0.89	96.75	0.01–3.26	0.80±0.04	80.00	0.06–3.30	1.05±0.11	68.57
Biological yield plant^{-1} (g)	6.35–13.90	8.92±1.70	38.22	0.50–18.50	3.73±0.15	54.95	0.25–10.55	5.45±0.29	35.59

Table 4. Range of variations for different quantitative agro-morphological traits in annual *Lens* species.

Character	*Lens culinaris* ssp. *tomentosus*			*Lens nigricans*			*Lens ervoides*			*Lens lamottei*		
	Range	Mean ± SE	CV (%)	Range	Mean ± SE	CV (%)	Range	Mean ± SE	CV (%)	Range	Mean ± SE	CV (%)
Days to 50% flowering	120.00–132.00	129.65±0.66	2.28	92.00–139.00	123.88±2.60	12.40	60.00–141.00	118.88±1.99	18.67	92.00–130.00	119.88±5.29	13.24
Days to maturity	20.00–173.00	162.75±0.78	2.16	130.00–176.00	160.42±2.41	8.90	93.00–173.00	153.63±1.97	14.38	130.00–165.00	154.11±4.65	9.05
Plant height (cm)	20.75–50.00	33.27±1.85	24.94	19.66–67.00	40.23±1.95	28.66	2.00–59.00	37.51±0.94	28.15	27.25–54.00	40.90±2.99	21.95
No. of branches plant^{-1}	10.50–19.50	15.26±0.60	17.62	2.00–22.75	13.57±0.85	37.14	4.25–52.00	16.99±1.67	109.53	6.00–19.80	14.20±1.54	32.53
No. of pods plant^{-1}	4.50–218.00	74.05±11.50	68.84	5.50–153.00	58.30±7.58	76.92	3.00–1002.00	111.78±11.74	116.94	8.00–185.50	63.97±18.80	88.19
No. of seeds pod^{-1}	1.00–1.97	1.70±0.06	17.05	1.00–2.00	1.64±0.04	14.63	1.00–2.22	1.72±0.02	14.53	1.68–2.10	1.85±0.04	7.02
No. of seeds plant^{-1}	6.00–420.00	133.00±22.30	74.97	7.50–275.00	100.45±14.01	82.46	3.00–1425.00	196.31±18.72	106.16	13.75–364.00	121.10±36.89	91.38
100-seed weight (g)	0.10–1.10	0.69±0.06	44.92	0.25–1.20	0.57±0.03	38.59	0.10–2.00	0.49±0.02	51.02	0.60–1.85	0.92±0.15	48.91
Seed yield plant^{-1} (g)	0.01–3.30	1.01±0.17	79.20	0.02–2.30	0.62±0.09	93.54	0.04–5.70	0.97±0.09	104.12	0.10–2.75	0.99±0.29	89.89
Biological yield plant^{-1} (g)	1.90–10.00	5.79±0.40	31.43	1.70–15.30	5.37±0.47	52.32	1.20–29.50	5.87±0.35	66.78	3.65–9.90	5.55±0.71	38.55

study was the same as established by the ICARDA Genetic Resources Unit.

Species characterization and agronomic evaluation

All 405 global annual *Lens* accessions including four commercial lentil cultivars were grown at two agro-ecological regions of India, the National Bureau of Plant Genetic Resources (NBPGR), New Delhi, India during the winter (*rabi*) season of 2011–2012, 2012–2013 and 2013–2014 for characterization and agro-morphological evaluation and the summer Himalayan nursery (*summer* 2012 and 2013) at Himachal Pradesh Agricultural University (HPAU), Research Station (RS), Sangla. The permission to conduct experiments at Sangla was granted by worthy Vice-Chancellor of the University. The location of NBPGR lies in the northern plain between 28^0 35′ N latitude, 70^0 18′ E longitude, 226 m asl altitude, whereas the HPAU, RS, Sangla lies between latitudes 31^0 55′ and 32^0 20′ N, and longitudes 77^0 00′ and 79^0 50′ E, in the northwestern Himalayas. Each accession was sown by hand in three replications of 3 m long rows spaced 30 cm apart. The soil type was sandy loam at both the locations. All these annual *Lens* species were characterized for qualitative morphological traits viz; seedling stem pigmentation, leaf pubescence, leaflet size, tendril length, pod shedding, pod dehiscence, flower colour, ground colour of testa, pattern of testa and cotyledon colour using standard lentil descriptors [32]. Observations were also recorded on ten plants accession^{-1} in each replication for ten quantitative morphological traits viz; days to flowering, days to maturity, plant height (cm), number of branches plant^{-1}, number of pods plant^{-1}, number of seeds pod^{-1}, number of seeds plant^{-1}, 100-seed weight (g), seed yield plant^{-1} (g) and biological yield plant^{-1}(g). The data were subjected to statistical analysis using MS Ofiice Excel and SAS software.

Screening against foliar biotic stresses

Rust (Uromyces fabae (Grev.) Fuckel). Reaction to *Uromyces fabae* was conducted at the Himachal Pradesh Agricultural University Research and Extension Centre (HPAUREC), Dhaulakuan (30^0 4′ N latitude, 77^0 5′ longitudes, 468 m asl) located in the foot hills of northwestern Himalayas. All the accessions were grown in 25 cm diameter plastic pots in soil and farmyard manure mixture (10:1). Each accession had two replications. Pots with plants at the vegetative stage were transferred in the field and were artificially inoculated by frequently spraying with ascosporic suspension (1×10^6 spores/ml) using the isolate of *Uromyces fabae*. The data were recorded on terminal disease reaction at vegetative and reproductive plant stages using 1–9 rating scale as suggested by [33]. Based on disease score, the wild accessions were categorized for their reaction to rust infection as: 1 = asymptomic (Free); 3 = resistant (R); 5 = moderately resistant (MR); 7 = susceptible (S); and 9 = highly susceptible (HS).

Powdery mildew (*Erysiphe polygoni DC.*). Powdery mildew is an economically important foliar disease of lentil especially in those areas where warm and humid weather persists. For the screening of wild *Lens* species, all accessions were grown in 25 cm diameter plastic pots in soil and farmyard manure mixture (10:1), and planted pots were kept and maintained in the glasshouse. The disease appeared at flower initiation stage, which was further allowed to develop. The pots with heavily infected plants were shifted near to test entries grown in the glass house. The heavily infected plants were shaken well over the test entries in the evening time to disperse conidia, which helps in spreading the disease on the test entries as suggested by [34]. Data on disease incidence were recorded using 1–9 rating scale at vegetative and reproduc-

Table 5. Species-wise representation of entire *Lens* collection and representation of accessions in core sets derived from different data sets.

Lens species	Accessions in entire collection	Accessions in core sets		
		I	II	III
L. culinaris ssp. *culinaris*	4	4	4	3
L. culinaris ssp. *orientalis*	171	21	19	7
L. culinaris ssp. *odemensis*	42	15	12	4
L. culinaris ssp. tomentosus	20	12	9	6
L. nigricans	35	18	17	7
L. ervoides	124	20	19	5
L. lamottei	9	6	6	4
Total	405	96 (23.7%)	86 (21%)	36 (8.9%)

Core sets based on I = combined data on both quantitative and qualitative variables; II = data only on quantitative variables, and III = data only on qualitative variables, respectively.

tive stages as: 1 = asymptomic (Free); 3 = resistant (R); 5 = moderately resistant (MR); 7 = susceptible (S); and 9 = highly susceptible (HS).

Development of core set

The entire set of *Lens* species was used for developing a core collection using "PowerCore" (http://genebank.rda.go.kr/powercore/) software developed at the Rural Development

Table 6. Comparison of SDI of entire collection to core sets derived from different data sets.

S. No.	Traits	Entire collection	Core sets (Using PowerCore)*			Core set derived through PCSS
			I	II	III	
Quantitative characters						
1	Days to flowering	0.733	0.819	0.843	0.750	0.691
2	Days to maturity	0.795	0.851	0.881	0.685	0.729
3	Plant height	0.964	0.777	0.725	0.874	0.664
4	No. of branches plant^{-1}	0.035	0.116	0.127	0.987	0.116
5	No. of pods plant^{-1}	0.200	0.406	0.440	0.760	0.511
6	No. of seeds plant^{-1}	0.297	0.533	0.635	0.632	0.716
7	No. of seeds pod^{-1}	0.806	0.817	0.830	0.958	0.837
8	100-seed weight	0.725	0.883	0.914	0.766	0.710
9	Seed yield plant^{-1}	0.735	0.745	0.698	0.816	0.698
10	Biological yield plant^{-1}	0.637	0.972	0.641	0.787	0.821
11	Resistance to rust	0.942	0.979	0.973	0.849	0.908
12	Resistance to powdery mildew	0.886	0.953	0.948	0.880	0.843
Qualitative characters						
13	Seedling stem pigmentation	0.475	0.613	0.613	0.687	0.525
14	Leaf pubescence	0.963	0.917	0.952	0.966	0.982
15	Leaflet size	0.871	0.926	0.858	0.935	0.890
16	Tendril length	0.614	0.584	0.581	0.654	0.562
17	Flower colour	0.235	0.391	0.359	0.572	0.311
18	Ground colour of testa	0.543	0.679	0.629	0.787	0.562
19	Pattern of testa	0.790	0.915	0.880	0.970	0.805
20	Cotyledon colour	0.500	0.621	0.622	0.637	0.562
21	Pod shedding	0.956	0.833	0.819	0.921	0.809
22	Pod dehiscence	0.837	0.835	0.833	0.870	0.970

Core sets based on I = combined data on both quantitative and qualitative variables; II = data only on quantitative variables, and III = data only on qualitative variables, respectively.

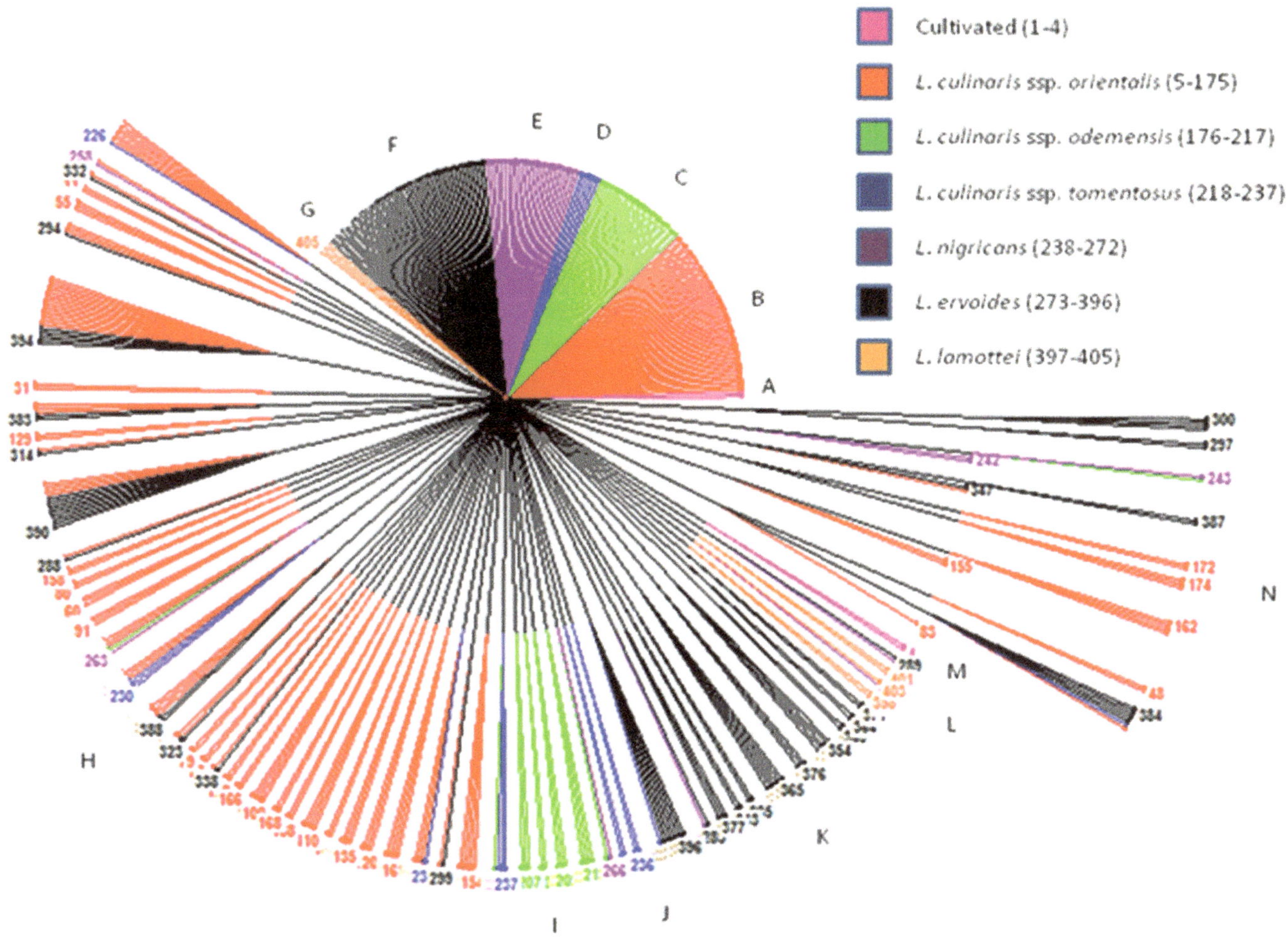

Figure 1. Hierarchical clustering of *Lens* accessions based on qualitative phenotypic data.

Administration (RDA), South Korea [22]. Separate sets were developed using both quantitative and qualitative data in combination as well as in isolation. "PowerCore" creates subsets representing all alleles or observation classes, with the least allelic redundancy, and ensures a highly reproducible list of entries. The Shannon Diversity Index [35] was used to compare the representativeness of the entire holding and the accessions selected as core entries.

Diversity analysis

Diversity analysis was conducted using both qualitative and quantitative phenotypic data of complete set of 405 accessions as well as of the core set developed from it. Genetic relationships among accessions were determined by means of Dice's coefficient [36] in case of qualitative data. Phenotypic relationships based on quantitative data were determined from the Euclidean distances. Trees were constructed using hierarchical clustering method by means of Jaccard's coefficient with UPGMA mode with the DARwin 5.0 software [37,38].

Results

All 405 accessions of global annual *Lens* taxa including four commercial cultivars were characterized and evaluated for morphological and agronomic traits as well as for two major fungal diseases under two agro-climatic regions in the north-western Indian conditions.

Species characterization using qualitative traits

The frequency distribution of seven annual *Lens* species showed a wide range of intraspecific variation for majority of the important morphological plant characteristics (Table 2). Seedling stem pigmentation exhibited variations in all the *Lens* species except, *L. culinaris* ssp. *culinaris*, where it was absent. Likewise, leaf pubescence, leaflet size and tendril length also revealed remarkable variation in all seven annual *Lens* taxa. There was substantial variation in pod shedding and dehiscence for majority of the *Lens* species except, *L. culinaris* ssp. *culinaris*. Flower colour was white and purple in majority of the *Lens* species like, *L. culinaris* ssp. *culinaris*, *L. culinaris* ssp. *orientalis*, *L. nigricans* and *L. lamottei* and in *L. culinaris* ssp. *odemensis*, *L. culinaris* ssp. *tomentosus*, and in *L. ervoides*, it was purple only. The ground colour of testa was mostly grey and brown in all the *Lens* species except, *L. culinaris* ssp. *culinaris*, where it was all brown. Substantial variations were also found in the pattern of testa for all the annual *Lens* species. All the annual *Lens* species had yellow and orange cotyledon colour except, *L. culinaris* ssp. *tomentosus*, where it was all orange (Table 2).

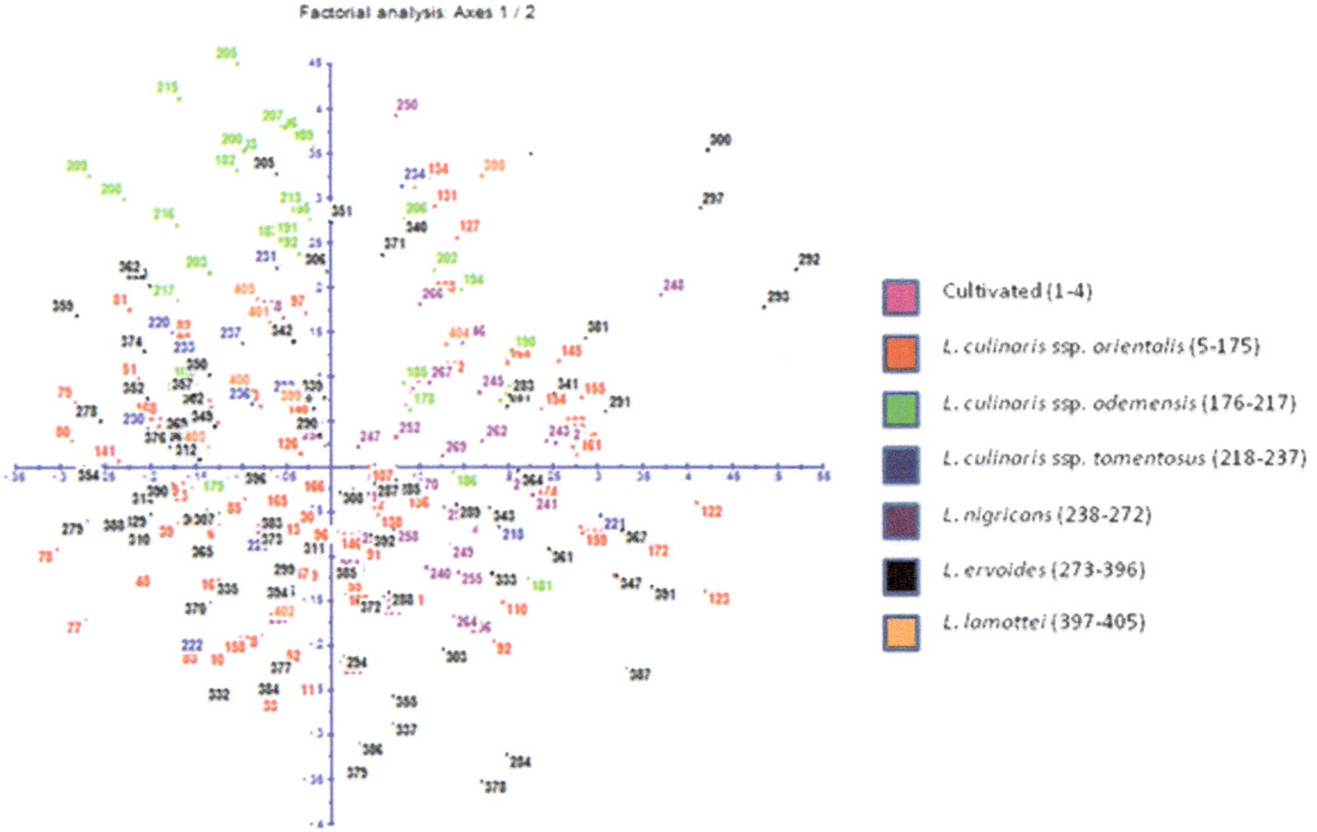

Figure 2. Factorial analysis of *Lens* accessions based on qualitative phenotypic data.

Agronomic evaluation using quantitative traits

All the annual *Lens* accessions were evaluated for various agro-morphological quantitative traits. The results revealed significant variations ($p = 0.05$) as was evident from the analysis of variance as also the range, mean and coefficient of variations for all traits studied (Table 3 and 4). In general, the coefficient of variation ranged from low (days to flowering and maturity) to high (number of branches^{-1}, number of pod plant^{-1} and number of seeds pod^{-1}) for important agro-morphological traits in all the cropping seasons. However, the mean performance for days to flowering and maturity was lowest in *L. culinaris* ssp. *culinaris*, followed by *L. culinaris* ssp. *orientalis*. Likewise, maximum plant height was recorded in *L. lamottei and L. nigricans* and lowest in *L. culinaris* ssp.*culinaris*. The mean number of branches plant^{-1} was highest in *L. ervoides* followed by *L. culinaris* ssp. *tomentosus, L. culinaris* ssp. *odemensis*, *L. lamottei* and *L. culinaris* ssp. *orientalis*, and lowest in *L. culinaris* ssp. *culinaris*. There was also substantial variation in the range for number of pods plant^{-1} with maximum number of pods plant^{-1} recorded in *L. ervoides* followed by *L. culinaris* ssp. *culinaris*, *L. culinaris* ssp. *odemensis*, *L. culinaris* ssp. *tomentosus* and *L. culinaris* ssp. *orientalis*. The number of seeds pod^{-1} was >1 in all the annual *Lens* species. As expected, cultivated species showed superior grain yield related agronomic performance as compared to the wild *Lens* species. But, several useful yield attributing traits like high number of pods plant^{-1}, high number of seeds plant^{-1}, high number of branches plant^{-1}, and greater plant height as well as resistance to major foliar diseases was observed in various wild annual *Lens* species.

Screening against foliar biotic stresses

Rust and powdery mildew. Many of the wild *Lens* accessions were resistant and moderately resistant to rust and powdery mildew during all the cropping seasons. Among accessions ILWL247 of *L. culinaris* ssp. *orientalis*; ILWL165, ILWL167 and ILWL238 of *L. culinaris* ssp. *odemensis*; ILWL37, ILWL22 and ILWL38 of *L. nigricans*; and ILWL40, ILWL41, ILWL54 and ILWL58 of *L. ervoides* recorded a score of 2 and identified as highly resistant to rust. While, ILWL350, ILWL369, and ILWL381 of *L. culinaris* ssp. *orientalis*; ILWL39 and ILWL81 of *L. culinaris* ssp. *odemensis*; ILWL191 of *L. nigricans,* and ILWL418, ILWL398, ILWL292, ILWL91 and ILWL294 of *L. ervoides* were found resistant to powdery mildew. Some of the accessions viz., ILWL37, ILWL38, ILWL22 and ILWL191 of *L. nigricans*, and ILWL269 of *L. ervoides* were rated as resistant to both rust and powdery mildew.

Development of core set

"PowerCore" statistical programme selected a set of 96 accessions (24% of the entire collection), when data on both quantitative and qualitative variables were used in combination (Table 5). A total of 86 accessions (21%) were selected, when data only on quantitative variables were used, whereas 36 accessions (about 9%) could be selected, when data only on qualitative characters were used for analysis. Compared with the whole collection, the Shannon Diversity Indices (SDIs) were always higher in the core sets derived from different data sets indicating better representation of the existing diversity in each core set. Better representation of diversity was observed using "Power-

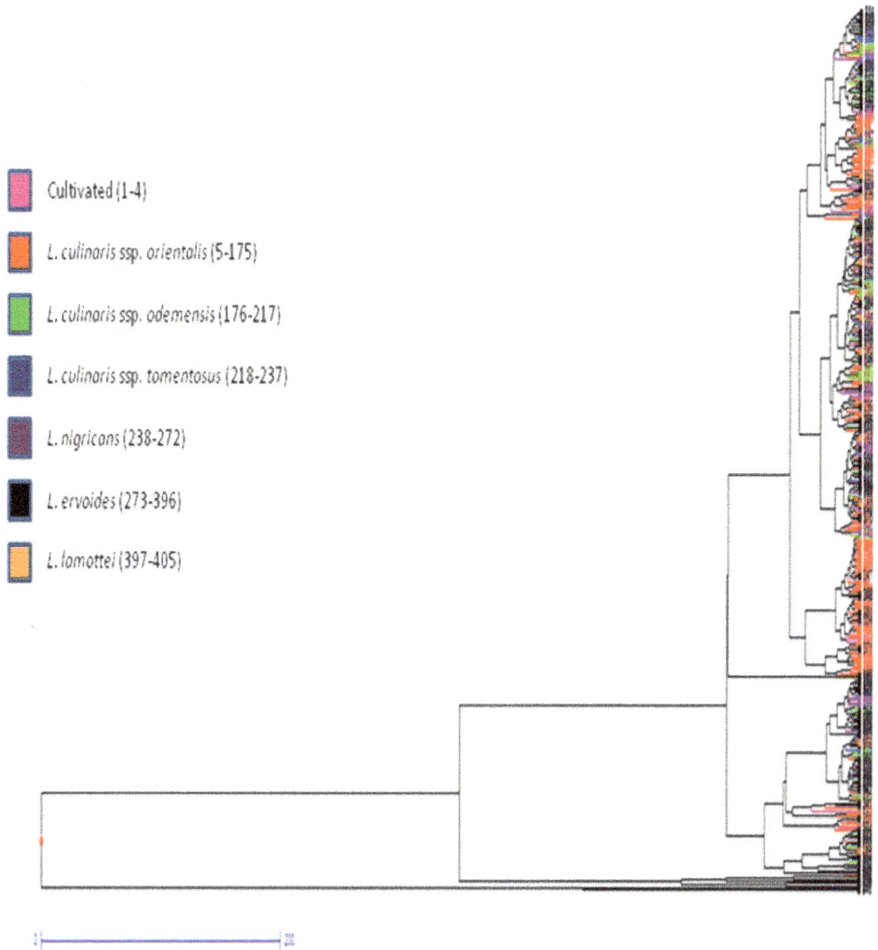

Figure 3. Hierarchical clustering of *Lens* accessions based on quantitative phenotypic data.

Core" as compared to Principal Components Score Strategy (PCSS, Table 6).

Diversity analysis of global *Lens* collection

A Dice similarity matrix was generated based on the morphological qualitative data, which was further used to construct phylogenetic tree for deciphering the diversity of wild *Lens* species. Based on the Dice dissimilarity matrix, an average dissimilarity coefficient value of 0.5 was recorded. The dissimilarity coefficient ranged from 0.0–1.0 indicating that accessions showed 100% similarity to 100% dissimilarity (Table S1). Maximum dissimilarity was assessed between ILWL429 (*L. lamottei*, Spain) and ILL10829 (*L. culinaris* ssp. *culinaris*, Syria); ILWL126 (*L. ervoides*, Syria) and ILL10829 (*L. culinaris* ssp. *culinaris,* Syria), whereas ILWL65 (*L. ervoides*, Turkey) showed 100% dissimilarity to the accessions of *L. culinaris* ssp. *orientalis*, namely ILWL378 (Turkmenistan), ILWL241 (Syria), ILWL231 (Syria) and ILWL230 (Syria); ILWL55 (*L. ervoides*, Israel) to *L. culinaris* ssp. *orientalis* accessions namely ILWL378 (Turkmenistan), ILWL241 (Syria), ILWL231 (Syria) and ILWL230 (Syria); ILWL55 (*L. ervoides*, Israel) to ILWL378 (*L. culinaris* ssp. *orientalis*, Turkmenistan); EC718264 (*L. nigricans*, Spain) to ILWL238 (*L. culinaris* ssp. *odemensis*, Syria); ILWL23 (*L. nigricans*, Italy) to ILWL241 (*L. culinaris* ssp. *orientalis*, Syria); ILWL97 (*L. culinaris* ssp. *tomentosus,* Turkey) to ILWL125 (*L. culinaris* ssp. *orientalis,* Syria); ILL10829 (*L. culinaris* ssp. *culinaris,* Syria) to *L. culinaris* ssp. *odemensis* from Syria including ILWL357, ILWL238 and ILWL167; ILWL371 (*L. culinaris* ssp. *orientalis*, Syria) to *L. culinaris* ssp. *orientalis* ILWL62 (11) and ILWL8 (9); ILWL227 (*L. culinaris* ssp. *orientalis*, Syria) to accessions of the same species ILWL62 (11) and ILWL8 (9). After exclusion of accessions, which showed 100% similarity to each other with dissimilarity matrix value of 0.0 (Table S1), the minimum dissimilarity was exhibited by ILWL247 (*L. culinaris* ssp. *orientalis,* Syria) to *L. ervoides* namely EC718432 (HRV), EC718429 (HRV), EC718427 (Ukraine), EC718422 (Turkey), ILWL67 (Turkey), and ILWL121 (*L. culinaris* ssp. *tomentosus*, Syria); ILWL443 (*L. culinaris* ssp. *orientalis,* Turkey) to other accessions of *L. culinaris* ssp. *orientalis,* ILWL469 (Syria), ILWL416 (Syria), ILWL402 (Lebanon), ILWL385 (Uzbekistan) and ILWL384 (Tajikistan); ILWL247 (*L. nigricans*, Syria) to *L. culinaris* ssp. *orientalis* accessions ILWL86 (Turkey), ILWL95 (Turkey) and ILWL117 (Syria), with dissimilarity coefficient of 0.053. The hierarchical clustering of accessions resulted into various clusters A–N, where clusters A–G, J and M were occupied by accessions of respective individual species, whereas other clusters included accessions from two or more species as shown in Fig. 1. The grouping pattern of different accessions in hierarchical clustering is listed along with their species and respective groups (Table S2). Clustering pattern of the various accessions obtained through factorial analysis, where

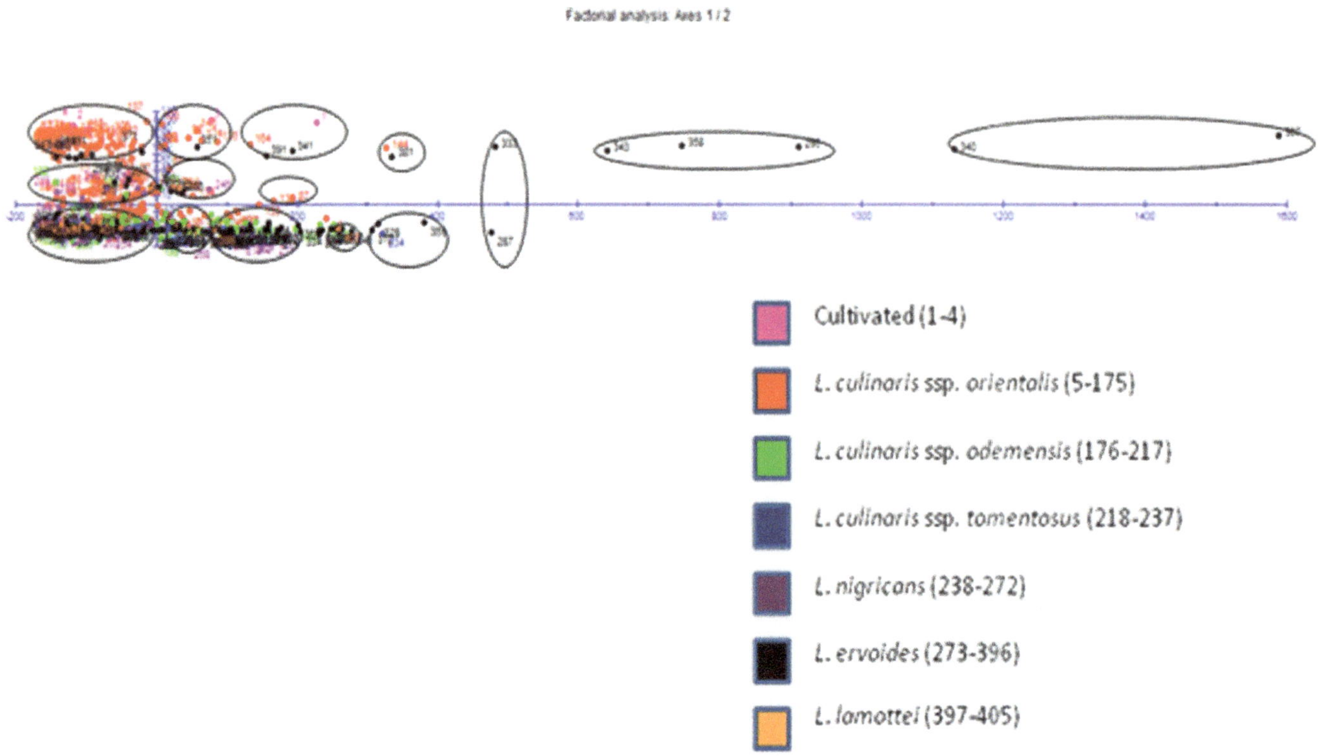

Figure 4. Factorial analysis of *Lens* accessions based on quantitative phenotypic data.

first and second principal coordinates explained only 10.17% and 8.8% variations, respectively (Fig. 2). In general, the clustering pattern of accessions, in both the hierarchical and factorial analysis was not according to their geographical origin.

In case of quantitative data, the Euclidean dissimilarity matrices varied from 2.67 between ILWL159 (Syria) and ILWL67 (Turkey) to 1740.53 between ILWL418 (Syria) and ILWL276 (Turkey) of *L. ervoides* (Table S3). In hierarchical clustering, two major clusters were formed in which major cluster I was occupied by only two accessions namely ILWL292 and ILWL418 of *L. ervoides* from Turkey and Syria, respectively, whereas cluster II included others. Major cluster II further divided into two clusters A and B, where A was occupied by accessions of *L. ervoides*, ILWL401 (Lebanon), ILWL321, ILWL58 and ILWL269 (Turkey), and ILWL51 (Montenegro), whereas cluster B was occupied by accessions from all seven species included in the study (Fig. 3). Cluster B was further divided into sub-clusters (B1 and B2), groups (a, b, c and d) outgroups (a1 and a2, b1 and b2, d1 and d2), clutches (b2a and b2b, d1a and d1b, d2a and d2b) and finally to sub-clutches (i-xxiii) as shown in Fig. 3 (Table S4). Cluster II was reported to be more heterogeneous with different sub-clusters and few individual accessions forming separate groups. Factorial analysis verified the results obtained *via* hierarchical clustering as it resulted into almost similar clustering pattern as shown in Fig. 4. In factorial analysis, first two principal coordinates explained 98.26 per cent cumulative variations with 92.94 and 5.32 per cent dissimilarity, respectively. The clustering pattern in both, the hierarchical and factorial analysis, grouping of accessions was not according to their geographical origin.

Diversity analysis of *Lens* core collection

Based on Dice dissimilarity matrix, the maximum dissimilarity matrix values ranged from 0.1–1.0 with an average of 0.55. No two genotypes showed 100% similarity to each other (Table S5). Minimum diversity was observed between accessions ILWL13 (*L. nigricans*, Italy) and ILWL20 (*L. culinaris* ssp. *odemensis*, Palestine); ILWL437 (*L. lamottei*, Turkey) and EC718266 (*L. nigricans*, Italy), with dissimilarity matrix value of 0.1. Maximum dissimilarity i.e. 100% was found between ILWL65 (*L. ervoides*, Turkey) and ILWL378 (*L. culinaris* ssp. *orientalis,* Turkmenistan); and ILWL227 (*L. culinaris* ssp. *orientalis*, Syria) and ILWL62 (*L. culinaris* ssp. *orientalis*, Turkey). In hierarchical clustering of the accessions, ILWL20 (*L. culinaris* ssp. *odemensis*, Palestine), ILWL13 (*L. nigricans*, Italy), ILWL15 (*L. lamottei,* Turkey) and ILWL18 (*L. nigricans*, France) each constitute a separate group alone as shown in Fig. 5. EC718266 of *L. nigricans* (Italy) clustered with ILWL437 of *L. lamottei* (Turkey), whereas ILWL97 of *L. culinaris* ssp. *tomentosus* (Turkey) clustered with EC718275 of *L. nigricans* (Turkey). ILWL203 (*L. culinaris* ssp. *odemensis*, Turkey), ILWL282, ILWL305 and ILWL307 (*L. culinaris* ssp. *tomentosus*, Turkey), and EC718267 (*L. nigricans*, Montenegro) occupied the same cluster. Further, ILWL343 and ILWL227 from *L. culinaris* ssp. *orientalis* (Syria) formed group with L830 of *L. culinaris* ssp. *culinaris* (India) and ILWL91 of *L. ervoides* (Turkey), respectively. Factorial analysis was also carried out as shown in Fig. 6, where PC1 and PC2 explained only 12.99 and 12.53% variations respectively. In general, first 12 principal coordinates had Eigen values >1.0. The clustering pattern in both, hierarchical and factorial analysis, did not reflect their geographical origin.

In case of quantitative data, the Euclidean dissimilarity matrices ranged from 7.15 between accessions ILWL282 (*L. culinaris* ssp.

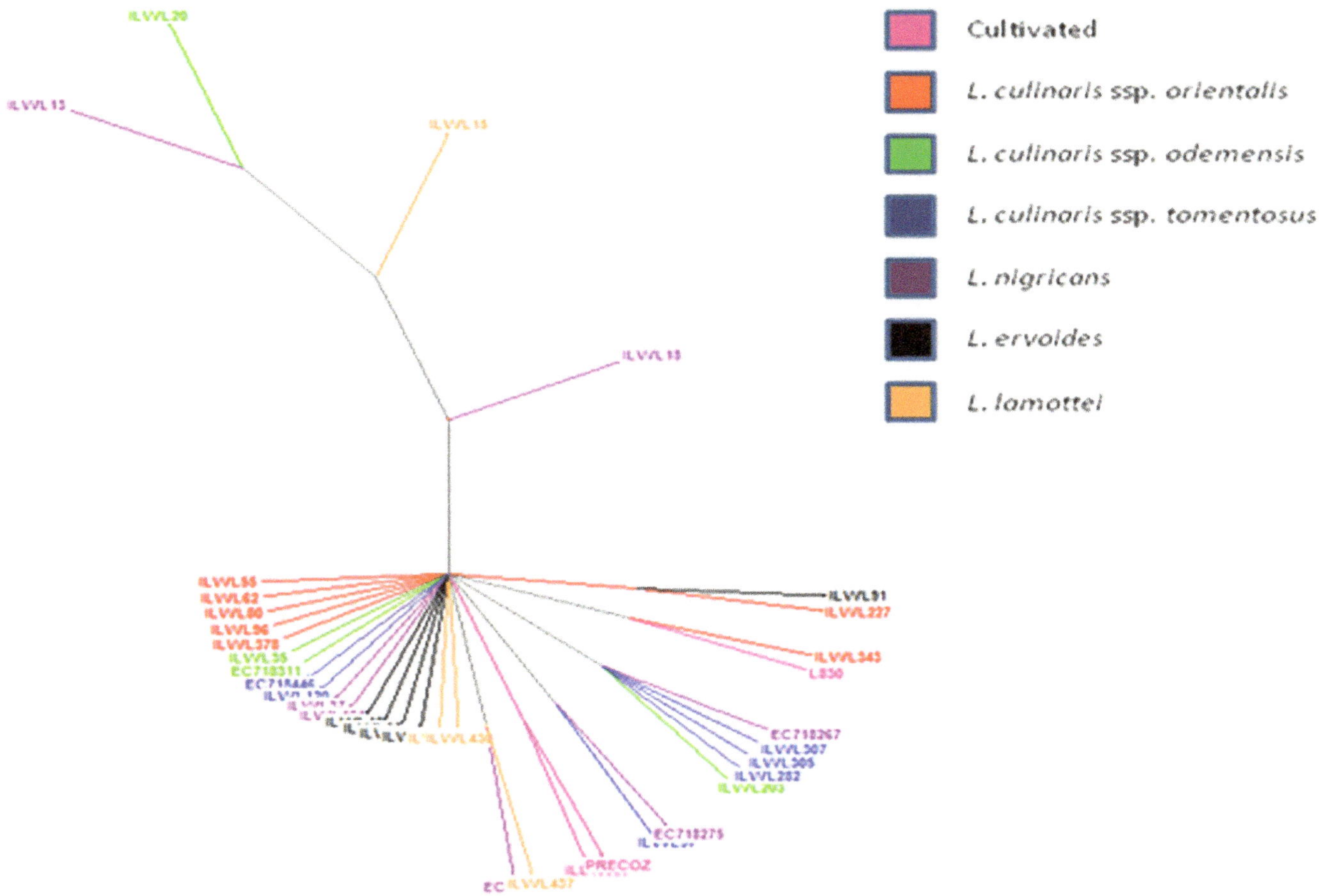

Figure 5. Hierarchical clustering of *Lens* core set accessions based on qualitative phenotypic data.

tomentosus, Turkey) and ILWL227 (*L. culinaris* ssp. *orientalis*, Syria) to 635.85 between accessions ILWL276 and ILWL269 of *L. ervoides* from Turkey (Table S6). The hierarchical clustering of accessions, where all the outgroups were occupied by accessions from different species excepting one containing Precoz (*L. culinaris* ssp. *culinaris*, Brazil) and ILL10829 (*L. culinaris* ssp. *culinaris*, Syria) as shown in Fig. 7. Factorial analysis resulted into first two principal coordinates having Eigen values >1 where first and second principal coordinates explained 89.6 and 8.85% variability respectively (Fig. 8). The clustering pattern in both, hierarchical and factorial analysis was not according to their geographical origin.

Discussion

Crop wild relatives (CWRs) can contain far greater diversity than their domestic counterparts and may hold increased potential to adapt against crop pests and diseases, adverse weather conditions and longer term changes in climate. One way to improve the resilience of our crop plants is to harness the genetic diversity found in their wild relatives. By introducing CWRs into breeding programs, the useful traits they contain, such as high yield and disease resistance, can be passed into cultivated crops. The introduction, characterization, evaluation, maintenance and utilization of CWRs for important morphological and agronomic traits, including adaptation to biotic and abiotic stresses, are often prerequisites for conducting a successful crop breeding research in annual plant species with a view to introgress novel trait of interest [39–41]. Effective evaluation and screening of diverse germplasm, especially unadapted gene pool [42–43] may prove useful to retrieve invaluable phenotypes to combat emerging agricultural challenges [44]. A strong pre-breeding program is needed to introgress genes from distant species and supply breeding pools with adequate parental germplasm.

Species characterization and agronomic evaluation

The results of the present study showed that global wild annual *Lens* collection of diverse origin had significant variation with respect to morphological and agronomic characters. Most of the wild annual *Lens* species had wide variations to seedling stem pigmentation and low to dense leaf pubescence. Dense leaf pubescence character can contribute towards better tolerance to insects and is a subject of further detailed investigation. More importantly to note that variation was observed within a species for leaflet size, tendril length, pod shedding, pod dehiscence, flower colour, ground colour of testa, pattern of testa and cotyledon colour (Table 2). These morphological plant characteristics can help to distinguish distinct phenotypic classes and their mode of inheritance can be studied using principles of classical genetics. Some of these traits are considered as descriptors for the taxonomic classification of taxon into their respective species and sub-species. It also allows phenotypic identification of specific alleles for specific loci [45–46]. Further, there were substantial variations exhibited in *Lens* species for important agro-morphological traits as reflected by their range, mean and coefficient of

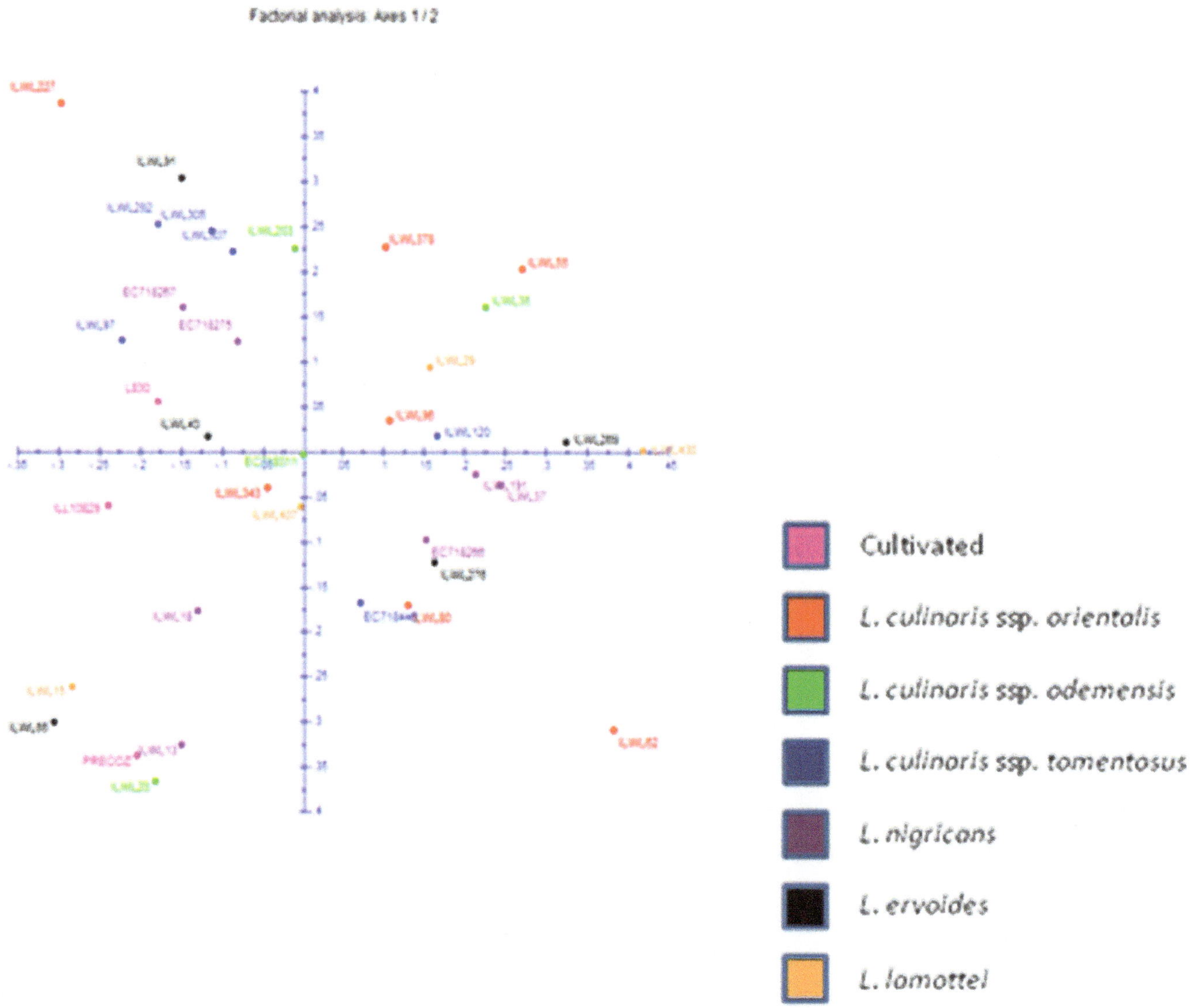

Figure 6. Factorial analysis of *Lens* core set accessions based on qualitative phenotypic data.

variation (Table 3 and 4), suggesting that it could be due to the diverse origin of these species and to their taxonomic identification. However, seed yield plant^{-1} and biological yield plant^{-1} - were higher in magnitude in the cultivated varieties, including Precoz, L830, ILL10829 and ILL8006 in both the agro-climatic regions. By contrast, in wild annual *Lens* species, the range and mean performance found promising for useful agro-morphological traits, including high number of pods plant^{-1} in accessions of *L. ervoides* viz., ILWL418 (1002 pods plant^{-1}), collected from Syria and ILWL321 (403 pods plant^{-1}), ILWL401(478 pods plant^{-1}), ILWL292 (697 pods plant^{-1}), ILWL369 (355 pods plant^{-1}) and ILWL308 (218 pods plant^{-1}) collected from Turkey. Other important accessions for high number of pods plant^{-1} and resistance to rust and powdery mildew in *L. nigricans* included, ILWL191 (141 pods plant^{-1}), ILWL40 (113 pods plant^{-1}), ILWL41 (147 pods plant^{-1}) and ILWL42 (143 pods plant^{-1}), also collected from Turkey. It is important to note that, wild accessions collected from Turkey exhibited substantial variations for useful agro-morphological traits including multiple resistance against major foliar diseases. These promising accessions of *L. nigricans* and *L. ervoides* from the secondary genepool can be exploited for broadening the genetic base of cultivated lentil. Further, rust and powdery mildew resistance gene sources identified can be of great significance for exploiting in breeding resistant lentil varieties, especially for the epidemiologically important regions in the lentil growing areas of the world. The list of some important accessions selected from different *Lens* species carrying useful traits of interest is presented in Table 7. The present study further strongly suggests revisiting of Turkey and other areas for more collections through explorations. In order to make quality *ex-situ* collection, it is necessary to know the geographical distribution of the target *Lens* taxa and where the gaps in the seed collection occur. Knowledge of current ICARDA genebank holding and understanding the gaps in collection would help prioritize the species and locations of wild *Lens*, that are in most need of collecting and conserving them in gene banks.

Assessment of among and within population diversity will also help to establish the potential of *in-situ* conservation of wild *Lens* species. Many areas of greatest interest for *in- situ* conservation (e.g. Turkey and other Mediterranean countries) are suffering

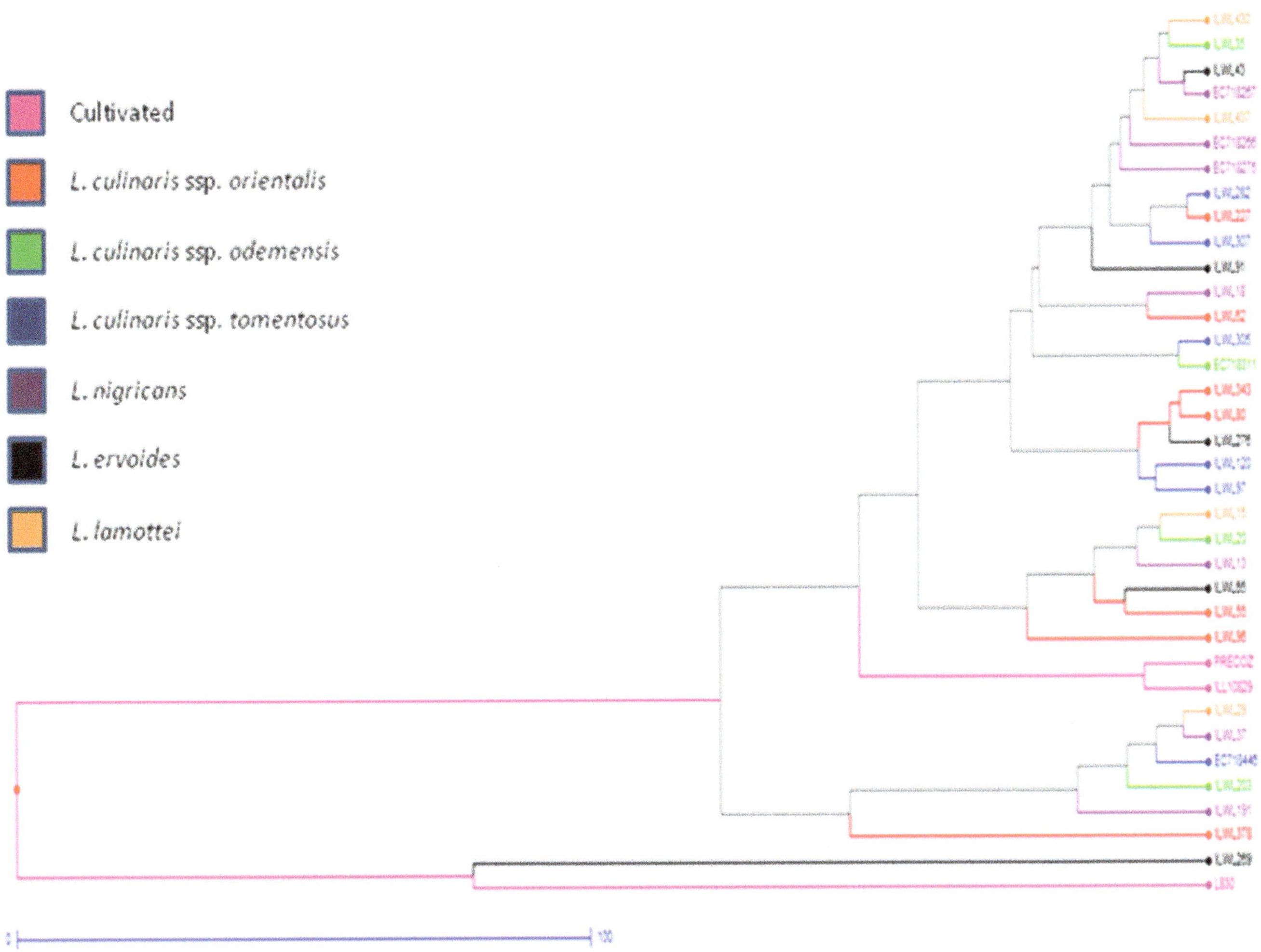

Figure 7. Hierarchical clustering of *Lens* core set accessions based on quantitative phenotypic data.

from rapid loss of invaluable genetic resources due to habitat destruction [47]. Important areas to target for *in-situ* conservation includes west Turkey for *L. nigricans*, Southwest Turkey, North-west Syria, south Syria and Jordan for *L. culinaris* spp. *orientalis*, south Syria for *L. culinaris* ssp. *odemensis* and the coastal border region between Turkey and Syria stretching along the Syrian coast for *L. ervoides* [48]. The accumulated information on this global collection of wild *Lens* will be used to conduct a gap analysis using appropriate approaches to guide future collecting missions to fill the gaps, but also to conduct traits-targeted collecting missions to sample new accessions with potential resistance/tolerance to major biotic and abiotic stresses. The same information can be used to define natural habitats that can allow effective *in-situ* conservation of the species richness as well as populations with potential sought traits. Maxted et al. (2012b) [49] proposed the Fertile Crescent, mainly the Aegean and South-western region of Turkey, North Syria and Lebanon as appropriate sites for *in-situ* conservation of most of temperate legume species.

Lens core collection

"PowerCore" is a rapid approach for developing core collection, which effectively simplifies the generation process of core set with reduced number of core entries, but maintaining high percent of diversity as compared to other methods used [22,50]. For larger genebank holdings, the core sets identified using "PowerCore" normally are small in size with greater diversity captured as compared to traditional clustering procedures. However, in the present study, as the number of accessions representing different *Lens* taxa in the entire collection were low, relatively more number of accessions were selected when quantitative variables were used either in combination with qualitative characters or alone. Using qualitative variables alone, a smaller core size of about 9% was achieved.

Majority of the wild *Lens* taxa in the entire collection belongs to Syria (37%) and Turkey (28%). However, maximum representation of accessions in core collections, derived from all data sets was from Turkey, followed by Syria, indicating rich diversity for many of the wild relatives of cultivated lentil. It is only in Aegean and the South-western region that the distribution of all four wild taxa of the genus *Lens* overlaps [51]. Unfortunately, Turkey, like other Mediterranean countries is suffering the rapid loss of many of its valuable genetic resources. The Aegean and South-western region of Turkey is, therefore, particularly important for exploration and collection of wild taxon for conservation *ex-situ* and also for analysis of within population variations in order to assess the threat of genetic erosion to each species and potential for *in-situ* conservation. Ferguson et al. (1996) [51] reported that *L. culinaris* ssp. *odemensis* and *L. ervoides* are the taxa most threatened by

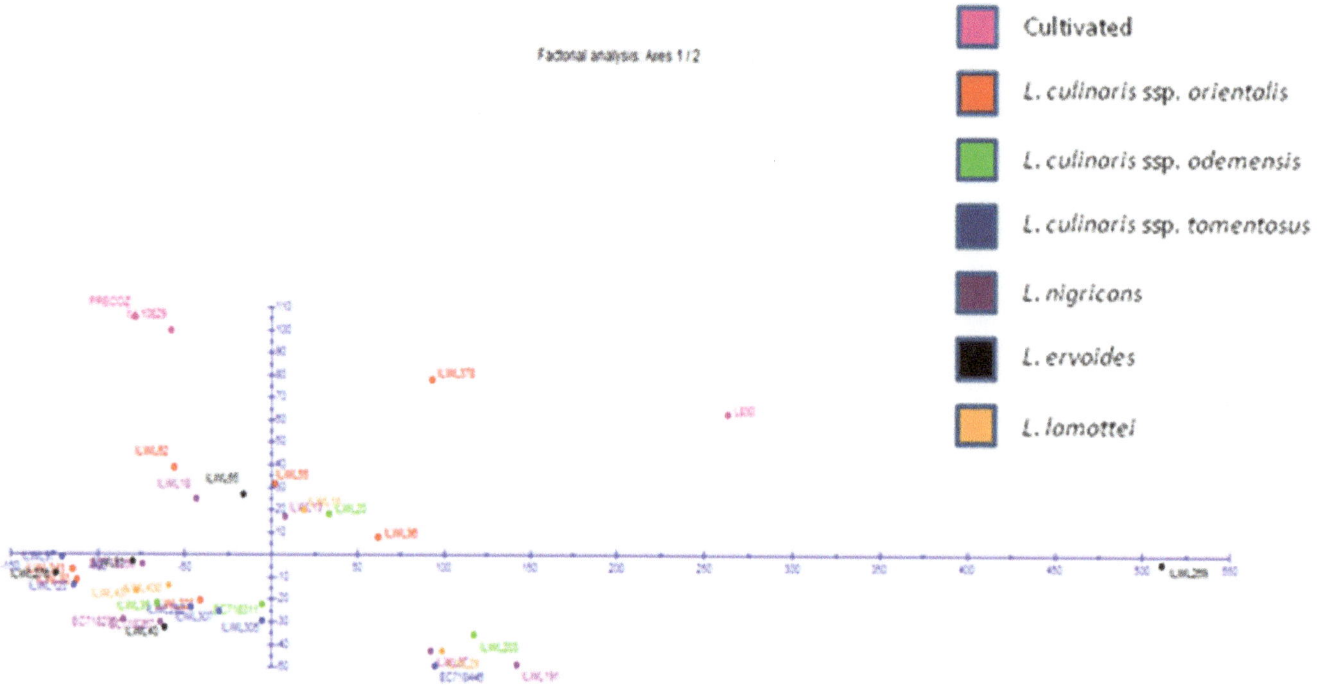

Figure 8. Factorial analysis of *Lens* core set accessions based on quantitative phenotypic data.

genetic erosion. Selection of core entries using different data sets gives a choice to the users as representation of common accessions across different core sets varies with the data sets used. A total of 66 and 25 accessions were common between the core sets developed using combined data on qualitative and quantitative variables and that obtained from quantitative and qualitative variables alone, respectively. Only 15 accessions were common between the core set obtained from quantitative and qualitative variables separately. The user, therefore, has a choice to use his experience and decide which core sets better serves the specific purpose.

Of the 46 accessions identified as promising for various desired traits from the entire *Lens* accessions (Table 7), a total of 21, 15 and 10 accessions, respectively, were represented in core sets derived from different data sets i.e. quantitative and qualitative traits combined, and quantitative and qualitative traits individually. Representation of proportionately greater number of desired accessions in the core set derived from qualitative morphological data could be the existence of good polymorphism with different descriptor states for qualitative traits with major gene effect. The core sets derived from different data sets gives us an opportunity to precisely estimate diversity at genetic/molecular level for use by the germplasm curators and the users. As majority of accessions in the core set derived from qualitative data originate from Turkey, this indirectly indicates greater polymorphism occurring in accessions originating from parts of Turkey (Table 8). Therefore, it is imperative that the wild gene pool of Turkey needs to be further explored, collected, conserved, characterized, and used in crop improvement. As has been emphasized, for broadening the genetic base of crop productions, where domestication is the bottleneck, protecting the wild gene pool is an utmost necessity in order to reconstructing the early evolutionary stages [52] to facilitate evolution. When selecting the species to conserve *in-situ*, the highest emphasis should be given to most genetically distinct groups of taxa (the widespread species with a wide range of adaptation). *L. culinaris* ssp. *orientalis* are distributed in Turkey throughout the northern belt of Fertile Crescent and share different types of habitat in different ecosystems, and they have a wide range of adaptability [53]. A few promising common accessions found across all core sets include ILWL29 (*L. lamottei*) from Spain; ILWL37 (*L. nigricans*) from Turkey; ILWL97 (*L. culinaris* ssp. *tomentosus*) from Turkey; ILWL269 (*L. ervoides*) from Turkey, and ILWL378 (*L. culinaris* ssp. *orientalis*) from Syria (Table 9). Besides possessing disease resistance, these accessions were also promising for yield related traits as high number of pods plant^{-1} and have significant breeding potential for yield enhancement.

Diversity assessment

The present study depicted a wide genetic variation in the wild annual *Lens* collection for agro-morphological traits as revealed by the dendrogram generated from the dissimilarity matrix and factorial analysis using DARwin 5 approach. In case of entire collection, qualitative data analysis revealed that maximum variability among accessions was interspecific and not limited to a particular country and this could be partly explained by the taxonomic differences. But in case of quantitative data, minimum as well as maximum diversity was reported among *L. ervoides* accessions collected from Turkey and Syria. Among core set accessions, qualitative data analysis revealed maximum interspecific as well as intraspecific variability between *L. ervoides* and *L. culinaris* ssp. *orientalis,* and among *L. culinaris* ssp. *orientalis*, accessions, respectively. Further, quantitative data revealed maximum diversity among *L. ervoides* accessions from Turkey. The diversity could mainly be attributed to diverse agro-ecological conditions as the current study included global wild annual *Lens* collection. The Euclidian distances calculated based on the quantitative data manifested large genetic distances among the accessions. Vieira et al. (2007) [54] analysed 19 wheat accessions for 17 phenotypic characters and reported genetic distances upto

Table 7. Identification of promising accessions for few agronomic and major biotic traits for their introgression in lentil enhancement selected from the entire *Lens* collection.

S. No.	Accessions	Species	Gene pool	Trait of Interest	Origin
1.	ILWL 247, 255, 476,	*L. culinaris* ssp. *orientalis*	Primary	High number of pods plant^{-1}, resistant to rust	Syria, Turkey
2.	ILWL 346, 350, 369, 378, 490	*L. culinaris* ssp. *orientalis*	Primary	High number of pods plant^{-1}, resistant to powdery mildew	Syria, Armenia
3.	ILWL 372	*L. culinaris* ssp. *orientalis*	Primary	High number of pods plant^{-1}, resistant to rust and powdery mildew	Uzbekistan
4.	ILWL 80, 381	*L. culinaris* ssp. *orientalis*	Primary	Resistant to rust and powdery mildew	Tajikistan
5.	ILWL 456	*L. culinaris* ssp. *orientalis*	Primary	High 100-seed weight and resistant to rust	Turkey
6.	ILWL 39	*L. culinaris* ssp. *odemensis*	Primary	Resistant to rust and powdery mildew	Turkey
7.	ILWL 165, 203, 237, 238	*L. culinaris* ssp. *odemensis*	Primary	High number of pods plant^{-1}, resistant to rust	Syria
8.	ILWL 167	*L. culinaris* ssp. *odemensis*	Primary	Resistant to rust and powdery mildew	Syria
9.	ILWL 97	*L. culinaris* ssp. *tomentosus*	Primary	Resistant to powdery mildew	Turkey
10.	ILWL 121, EC-718446	*L. culinaris* ssp. *tomentosus*	Primary	High number of pods plant^{-1}	Syria
11.	ILWL 23, 24, 25	*L. nigricans*	Secondary/ Tertiary	Resistant to rust	Italy, HRV
12.	ILWL 22	*L. nigricans*	Secondary/ Tertiary	Resistant to rust and powdery mildew	Italy
13.	ILWL 37, 191	*L. nigricans*	Secondary/ Tertiary	High number of pods plant^{-1}, resistant to rust and powdery mildew	Turkey
14.	ILWL 40, 41, 42, 159, 269, 292, 294, 321, 398, 418, EC-718428, 718439	*L. ervoides*	Secondary/ Tertiary	High number of pods plant^{-1}, resistant to rust and powdery mildew	Ukraine, Turkey, Italy, Syria,
15.	ILWL 50, 54, 56, 58	*L. ervoides*	Secondary/ Tertiary	High number of pods plant^{-1}, resistant to rust	HRV, Israel, PSE, Turkey
16.	ILWL 91	*L. ervoides*	Secondary/ Tertiary	Resistant to powdery mildew	Turkey
17.	ILWL 29, EC-718692	*L. lamottei*	Secondary/ Tertiary	High number of pods plant^{-1}	Spain, France

196.61. A large genetic distance between heterotic germplasm can be useful for developing lines with good combining ability in heterosis breeding [55–56]. The knowledge of genetic diversity among accessions provides clues about the heterotic potential that can be exploited while making crosses in order to create polymorphic populations. The clustering of accessions on the basis of phenotypic multivariate data, as compared to the molecular classification, was observed to be a much better procedure to explain the genotypic effects, since, the percentage of sum of squares between phenotypic sub-populations was higher in magnitude than that of between genetic sub-populations for all the traits analysed [57].

When genetic diversity was compared based on qualitative traits between entire set and core set, the core set included accessions which explained reasonable variability. In case of wild taxa namely *L. culinaris* ssp. *culinaris*, *L. culinaris* ssp. *orientalis*, *L. culinaris* ssp. *odemensis* and *L. lamottei*, core set included representative accessions from each group categorised based on entire set qualitative data excluding two to three accessions from *L. nigricans* and *L. culinaris* ssp. *tomentosus*, each. Further, accessions of various species namely *L. culinaris* ssp. *odemensis* from Palestine, *L. nigricans* from Italy and France, and *L. lamottei* from Turkey geographical origins constituted separate groups alone. Thus overall, it can be suggested that based on qualitative data, core set accessions revealing substantial diversity as available in the entire global collection. In case of quantitative data, core set included accessions from all the species and from majority of outgroups formed in genetic diversity analysis of entire set. Wang et al. (2006) [58] developed a core set capturing the genetic diversity of large soybean collection using 2% of total accessions to represent about 70% of the diversity from a whole sample set.

The tree formed from Dice Co-efficient and Euclidean distances of the standardized phenotypic means showed that genetic relatedness rarely matched analogous geographical origins in general. Plants that are phenotypically similar are not necessarily genetically so, as different gene pools could result in similar phenotypes. The reason might be continuous gene flow among wild species occurring in nature. Among wild taxa, one important aspect for species prioritization is the degree of relatedness, as it determines the actual potential for introducing useful traits from

Table 8. Agro-ecological representation of species-wise accessions in entire *Lens* collection and core sets derived from different data sets.

S. No.	Origin	Accessions in entire collection	Accessions in core sets		
			I	II	III
1	Alpes-Cote d'Azur	1	1	1	–
2	Armenia	2	1	1	–
3	Azerbaijan	1	–	–	–
4	Bahrain	4	–	1	–
6	Cyprus	3	–	–	–
7	Former Soviet Union	2	–	–	–
8	Former Yugoslavia	2	–	–	–
9	France	3	3	2	1
10	HRV	14	2	2	1
11	Iran	2	–	–	–
12	Israel	10	2	1	1
13	Italy	8	2	3	2
14	Jordan	11	1	1	–
15	Lebanon	9	2	2	–
16	Montenegro	3	–	1	1
17	Palestine	1	1	–	1
18	PSE	3	–	–	–
19	Russia	1	–	–	–
20	Spain	12	6	5	2
21	Syria	152	20	24	3
22	Tajikistan	4	2	1	–
23	Turkmenistan	2	1	1	1
24	Turkey	113	28	22	14
25	UBA	1	–	–	–
26	Ukraine	3	1	1	–
27	Uzbekistan	8	2	–	–
28	Unknown	30	21	17	9
	Total	**405**	**96**	**86**	**36**

Table 9. Identification of some promising accessions for few agronomic and major biotic traits for their introgression in lentil enhancement selected from the qualitative core set.

S. No.	Accessions	Species	Gene pool	Trait of Interest	Origin
1.	ILWL 80	*L. culinaris* ssp. *orientalis*	Primary	Resistant to rust and powdery mildew	Tajikistan
2.	ILWL 378	*L. culinaris* ssp. *orientalis*	Primary	High number of pods plant^{-1}, resistant to powdery mildew	Turkmenistan
3.	ILWL 203	*L. culinaris* ssp. *odemensis*	Primary	High number of pods plant^{-1}, resistant to rust	Syria
4.	EC-718446	*L. culinaris* ssp. *tomentosus*	Primary	High number of pods plant^{-1}	Syria
5.	ILWL 97	*L. culinaris* ssp. *tomentosus*	Primary	Resistant to powdery mildew	Turkey
6.	ILWL 37	*L. nigricans*	Secondary/Tertiary	High number of pods plant^{-1}, resistant to rust and powdery mildew	Turkey
7.	ILWL 191	*L. nigricans*	Secondary/Tertiary	High number of pods plant^{-1}, resistant to rust and powdery mildew	Turkey
8.	ILWL 91	*L. ervoides*	Secondary/Tertiary	Resistant to powdery mildew	Turkey
9.	ILWL 269	*L. ervoides*	Secondary/Tertiary	High number of pods plant^{-1}, resistant to rust and powdery mildew	Turkey
10.	ILWL 29	*L. lamottei*	Secondary/Tertiary	High number of pods plant^{-1}	Spain

the wild into cultigen. Relatedness information derived from a combination of biosystematics based upon traditional morphological and genotypic data, which are being used increasingly as well as information derived from plant breeders attempting crosses between CWRs and crops [59] are extremely useful for planning future study for the introgression of crop wild relatives.

Conclusions

The present study has helped in identifying the useful gene sources viz, earliness, high number of branches plant^{-1}, high number of pods plant^{-1} and, multiple disease resistance etc. in different *Lens* taxa across gene pools. Diversity analysis suggested ample scope for future germplasm collecting from the hotspots in Turkey and Syria. The promising trait-specific accessions reported particularly in *L. nigricans* and *L. ervoides* need to be considered while planning future lentil breeding programme for introgressing gene of interests from wild *Lens*. Further, certain accessions belonging to the secondary gene pool of *Lens* and also extracted in core set developed, can be used as a starting material aimed at large scale base broadening of cultivated lentil.

Acknowledgments

The authors acknowledge the Genetic Resources Unit at the International Centre for Agricultural Research in Dry Areas (ICARDA), Aleppo, Syria, for providing global wild *Lens* collection. The authors also thank the Vice Chancellor, HPAU, Palampur, Himachal Pradesh, India for allowing use of the hot spot facility at Sangla and Dhaulakuan centre.

Author Contributions

Conceived and designed the experiments: MS. Performed the experiments: MS ISB. Analyzed the data: Sandeep Kumar MD KCB MK. Contributed reagents/materials/analysis tools: AS AA Shiv Kumar. Contributed to the writing of the manuscript: MS ISB Sandeep Kumar. Provided necessary support for inter-institutional collaboration: SKD.

References

1. Lane A, Jarvis A (2007) Changes in climate will modify the geography of crop suitability: Agricultural biodiversity can help with adaptation. Paper presented at ICRISAT/CGIAR 35th Anniversary Symposium. "Climate Proofing Innovation for Poverty Reduction and Food Security" 22–24 November 2007, ICRISAT, Patancheru, India.
2. Maxted N, Ford BV, Kell SP, Iriondo JM, Dulloo ME, et al. (2008) Crop wild relative conservation and use. Wallingford, UK, CABI publishing.
3. Soltis DE, Smith SA, Cellinese N, Wurdack KJ, Tank DC, et al. (2011) Angiosperm phylogeny: 17 genus, 640 taxa. Amer J Bot 98 (4): 704–730.
4. Sonnante G, Hammer K, Pignone D (2009) From the cradle of agriculture a handful of lentils. History of domestication. Rendiconti. Lincei 20 (1): 21–37.
5. Arumuganathan K, Earle ED (1991) Nuclear DNA content of some important plant species. Plant Mol Biol Rep 9(3): 208–218.
6. Ferguson ME, Maxted N, Slageren MV, Robertson LD (2000) A re-assessment of the taxonomy of *Lens* Mill. (Leguminosae, Papilionoideae, Vicieae). Bot J Linn Soc 133(1): 41–59.
7. Erskine W, Muehlbauer FJ, Sarker A, Sharma B (2009) The Lentil: Botany, Production and Uses. Wallingford, UK, CABI publishing.
8. Muench DG, Slinkard AE, Scoles GJ (1991) Determination of genetic variation and taxonomy in lentil species by chloroplast DNA polymorphism. Euphytica 56(3): 213–218.
9. Alveraz MT, Garcia P, Perez de la Vega M (1997) RAPD polymorphism in Spanish lentil landraces and cultivars. J Genet Breed 51(2): 91–96.
10. Ford R, Pang ECK, Taylor PWJ (1997) Diversity analysis and species identification in *Lens* using PCR generated markers. Euphytica 96(2): 247–255.
11. Duran Y, Fratini R, Garcia P, Perez de la Vega M (2004) An inter sub-specific genetic map of *Lens*. Theor Appl Genet 108: 1265–1273.
12. Gupta D, Sharma SK (2006) Evaluation of wild *Lens* taxa for agro-morphological traits, fungal diseases and moisture stress in North-western Indian Hills. Genet Resour Crop Evol 53: 1233–1241.
13. Singh M, Bisht IS, Dutta M, Kumar K, Gautam NK, et al. (2011) Widening genetic base through pre-breeding in lentil and chickpea for faster utilization of genetic resources at the National Seminar on contemporary approaches to crop improvement, May 22–25 at Bangalore, India pp 365–366.
14. Scott JM, Davis F, Csuti B, Noss R, Butterfield B, et al. (1993) Gap analysis: a geographic approach to protect of biological diversity. Wildlife Monographs 123: 3–41.
15. Maxted N, Hargreaves S, Kell SP, Amri A, Street K, et al. (2012a) Temperate forage and pulse legume genetic gap analysis. Bocconea 24: 5–36.
16. Frankel OH (1984) Genetic perspectives of germplasm conservation. In: Arber WK, Llimensee K, Peacock WJ, Starlinger P, ed. Genetic Manipulation: Impact on Man and Society. Cambridge: Cambridge University Press. pp 161–170.
17. Frankel OH, Speer AHD (1984) Current plant genetic resources: a critical appraisal. In: Chopra VL, Joshi BC, Sharma RP, Sharma HC, eds. Genetics: New Frontiers. Vol. IV. Oxford & IBH Publ. Co. New Delhi, India.
18. Brown AHD (1989a) Core collections: A practical approach to genetic resources management. Genome 3: 818–824.
19. Brown AHD (1989b) The case for core collections. In: Brown AHD, ed. The use of plant genetic resources. Cambridge University Press, Cambridge, England. pp 136–155.
20. Gouesnard B, Bataillon TM, Decoux G, Rozale C, Schoen DJ, et al. (2001) M-Strat: an algorithm for building germplasm core collection by maximizing allelic or phenotypic richness. J Hered 92: 93–94.
21. Jansen J, van Hintum TJL (2007) Genetic distance sampling: a novel sampling method for obtaining core collections using genetic distances with an application to cultivated lettuce. Theor Appl Genet 114: 421–428.
22. Kim KW, Chung HK, Cho GT, Ma KH, Chandrabalan D, et al. (2007) Power Core: a program applying the advanced M strategy with a heuristic search for establishing core sets. Bioinformatics 23: 2155–2162.
23. Thachuk C, Crossa J, Franco J, Dreisigacker S, Warburton M, et al. (2009) Core Hunter: An algorithm for sampling genetic resources based on multiple genetic measures. BMC Bioinformatics pp 10: 243.
24. Zdenik S, Iva F, Ladislav D, Sergey M, Dobrotvorskaya T (2006) Genealogical analysis in the Czech Spring Wheat collection and its use for the creation of core collection. Czech J Genet Breed 42: 117–125.
25. Upadhyaya HD (2003) Phenotypic diversity in groundnut core collection assessed by morphological and agronomical evaluations. Genet Resour Crop Evol 50: 539–550.
26. Upadhyaya HD, Ortiz R, Bramel PJ (2003) Development of a groundnut core collection using taxonomical, geographical and morphological descriptors. Genet Resour Crop Evol 50: 139–148.
27. Francois B, Valérie R, Pjotr S, Florence EV, Pierre S, et al. (2007) A worldwide bread wheat core collection arrayed in a 384-well plate. Theor Appl Genet 114: 1265–1275.
28. Borba TCO, Brondani RPV, Rangel PHN, Brondani C (2009) Microsatellite marker– mediated analysis of the EMBRAPA rice core collection genetic diversity. Genetica 137: 293–304.
29. Yan W, Rutger JN, Bryant RJ, Bockelman HE, Fjellstrom RG, et al. (2007) Development and evaluation of a core subset of the USDA rice germplasm collection. Crop Sci 47: 869–878.

30. Bari A, Street K, Mackay M, Endresen DTF, De Pauw E, et al. (2012) Focused Identification of germplasm strategy (FIGS) detects wheat stem rust resistance linked to environmental variables. Genet Resour Crop Evol 59: 1465–1481.
31. Khazaei H, Street K, Bari A, Mackay M, Stoddard FL (2013) The FIGS (Focused Identification of Germplasm Strategy) approach identifies traits related to drought adaptation in *Vicia faba* genetic resources. PLoSONE 8(5): e63107. doi:10.1371/journal.pone.0063107.
32. IBPGR/ICARDA (1993) Lentil descriptors: IBPGR Secretariat, Rome, Italy pp 1–15.
33. Mayee CD, Datar VV (1986) Phytopathometry. Technical Bulletin-1, Marathwada Agricultural University, Parbhani, p 218.
34. Tiwari KR, Penner GA, Warkentin TD, Rashid KY (1997) Pathogenic variation in *Erisiphe polygoni*, the casual organism of powdery mildew of lentil. Can J Plant Pathol 19: 267–271.
35. Shannon CE, Weaver W (1949) The Mathematical Theory of Communication. University of Illinois Press, Urbana.
36. Dice LR (1945) Measurements of the amount of ecologic association between species. Ecology 26: 297–302.
37. Perrier X, Flori A, Bonnot F (2003) Data analysis methods In: P. Hamon, M. Seguin, X. Perrier, J.C. Glaszmann, eds. Genetic diversity of cultivated tropical plants. Enfield, Science Publishers. Montpellier pp 43–76.
38. Perrier X, Jacquemoud-Collet JP (2006) DARwin software. Available: http://darwin.cirad.fr/darwin.
39. Duvick DN (1984) Genetic diversity in major farm crops on the farm and in reverse. Eco Bot 38: 151–178.
40. Lazaro A, Ruiz M, Rosa L, Martin I (2001) Relationship between agro-morphological characters and climate parameters in Spanish landraces of lentil. Genet Resour Crop Evol 48: 239–249.
41. Naghavi MR, Jahansouz MR (2005) Variation in the agronomic and morphological traits of lentil accessions. J Integ Plant Biol 47: 375–379.
42. Tanksley SD, McCouch SR (1997) Seed banks and Molecular maps: unlocking genetic potentials from the wild. Science 277: 418–423.
43. Gur A, Zamir D (2004) Unused natural variation can lift yield barriers in Plant Breeding. PLoS Biol 2: e245.
44. Kumar GR, Sakthivel K, Sundaram RM, Neeraja CN, Balachandran SM, et al. (2010) Allele mining in crops: prospects and potential. Biotech Adv 28: 451–461.
45. Muehlbauer FJ, Singh KB (1987) Genetics of chickpea. In: Saxena MC, Singh KB, ed. The chickpea. CAB International, Cambridge pp 99–126.
46. Pundir RPS, Rao NK, van der Maesen LJG (1985) Distribution of qualitative traits in the world germplasm of chickpea. Euphytica 34(3): 697–703.
47. Solh M, Erskine W (1981) In: Webb C, Hawtin GC, eds. Commonwealth Agricultural Bureaux, Slough, UK. Plant Genet Resour pp 54–67.
48. Ferguson ME, Newbury HJ, Maxted N, Ford-Lloyd BV, Robertson LD (1998) Population genetic structure in *Lens* taxa revealed by isozyme and RAPD analysis. Genet Resour Crop Evol 45: 549–559.
49. Maxted N, Kell S, Ford-Lloyd B, Dulloo E, Toledo A (2012b) Towards the systematic conservation of global crop wild relative diversity. Crop Sci 52(2): 774–785.
50. Gowda J, Krishanappa M, Pathak N, Mathur PN, Seetharam A (2012) Use of Heuristic Approach for the Development of a Core Set from Large Germplasm Collection of Foxtail Millet. Indian J Plant Genet Resour 26: 13–18.
51. Ferguson MN, Açikgöz A, Ýsmaýl AS, Cýnsoy B (1996) An eco-geographic survey of wild *Lens* species in Aegean and South West Turkey. Ana J Agric Sci 6: 159–166.
52. Cooper DC, Spillane C, Hodgkin T (2000) Broadening the genetic bases of crop production, IPGRI, FAO: Rome, CABI: Wallingford 480.
53. Tan A, Tan AS (2002) *In-situ* conservation of wild species related to crop plants: the case of Turkey. Managing Plant Genetic Diversity. In: Engels JMM, Ramanatha Rao V, Brown AHD, Jackson MT, eds. CABI Publishing, UK pp 195–204.
54. Vieira EA, de Carvalho FIF, Bertan I, Kopp M, Zimmer PD, et al. (2007) Association between genetic distances in wheat (*Triticum aestivum* L.) as estimated by AFLP and morphological markers. Genet Mol Biol 30(2): 392–399.
55. Melchinger AE (1999) Genetic diversity and heterosis. In: JG Coors, and S Pandey, eds. The genetics and exploitation of heterosis in crops, ASA, CSSA, and SSSA, Madison, WI. pp 99–118.
56. Moll RH, Longquist JH, Fortuna JV, Johnson EC (1965) The relation of heterosis and genetic divergence in maize. Genetics 52: 139–144.
57. Royo C, Maccaferri M, Alvaro F, Moragues M, Sanguineti MC, et al. (2010) Understanding the relationships between genetic and phenotypic structures of a collection of elite durum wheat accessions. Field Crops Res 119: 91–105.
58. Wang L, Guan Y, Guan R, Li Y, Ma Y, et al. (2006) Establishment of Chinese soybean (*Glycine max*) core collections with agronomic traits and SSR markers. Euphytica 151: 215–223.
59. Singh M, Rana MK, Kumar K, Bisht IS, Dutta M, et al. (2013) Broadening the genetic base of lentil cultivars through inter-sub-specific and interspecific crosses of *Lens* taxa. Plant Breed 132 (6): 667–675.

Development of DArT Marker Platforms and Genetic Diversity Assessment of the U.S. Collection of the New Oilseed Crop Lesquerella and Related Species

Von Mark V. Cruz[1,3], Andrzej Kilian[2], David A. Dierig[1]*

1 USDA-ARS National Center for Genetic Resources Preservation, Fort Collins, Colorado, United States of America, **2** Diversity Arrays Technology Pty. Ltd., Yarralumla, ACT, Australia, **3** Department of Bioagricultural Sciences and Pest Mgt., Colorado State University, Fort Collins, Colorado, United States of America

Abstract

The advantages of using molecular markers in modern genebanks are well documented. They are commonly used to understand the distribution of genetic diversity in populations and among species which is crucial for efficient management and effective utilization of germplasm collections. We describe the development of two types of DArT molecular marker platforms for the new oilseed crop lesquerella (*Physaria* spp.), a member of the Brassicaceae family, to characterize a collection in the National Plant Germplasm System (NPGS) with relatively little known in regards to the genetic diversity and traits. The two types of platforms were developed using a subset of the germplasm conserved *ex situ* consisting of 87 *Physaria* and 2 *Paysonia* accessions. The microarray DArT revealed a total of 2,833 polymorphic markers with an average genotype call rate of 98.4% and a scoring reproducibility of 99.7%. On the other hand, the DArTseq platform developed for SNP and DArT markers from short sequence reads showed a total of 27,748 high quality markers. Cluster analysis and principal coordinate analysis indicated that the different accessions were successfully classified by both systems based on species, by geographical source, and breeding status. In the germplasm set analyzed, which represented more than 80% of the *P. fendleri* collection, we observed that a substantial amount of variation exists in the species collection. These markers will be valuable in germplasm management studies and lesquerella breeding, and augment the microsatellite markers previously developed on the taxa.

Editor: Matthew E. Hudson, University of Illinois, United States of America

Funding: Funding for the study was provided by U.S. Department of Agriculture (USDA) - Agricultural Research Service Project No. 5402-21000-007-00. The acquisition of new germplasm in 2010 was supported by the NPGS Plant Exploration/Exchange Program. No additional external funding was received or used for this study. USDA determines the research priorities of all research that it funds. Publication of research is subject to review of USDA officials. Specific study design, data collection, data analysis, and preparation of the manuscript is under the control of the individual researchers.

Competing Interests: The authors acknowledge the help and technical support from the employees of Diversity Arrays Technology Pty. Ltd. AK is an employee of Diversity Arrays Technology Pty Ltd, which offers genome profiling service using the technologies described in this report. This fact, however, has not interfered whatsoever with the full, objective, transparent and unbiased presentation of the research results described in the manuscript nor alters the authors' adherence to all the PLOS ONE policies on sharing data and materials.

* E-mail: david.dierig@ars.usda.gov

Introduction

Lesquerella (*Physaria* sp.), a member of the Brassicaceae, is a North American genus which has more than 90 member species thriving mostly in dry and arid habitats usually mixed with sparse vegetation [1], [2]. Commonly known as bladderpod, it has been identified by the U.S. Department of Agriculture and the U.S. Department of Energy as very promising species for the production of lubricants, engine oils, waxes, coatings and a source of natural estolides with valuable application in the automobile and biofuel industries [3], [4], [5]. There has been an increasing trend in research activities in the crop since its oil was first characterized for novel properties [6]. The fatty acid profile of lesquerella oil has been found to vary. Species found east of the Mississippi have mostly densipolic acid, those in the western U.S. mostly lesquerolic acid, and a species from Oklahoma predominantly with auricolic acid [7]. Lesquerella could be grown successfully as a winter annual in the southwest U.S. producing an average seed yield of 1.7 tons/ha [8]. At present, there are more than eight advanced breeding lines of *P. fendleri* ready for commercialization and a corresponding substantial germplasm collection has been assembled in the U.S. National Plant Germplasm System [9].

Molecular characterization of germplasm collections supplements phenotypic assessment of diversity and is important in the effective management of genetic resources. Molecular markers have been very useful in efforts to accurately identify gaps and redundancy within and among individual germplasm collections and have helped resolve important genebank management issues [10], [11], [12]. Examples of markers used in specific Brassicaceae collections include microsatellites for examining diversity in *B. napus* [13], [14], [15], lesquerella [16] and wild relatives such as *Capsella*, *Crambe* and *Sinapis* [17]; AFLPs in *B. oleracea* [18] and *Lepidium* [19]; and RAPD markers in *Raphanus* [20].

Compared to the previously mentioned molecular marker systems, Diversity Array Technology (DArT) markers are relatively new and were developed only in the early 2000. DArT markers overcame the difficulties in correlating bands on gels with allelic variants by utilizing hybridization-based methods and solid state surfaces [21]. They have been increasingly utilized as evidenced by the number of publications reporting successful implementation [22]. Among its advantages over other marker systems include high throughput capability allowing rapid germplasm characterization in a single experiment, independence of sequence data, and ability to detect single base changes and indels [23]. To date, DArT markers have been successfully applied in genetic diversity analysis, linkage mapping and in finding out population structure of collections in various crop species [24]. Their application in minor crops are likewise increasing due to their potential to accelerate gene discovery and initiate molecular breeding because of their whole genome coverage without relying on prior sequence data information [25], [26].

In lesquerella, allozymes have previously been used [27] and microsatellite markers have been developed [16]. But the small number of markers that are available presents a limitation in linkage mapping and in the study of genetic resources collections. At present, only fifteen microsatellites have demonstrated utility across different *Physaria* species. We present in this paper the development of two platforms of high density DArT markers for lesquerella and the results obtained after testing them to analyze the genetic diversity of *Physaria* and *Paysonia* national germplasm collection. This new molecular marker system for the new crop species will serve as an additional resource to augment the existing systems to assist crop improvement efforts, germplasm management activities and genetic studies.

Materials and Methods

Tissue Sampling

The lesquerella germplasm set was obtained from collections of the USDA-ARS National Arid Land Plant Genetic Resources Unit, Parlier, CA and the USDA-ARS Arid Land Agricultural Research Center, Maricopa, AZ. The samples used in this study were selected based on each accession's geographic location, sampling within distinct counties in each State when possible. Seven advanced *P. fendleri* breeding lines, WCL-LH1, WCL-LO1, WCL-LO2, WCL-LO4, WCL-LY1, WCL-SL1 and WCL-YS1 were included among the samples. DNA was extracted from fresh leaf tissue obtained from five week old seedlings using Qiagen DNeasy 96 Plant Kits (Qiagen Inc., Valencia, CA). Two DArT platforms were developed for lesquerella as described below. The platform development utilized 86 *Physaria* accessions (see Table 1), representing 11 species and one *Paysonia* accession since the latter is a sister genus of *Physaria* and only recently was there a recognized taxonomic delineation between them [2].

Microarray DArT Platform Development

The microarray DArT was developed by first testing combinations of the rare-cutting restriction enzyme *PstI* with several restriction endonucleases that cut frequently on DNA samples from 8 representative accessions to determine the restriction enzyme combination that provided the best complexity reduction. Final genomic representations were prepared using *PstI/BstNI* combinations. Approximately 50 ng of genomic DNA was digested with *PstI/BstNI* combinations and the resulting fragments ligated to a *PstI* overhang compatible oligonucleotide adapter. A primer annealing to this adapter was used in PCR reaction to amplify complexity-reduced representation of a sample. Amplification products were either used for cloning a in marker development process or labeled with fluorescent dyes and hybridized to DArT array in the genotyping process. Library construction was subsequently performed using 80 *P. fendleri* accessions and 16 accessions of wild related *Brassica* species. The amplified *PstI* restriction fragments from all accessions were cloned into pCR2.1-TOPO vector (Invitrogen, Australia) as described by Jaccoud et al. [21] and four libraries were generated. The white colonies containing genomic fragments inserted into pCR2.1-TOPO vector were picked into individual wells of 384-well microtiter plates filled with ampicillin/kanamycin-supplemented freezing medium. A total of 3,456 clones from *Physaria* were obtained –1,920 from three libraries of *P. fendleri* and 1,536 from the related species. Inserts from these clones were amplified using M13F and M13R primers in 384 plate format, a subset of PCR products were assessed for quality (10% of 25 μl PCR reaction) through gel electrophoresis, and all remaining PCR products dried, washed and dissolved in a spotting buffer. A total of 6,144 clones were printed with spot duplication on SuperChip poly-L-lysine slides (Thermo Scientific, Australia) using a MicroGrid arrayer (Genomics Solutions, UK). The microarrays included 1,920 *Brassica* clones and 768 *Arabidopsis* clones in addition to the *Physaria* clones.

Each sample was assayed using methods described above for library construction. Genomic representations were assessed for quality through gel electrophoresis in 1.2% agarose and labeled with fluorescent dyes (Cy3 and Cy5). Labelled targets were then hybridized to printed DArT arrays for 16 hours at 62°C in a water bath. Slides were washed as described by Kilian et al. [28], dried initially by centrifugation at 500 × g for 7 min and later by a desiccator under vacuum for 30 min. The slides were scanned using Tecan LS300 scanner (Tecan Group Ltd, Männedorf, Switzerland) generating three images per array: one image scanned at 488 nm for reference signal measures the amount of DNA within the spot based on hybridization signal of FAM-labeled fragment of a TOPO vector multiple cloning site fragment and two images for "target" signal measurement: one scanned at 543 nM (for Cy3 labeled targets) and one at 633 nM (for Cy5 labeled targets). All DArT techniques applied for this work were recently described in much more detail by Kilian et al. [28].

DArTseq Platform Development

For the sequencing-based DArT genotyping, four complexity reduction methods optimized for several other plant species at DArT P/L were used. *PstI*-RE site specific adapter was tagged with 96 different barcodes enabling encoding a plate of DNA samples to run within a single lane on an Illumina Genome Analyzer IIx (Illumina Inc., San Diego, CA). *PstI* adapter included also a sequencing primer site, so that the tags generated were always reading into the genomic fragments from the *PstI* sites. After the sequencing run, the FASTQ files were quality filtered using the threshold of 90% confidence for at least 50% of the bases and in addition filtered more stringently for barcode sequences. Two lanes of GAIIx were run with all samples providing fully replicated sequencing data. The filtered data were split into their respective target (individual) data using barcode splitting script. Each sample had on average 500,000 counts per replicate. After producing various QC statistics and trimming of the barcode the sequences were aligned against the reference created from the tags identified in the sequence reads generated from all the samples. In addition the short sequence tags were aligned against *Arabidopsis thaliana*'s genome available in Genbank. *Arabidopsis* is a close relative of *Physaria* in the Brassicaceae [29]. The output files from alignment generated using Bowtie software [30] were processed using an analytical pipeline developed by DArT P/L to produce "DArT score" tables and "SNP" tables.

Genotyping

Both DArT on array platform and DArTseq use a set of quality parameters to select markers which are of use for a specific

Table 1. Passport information of accessions used in *Physaria* DArT and DArTseq platform development.

Collection No. and ID	Accession No.	Species	Country	State/ Prov.	County	Latitude	Longitude	Elevation (m)
1809	W6 20822	*P. fendleri*	United States	AZ	Cochise	31.75600	−110.05500	134.15
1817	W6 20823	*P. fendleri*	United States	AZ	Cochise	31.74100	−109.96800	1493.90
1818	W6 20824	*P. fendleri*	United States	AZ	Cochise	31.50000	−110.00000	1432.56
1819	W6 20825	*P. fendleri*	United States	AZ	Santa Cruz	31.72400	−110.52600	1493.90
1826	W6 20856	*P. fendleri*	United States	AZ	Santa Cruz	31.72300	−110.52400	1463.41
1833	W6 20833	*P. gordonii*	United States	NM	Hidalgo	31.93333	−108.95000	1024.13
1834	PI 641918	*P. fendleri*	United States	NM	Hidalgo	31.93333	−108.93333	1292.35
1835	–	*P. fendleri*	United States	NM	Luna	32.21667	−107.41667	1277.11
1840	W6 20858	*P. fendleri*	United States	NM	Eddy	32.61667	−104.40000	1042.42
1841	PI 596417	*P. fendleri*	United States	NM	Eddy	32.31667	−104.25000	1130.81
1842	W6 20860	*P. fendleri*	United States	NM	Eddy	32.53333	−103.98333	1060.70
1843	PI 596418	*P. fendleri*	United States	NM	Eddy	32.75000	−104.10000	1082.04
1844	PI 596419	*P. fendleri*	United States	NM	Lincoln	33.68333	−105.83333	1636.78
1845	PI 596420	*P. fendleri*	United States	NM	Lincoln	33.68333	−105.91667	1572.77
1848	PI 596421	*P. fendleri*	United States	AZ	Graham	33.25000	−110.28333	795.53
1851	W6 20826	*P. fendleri*	United States	AZ	Cochise	31.91667	−109.16667	1493.52
1852	W6 20861	*P. fendleri*	United States	AZ	Cochise	31.65000	−110.01667	1447.80
1874	PI 596423	*P. fendleri*	United States	AZ	Apache	34.56667	−109.50000	1755.65
1889	PI 596425	*P. fendleri*	United States	NM	Socorro	33.88333	−106.40000	1661.16
1904	W6 20865	*P. fendleri*	United States	AZ	Apache	34.35000	−109.36667	1859.28
1906	PI 596426	*P. fendleri*	United States	NM	Socorro	34.41667	−106.65000	1645.92
1909	W6 20868	*P. fendleri*	United States	NM	Guadalupe	33.56667	−105.05000	1697.74
1910	PI 596428	*P. fendleri*	United States	NM	De Baca	34.46667	−104.30000	1149.10
1912	PI 596429	*P. fendleri*	United States	NM	Guadalupe	34.73333	−104.46667	1359.41
1919	PI 596430	*P. fendleri*	United States	NM	San Juan	36.60000	−108.91667	1758.70
1920	PI 596431	*P. fendleri*	United States	NM	San Juan	36.58333	−108.96667	1932.43
1924	W6 20851	*P. rectipes*	United States	AZ	Luna	35.86667	−109.60000	1912.00
1932	PI 596433	*P. fendleri*	United States	AZ	Graham	33.28333	−110.38333	877.82
2212	W6 20863	*P. argyraea*	United States	TX	Uvalde	29.15000	−99.70000	228.60
2226	W6 20828	*P. fendleri*	United States	TX	Webb	27.36667	−98.98333	243.84
2232	PI 643174	*P. lindheimeri*	United States	TX	Nueces	27.06667	−97.06667	107.00
2243	W6 20835	*Paysonia grandiflora*	United States	TX	Wilson	29.16667	−97.93333	91.44
2250	W6 20854	*P. recurvata*	United States	TX	Medina	29.53333	−99.21667	434.34
2255	PI 596435	*P. fendleri*	United States	TX	El Paso	31.83333	−106.01667	1341.12
2256	PI 596436	*P. fendleri*	United States	TX	Hudspeth	31.70000	−105.41667	1240.54
2260	PI 596439	*P. fendleri*	United States	TX	Hudspeth	31.81667	−105.13333	1219.20
2262	PI 596441	*P. fendleri*	United States	TX	Culberson	31.36667	−104.81667	944.88
2269	PI 596447	*P. fendleri*	United States	TX	Pecos	30.71667	−103.18333	975.36
2273	PI 596450	*P. fendleri*	United States	TX	Brewster	30.20000	−103.56667	1554.48
2277	PI 596452	*P. fendleri*	United States	TX	Brewster	29.78333	−103.18333	853.44
2278	PI 596453	*P. fendleri*	United States	TX	Brewster	30.20000	−103.20000	1310.64
2281	PI 596454	*P. fendleri*	United States	TX	Crockett	30.93333	−101.86667	640.08
2282	PI 596455	*P. fendleri*	United States	TX	Pecos	30.96667	−101.96667	563.88
2286	PI 596456	*P. fendleri*	United States	TX	Andrews	32.31667	−102.50000	807.72
2290	PI 596457	*P. fendleri*	United States	TX	Winkler	31.78333	−103.26667	853.44
2291	PI 596458	*P. fendleri*	United States	TX	Reeves	31.11667	−103.71667	944.88
2292	PI 596459	*P. fendleri*	United States	TX	Reeves	31.01667	−103.73333	1005.84
2294	PI 596461	*P. fendleri*	United States	TX	Jeff Davis	30.53333	−104.10000	1722.12
2295	PI 596462	*P. fendleri*	United States	TX	Jeff Davis	30.55000	−103.91667	1676.40

Table 2. Passport information of additional *Physaria* and *Paysonia* accessions genotyped using DArTseq.

Collection No. and ID	Accession No.	Species	Country	State/ Prov.	County	Latitude	Longitude	Elevation (m)
715	PI 293027	*P. fendleri*	United States	TX	Brewster	29.32768	−103.20527	1139.00
720	PI 293028	*P. fendleri*	United States	TX	Val Verde	29.84683	−101.68152	519.00
Barclay 998	PI 279650	*P. fendleri*	United States	NM	–	–	–	–
1836	W6 20857	*P. fendleri*	United States	NM	Luna	32.23333	−107.38333	1264.92
1838	PI 596415	*P. fendleri*	United States	NM	Chaves	32.86667	−105.03333	1621.54
1839	PI 596416	*P. fendleri*	United States	NM	Eddy	32.83333	−104.80000	1301.50
1854	W6 23374	*P. gordonii*	United States	AZ	Cochise	31.85000	−109.96667	1732.79
1855	PI 596422	*P. fendleri*	United States	AZ	Cochise	31.96667	−109.46667	1548.38
1871	W6 20862	*P. fendleri*	United States	AZ	Navajo	34.88333	−110.11667	1546.86
1879	W6 20838	*P. intermedia*	United States	AZ	Navajo	34.35000	−110.16667	1828.80
1880	PI 596424	*P. fendleri*	United States	AZ	Navajo	34.36667	−110.15000	1825.75
1902		*P. kaibabensis*	United States	AZ	Coconino	36.73333	−112.11667	2237.23
1907	PI 596427	*P. fendleri*	United States	NM	Torrance	34.55000	−106.13333	1850.14
1926	PI 596432	*P. fendleri*	United States	AZ	Navajo	34.75000	−110.16667	1615.44
1934	W6 20869	*P. fendleri*	United States	AZ	Graham	33.23333	−110.25000	908.30
2097	PI 331165	*P. fendleri*	United States	AZ	Graham	33.38333	−110.38333	917.00
2202	W6 20819	*P. densiflora*	United States	TX	Gillespie	30.20000	−98.93333	434.34
2203		*P. recurvata*	United States	TX	Kerr	30.00000	−99.36667	518.16
2217	PI 643172	*Paysonia lasiocarpa*	United States	TX	Frio	28.63333	−99.38333	198.12
2228	W6 20840	*Paysonia lasiocarpa*	United States	TX	Duval	27.58333	−98.56667	137.16
2246	W6 20836	*Paysonia grandiflora*	United States	TX	Atascosa	28.85000	−98.35000	121.92
2247	W6 20837	*Paysonia grandiflora*	United States	TX	Frio	29.06667	−98.85000	121.92
2257	PI 596437	*P. fendleri*	United States	TX	Hudspeth	31.73333	−105.21667	1097.28
2258	W6 20829	*P. fendleri*	United States	TX	Hudspeth	31.83333	−105.20000	1219.20
2259	PI 596438	*P. fendleri*	United States	TX	Hudspeth	31.81667	−105.13333	1188.72
2264	W6 20870	*P. fendleri*	United States	TX	Reeves	30.91667	−103.78333	914.40
2265	PI 596443	*P. fendleri*	United States	TX	Jeff Davis	30.86667	−103.78333	1127.76
2267	PI 596445	*P. fendleri*	United States	TX	Jeff Davis	30.68333	−103.78333	1234.44
2268	PI 596446	*P. fendleri*	United States	TX	Jeff Davis	30.48333	−103.75000	1554.48
2270	PI 596448	*P. fendleri*	United States	TX	Brewster	30.38333	−103.58333	1402.08
2274		*P. fendleri*	United States	TX	Brewster	30.03333	−103.56667	1280.16
2276	W6 20871	*P. fendleri*	United States	TX	Brewster	29.66667	−103.11667	822.96
2279	W6 20844	*P. mcvaughiana*	United States	TX	Pecos	30.46667	−102.91667	1432.56
2293	PI 596460	*P. fendleri*	United States	TX	Jeff Davis	31.00000	−104.18333	1310.64
2297	W6 20859	*P. fendleri*	United States	TX	Presidio	30.28333	−104.01667	1463.04
2401		*P. douglasii*	United States	WA	Grant	46.63333	−119.73333	89.61
2819	W6 20873	*P. fendleri*	–	–	–	–	–	–
2894		*Paysonia densipila*	United States	TN	Williamson	35.82861	−86.69722	208.00
2997	W6 20859	*P. fendleri*	United States	AZ	Apache	34.48333	−109.51667	1899.00
3029		*P. intermedia*	United States	UT	Garfield	38.02044	−112.36819	1944.00
3011	W6 20320	*Paysonia auriculata*	United States	OK	Blaine	35.79325	−98.42029	454.00
3000		*Paysonia lyrata*	United States	AL	Colbert	–	–	–
3030	PI 337050	*P. fendleri*	United States	NM	Sierra	33.36355	−107.28041	1500.00
3042	W6 20843	*P. ludoviciana*	United States	UT	Garfield	37.76753	−112.06660	1791.00
3050		*P. acutifolia*	United States	UT	San Juan	37.50160	−109.64779	1410.00
3058	PI 345712	*Paysonia auriculata*	United States	OK	Garfield	36.43716	−97.58647	360.00
3083	PI 355037	*P. fendleri*	United States	TX	Kinney	29.15197	−100.38664	288.00
3085		*P. acutifolia*	United States	CO	Mesa	38.83627	−108.56218	2073.00
3091	PI 355041	*Paysonia perforata*	United States	TN	Wilson	36.23255	−86.38455	182.00

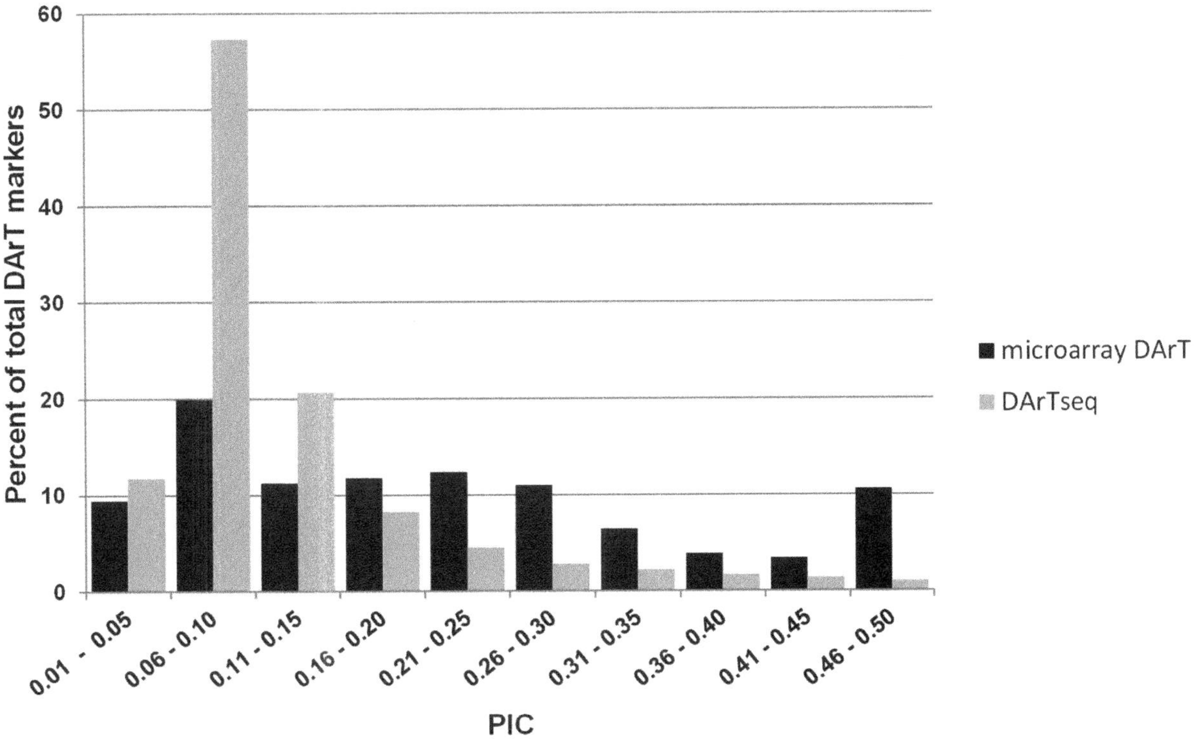

Figure 1. Distribution of marker PIC values by DArT platform.

application. One of these parameters is reproducibility of markers in technical replicates for a subset of samples. In diversity analysis, the reproducibility parameter threshold is set usually at 97% which translates to average reproducibility of the dataset around 99.7%. A total of 87 common accessions were genotyped using both the microarray DArT and DArTseq platforms. A total of 11 *Physaria* species, 1 interspecific hybrid, and 2 *Paysonia* species were analyzed by microarray DArT. Additional accessions were genotyped using the DArTseq platform comprised of 17 *Physaria* and 7 *Paysonia* species. Overall, there was a total of 177 accessions represented by single plant samples genotyped using the two platforms, majority of which are *P. fendleri* (Tables 1 and 2). The marker sequences and genotype data will be stored in the U.S. Germplasm Resources Information Network (GRIN) database (http://www.ars-grin.gov) along with the accessions' phenotypic observations and germplasm passport data curated by the U.S. National Plant Germplasm System.

Data Analysis

All the images from DArT platforms were analyzed using DArTsoft v.7.4.7 (DArT P/L, Canberra, Australia). The markers were scored as binary data (1/0), indicating presence or absence of a marker in genomic representation of each sample as described by Wenzl et al. [31]. For quality control, 30% of genotypes were genotyped in full technical replication. Clones with P>77%, a call rate >97% and >98% allele-calling consistency across the replicates were selected as markers. P value represents the allelic-states variance of the relative target hybridization intensity as a percentage of the total variance. The informativeness of the DArT markers was determined by calculating the polymorphism information content (PIC) within the panel of diverse accessions according to Anderson et al. [32]. The *P. fendleri* data were used in GenAlEx v.6.41 [33] to determine 2D spatial autocorrelation of the DArT microarray and Arlequin v.3.5.1.3 [34] to assess the amount of variation among the assigned regional groupings by AMOVA. To summarize the relationships among all examined accessions, cluster analysis was performed on Dice similarity values with the SAHN procedure using the unweighted pair-group method done using NTSYS-pc v. 2.21 m [35]. The Dice coefficient was preferred over simple matching coefficient because DArT is a dominant marker system and there were several non-*Physaria* species in the sample set [36], [37]. In addition, a Bayesian model-based clustering was performed on microarray DArT markers using STRUCTURE v.2.3.4 [38] testing 3 independent runs with *K* from 1 to 8, each run with a burn-in period of 50,000 iterations and 300,000 Monte Carlo Markov Chain (MCMC) iterations, assuming an admixture model and correlated allele frequencies. The STRUCTURE data was subsequently analyzed by HARVESTER v.06.92 [39]. Mantel tests were made to determine if there are significant correlations between the dendrogram representations and the distance matrices, between the matrices of geographic and genetic distances, and between the distance matrices from the two DArT platforms used in the *P. fendleri*. The missing geographic coordinates of collection sites of ten accessions (PI 293027, PI 293028, PI 337050, PI 345712, PI 355037, PI 355041, PI 355042, PI 275771, PI 283700 and PI 299412) in the GRIN database were estimated using Google Earth v. 6.1.0.5001 based on available locality description and collectors' notes and the information included in the analysis. Map projections of the analyzed *P. fendleri* accessions were made using ArcGIS Explorer v.2.0.0.1750 (ESRI, Redlands, CA) and non-parametric correlation tests done using JMP v.9 (SAS Institute, Cary, NC).

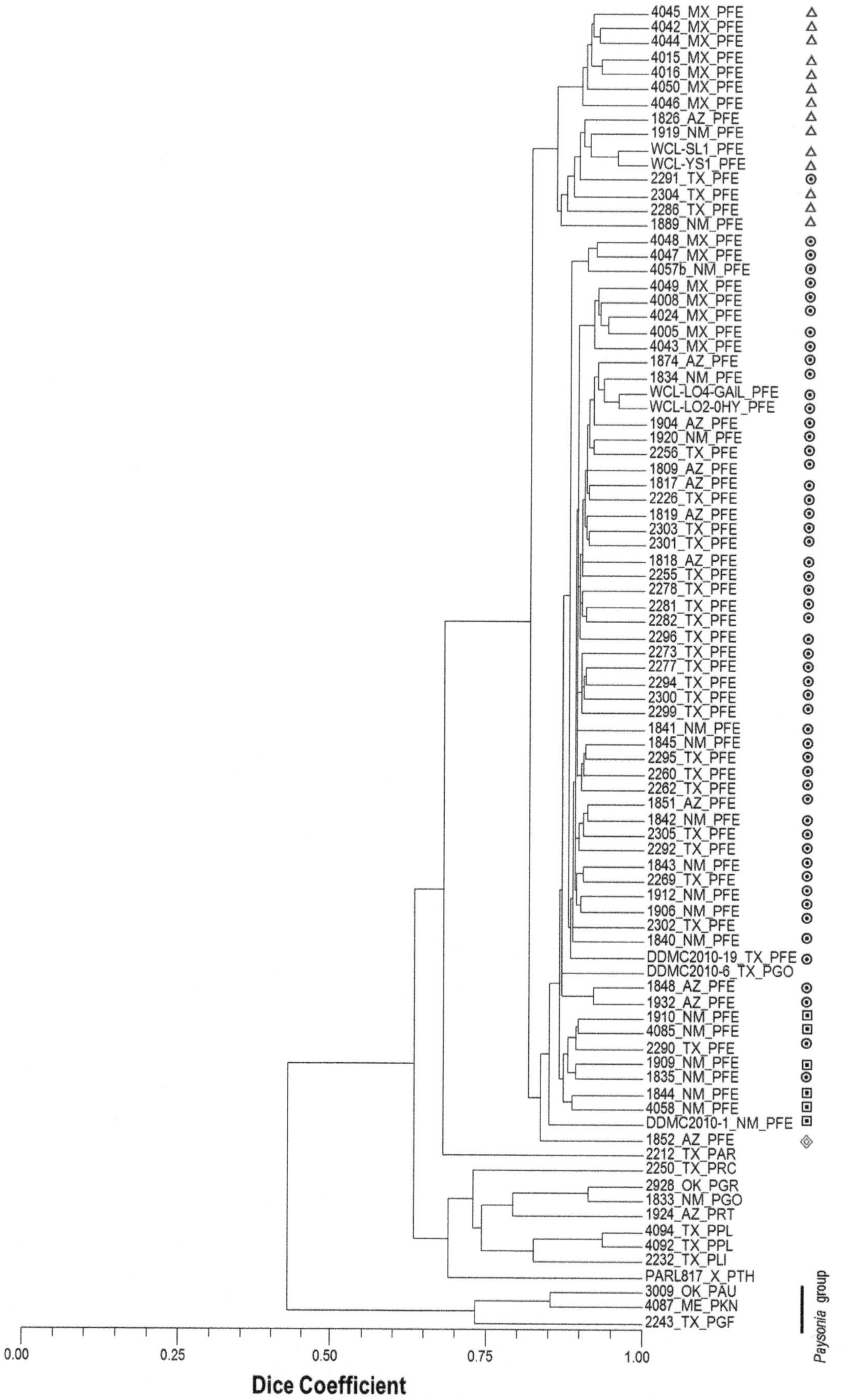
4045_MX_PFE
4042_MX_PFE
4044_MX_PFE
4015_MX_PFE
4016_MX_PFE
4050_MX_PFE
4046_MX_PFE
1826_AZ_PFE
1919_NM_PFE
WCL-SL1_PFE
WCL-YS1_PFE
2291_TX_PFE
2304_TX_PFE
2286_TX_PFE
1889_NM_PFE
4048_MX_PFE
4047_MX_PFE
4057b_NM_PFE
4049_MX_PFE
4008_MX_PFE
4024_MX_PFE
4005_MX_PFE
4043_MX_PFE
1874_AZ_PFE
1834_NM_PFE
WCL-LO4-GAIL_PFE
WCL-LO2-0HY_PFE
1904_AZ_PFE
1920_NM_PFE
2256_TX_PFE
1809_AZ_PFE
1817_AZ_PFE
2226_TX_PFE
1819_AZ_PFE
2303_TX_PFE
2301_TX_PFE
1818_AZ_PFE
2255_TX_PFE
2278_TX_PFE
2281_TX_PFE
2282_TX_PFE
2296_TX_PFE
2273_TX_PFE
2277_TX_PFE
2294_TX_PFE
2300_TX_PFE
2299_TX_PFE
1841_NM_PFE
1845_NM_PFE
2295_TX_PFE
2260_TX_PFE
2262_TX_PFE
1851_AZ_PFE
1842_NM_PFE
2305_TX_PFE
2292_TX_PFE
1843_NM_PFE
2269_TX_PFE
1912_NM_PFE
1906_NM_PFE
2302_TX_PFE
1840_NM_PFE
DDMC2010-19_TX_PFE
DDMC2010-6_TX_PGO
1848_AZ_PFE
1932_AZ_PFE
1910_NM_PFE
4085_NM_PFE
2290_TX_PFE
1909_NM_PFE
1835_NM_PFE
1844_NM_PFE
4058_NM_PFE
DDMC2010-1_NM_PFE
1852_AZ_PFE
2212_TX_PAR
2250_TX_PRC
2928_OK_PGR
1833_NM_PGO
1924_AZ_PRT
4094_TX_PPL
4092_TX_PPL
2232_TX_PLI
PARL817_X_PTH
3009_OK_PAU
4087_ME_PKN
2243_TX_PGF
Paysonia group
0.00
0.25
0.50
0.75
1.00
Dice Coefficient

Figure 2. Cluster analysis of *Physaria* and *Paysonia* accessions based on 2,833 DArT markers. The labels denote the germplasm collection numbers and origin. Suffixes indicate respective species (PAR: *P. argyraea*, PAU: *P. auriculata*, PFE: *P. fendleri*, PGO: *P. gordonii*, PGF: *P. grandiflora*, PGR: *P. gracilis*, PKN: *P.* 'kathryn', PLI: *P. lindheimeri*, PPL: *P. pallida*, PRC: *P. recurvata*, PRT: *P. rectipes*, and PTH: *P. thamnophila*).

Results

Relationships among Accessions from Microarray DArT Analysis

A total of 2,833 polymorphic markers were found using microarray DArT, with an average genotype call rate of 98.4% and a scoring reproducibility of 99.7%. The average PIC value was 0.21 and the median 0.19. About 20% of the markers have values in the range of 0.06 to 0.10, and almost an equal proportion of 12% of markers on the following PIC classes - 0.11 to 0.15, 0.16 to 0.20, and 0.21 to 0.25 (Figure 1). Overall, the distribution of PIC values was asymmetrical and skewed towards the lower values.

Cluster analysis and principal coordinate analysis (PC plot not shown) indicated that the different accessions were successfully classified by the marker system based on species, by geographical source, and breeding status (Figure 2), except for one new collection of *P. gordonii*, DDMC2010-6, which clustered with the *P. fendleri* accessions. The cophenetic correlation coefficient between the dendrogram and the distance matrix was highly significant ($r = 0.98$, $t = 10.30$, prob random $Z < obs.$ $Z = 1.00$, 3000 permutations) indicating that the tree is a very good representation of the distance matrix. All *P. fendleri* accessions grouped in a separate cluster from the other species. The main cluster has two subgroups from Mexico, and a subgroup with a majority of accessions from Texas. There was no single group of *P. fendleri* accessions from Arizona and New Mexico. Most accessions from these States were associated with other accessions from Texas.

The four advanced *P. fendleri* breeding lines were partitioned into two clusters. The breeding lines WCL-SL1 and WCL-YS1 were found to be more genetically distant to WCL-LO2 and WCL-LO4. The last two lines were determined to be more similar to the rest of the *P. fendleri* from North America than WCL-SL1 and WCL-YS1.

Among the different species, the most similar to *P. fendleri* was determined to be *P. argyraea* while the least similar was *P. thamnophila*. The *P. pallida* accessions grouped in one cluster along with *P. lindheimeri* and other accessions representing the species, *P. gracilis*, *P. gordonii*, *P. recurvata*, and *P. rectipes*. The two accessions of *Paysonia auriculata* (3009) and *Paysonia grandiflora* (2243) grouped together in a separate cluster along with the interspecific hybrid swarm 'Kathryn' (4087) from five *Paysonia* species.

The analysis of molecular variance considering *P. fendleri* accessions only, showed that there was a much greater proportion of variation within groups (90%) than among groups (10%) in the species (Table 3). Among the unimproved germplasm set, the pairwise Fst values that showed the greatest differentiation was between Arizona accessions and those from New Mexico and the least amount of differentiation between Mexico and Arizona (Table 4). There was a significant correlation found between the computed genetic distance and geographic distance matrices from Mantel test ($r = 0.33$, $t = 5.94$, prob random $Z < obs$ $z = 1.00$, 3000 permutations), indicating that distant accession pairs are more

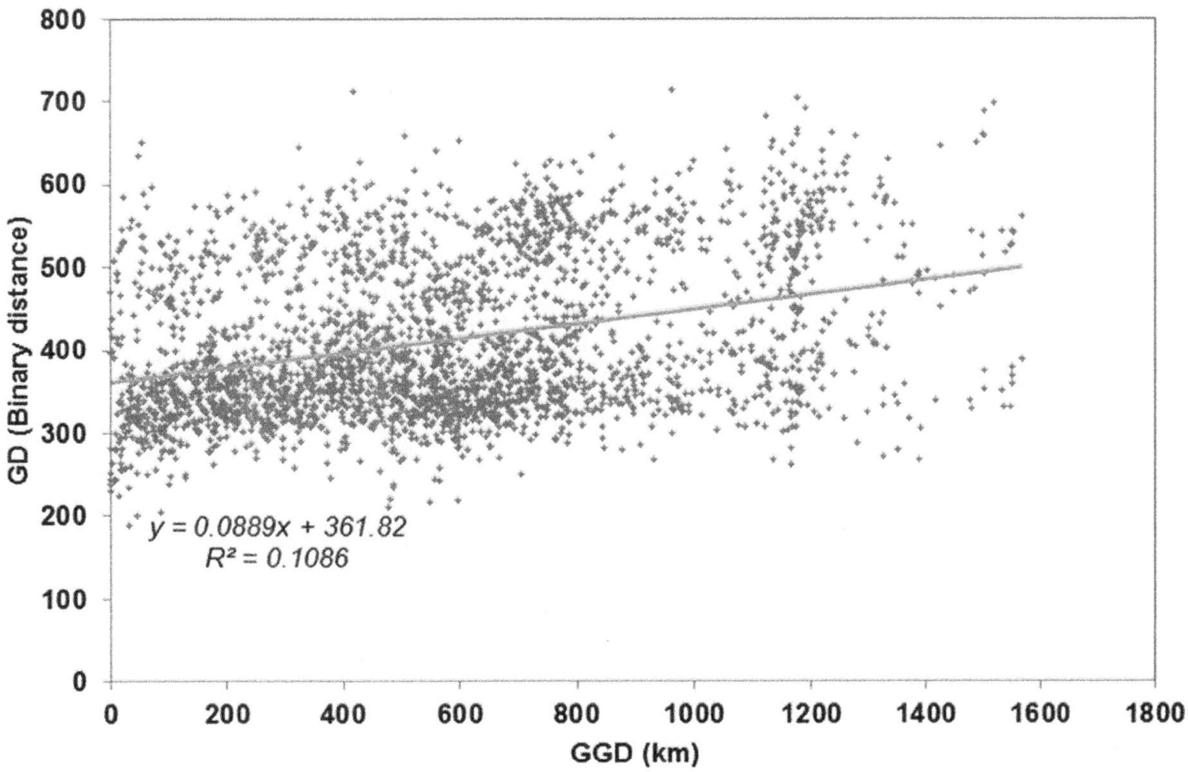

Figure 3. GenAlEx plot of geographic distance (GGD) and genetic distance (GD) from microarray DArT markers.

Table 3. Analysis of molecular variance in *P. fendleri* accessions.

Source	df	SS	Variance components	%
A. Microarray DArT				
Among Groups	4	1597.506	17.19956	9.66%
Within Groups	69	11094.561	160.79074	90.34%
Total	73	12692.068	177.99031	100%
B. DArTseq				
Among Groups	4	14299.235	98.59559	7.22%
Within Groups	123	155895.234	1267.44093	92.78%
Total	127	170194.469	1366.03651	100%

Table 4. Comparison of population pairwise Fst values using microarray DArT and DArTseq.

	Arizona	Advanced lines	Mexico	New Mexico	Texas
Arizona		**0.10717**	**0.13076**	**0.03055**	**0.04049**
Advanced lines	**0.13854**		**0.22289**	**0.10951**	**0.08193**
Mexico	0.00448	**0.13043**		**0.12926**	**0.10552**
New Mexico	**0.16823**	**0.17758**	**0.16280**		**0.03056**
Texas	0.01547	**0.12599**	**0.02649**	**0.16289**	

*Values in bold are significant at $\alpha = 0.05$. Values below diagonal are from microarray DArT and those above diagonal are from DArTseq.

different genetically, supporting the previously mentioned result (Figure 3).

Relationships among Accessions from DArTseq Analysis

There was a total of 27,748 markers obtained using the *Physaria* DArTseq platform. The average genotype call rate was 98.8% and a scoring reproducibility of 99.7%. The average PIC value was 0.12 and the median was 0.09. About 57% of markers have PIC values in the range of 0.06 to 0.10, while 20% and 11% on PIC classes 0.11–0.15 and 0.01–0.05, respectively (Figure 1). The distribution of the PIC values of DArTseq markers follows the same skewed pattern as the DArT microarray markers presented earlier.

Cluster analysis using the markers from this platform resulted in a relationship that follows that of the taxonomic groupings based on general morphological affinities presented by Rollins and Shaw [1]. The results did not deviate from those when the first platform with fewer markers used. Four accessions of *P. fendleri* (2258, 2274, 3083, and DDMC2010-8) clustered with the other *Physaria* species – *P. gordonii* and *P. gracilis*, indicating higher genetic similarity to representative accessions of these species than the rest of *P. fendleri*. These four *P. fendleri* accessions will be further examined for misidentification and for oil and morphological trait variation when verified.

The cophenetic correlation coefficient between the dendrogram and the distance matrix was highly significant (r = 0.94, t = 49.20, prob random Z<obs. Z = 1.00, 3,000 permutations) indicating that the tree is a very good representation of the distance matrix. The cluster of *P. fendleri* showed a distinct group of accessions from Mexico and a cluster comprised of all other germplasm from North America (Figure 4). The DArTseq platform indicated two separate clusters for the advanced lines. The breeding lines WCL-LH1, WCL-LO1, and WCL-LY1 are all in a group with greater similarity to accessions from Texas. WCL-LO2, WCL-LO4, WCL-SL1, and WCL-YS1 grouped together in one cluster indicating greater genetic similarity to accessions from Arizona and Mexico. WCL-LO2 was derived from the WCL-LO1 and the remaining three from WCL-LO2. It appears from these data that perhaps more genotypes from the Arizona accession were integrated into WCL-LO2.

Similar to results from the microarray DArT, the accessions of the following species formed distinct groups consistent with the classification by Rollins and Shaw [1]: a) *P. intermedia*, *P. valida*, and *P. rectipes*, b) *P. gordonii*, *P. gracilis*, *P. rectipes*, and *P. lindheimeri*, and c) *P. auriculata*, *P. densipila*, *P. lyrata*, *P. stonensis*, *P. grandiflora*, and *P. perforata*. The two accessions of *P. lasiocarpa* (2217 and 2228) were most genetically similar to the accessions of *P. grandiflora*. This last group of species comprised of annual, auriculate-leaved types whose taxonomic nomenclature was segregated from *Physaria* and transferred to *Paysonia* based on leaf-trichome morphology, chromosome number, and molecular data from analyzing internal transcribed spacers of nuclear ribosomal DNA [40]. Apart from the accession of *Physaria mexicana* nested within the *Paysonia* cluster, the results of this genetic analysis using both microarray DArT and DArTseq platforms support the previous segregation of *Physaria* from *Paysonia* as proposed by O'Kane and Al-Shehbaz [40] indicating that the group of *Paysonia* species to be the least genetically similar to the *P. fendleri* and other *Physaria* accessions.

Results of the analysis of molecular variance when DArTseq markers were used correspond to that from microarray DArT. A greater proportion of variation within groups (93%) than among groups (7%) in the species was found (Table 3).

The average genetic similarity in the *P. fendleri* group when using microarray DArT was 0.86, while only 0.44 when DArTseq was used. This is in line with the assumption that more differences may be found when more markers are used because of the increased sensitivity and resolution to detect genetic distinctiveness [41]. The genetic similarity matrices of *P. fendleri* obtained using the two platform systems showed a good fit when compared by a Mantel test (r = 0.48, t = 11.28, prob random Z<obs z = 1.00, 3,000 permutations).

Population Structure Analysis of *P. fendleri*

Using results from microarray DArT, the population structure of the *P. fendleri* samples was determined. The plot of ΔK for each K value is shown in Figure 5a. It was estimated through the method of Evanno et. al. [42] that there are 4 groups contributing significant genetic information in the *P. fendleri* collection. The bar plot of the population assignment test when $K = 4$ is shown in Figure 5b. Three of the four accessions of *P. fendleri* breeding lines are shown to have mixed backgrounds. Of the other *P. fendleri* accessions, twelve (16%) have close to homogeneous genetic background (>98% probability) while 63 accessions (84%) are highly heterogeneous showing intermediate and/or highly mixed composition. A majority of the accessions from Texas are assigned to one cluster, while those from New Mexico have myriad cluster assignments suggesting the greater diversity of *P. fendleri* in this U.S. state. When the cluster assignments of the *P. fendleri* accessions were projected on a map, the segregation among clusters was evident in their geographic location (Figure 6). Further testing for association between the assigned clusters and available site elevation data by computing Spearman's rank correlation

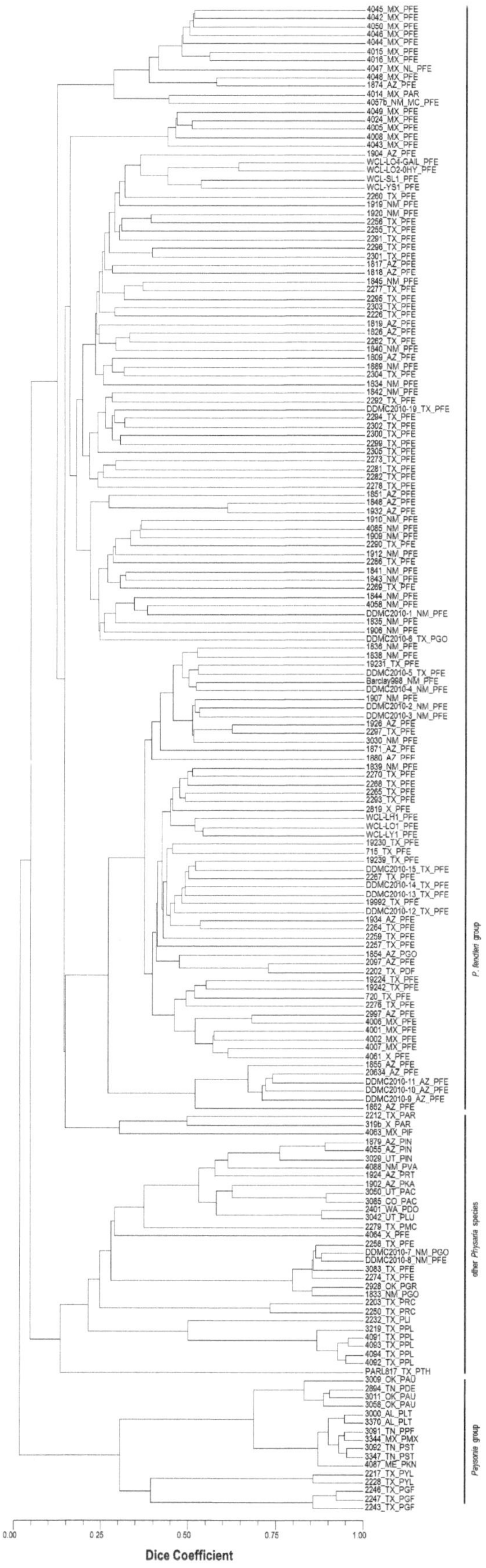

Figure 4. Cluster analysis of *Physaria* and *Paysonia* accessions based on 27,748 DArTseq markers. The labels denote the germplasm collection numbers and origin. Suffixes indicate respective species (PAC: *P. acutifolia*, PAR: *P. argyraea*, PAU: *P. auriculata*, PDE: *P. densipila*, PDF: *P. densiflora*, PDO: *P. douglasii*, PFE: *P. fendleri*, PGO: *P. gordonii*, PGF: *P. grandiflora*, PGR: *P. gracilis*, PIF: *P. inflata*, PIN: *P. intermedia*, PKA: *P. kaibabensis*, PKN: *P.* 'kathryn', PYL: *P. lasiocarpa*, PLI: *P. lindheimeri*, PLT: *P. lyrata*, PLU: *P. ludoviciana*, PMC: *P. mcvaughiana*, PMX: *P. mexicana*, PPF: *P. perforata*, PPL: *P. pallida*, PRC: *P. recurvata*, PRT: *P. rectipes*, PST: *P. stonensis*, PTH: *P. thamnophila* and PVA: *P. valida*).

coefficient showed a very weak positive correlation (Spearman $\rho = 0.11$, $p = 0.33$).

Discussion

The importance of understanding genetic diversity in germplasm collections is critical for the effective management of accessions in genebanks. Molecular characterization supplements morphological evaluation of germplasm and allows measurements to help resolve numerous operational, logistical, and biological questions that face genebank managers and conservation biologists [43], [44]. The *Physaria* collection in the U.S. NPGS has not been well characterized before for genetic diversity, though there have been preliminary studies on subsets of accessions using a limited number of microsatellite markers [16], and an extensive evaluation for diversity in oil characteristics and other morphological characters [45]. In this study, the new DArT platforms for *Physaria* were found both acceptable and provided robust information about the genetic variability of the collection.

The development and utilization of DArT markers allowed us to determine the genetic diversity of the *Physaria* collection. The 2,833 microarray DArT markers were found to be useful in providing a picture of genetic diversity in the *Physaria* germplasm collection using a large set of accessions. Overall, the average PIC of the *Physaria* and *Paysonia* microarray DArT markers was found to be lower than that observed in other species where similar markers were developed, like wheat (0.44) [46], cassava (0.42) [47], and sorghum (0.41) [48], but comparable to that observed in sugar beet (0.28) [49] and *Asplenium* fern (0.21) [50]. The average PIC of the DArTseq markers is much less than that of the microarray DArT. However, the more numerous DArTseq markers may have the capability of providing a better picture of diversity by sampling more points in the genome. The distribution of these almost 30,000 new DArT markers in the *Physaria* and *Paysonia* genome remains to be determined. However, based on the information from a large number of organisms in which DArT system was applied more broadly including genetic mapping and/or sequence-based physical mapping, we can assume that DArT marker from both platforms will be distributed throughout the genome with marker density highly correlated to gene density [28], [51]. Compared to microsatellite markers, DArT markers are very suitable for high-throughput work and previously have been determined to have clear advantages in cost and time aspects of genotyping as demonstrated in other crops [52]. Both microarray DArT and DArTseq platforms have the same development costs. However, the higher number of markers obtained in DArTseq resulted to an overall lower cost per datapoint than microarray DArT. This higher cost effectiveness of DArTseq is in parallel to other sequenced-based genotyping strategies which can provide substantial cost savings compared to microarrays when conducting genetic diversity studies [53], [54]. Importantly, when comparing *effectiveness* of the two platforms one has to keep in mind that it may vary according to specific *application*: in genetic ID and product quality testing (i.e. seed

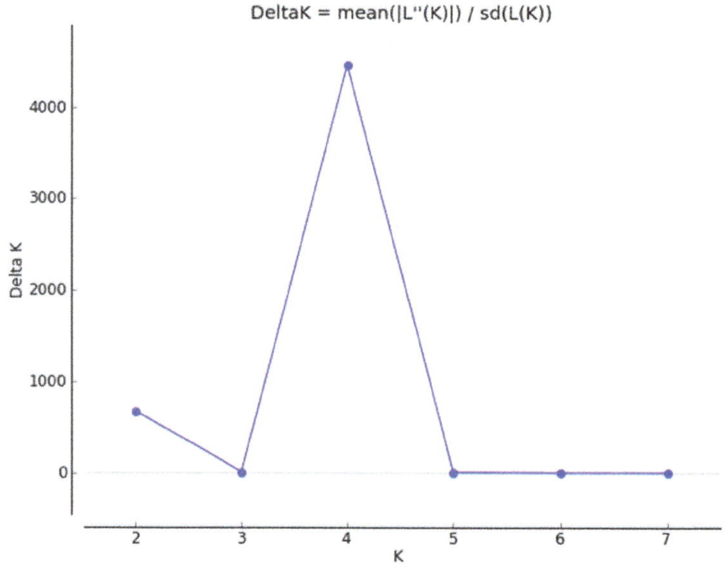

A

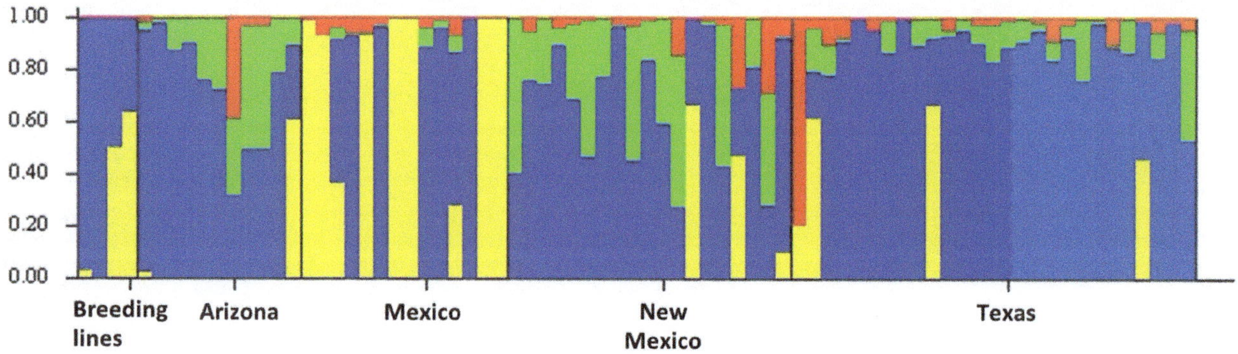

B

Figure 5. Plot of ΔK from K=2 to 7 (a) and the population structure of 75 *P. fendleri* accessions at K=4 (b).

purity) modest number of array-based DArT markers may perform as well as DArTseq platform and currently for a better price. A further cost reduction of sequencing may however push even this balance towards DArTseq platform in the future.

The relationships found among the accessions are in line with the previously proposed evolution within the genus. Taxonomists assert that *P. auriculata* is the most primitive species due to its very distinct evolutionarily primitive characters such as large siliques, large number of ovules around the replum, and predominance of simple trichomes [1]. *P. auriculata* has been proposed to be closely related to *P. grandiflora* and this relationship is supported by the results of the molecular marker analysis between the accessions representative of these species (2243 and 3009) showing high genetic similarities. Likewise, *P. gordonii* and *P. gracilis* grouped in the same cluster with *P. rectipes*, *P. recurvata*, and *P. thamnophila* which is in agreement with their previously known taxonomic groupings based on very general morphological similarities. The high genetic similarity between the *P. argyraea* accessions (2212 and 319 b) to the *P. fendleri* group supports phylogenetic subsectional grouping based on pod morphology.

Overall, there was higher genetic similarity found among accessions of the other *Physaria* and *Paysonia* species than among accessions of *P. fendleri*. In particular, species that are included in the federal or state threatened and endangered species list, like *P. pallida* and *P. stonensis* [55], have very low genetic diversity as indicated by results of DArTseq markers. In *P. pallida*, the representative accessions (4091 and 4093) were found to be highly similar. The limited geographic range and the proximity of the collection sites of these two samples suggest that they might have been from just one population. Likewise for *P. stonensis*, there was a very high genetic similarity found on both of the representative accessions (3092 and 3347). Because only a limited number of accessions were included in these other species, a follow up study that includes additional accessions is recommended to validate if this is a general trend.

Physaria 'Kathryn' (4087) is a cultivar from interspecific hybridization developed by allowing five species (*P. densipila*, *P. lescurii*, *P. lyrata*, *P. perforata*, and *P. stonensis*) to intermate for twenty two generations [56]. *P.* 'Kathryn' was determined to be most closely related to *P. auriculata* in our analysis using microarray DArT. However, with the expanded set of accessions included in

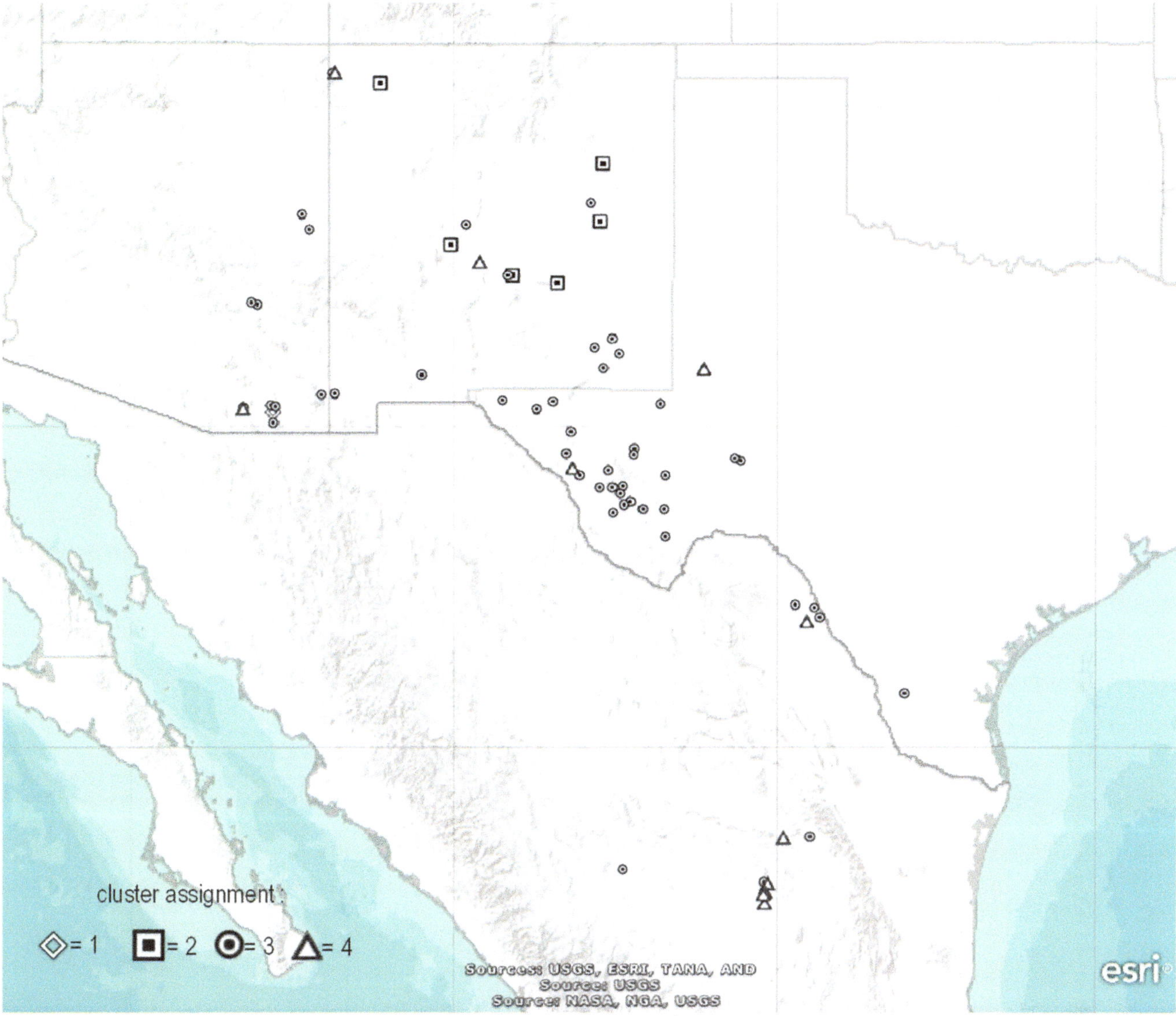

Figure 6. Geographic locations of the *P. fendleri* accessions and their respective cluster assignments (indicated by different icons) based on Bayesian model-based clustering methods.

the DArTseq analysis, it grouped with the representative samples of its parent species – *P. lyrata* (3000 and 3370), *P. perforata* (3091), and *P. stonensis* (3092, 3347). These species were also found most genetically similar to *P. mexicana* (3344) which is the only perennial type in the species cluster. *P. mexicana* is a previously undescribed species in Mexico and is among the more recent species reported by Rollins [57].

DDMC2010-6 was from a more recent germplasm collecting trip and it was entered in the database as *P. gordonii*. However, based on results of DArT markers, it clustered with the *P. fendleri* accessions after using both DArT platforms, indicating the need to review its species assignment. The species identity of this particular accession will again be verified using its plant voucher specimen as well as in the NPGS site handling the germplasm when it is regenerated in the future.

The two clusters of *P. fendleri* breeding lines - WCL-LO2, WCL-LO4, WCL-SL1, and WCL-YS1 in one cluster, while the other set of breeding lines WCL-LO1, WCL-LH1, and WCL-LY1 in another, indicates the possibility of developing genetically differentiated lines for crop improvement and may have applications in future hybrid development work. The lines WCL-LH1, WCL-LO1 and WCL-LY1 are the first three germplasm lines that were publicly released in 1996. These were developed using recurrent selection on a population made by bulking seeds of one accession that came from Arizona and nine from Texas in 1986 [58]. The bulked seeds were also the starting material for other breeding lines. WCL-LO4 was derived from mass selection from WCL-LO3 and WCL-LY2 (both not included in this study) which has WCL-LY1 as the source population [59], [60]. The other breeding lines were developed through phenotypic selection: WCL-YS1 was a selection from PI 311165, one of the initial accession from Arizona that comprise the original bulk in 1986 [61], while WCL-SL1 came from plants that survived at the highest salinity levels during a salt tolerance screening study when seeds from the original bulked seeds were planted [62].

The accessions of *P. fendleri* from Mexico are genetically similar as indicated by the cluster analyses. The range of *Physaria* species has been reported as limited to the northeastern part of country and concentrated on mountain and high plains of Coahuila, Nuevo León, and Zacatecas [57]. This limited geographic distribution likely prevented their further genetic differentiation. A similar investigation focusing on the other *Physaria* species in this region could confirm this trend.

The array of *P. fendleri* germplasm consisted of 75 accessions analyzed using microarray DArT and 128 accessions using DArTseq. There is ample genetic variability in the *P. fendleri* collection found, as indicated by the cluster analysis as well as analysis of population structure. This is expected for a cross pollinating species that has not been fully domesticated [63]. However, the higher within group variation detected by AMOVA using data from both DArT platforms suggests that there is only a small amount of genetic differentiation among groups in the sample as a whole. Results from the Bayesian clustering approach when comparing the geographical sources of the accessions suggested that there is more variability in New Mexico than the other *P. fendleri* locations and that there is a spatial pattern evident in the microarray DArT results. This pattern of genetic differentiation occurs outward from central Texas, the region identified as the putative origin of *Physaria* as proposed earlier by Payson [64].

An increased population differentiation has been reported in many plant species between source populations and new ones when plants colonize new habitats [65]. The more dynamic nature of *Physaria* populations in distant locations from Texas has been reported by Payson [64] and he attributed part of the process as caused by barriers that separate populations, such as soil properties and moisture availability which are very important to survival of ephemeral populations of the taxon. In *P. fendleri*, Dierig et al. [66] stated that temperature and elevation effects can also account for significant differences in reproductive capacity. Selection patterns have also been investigated by past studies reporting that non-random mating, sexual selection and dormancy characteristics play a significant role in how traits evolved in the species [67], [68]. Seed dormancy characteristics in particular cause a persistent soil seed bank which may prevent genetic differentiation in the species because certain genotypes are reintroduced back during subsequent seasons [69], [70]. In terms of genetic resources conservation of *P. fendleri* germplasm, the DArT results suggest the existence of more variable genotypes in New Mexico and it is recommended that future collecting missions select this geographic area to possibly expand the genetic diversity in the species collection.

Conclusion

The availability of genetic diversity information of the *P. fendleri* collection will enable better germplasm management and conservation of the species. In this study we report the successful development of two DArT marker platforms that were utilized for genotyping *Physaria* and *Paysonia* accessions. This marker system complements the microsatellite markers developed previously for *Physaria*. The high number of DArT markers allows a greater resolution of genetic differences among accessions and enabled us to examine the extent of variation in the *P. fendleri* collection, as well as provide support to known taxonomic classification and recent nomenclatural changes of certain *Physaria* species to *Paysonia*. We intend to further utilize the DArT markers in developing a linkage map in *Physaria* to assist breeding efforts and for future genetic mapping studies.

Acknowledgments

USDA is an equal opportunity provider and employer. Mention of companies or commercial products does not imply recommendation or endorsement by the USDA over others not mentioned. USDA neither guarantees nor warrants the standard of any product mentioned. Product names are mentioned solely to report factually on available data and to provide specific information. We would like to thank David Ellis and Chris Richards for the valuable suggestions during the conduct of the study.

Author Contributions

Conceived and designed the experiments: VMVC AK DAD. Performed the experiments: VMVC AK DAD. Analyzed the data: VMVC AK. Wrote the paper: VMVC AK DAD. Participated in germplasm collection: VMVC DAD.

References

1. Rollins RC, Shaw EA (1973) The Genus *Lesquerella* (Cruciferae) in North America. Harvard Univ. Press, Cambridge, MA. 288 p.
2. Al-Shehbaz IA, O'Kane SL Jr (2002) *Lesquerella* is united with *Physaria* (Brassicaceae). Novon 12: 319–329.
3. Kish S (2008) *Lesquerella*: The next source of biofuel. USDA Cooperative State Research, Education, and Extension Service (CSREES). Available: http://www.csrees.usda.gov/newsroom/impact/2008/nri/pdf/lequerella.pdf. Accessed 2012 March 29.
4. Moser BR, Cermak SC, Isbell TA (2008) Evaluation of castor and lesquerella oil derivatives as additives in biodiesel and ultralow sulfur diesel fuels. Energy & Fuels 22: 1349–1352.
5. Dierig DA, Ray DT (2009) New Crops Breeding: *Lesquerella*. In: Vollman J, Rajcan I (eds). Oil Crops, Handbook of Plant Breeding 4. Springer Science Media, NY. 507–516.
6. Cruz VMV, Dierig DA (2012) Trends in literature on new oilseed crops and related species: Seeking evidence of increasing or waning interest. Industrial Crops and Products 37: 141–148.
7. Jenderek MM, Dierig DA, Isbell TA (2009) Fatty-acid profile of *Lesquerella* germplasm in the National Plant Germplasm System collection. Industrial Crops and Products 29: 154–164.
8. Wang GS, McCloskey W, Foster M, Dierig D (2010) Lesquerella: A winter oilseed crop for the Southwest. Arizona Cooperative Extension. The University of Arizona, Tucson, AZ. 4 p.
9. Cruz VMV, Dierig DA (2010) National conservation activities on *Lesquerella* spp. for ensuring species survival and germplasm availability for new crops research and development. Proc Colorado Native Plant Society Annual Meeting. Sept. 10–12, 2010. Denver, CO.
10. Mondini L, Noorani A, Pagnotta MA (2009) Assessing plant genetic diversity by molecular tools. Diversity 1: 19–35.
11. FAO (2010) The Second Report on the State of the World's Plant Genetic Resources for Food and Agriculture. Commission on Genetic Resources for Food and Agriculture. Food and Agriculture Organization of the United Nations. Rome, Italy. 69–70.
12. Börner A, Khlestkina EK, Pshenichnikova TA, Osipova SV, Kobiljski B, et al. (2012) Genetics and genomics of plant genetic resources. Journal of Stress Physiology & Biochemistry 8 suppl. vol. 3, S10.
13. Cruz VMV, Nason J, Luhman R, Marek LF, Shoemaker RC, et al. (2006) Analysis of bulked and redundant accessions of *Brassica* germplasm using assignment tests of microsatellite markers. Euphytica 152: 339–349.
14. Cruz VMV, Luhman R, Marek LF, Rife CL, Shoemaker RC, et al. (2007) Characterization of flowering time and SSR marker analysis of spring and winter type *Brassica napus* L. germplasm. Euphytica 153: 43–57.
15. Hasan M, Seyis F, Badani AG, Pons-Kühnemann J, Friedt W, et al. (2006) Analysis of genetic diversity in the *Brassica napus* L. gene pool using SSR markers. Genetic Resources and Crop Evolution 53: 793–802.
16. Salywon AM, Dierig DA (2006) Isolation and characterization of microsatellite loci in *Lesquerella fendleri* (Brassicaceae) and cross-species amplification. Mol Ecol Notes 6: 382–384.
17. Redden R, Vardy M, Edwards D, Raman H, Batley J (2009) Genetic and morphological diversity in the Brassicas and wild relatives. Proc. 16th Australian Research Assembly on Brassicas. Sept. 14–16, 2009. Ballarat, Victoria. 5 p.
18. Faltusová Z, Kučera L, Ovesná J (2011) Genetic diversity of *Brassica oleracea* var. *capitata* Gene Bank accessions assessed by AFLP. Electronic Journal of Biotechnology. http://dx.doi.org/10.2225/vol14-issue3-fulltext-4http://dx.doi.org/10.2225/vol14-issue3-fulltext-4. Accessed 2012 March 29.

19. Toledo J, Dehal P, Jarrin F, Hu J, Hermann M, et al. (1998) Genetic variability of *Lepidium meyenii* and other Andean *Lepidium* species (Brassicaceae) assessed by molecular markers. Ann Bot 82: 523–530.
20. Kamel EA, Hassan HZ, El-Nahas AI, Ahmed SM (2004) Molecular characterization of some taxa of the genus *Raphanus* L. (Cruciferae = Brassicaceae). Cytologia 69: 249–260.
21. Jaccoud D, Peng K, Feinstein D, Kilian A (2001) Diversity Arrays: a solid state technology for sequence information independent genotyping. Nucleic Acids Res 29: e25.
22. Cruz, unpublished data.
23. Jones N, Ougham H, Thomas H, Pasakinskiene I (2009) Markers and mapping revisited: finding your gene. New Phytologist 137: 165–177.
24. DArT (2012) Papers about DArt. Diversity Arrays Technology, Pty Ltd. Available: http://www.diversityarrays.com/publications.html. Accessed 2012 March 30.
25. Varshney RK, Glaszmann JC, Leung H, Ribaut JM (2010) More genomic resources for less-studied crops. Trends in Biotechnology 28: 452–460.
26. Howard EL, Whittock SP, Jakše J, Carling J, Matthews PD, et al. (2011) High-throughput genotyping of hop (*Humulus lupulus* L.) utilising diversity arrays technology (DArT). Theor Appl Genet 122: 1265–1280.
27. Cabin RJ, Mitchell RJ, Marshall DL (1998) Do surface plant and soil seed bank populations differ genetically? A multipopulation study of the desert mustard *Lesquerella fendleri* (Brassicaceae). Am J Bot 85: 1098–1109.
28. Kilian A, Wenzl P, Huttner E, Carling J, Xia L, et al. (2012) Diversity Arrays Technology: A Generic Genome Profiling Technology on Open Platforms. Methods Mol Bio 888: 67–88.
29. O'Kane SL Jr, Al-Shehbaz IA (2003) Phylogenetic position and generic limits of *Arabidopsis* (Brassicaceae) based on sequences of nuclear ribosomal DNA. Ann. Missouri Bot. Gard. 90: 603–612.
30. Langmead B, Trapnell C, Pop M, Salzberg SL (2009) Ultrafast and memory-efficient alignment of short DNA sequences to the human genome. Genome Biol 10: R25.
31. Wenzl P, Carling J, Kudrna D, Jaccoud D, Huttner E, et al. (2004) Diversity arrays technology (DArT) for whole genome profiling of barley. Proc Natl Acad Sci (USA) 101: 9915–9920.
32. Anderson JA, Churchill GA, Sutrique JE, Tanksley SD, Sorrells ME (1993) Optimizing parental selection for genetic linkage maps. Genome 36: 181–186.
33. Peakall R, Smouse P (2006) GenalEx 6: Genetic Analysis in Excel. Population genetic software for teaching and research. Mol Ecol Notes 6, 288–295.
34. Excoffier L, Laval G, Schneider S (2005) Arlequin ver. 3.0: An integrated software package for population genetics data analysis. Evolutionary Bioinformatics Online 1: 47–50.
35. Rohlf FJ (2011) NTSYSpc: Numerical Taxonomy System, ver. 2.21 m. Exeter Publishing, Ltd., Setauket, NY.
36. Dalirsefat S, Meyer A, Mirhoseini S (2009) Comparison of similarity coefficients used for cluster analysis with amplified fragment length polymorphism markers in the silkworm, *Bombyx mori*. Journal of Insect Science 9: 71.
37. Sesli M, Yegenoglu ED (2010) Comparison of similarity coefficients used for cluster analysis based on RAPD markers in wild olives. Genet Mol Res 9: 2248–2253.
38. Pritchard JK, Stephens M, Donnelly P (2000) Inference of population structure using multilocus genotype data. Genetics 155: 945–959.
39. Earl DA, vonHoldt BM (2012) STRUCTURE HARVESTER: a website and program for visualizing STRUCTURE output and implementing the Evanno method. Conservation Genetics Resources 4: 359–361.
40. O'Kane SL Jr, Al-Shehbaz IA (2002) *Paysonia*, a new genus segregated from *Lesquerella* (Brassicaceae). Novon 12: 379–381.
41. Agarwal M, Shrivastava N, Padh H (2008) Advances in molecular marker techniques and their applications in plant sciences. Plant Cell Rep 27: 617–631.
42. Evanno G, Regnaut S, Goudet J (2005) Detecting the number of clusters of individuals using the software STRUCTURE: a simulation study. Mol Ecol 14: 2611–2620.
43. Hedrick PW (2000) Applications of population genetics and molecular techniques to conservation biology. In: Conservation Biology 4. Genetics, Demography and Viability of Fragmented Populations. Young, A.G, Clarke, G.M. (eds.). Cambridge Univ. Press, Cambridge, UK. 113–125.
44. Karp A, Kresovich S, Bhat KV, Ayad WG, Hodgkin T (1997) Molecular tools in plant genetic resources conservation: a guide to the technologies. IPGRI Technical Bulletin No. 2. International Plant Genetic Resources Institute, Rome, Italy. 47 p.
45. Salywon AM, Dierig DA, Rebman JP, Jasso de Rodriguez D (2005) Evaluation of new *Lesquerella* and *Physaria* (Brassicaceae) oilseed germplasm. Am J Bot 92: 53–62.
46. Raman H, Stodart BJ, Cavanagh C, Mackay M, Morell M, et al. (2010) Molecular diversity and genetic structure of modern and traditional landrace cultivars of wheat (*Triticum aestivum* L.). Crop and Pasture Science 61: 222–229.
47. Xia L, Peng KM, Yang SY, Wenzl P, de Vicente MC, et al. (2005) DArT for high-throughput genotyping of cassava (*Manihot esculenta*) and its wild relatives. Theor Applied Genet 110: 1092–1098.
48. Mace ES, Xia L, Jordan DR, Halloran K, Path DK, et al. (2008) DArT markers: diversity analyses and mapping in *Sorghum bicolor*. BMC Genomics. 9, 26. doi: 10.1186/1471-2164-9-26. Accessed 2012 March 29.
49. Simko I, Eujayl I, van Hintum TJL (2012) Empirical evaluation of DArT, SNP, and SSR marker-systems for genotyping, clustering, and assigning sugar beet hybrid varieties into populations. Plant Science 184: 54–62.
50. James KE, Schneider H, Ansell SW, Evers M, Robba L, et al. (2008) Diversity Arrays Technology (DArT) for pan-genomic evolutionary studies of non-model organisms. PLoS ONE 3: e1682.
51. Petroli CD, Sansaloni CP, Carling J, Steane DA, Vaillancourt RE, et al. (2012) Genomic characterization of DArT markers based on high-density linkage analysis and physical mapping to the *Eucalyptus* genome. PLoS ONE 7: e44684.
52. Kilian A, Huttner E, Wenzl P, Jaccoud D, Carling J, et al. (2005) The fast and the cheap: SNP and DArT-based whole genome profiling for crop improvement. In: Tuberosa R, Phillips RL, Gale M (eds) Proceedings of the international congress in the wake of the double helix: from the green revolution to the gene revolution. Avenue Media, Bologna, Italy, 27–31 May 2003, 443–461.
53. Illumina (2013) Agrigenomics genotyping decisions reach a crossroads. Application Spotlight: Analyzing Genetic Variation. Available: http://www.illumina.com/Documents/products/appspotlights/app_spotlight_ngg_ag.pdf. Accessed 2013 February 13.
54. Ayling S (2012) Technical appraisal of strategic approaches to large-scale germplasm evaluation. The Genome Analysis Center, Norwich, UK. Available: http://agro.biodiver.se/wp-content/uploads/2012/12/Technical-appraisal-NGS-for-genebanks-please-comment.pdf. Accessed 2013 February 13.
55. USDA-NCRS (2012) The PLANTS Database. National Plant Data Team, Greensboro, NC. Available: http://plants.usda.gov. Accessed 2012 September 24.
56. Rollins RC (1988) A population of interspecific hybrids of *Lesquerella* (Cruciferae). Systematic Botany 13: 60–63.
57. Rollins RC (1958) Notes on *Lesquerella* (Cruciferae) in Mexico. Boletin de la Sociedad Botanica de Mexico 23: 42–47.
58. Dierig DA, Thompson AE, Coffelt TA (1998) Registration of three *Lesquerella fendleri* germplasm lines selected for improved oil traits. Crop Sci 38: 287.
59. Dierig DA, Dahlquist GH, Coffelt TA, Ray DT, Isbell TA, et al. Registration of WCL-LO4-Gail lesquerella with improved harvest index. J Plant Reg "submitted".
60. Dierig DA, Dahlquist GA, Tomasi PM (2006b) Registration of WCL-LO3 high oil *Lesquerella fendleri* germplasm. Crop Sci 46: 1832–1833.
61. Dierig DA, Tomasi PM, Coffelt TA, Rayford WE, Lauver L (2000) Yellow seed coat *Lesquerella*. Crop Sci 40: 865.
62. Dierig DA, Shannon MC, Grieve CM (2001) Salt tolerant *Lesquerella*. Crop Sci 41: 604.
63. Rauf S, Teixeira da Silva JA, Khan AA, Naveed A (2010) Consequences of plant breeding on genetic diversity. Intl J Plant Breeding 4: 1–21.
64. Payson EB (1921) A Monograph of the Genus *Lesquerella*. Annals of the Missouri Botanical Garden 8: 103–236.
65. Barrett SCH, Husband BC (1990) The Genetics of Plant Migration and Colonization. In: Brown AHD, Clegg MT, Kahler AL, Weir BS (eds). Plant population genetics, breeding, and genetic resources. Sinauer Associates Inc., Sunderland, MA. 254–277.
66. Dierig DA, Adam NR, Mackey NR, Dahlquist GH, Coffelt TA (2006a) Temperature and elevation effects on plant growth, development, and seed production of two *Lesquerella* species. Industrial Crops and Products 24: 17–25.
67. Cabin RJ, Evans AS, Mitchell RJ (1997) Do plants derived from seeds that readily germinate differ from plants derived from seeds that require forcing to germinate? A case study of the desert mustard *Lesquerella fendleri*. Ann Midl Nat 138: 121–133.
68. Mitchell RJ, Marshall DL (1998) Nonrandom mating and sexual selection in a desert mustard: an experimental approach. Am J Bot 85: 48–55.
69. Cabin RJ (1996) Genetic comparisons of seed bank and seedling populations of a perennial desert mustard, *Lesquerella fendleri*. Evolution 50: 1830–1841.
70. Freeland J (2005) Molecular Ecology. John Wiley & Sons, Ltd., Hoboken, NJ. 109–154.

12

A Meta-Analysis of the Impacts of Genetically Modified Crops

Wilhelm Klümper, Matin Qaim*

Department of Agricultural Economics and Rural Development, Georg-August-University of Goettingen, Goettingen, Germany

Abstract

Background: Despite the rapid adoption of genetically modified (GM) crops by farmers in many countries, controversies about this technology continue. Uncertainty about GM crop impacts is one reason for widespread public suspicion.

Objective: We carry out a meta-analysis of the agronomic and economic impacts of GM crops to consolidate the evidence.

Data Sources: Original studies for inclusion were identified through keyword searches in ISI Web of Knowledge, Google Scholar, EconLit, and AgEcon Search.

Study Eligibility Criteria: Studies were included when they build on primary data from farm surveys or field trials anywhere in the world, and when they report impacts of GM soybean, maize, or cotton on crop yields, pesticide use, and/or farmer profits. In total, 147 original studies were included.

Synthesis Methods: Analysis of mean impacts and meta-regressions to examine factors that influence outcomes.

Results: On average, GM technology adoption has reduced chemical pesticide use by 37%, increased crop yields by 22%, and increased farmer profits by 68%. Yield gains and pesticide reductions are larger for insect-resistant crops than for herbicide-tolerant crops. Yield and profit gains are higher in developing countries than in developed countries.

Limitations: Several of the original studies did not report sample sizes and measures of variance.

Conclusion: The meta-analysis reveals robust evidence of GM crop benefits for farmers in developed and developing countries. Such evidence may help to gradually increase public trust in this technology.

Editor: emidio albertini, University of Perugia, Italy

Funding: This research was financially supported by the German Federal Ministry of Economic Cooperation and Development (BMZ) and the European Union's Seventh Framework Programme (FP7/2007-2011) under Grant Agreement 290693 FOODSECURE. The funders had no role in study design, data collection and analysis, decision to publish, or preparation of the manuscript. Neither BMZ nor FOODSECURE and any of its partner organizations, any organization of the European Union or the European Commission are accountable for the content of this article.

Competing Interests: The authors have declared that no competing interests exist.

* Email: mqaim@uni-goettingen.de

Introduction

Despite the rapid adoption of genetically modified (GM) crops by farmers in many countries, public controversies about the risks and benefits continue [1–4]. Numerous independent science academies and regulatory bodies have reviewed the evidence about risks, concluding that commercialized GM crops are safe for human consumption and the environment [5–7]. There are also plenty of studies showing that GM crops cause benefits in terms of higher yields and cost savings in agricultural production [8–12], and welfare gains among adopting farm households [13–15]. However, some argue that the evidence about impacts is mixed and that studies showing large benefits may have problems with the data and methods used [16–18]. Uncertainty about GM crop impacts is one reason for the widespread public suspicion towards this technology. We have carried out a meta-analysis that may help to consolidate the evidence.

While earlier reviews of GM crop impacts exist [19–22], our approach adds to the knowledge in two important ways. First, we include more recent studies into the meta-analysis. In the emerging literature on GM crop impacts, new studies are published continuously, broadening the geographical area covered, the methods used, and the type of outcome variables considered. For instance, in addition to other impacts we analyze effects of GM crop adoption on pesticide quantity, which previous meta-analyses could not because of the limited number of observations for this particular outcome variable. Second, we go beyond average impacts and use meta-regressions to explain impact heterogeneity and test for possible biases.

Our meta-analysis concentrates on the most important GM crops, including herbicide-tolerant (HT) soybean, maize, and cotton, as well as insect-resistant (IR) maize and cotton. For these crops, a sufficiently large number of original impact studies have been published to estimate meaningful average effect sizes. We estimate mean impacts of GM crop adoption on crop yield, pesticide quantity, pesticide cost, total production cost, and farmer profit. Furthermore, we analyze several factors that may influence outcomes, such as geographic location, modified crop trait, and type of data and methods used in the original studies.

Materials and Methods

Literature search

Original studies for inclusion in this meta-analysis were identified through keyword searches in relevant literature databanks. Studies were searched in the ISI Web of Knowledge, Google Scholar, EconLit, and AgEcon Search. We searched for studies in the English language that were published after 1995. We did not extend the review to earlier years, because the commercial adoption of GM crops started only in the mid-1990s [23]. The search was performed for combinations of keywords related to GM technology and related to the outcome of interest. Concrete keywords used related to GM technology were (an asterisk is a replacement for any ending of the respective term; quotation marks indicate that the term was used as a whole, not each word alone): GM*, "genetically engineered", "genetically modified", transgenic, "agricultural biotechnology", HT, "herbicide tolerant", Roundup, Bt, "insect resistant". Concrete keywords used related to outcome variables were: impact*, effect*, benefit*, yield*, economic*, income*, cost*, soci*, pesticide*, herbicide*, insecticide*, productivity*, margin*, profit*. The search was completed in March 2014.

Most of the publications in the ISI Web of Knowledge are articles in academic journals, while Google Scholar, EconLit, and AgEcon Search also comprise book chapters and grey literature such as conference papers, working papers, and reports in institutional series. Articles published in academic journals have usually passed a rigorous peer-review process. Most papers presented at academic conferences have also passed a peer-review process, which is often less strict than that of good journals though. Some of the other publications are peer reviewed, while many are not. Some of the working papers and reports are published by research institutes or government organizations, while others are NGO publications. Unlike previous reviews of GM crop impacts, we did not limit the sample to peer-reviewed studies but included all publications for two reasons. First, a clear-cut distinction between studies with and without peer review is not always possible, especially when dealing with papers that were not published in a journal or presented at an academic conference [24]. Second, studies without peer review also influence the public and policy debate on GM crops; ignoring them completely would be short-sighted.

Of the studies identified through the keyword searches, not all reported original impact results. We classified studies by screening titles, abstracts, and full texts. Studies had to fulfill the following criteria to be included:

- The study is an empirical investigation of the agronomic and/or economic impacts of GM soybean, GM maize, or GM cotton using micro-level data from individual plots and/or farms. Other GM crops such as GM rapeseed, GM sugarbeet, and GM papaya were commercialized in selected countries [23], but the number of impact studies available for these other crops is very small.
- The study reports GM crop impacts in terms of one or more of the following outcome variables: yield, pesticide quantity (especially insecticides and herbicides), pesticide costs, total variable costs, gross margins, farmer profits. If only the number of pesticide sprays was reported, this was used as a proxy for pesticide quantity.
- The study analyzes the performance of GM crops by either reporting mean outcomes for GM and non-GM, absolute or percentage differences, or estimated coefficients of regression models that can be used to calculate percentage differences between GM and non-GM crops.
- The study contains original results and is not only a review of previous studies.

In some cases, the same results were reported in different publications; in these cases, only one of the publications was included to avoid double counting. On the other hand, several publications involve more than one impact observation, even for a single outcome variable, for instance when reporting results for different geographical regions or derived with different methods (e.g., comparison of mean outcomes of GM and non-GM crops plus regression model estimates). In those cases, all observations were included. Moreover, the same primary dataset was sometimes used for different publications without reporting identical results (e.g., analysis of different outcome variables, different waves of panel data, use of different methods). Hence, the number of impact observations in our sample is larger than the number of publications and primary datasets (Data S1). The number of studies selected at various stages is shown in the flow diagram in Figure 1. The number of publications finally included in the meta-analysis is 147 (Table S1).

Effect sizes and influencing factors

Effect sizes are measures of outcome variables. We chose the percentage difference between GM and non-GM crops for five different outcome variables, namely yield, pesticide quantity, pesticide cost, total production cost, and farmer profits per unit area. Most studies that analyze production costs focus on variable costs, which are the costs primarily affected through GM technology adoption. Accordingly, profits are calculated as revenues minus variable production costs (profits calculated in this way are also referred to as gross margins). These production costs also take into account the higher prices charged by private companies for GM seeds. Hence, the percentage differences in profits considered here are net economic benefits for farmers using GM technology. Percentage differences, when not reported in the original studies, were calculated from mean value comparisons between GM and non-GM or from estimated regression coefficients.

Since we look at different types of GM technologies (different modified traits) that are used in different countries and regions, we do not expect that effect sizes are homogenous across studies. Hence, our approach of combining effect sizes corresponds to a random-effects model in meta-analysis [25]. To explain impact heterogeneity and test for possible biases, we also compiled data on a number of study descriptors that may influence the reported effect sizes. These influencing factors include information on the type of GM technology (modified trait), the region studied, the type of data and method used, the source of funding, and the type of publication. All influencing factors are defined as dummy variables. The exact definition of these dummy variables is given in Table 1. Variable distributions of the study descriptors are shown in Table S2.

Statistical analysis

In a first step, we estimate average effect sizes for each outcome variable. To test whether these mean impacts are significantly

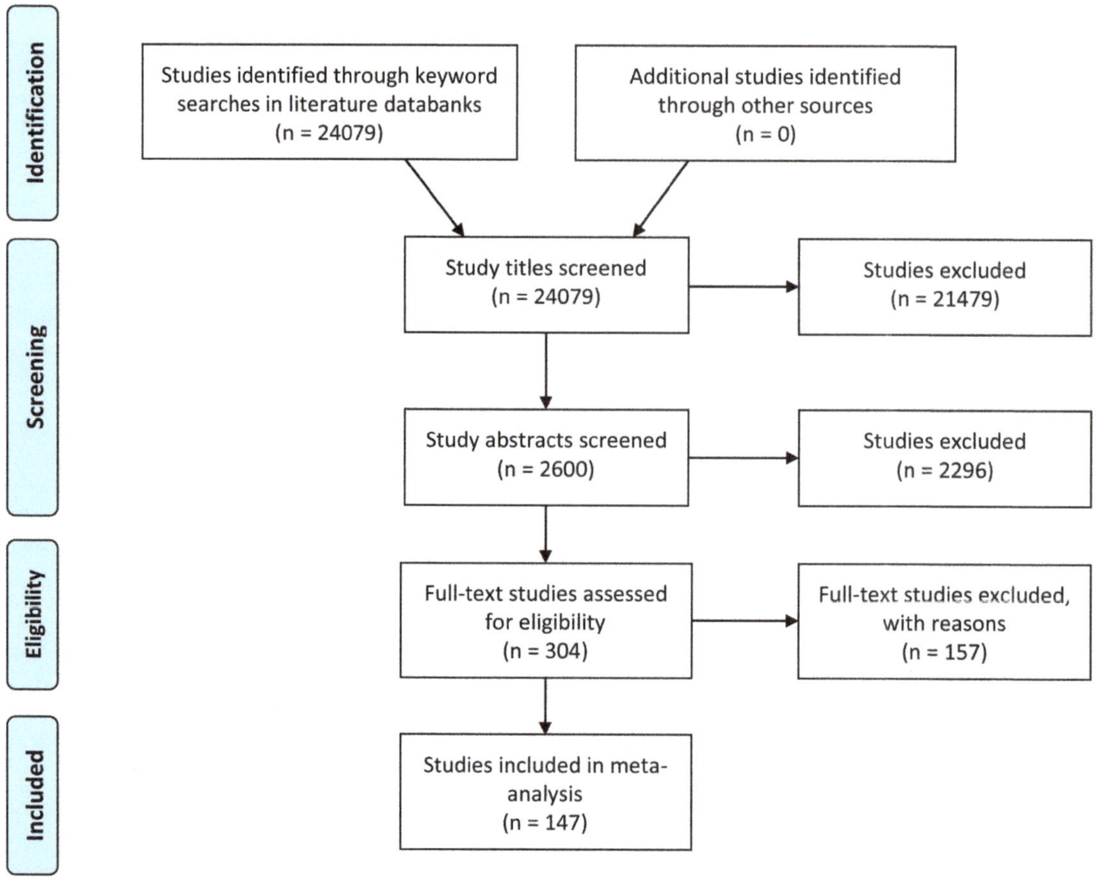

Figure 1. Selection of studies for inclusion in the meta-analysis.

different from zero, we regress each outcome variable on a constant with cluster correction of standard errors by primary dataset. Thus, the test for significance is valid also when observations from the same dataset are correlated. We estimate average effect sizes for all GM crops combined. However, we expect that the results may differ by modified trait, so that we also analyze mean effects for HT crops and IR crops separately.

Meta-analyses often weight impact estimates by their variances; estimates with low variance are considered more reliable and receive a higher weight [26]. In our case, several of the original studies do not report measures of variance, so that weighting by variance is not possible. Alternatively, weighting by sample size is common, but sample sizes are also not reported in all studies considered, especially not in some of the grey literature publications. To test the robustness of the results, we employ a

Table 1. Variables used to analyze influencing factors of GM crop impacts.

Variable name	Variable definition
Insect resistance (IR)	Dummy that takes a value of one for all observations referring to insect-resistant GM crops with genes from *Bacillus thuringiensis* (Bt), and zero for all herbicide-tolerant (HT) GM crops.
Developing country	Dummy that takes a value of one for all GM crop applications in a developing country according to the World Bank classification of countries, and zero for all applications in a developed country.
Field-trial data	Dummy that takes a value of one for all observations building on field-trial data (on-station and on-farm experiments), and zero for all observations building on farm survey data.
Industry-funded study	Dummy that takes a value of one for all studies that mention industry (private sector companies) as source of funding, and zero otherwise.
Regression model result	Dummy that takes a value of one for all impact observations that are derived from regression model estimates, and zero for observations derived from mean value comparisons between GM and non-GM.
Journal publication	Dummy that takes a value of one for all studies published in a peer-reviewed journal, and zero otherwise.
Journal/academic conference	Dummy that takes a value of one for all studies published in a peer-reviewed journal or presented at an academic conference, and zero otherwise.

different weighting procedure, using the inverse of the number of impact observations per dataset as weights. This procedure avoids that individual datasets that were used in several publications dominate the calculation of average effect sizes.

In a second step, we use meta-regressions to explain impact heterogeneity and test for possible biases. Linear regression models are estimated separately for all of the five outcome variables:

$$\%\Delta Y_{hij} = \alpha_h + \mathbf{X}_{hij}\boldsymbol{\beta}_h + \varepsilon_{hij}$$

$\%\Delta Y_{hij}$ is the effect size (percentage difference between GM and non-GM) of each outcome variable h for observation i in publication j, and $\mathbf{X}_{hij}$ is a vector of influencing factors. α_h is a coefficient and $\boldsymbol{\beta}_h$ a vector of coefficients to be estimated; ε_{hij} is a random error term. Influencing factors used in the regressions are defined in Table 1.

Results and Discussion

Average effect sizes

Distributions of all five outcome variables are shown in Figure S1. Table 2 presents unweighted mean impacts. As a robustness check, we weighted by the inverse of the number of impact observations per dataset. Comparing unweighted results (Table 2) with weighted results (Table S3) we find only very small differences. This comparison suggests that the unweighted results are robust.

On average, GM technology has increased crop yields by 21% (Figure 2). These yield increases are not due to higher genetic yield potential, but to more effective pest control and thus lower crop damage [27]. At the same time, GM crops have reduced pesticide quantity by 37% and pesticide cost by 39%. The effect on the cost of production is not significant. GM seeds are more expensive than non-GM seeds, but the additional seed costs are compensated through savings in chemical and mechanical pest control. Average profit gains for GM-adopting farmers are 69%.

Results of Cochran's test [25], which are reported in Figure S1, confirm that there is significant heterogeneity across study observations for all five outcome variables. Hence it is useful to further disaggregate the results. Table 2 shows a breakdown by modified crop trait. While significant reductions in pesticide costs are observed for both HT and IR crops, only IR crops cause a consistent reduction in pesticide quantity. Such disparities are expected, because the two technologies are quite different. IR crops protect themselves against certain insect pests, so that spraying can be reduced. HT crops, on the other hand, are not protected against pests but against a broad-spectrum chemical herbicide (mostly glyphosate), use of which facilitates weed control. While HT crops have reduced herbicide quantity in some situations, they have contributed to increases in the use of broad-spectrum herbicides elsewhere [2,11,19]. The savings in pesticide costs for HT crops in spite of higher quantities can be explained by the fact that broad-spectrum herbicides are often much cheaper than the selective herbicides that were used before. The average farmer profit effect for HT crops is large and positive, but not statistically significant because of considerable variation and a relatively small number of observations for this outcome variable.

Impact heterogeneity and possible biases

Table 3 shows the estimation results from the meta-regressions that explain how different factors influence impact heterogeneity. Controlling for other factors, yield gains of IR crops are almost 7 percentage points higher than those of HT crops (column 1). Furthermore, yield gains of GM crops are 14 percentage points higher in developing countries than in developed countries. Especially smallholder farmers in the tropics and subtropics suffer from considerable pest damage that can be reduced through GM crop adoption [27].

Most original studies in this meta-analysis build on farm surveys, although some are based on field-trial data. Field-trial results are often criticized to overestimate impacts, because farmers may not be able to replicate experimental conditions. However, results in Table 3 (column 1) show that field-trial data do not overestimate the yield effects of GM crops. Reported yield gains from field trials are even lower than those from farm surveys. This is plausible, because pest damage in non-GM crops is often more severe in farmers' fields than on well-managed experimental plots.

Table 2. Impacts of GM crop adoption by modified trait.

Outcome variable	All GM crops	Insect resistance	Herbicide tolerance
Yield	21.57*** (15.65; 27.48)	24.85*** (18.49; 31.22)	9.29** (1.78; 16.80)
n/m	451/100	353/83	94/25
Pesticide quantity	-36.93*** (-48.01; −25.86)	-41.67*** (-51.99; −31.36)	2.43 (-20.26; 25.12)
n/m	121/37	108/31	13/7
Pesticide cost	-39.15*** (-46.96; −31.33)	-43.43*** (-51.64; −35.22)	-25.29*** (-33.84; −16.74)
n/m	193/57	145/45	48/15
Total production cost	3.25 (-1.76; 8.25)	5.24** (0.25; 10.73)	-6.83 (-16.43; 2.77)
n/m	115/46	96/38	19/10
Farmer profit	68.21*** (46.31; 90.12)	68.78*** (46.45; 91.11)	64.29 (-24.73; 153.31)
n/m	136/42	119/36	17/9

Average percentage differences between GM and non-GM crops are shown with 95% confidence intervals in parentheses. *, **, *** indicate statistical significance at the 10%, 5%, and 1% level, respectively. *n* is the number of observations, *m* the number of different primary datasets from which these observations are derived.

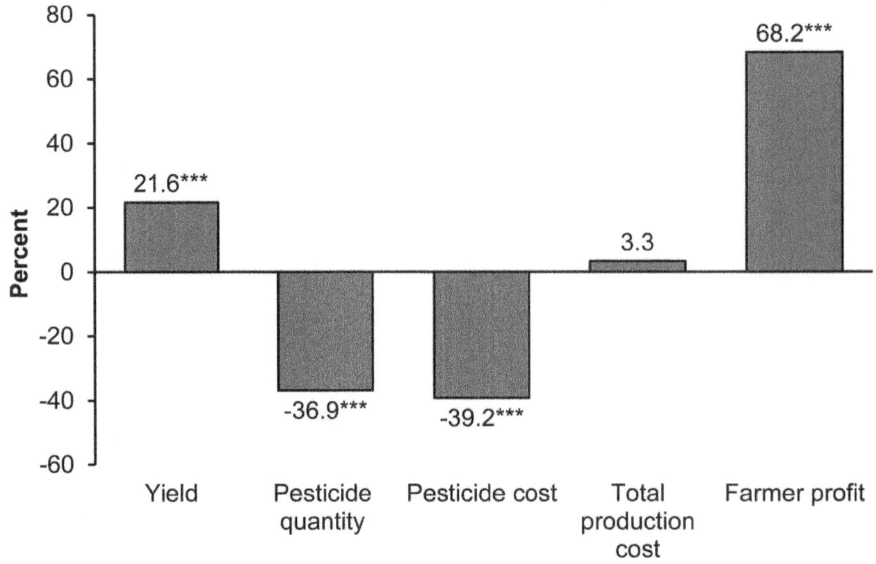

Figure 2. Impacts of GM crop adoption. Average percentage differences between GM and non-GM crops are shown. Results refer to all GM crops, including herbicide-tolerant and insect-resistant traits. The number of observations varies by outcome variable; yield: 451; pesticide quantity: 121; pesticide cost: 193; total production cost: 115; farmer profit: 136. *** indicates statistical significance at the 1% level.

Another concern often voiced in the public debate is that studies funded by industry money might report inflated benefits. Our results show that the source of funding does not significantly influence the impact estimates. We also analyzed whether the statistical method plays a role. Many of the earlier studies just compared yields of GM and non-GM crops without considering possible differences in other inputs and conditions that may also affect the outcome. Net impacts of GM technology can be estimated with regression-based production function models that control for other factors. Interestingly, results derived from regression analysis report higher average yield effects.

Finally, we examined whether the type of publication matters. Controlling for other factors, the regression coefficient for journal publications in column (1) of Table 3 implies that studies published in peer-reviewed journals show 12 percentage points higher yield gains than studies published elsewhere. Indeed, when only including observations from studies that were published in journals, the mean effect size is larger than if all observations are included (Figure S2). On first sight, one might suspect publication bias, meaning that only studies that report substantial effects are accepted for publication in a journal. A common way to assess possible publication bias in meta-analysis is through funnel plots [25], which we show in Figure S3. However, in our case these funnel plots should not be over-interpreted. First, only studies that report variance measures can be included in the funnel plots, which holds true only for a subset of the original studies used here. Second, even if there were publication bias, our mean results would be estimated correctly, because we do include studies that were not published in peer-reviewed journals.

Further analysis suggests that the journal review process does not systematically filter out studies with small effect sizes. The journal articles in the sample report a wide range of yield effects, even including negative estimates in some cases. Moreover, when combining journal articles with papers presented at academic conferences, average yield gains are even higher (Table 3, column 2). Studies that were neither published in a journal nor presented at an academic conference encompass a diverse set of papers, including reports by NGOs and outspoken biotechnology critics. These reports show lower GM yield effects on average, but not all meet common scientific standards. Hence, rather than indicating publication bias, the positive and significant journal coefficient may be the result of a negative NGO bias in some of the grey literature.

Concerning other outcome variables, IR crops have much stronger reducing effects on pesticide quantity than HT crops (Table 3, column 3), as already discussed above. In terms of pesticide costs, the difference between IR and HT is less pronounced and not statistically significant (column 4). The profit gains of GM crops are 60 percentage points higher in developing countries than in developed countries (column 6). This large difference is due to higher GM yield gains and stronger pesticide cost savings in developing countries. Moreover, most GM crops are not patented in developing countries, so that GM seed prices are lower [19]. Like for yields, studies published in peer-reviewed journals report higher profit gains than studies published elsewhere, but again we do not find evidence of publication bias (column 7).

Conclusion

This meta-analysis confirms that – in spite of impact heterogeneity – the average agronomic and economic benefits of GM crops are large and significant. Impacts vary especially by modified crop trait and geographic region. Yield gains and pesticide reductions are larger for IR crops than for HT crops. Yield and farmer profit gains are higher in developing countries than in developed countries. Recent impact studies used better data and methods than earlier studies, but these improvements in study design did not reduce the estimates of GM crop advantages. Rather, NGO reports and other publications without scientific peer review seem to bias the impact estimates downward. But even with such biased estimates included, mean effects remain sizeable.

One limitation is that not all of the original studies included in this meta-analysis reported sample sizes and measures of variance. This is not untypical for analyses in the social sciences, especially when studies from the grey literature are also included. Future

Table 3. Factors influencing results on GM crop impacts (%).

	(1)	(2)	(3)	(4)	(5)	(6)	(7)
Variables	**Yield**	**Yield**	**Pesticide quantity**	**Pesticide cost**	**Total cost**	**Farmer profit**	**Farmer profit**
Insect resistance (IR)	6.58** (2.85)	5.25* (2.82)	−37.38*** (11.81)	−7.28 (5.44)	5.63 (5.60)	−22.33 (21.62)	−33.41 (21.94)
Developing country	14.17*** (2.72)	13.32*** (2.65)	−10.23 (8.99)	−19.16*** (5.35)	3.43 (4.78)	59.52*** (18.02)	60.58*** (17.67)
Field-trial data	−7.14** (3.19)	−7.81** (3.08)	–#	−17.56 (11.45)	−10.69* (5.79)	–#	–#
Industry-funded study	1.68 (5.30)	1.05 (5.21)	37.04 (23.08)	−7.77 (10.22)	–#	–#	–#
Regression model result	7.38* (3.90)	7.29* (3.83)	9.67 (10.40)	–#	–#	−11.44 (24.33)	−9.85 (24.03)
Journal publication	12.00*** (2.52)	–	9.95 (6.79)	−3.71 (4.09)	−3.08 (3.30)	48.27*** (15.48)	–
Journal/academic conference	–	16.48*** (2.64)	–	–	–	–	65.29*** (17.75)
Constant	−0.22 (2.84)	−2.64 (2.86)	−4.44 (10.33)	−16.13 (4.88)	−1.02 (4.86)	8.57 (24.33)	−1.19 (24.53)
Observations	451	451	121	193	115	136	136
R^2	0.23	0.25	0.20	0.14	0.12	0.12	0.14

Coefficient estimates from linear regression models are shown with standard errors in parentheses. Dependent variables are GM crop impacts measured as percentage differences between GM and non-GM. All explanatory variables are 0/1 dummies (for variable definitions see Table 1). The yield models in columns (1) and (2) and the farmer profit models in columns (6) and (7) have the same dependent variables, but they differ in terms of the explanatory variables, as shown. *, **, *** indicate statistical significance at the 10%, 5%, and 1% level, respectively. # indicates that the variable was dropped because the number of observations with a value of one was smaller than 5.

impact studies with primary data should follow more standardized reporting procedures. Nevertheless, our findings reveal that there is robust evidence of GM crop benefits. Such evidence may help to gradually increase public trust in this promising technology.

Acknowledgments

We thank Sinja Buri and Tingting Xu for assistance in compiling the dataset. We also thank Joachim von Braun and three reviewers of this journal for useful comments.

Author Contributions

Conceived and designed the research: WK MQ. Analyzed the data: WK MQ. Contributed to the writing of the manuscript: WK MQ. Compiled the data: WK.

References

1. Gilbert N (2013) A hard look at GM crops. Nature 497: 24–26.
2. Fernandez-Cornejo J, Wechsler JJ, Livingston M, Mitchell L (2014) Genetically Engineered Crops in the United States. Economic Research Report ERR-162 (United Sates Department of Agriculture, Washington, DC).
3. Anonymous (2013) Contrary to popular belief. Nature Biotechnology 31: 767.
4. Andreasen M (2014) GM food in the public mind–facts are not what they used to be. Nature Biotechnology 32: 25.
5. DeFrancesco L (2013) How safe does transgenic food need to be? Nature Biotechnology 31: 794–802.
6. European Academies Science Advisory Council (2013) Planting the Future: Opportunities and Challenges for Using Crop Genetic Improvement Technologies for Sustainable Agriculture (EASAC, Halle, Germany).
7. European Commission (2010) A Decade of EU-Funded GMO Research 2001–2010 (European Commission, Brussels).
8. Pray CE, Huang J, Hu R, Rozelle S (2002) Five years of Bt cotton in China - the benefits continue. The Plant Journal 31: 423–430.
9. Huang J, Hu R, Rozelle S, Pray C (2008) Genetically modified rice, yields and pesticides: assessing farm-level productivity effects in China. Economic Development and Cultural Change 56: 241–263.
10. Morse S, Bennett R, Ismael Y (2004) Why Bt cotton pays for small-scale producers in South Africa. Nature Biotechnology 22: 379–380.
11. Qaim M, Traxler G (2005) Roundup Ready soybeans in Argentina: farm level and aggregate welfare effects. Agricultural Economics 32: 73–86.
12. Sexton S, Zilberman D (2012) Land for food and fuel production: the role of agricultural biotechnology. In: The Intended and Unintended Effects of US Agricultural and Biotechnology Policies (eds. Zivin, G. & Perloff, J.M.), 269–288 (University of Chicago Press, Chicago).
13. Ali A, Abdulai A (2010) The adoption of genetically modified cotton and poverty reduction in Pakistan. Journal of Agricultural Economics 61, 175–192.
14. Kathage J, Qaim M (2012) Economic impacts and impact dynamics of Bt (*Bacillus thuringiensis*) cotton in India. Proceedings of the National Academy of Sciences USA 109: 11652–11656.
15. Qaim M, Kouser S (2013) Genetically modified crops and food security. PLOS ONE 8: e64879.
16. Stone GD (2012) Constructing facts: Bt cotton narratives in India. Economic & Political Weekly 47(38): 62–70.
17. Smale M, Zambrano P, Gruere G, Falck-Zepeda J, Matuschke I, et al. (2009) Measuring the Economic Impacts of Transgenic Crops in Developing Agriculture During the First Decade: Approaches, Findings, and Future Directions (International Food Policy Research Institute, Washington, DC).
18. Glover D (2010) Is Bt cotton a pro-poor technology? A review and critique of the empirical record. Journal of Agrarian Change 10: 482–509.
19. Qaim M (2009) The economics of genetically modified crops. Annual Review of Resource Economics 1: 665–693.
20. Carpenter JE (2010) Peer-reviewed surveys indicate positive impact of commercialized GM crops. Nature Biotechnology 28: 319–321.
21. Finger R, El Benni N, Kaphengst T, Evans C, Herbert S, et al. (2011) A meta analysis on farm-level costs and benefits of GM crops. Sustainability 3: 743–762.
22. Areal FJ, Riesgo L, Rodríguez-Cerezo E (2013) Economic and agronomic impact of commercialized GM crops: a meta-analysis. Journal of Agricultural Science 151: 7–33.
23. James C (2013) Global Status of Commercialized Biotech/GM Crops: 2013. ISAAA Briefs No.46 (International Service for the Acquisition of Agri-biotech Applications, Ithaca, NY).
24. Rothstein HR, Hopewell S (2009) Grey literature. In: Handbook of Research Synthesis and Meta-Analysis, Second Edition (eds. Cooper, H., Hegdes, L.V. & Valentine, J.C.), 103–125 (Russell Sage Foundation, New York).
25. Borenstein M, Hedges LV, Higgins JPT, Rothstein HR (2009) Introduction to Meta-Analysis (John Wiley and Sons, Chichester, UK).
26. Shadish WR, Haddock CK (2009) Combining estimates of effect size. In: Handbook of Research Synthesis and Meta-Analysis, Second Edition (eds. Cooper, H., Hegdes, L.V. & Valentine, J.C.), 257–277 (Russell Sage Foundation, New York).
27. Qaim M, Zilberman D (2003) Yield effects of genetically modified crops in developing countries. Science 299: 900–902.

13

MSH1-Induced Non-Genetic Variation Provides a Source of Phenotypic Diversity in *Sorghum bicolor*

Roberto de la Rosa Santamaria[1,2], Mon-Ray Shao[1,2], Guomei Wang[3], David O. Nino-Liu[3], Hardik Kundariya[1], Yashitola Wamboldt[1], Ismail Dweikat[2], Sally A. Mackenzie[1,2]*

1 Center for Plant Science Innovation, University of Nebraska Lincoln, Lincoln, Nebraska, United States of America, **2** Department of Agronomy and Horticulture, University of Nebraska Lincoln, Lincoln, Nebraska, United States of America, **3** Monsanto, Chesterfield Village Research Center, Chesterfield, Missouri, United States of America

Abstract

MutS Homolog 1 (*MSH1*) encodes a plant-specific protein that functions in mitochondria and chloroplasts. We showed previously that disruption or suppression of the *MSH1* gene results in a process of developmental reprogramming that is heritable and non-genetic in subsequent generations. In Arabidopsis, this developmental reprogramming process is accompanied by striking changes in gene expression of organellar and stress response genes. This developmentally reprogrammed state, when used in crossing, results in a range of variation for plant growth potential. Here we investigate the implications of *MSH1* modulation in a crop species. We found that *MSH1*-mediated phenotypic variation in *Sorghum bicolor* is heritable and potentially valuable for crop breeding. We observed phenotypic variation for grain yield, plant height, flowering time, panicle architecture, and above-ground biomass. Focusing on grain yield and plant height, we found some lines that appeared to respond to selection. Based on amenability of this system to implementation in a range of crops, and the scope of phenotypic variation that is derived, our results suggest that *MSH1* suppression provides a novel approach for breeding in crops.

Editor: Mark Gijzen, Agriculture and Agri-Food Canada, Canada

Funding: This work was supported by grants from the Bill & Melinda Gates Foundation (OPP1059119, http://www.gatesfoundation.org) and the National Science Foundation (IOS-1126935, http://www.nsf.gov). The funders had no role in study design, data collection and analysis, decision to publish, or preparation of the manuscript.

Competing Interests: The manipulation of MSH1 for use in plant breeding has been submitted for U.S. patent (Mackenzie, S. and De La Rosa Santamaria, R., Plants with Useful Traits and Related Methods, Pending Appl. No. 61/481,519). Two authors in this study (GW and DON) were employed by the company Monsanto. This does not alter the authors' adherence to PLOS ONE policies on sharing data and materials.

* Email: smackenzie2@unl.edu

These authors contributed equally to this work.

Introduction

One increasingly problematic threat to plant improvement is the depletion of natural stores of genetic diversity for most of our major crop species [1]. Centers of diversity for many species have been encroached by man-made or natural influences, limiting our ability to diversify germplasm appropriate for breeding efforts. Moreover, integration of unselected germplasm to a breeding program is laborious in the early selection process required to eliminate undesirable genetic linkages [2].

For years, breeders have speculated that non-genetic variation plays a key role in conventional crop improvement strategies [3]. For example, epigenetic influences have been implicated in heterosis [4], [5], flowering time and maturation [6], [7], inbreeding depression and its circumvention [8], and genotype×environmental interactions [9], [10]. Still, there has been no straightforward means of directly accessing such variation for plant improvement purposes.

MSH1 is a plant-specific gene that encodes a mitochondrial and plastid-localized protein [11], [12]. The expression level of *MSH1* appears to be influenced by environmental stress [12], [13]. In Arabidopsis, *msh1* mutants are characterized by variable phenotypes including dwarfing, variegation, delay in maturity transition and flowering, altered branching, and woody growth with aerial rosettes at short day length growth [14]. This developmental reprogramming (MSH1-dr) is associated with large changes in gene expression, particularly genes involved in organelle and stress response functions [14]. RNAi suppression of *MSH1* in crop plants, including tomato, soybean, tobacco, millet and sorghum produces a similar MSH1-dr phenotypic range in each that is subsequently inherited independent of the RNAi transgene [14]. These observations suggest that the MSH1-dr phenotype is both programmed and non-genetic.

Here we investigate the consequences of incorporating the MSH1-dr condition to plant selection, using sorghum as a model. We show that crossing with a transgene-null MSH1-dr line produces an unexpected range of phenotypic variation that is both heritable and responsive to selection. This variation appears to be stable over at least four generations. We also show evidence of line×environment interactions. Finally, we demonstrate gains in grain yield over two generations of selection, suggesting that this non-genetic variation may prove valuable for agricultural production as a potential crop breeding strategy.

Materials and Methods

Plant materials and growth conditions

Sorghum MSH1-dr plants used in these experiments were derived as described in [14]. Six T_3 individuals displaying the MSH1-dr phenotype but null for the MSH1-RNAi transgene were used as females in crosses to wild type inbred Tx430 to derive F_1 seed. Another three T_3 individuals were used as males in the reciprocal crosses to Tx430. The number of F_1 plants derived from each cross ranged from 5 to 19 individuals. Parents and F_1 progeny were grown under greenhouse conditions on a 14 hr/10 hr day-night cycle with 28°C/22°C day-night temperatures. Self-pollinated seed of F_1 plants was harvested individually to generate corresponding F_2 families.

Field experiments and phenotyping

In all field plots, plants were thinned to a final density of 15 plants/m^2 and fertilized according to standard growing practices. The 2010 field experiment was used to propagate F_2 lines, and contained F_2 and wild type Tx430. The 2011 field experiment contained F_2, F_3, and F_4 lines randomized across seven blocks with 28 rows per block (alpha lattice design) and two field replicates. Replicates were augmented with wild type Tx430 (16 rows total).

For estimating grain yield, threshed panicles from three plants were pooled and converted to grams/m^2 based on final plant density, with 2–3 such measurements taken per row. For comparison of panicle yield distributions in F_2 versus in wild type Tx430, individual panicle grain yield (i.e., prior to pooling) was used. For flowering time, plant height, and rachis length, measurements were taken on individual plants. For each dry biomass measurement, three fully dried plants were pooled together then converted to grams/plant. Plants showing the DR phenotype were not included in phenotypic variation analysis.

The 2012 multi-location experiment included Lincoln, NE (40° 51′N, 96° 35′W) and Mead, NE (41° 9′N, 96° 24′W) sites, which received 178 mm and 158 mm of precipitation over the growing season, respectively. Within each location, lines were grown in two-row plots arranged in a randomized complete block design with two replicates. For this experiment, grain yield was estimated by taking threshed panicles from a meter-length area of each row and converting to grams/m^2.

Statistical analysis

For evaluations in a single environment, mean phenotypic values and confidence intervals for each line were estimated using the linear mixed model $y_{ijk} = \mu+\alpha_i+r_k+(b/r)_{jk}+\varepsilon_{ijk}$ where y_{ijk} is the trait response, μ is the population mean, α_i is the effect of line i, r_k is the effect of replicate k, $(b/r)_{jk}$ is the effect of block j nested within replicate k, and ε_{ijk} is the residual error. For evaluations over multiple environments, mean phenotypic values and confidence intervals for each line were estimated using the linear mixed model $y_{ijkm} = \mu+\alpha_i+e_m+(r/e)_{km}+(b/r/e)_{jkm}+(\alpha e)_{im}+\varepsilon_{ijkm}$ where y_{ijkm} is the trait response, μ is the population mean, α_i is the effect of line i, e_m is the effect of environment m, $(r/e)_{km}$ is the effect of replicate k nested within environment m, $(b/r/e)_{jkm}$ is the effect of block j nested within replicate k of environment m, $(\alpha e)_{im}$ is the interaction between line i and environment m, and ε_{ijkm} is the residual. Line, environment, and line×environment effects were treated as fixed while block and replicate effects were treated as random. Models were fit by restricted maximum likelihood using the R package "nlme" [15]. When deemed appropriate, Box-cox transformations were performed. F_4 models for plant height and biomass excluded lines exhibiting mixed heights to avoid heteroscedasticity.

PCR assay for RNAi transgene and SSR marker analysis

PCR assay for MSH1-RNAi transgene presence in sorghum materials used primers RNAi-F 5′-GTGTACT CATCTG-GATCTGTATTG-3′ and RNAi-R 5′-GGTTGAGGAGCCT-GAATCTCTGAAC-3′. Positive and negative controls were included from a confirmed transgenic line and wild type Tx430, respectively.

SSR marker analysis used SSR primers that were developed and mapped previously [16], [17]. Fragments were assayed by capillary electrophoresis on an Advanced Analytical Fragment Analyzer (Advanced Analytical Technologies, Inc. Ames, IA) using the dsDNA Reagent kit, 35–1,500 bp 500S that separates DNA in the size range of 35–1,500 bp. Of the 136 primers that were tested, 43 produced unambiguous polymorphisms between Tx430 and the sweet sorghum control line Wray and were used for testing the epi-lines.

Sorghum SNP survey

Leaf tissue sample was collected from plants grown under controlled greenhouse conditions three weeks after germination. Genomic DNA was extracted from freeze-dried leaf tissue and processed following manufacturer's recommendations prior to Infinium beadchip hybridization (Illumina, San Diego, CA). The genotyping of five F_4 lines and wild type Tx430 was carried out at the Monsanto Applied Genotyping Labs (Chesterfield, MO). The platform used was an exclusive custom-designed *Sorghum bicolor* Infinium high-density beadchip containing 1,885 internally validated SNP markers.

For the six samples, 107 of the 1,885 SNP markers, ca 5.68%, provided invalid data due to one of the following: low marker signal intensity, marker failed data QC, or unscorable allele calls. The remaining 1,778 SNP markers were used for the analysis. These 1,778 SNP markers are distributed across all 10 sorghum chromosomes with genome coverage approximating 90%. The number of heterozygotes (# Het) and percentage of heterozygotes (% Het) were calculated based on the 1778 SNP markers.

Results

MSH1-altered lines and reciprocal crosses

Previously, we described MSH1-RNAi lines displaying numerous physiological changes, a condition of developmental reprogramming that was termed MSH1-dr [14]. Segregation of the MSH1-RNAi transgene gave rise to some *MSH1* +/+ individuals that retained the characteristic *msh1* phenotype despite having normal *MSH1* transcript levels [14]. These plants maintain the altered MSH1-dr growth phenotype through multiple (at least nine, to date) generations of self-pollination.

To investigate the mechanism of inheritance, we performed reciprocal crosses in sorghum of MSH1-dr individuals to their wild type counterpart. Figure 1 illustrates the transgene and crossing process used in this study, with all sorghum materials generated from the inbred line Tx430 [18]. When crossed to the wild type inbred Tx430 line, the transgene-null MSH1-dr individuals produced progeny that were restored to normal phenotype (Figure 1A). The derived F_1 progeny no longer showed the dwarfed, tillering, and late flowering phenotype; instead, many of the plants grew taller and produced more seed than the wild type. This was repeatedly observed in F_1 populations derived from nine separate crosses, three of which used an MSH1-dr plant as the pollen donor [14].

Lack of the MSH1-dr phenotype in the F_1 generation from either direct or reciprocal crosses argues against the observed phenotypes in this sorghum material being inherited via cytoplas-

Figure 1. Enhanced growth phenotype of MSH1 lines in sorghum. The transgene and crossing procedure used to derive sorghum populations is indicated. (A) The phenotype of the F_1 progeny derived from crossing Tx430×MSH1-dr. (B) Field grown F_2, F_3 and F_4 sorghum lines show variation in plant architecture and height. (C) Panicles from Tx430 (on left, 66 g, 8 mm stem) versus a larger F_2 individual (on right, 112 g, 11 mm stem), and (D) seed yield after threshing. (E) The MSH1-dr sorghum phenotype under field conditions. (F) Sorghum MSH1 F_2, F_3 and F_4 populations grown in progeny rows in a 2011 field experiment. Wild type inbred Tx430 is indicated. Variation in plant height, flowering time and plant architecture is apparent; all plants shown are non-transgenic and Tx430 genotype.

mic organellar genomes. Analogously generated crosses in Arabidopsis with *msh1* point or T-DNA insertion mutations also display enhanced vigor; in other species, including tomato, soybean and tobacco, heritable MSH1-dr phenotypes also persist despite restored *MSH1* expression following RNAi silencing, and crosses in those species to their respective wild type counterparts similarly produce progeny with enhanced growth phenotypes [14], (unpublished data). Taken together, the evidence suggests that the MSH1-dr and F_1 observations involve a conserved, programmed pathway.

MSH1 F_2 populations show enhanced variation

Self-pollination of the F_1 plants produced an F_2 population variable in plant phenotype (Figure 1B–F, Figure 2, Table S1), with a minority exhibiting the MSH1-dr phenotype (Figure 1E). This was initially apparent in several F2 families as an elongated tail in the distributions for panicle weight, suggesting a higher proportion of individuals with extreme values (Figure S1). Further analysis detected increased variation in the F_2 for plant height and grain yield (Figure 2A, Table S1), which although more prominent in the 2010 planting than the 2011 planting (Figure 2B–C), was still significant (Table S1). Although we did not detect a very significant increase in variance for flowering time or panicle length in the F_2, by the F_4 we were able to observe lines diverging from wild type Tx430 for those traits (Figure S2), suggesting modest but heritable variation for flowering time and panicle length.

A small proportion of greenhouse-grown MSH1 F_3 families also showed the MSH1-dr phenotype, with an overall frequency of ca. 8% (Table S2). By the F_4 generation, we estimate that the overall frequency drops to below 2%. Although the progeny from these sporadic MSH1-dr types in advanced generations have not been thoroughly investigated, some families appear more likely than others to produce this phenotype. When MSH1-dr frequencies were compared between parental and progeny generations, each derived from a single individual, the phenotype was only observed in progeny generations whose parental generation had some incidence of the phenotype (Table S3). Currently, we cannot rule out that the overall rarity of the MSH1-dr phenotype by the F_4 generation may be the consequence of inadvertent selection rather than a natural tendency to gradually stabilize away from the phenotype.

To ensure that the observed variation was not the consequence of inadvertent seed contamination or outcrossing, 50 SSR markers were used to test a number of derived lines, which produced no evidence of polymorphism (Figure S3; Table S4). This analysis was extended with 1778 SNP markers that, when assayed across five different MSH1 F_2 individuals and the wild type Tx430, detected less than 0.8% variation (Table S5, Figure S4). In Arabidopsis, the *msh1* mutant genome was DNA sequenced, with genome alignment and *de novo* assembly producing no evidence of unexplained genome rearrangement or unusual mutation frequency (unpublished). These data, together with reproducibility of

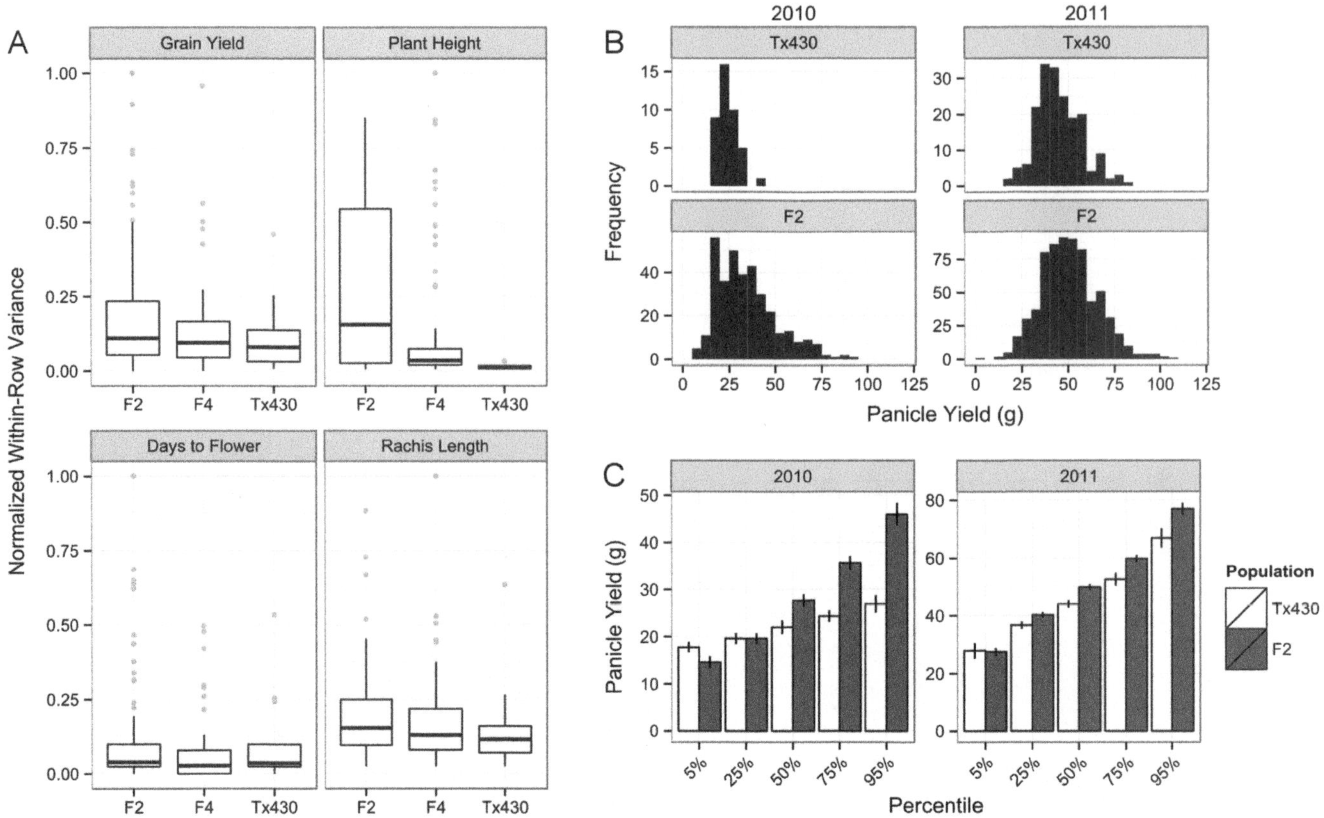

Figure 2. Increased phenotypic variation in MSH1 F2 lines. (A) Boxplots of within-row field variance for indicated traits, with values normalized as a proportion of the maximum observed row variance for that trait. Differences in variances between the F_2 and wild type populations were significant for plant height (Brown-Forsythe test, $p<0.001$) and grain yield ($p<0.01$). (B) Histograms for yield per panicle in the F_2 population compared to wild type, from the two field plantings. (C) Percentile values for yield per panicle in the F_2 population compared to wild type, estimated from bootstrapping; error bars represent standard deviation.

the phenomenon, argue against the developmental reprogramming phenotype as a consequence of genome hypermutability.

Significant increases in trait values persist for multiple generations

From the MSH1 F_2 families, individuals were self-pollinated and selected for grain yield and plant height to the F_3 and F_4 generations. F_4 lines, along with F_3 and F_2 lines from remnant seed, were evaluated together in a 2011 field experiment. Despite weak selection intensity (33% and 38% of phenotyped plants were propagated to F_3 and F_4, respectively, based on grain yield), derived F_3 and F_4 lines showed differences in grain yield and plant height, as well as differences in dry biomass and panicle length (Figure 3, Figure S2, Table S6). Differences were detectable even when F_3 and F_4 lines were analyzed separately or when a model term for generation was included, indicating that the variation did not simply come from maternal effects. While some traits appeared to be correlated, such as flowering time and grain yield, no correlation was detected between plant height and grain yield, indicating that height was not pleiotropically affecting grain yield (Figure S5).

Although the F_3 generation showed higher variance for some traits compared to the F_2 generation, for all measured traits the F_4 generation showed lower variance compared to the F_2 generation (Figure 2A). Furthermore, in contrast to the F_2 generation, we did not find significant heterogeneity for variance in grain yield among wild type, F_3 and F_4 lines ($p>0.1$, Brown-Forsythe test; $p<0.01$ among F_2 lines and wild type).

Analysis of several direct lineages from F_2 to F_4 showed high response to selection for plant height but variable response for grain yield (Figures 3A, S6). Overall, gains in the F_4 were more modest compared to the F_3, implying progress may taper off by F_4 in self-pollinated lineages. Indeed, there is evidence that the F_3 generation may be the most vigorous. As a population, it appears to have slightly higher overall grain yield than the F_2 or F_4. Nevertheless, the population mean for grain yield in the F_4 remains higher than that of wild type Tx430 (Figure 3B).

Line×environment interactions suggest an additional component to G×E

As plant development is heavily influenced by the surrounding conditions, genotype×environment interactions (G×E) have major impacts on phenotype. The causes underlying G×E effects can potentially come from multiple sources, both genetic and non-genetic [19]. We evaluated the yield performance of three F_5 families alongside wild type Tx430 at two different locations, which displayed a large difference in environmental means. Although the lines showed little between-line difference at the site of the earlier experiments (which may be a consequence of year-to-year climate effects), they showed large differences at the second site, which was more drought-stressed, demonstrating a line× environment effect (Figures 4, S7; Table S7). Results at the first

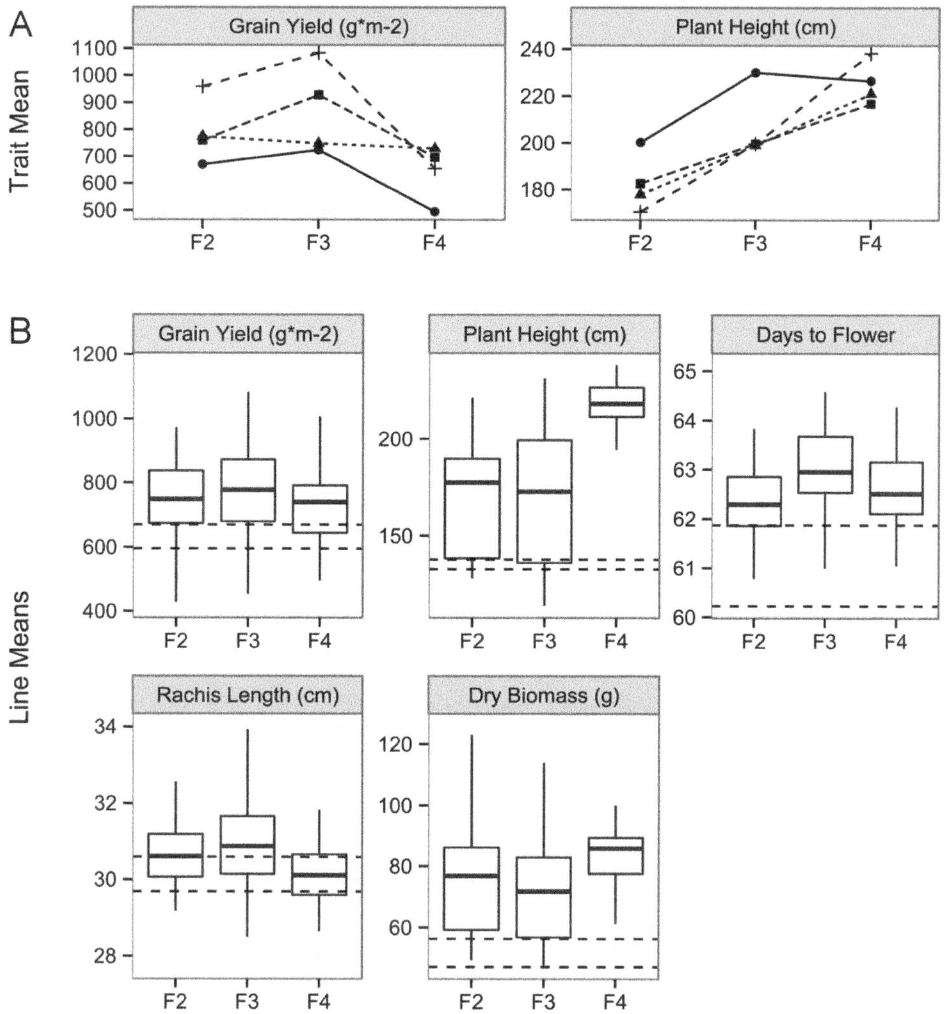

Figure 3. Phenotypic changes over MSH1 F2, F3 and F4 generations. (A) Selection had varying results, with response for yield into the F_3 generation, but not into the F_4 generation. For each lineage, the mean generation performance is represented as a point. (B) Boxplots of F_2, F_3, and F_4 line means for various traits, giving a population-wide view of line performance. Dashed lines indicate the 95% confidence interval for wild type Tx430 mean.

site also suggest that, depending on conditions, variation in these materials could begin to dissipate at around the F_5 generation. The outcomes of these experiments indicate that plant materials with little to no genetic variation have the potential to exhibit substantial variation in response to environmental influences, which may reflect epigenetic×environmental interactions.

Discussion

A substantial range of sorghum phenotypic variation observed in this study appears to be primarily non-genetic, and is induced by crossing to a MSH1-dr line, altered through *MSH1* suppression in a previous generation. The MSH1-dr lines used in this study were maintained as transgene-nulls seven generations following segregation of the transgene, suggesting that the non-genetic properties of the MSH1-dr line are stable through multiple rounds of self-pollination [14]. We do not presume that all of the variation observed is non-genetic; the observed bimodal distribution for plant height could support an alternative hypothesis of markedly enhanced reversion frequency for the dwarfing gene, *dw3*, in line Tx430 [20]. If this is the case, the unusually high reversion rate may be the consequence of increased local recombination, possibly due to cytosine methylation redistribution [21], [22]. We are investigating this possibility presently. Nevertheless, we see additional height variation within short and tall plants, indicating variation beyond a single-locus.

The range of phenotypic variation observed is surprising. While we were not able to take measurements of all parameters for this initial study, the F_3 and the F_4 generations showed highly significant increases in above-ground biomass and grain yield over Tx430 wild type. One interpretation of these increases would be that *dw3* reversion could cause pleiotropic changes in plant architecture. However, the greater range of plant height, panicle architecture and yield variation observed in this study appears to exclude that possibility [23].

The observation of line×environment interaction in test plots suggests that at least some portion of the genotype×environment interaction that is commonly observed in varietal studies may be non-genetic, which is supported by other recent studies [24]. The *MSH1* system may be useful in understanding this type of

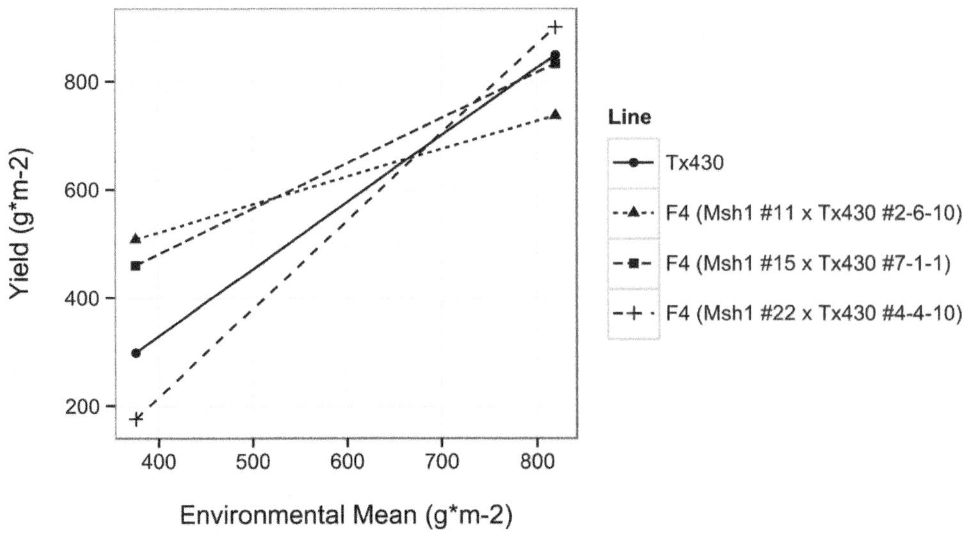

Figure 4. Line performance shows environmental interactions. Joint regression (with Wright modification) indicates differential response between lines to location.

environmental influence and selecting for enhanced stability of crop performance.

MSH1-dr transgene null lines developed on elite inbred genetic backgrounds would permit direct incorporation of the MSH1-enhanced growth phenomenon to hybrid production. However, studies to date have not observed the greatest gain in growth to occur in the derived F_1 populations, suggesting that the effects we observe in this system may be distinct from heterosis. It is possible that self- or open-pollination breeding will prove more effective at capturing maximal growth gain derived from *MSH1* manipulation. The transgene-null MSH1-dr line crossed to its wild type counterpart produces maximum variation in the F_2 population, at which point selection appears to be most effective. Large-scale seed increase in F_3 and F_4 generations permits rapid capture of the growth enhancement as variation tapers off. In our experience with this system, variation observed in the F_2 population tends to produce above wild type performance more often than below (Figure 3B). Consequently, development of MSH1-dr in an elite line followed by selection in the F_2, appears to result in, by the F_4, a population that is uniform genetically yet enhanced in growth vigor and productivity.

The progress, response to selection, and final phenotypic outcomes observed in this study are of sufficient magnitude to suggest that untapped non-genetic potential resides within crops. One possibility is that epigenetic changes such as DNA methylation may either directly cause or are indicators of such variation. In Arabidopsis, mutation of genes that comprise the DNA methylation machinery, followed by crossing to wild type for development of recombinant inbred lines, has provided valuable information on the phenotypic consequences of epigenomic perturbation, as well as heritability and stability of epigenetic changes [25], [26]. It has been suggested that doubled haploids, subjected to recursive selection for mitochondrial behavior, can produce epigenetic variation that may be amenable to selection [27]. Somaclonal variation derived from plant tissue culture has also been associated with epigenetic changes [28]. Whether crop enhancement using *MSH1* manipulation will produce crop vulnerabilities not yet considered is under investigation. However, the performance of these plant materials under low rainfall conditions suggests that this methodology holds significant promise.

Supporting Information

Figure S1 F2 families generally exhibit wider distributions for panicle weight. The majority of F2 families have an altered distribution of values for panicle weight, leading to generally flatter distributions with longer tails compared to wild type Tx430, and indicating that a higher proportion of individuals have unusually large values. Data presented are from 2010 field planting.

Figure S2 F4 lines show trait differences compared to wild type. Lines from the MSH1 F_4 generation show differences between each other and compared to wild type Tx430. Points represent line means; error bars represent 95% confidence intervals, which are shorter in wild type Tx430 due to higher sample size.

Figure S3 Sample SSR marker analysis. Sorghum genomic DNAs were prepared from wild type Tx430, Tx430 MSH1-DR line (transgene-null, displaying the dwarfed, tillered, delayed flowering phenotype), one F_2 and seven F_4 lines selected for phenotypic diversity. The sweet sorghum variety Wray was included as a control. The SSR marker shown is generated with SAM16073 primers. Arrow shows detected DNA polymorphism. M designates marker lane, with fragment sizes (bp) shown at left. The 1500 and 35 bp fragments are internal markers used to calibrate each lane.

Figure S4 Sorghum genetic map with markers displaying heterozygous genotype.

Figure S5 No correlation was found between plant height and grain yield. Each point represents a line mean. Spearman's rank correlation coefficient = 0.02 (p-value = 0.83).

Figure S6 Shifts in phenotypic means by individual lineages. Several lineages from F2 to F4 were (re)planted within the same field experiment and measured for (A) yield per panicle and (B) plant height, with varying results in terms of response to selection. Representative lineages from independent crosses are shown.

Figure S7 Line×environment effects were detected from a multiple location experiment. Although three F_4 lines (bulked from indicated F3 parent) were similar to wild type Tx430 in grain yield when grown in one location (Havelock), significant differences emerged when grown another location with a more challenging environment (Mead). Data were collected from a field experiment in 2012.

Table S1 MSH1-dr×Tx430 derived populations show more variation compared to wild type Tx430. Data were acquired from 2011 field experiment. Brown-Forsythe tests for homogeneous variances were performed between all individuals of indicated generation versus wild type (e.g. all F_2 vs Tx430, all F_3 vs Tx430).

Table S2 Frequency of MSH1-dr phenotype (8.4%) in F3 families derived from sorghum MSH1-dr×Tx430. Data were acquired from plants grown in greenhouse conditions.

Table S3 Msh1-dr phenotype shows a partially heritable or metastable component. From each of ten lines, a single individual that did not display the MSH1-dr phenotype was grown along with its parental generation. Parental and progeny generation frequencies were then counted with $N \geq 105$ in each generation.

Table S4 SSR marker polymorphism data for 43 markers. Markers were scored as + or − relative the pattern of Tx430 wild type. SSR markers were selected based on their polymorphic behavior in comparisons of Tx430 and a sweet sorghum variety, Wray. Assays included a transgene-null Tx430 line displaying the developmental reprogramming phenotype (DR), one F_2, two F_3, and seven F_4 lines.

Table S5 SNP marker analysis. (A) Summary of Het %. (B) A list of all the markers with heterozygous genotype. Markers with heterozygous genotypes are ordered by chromosome and genetic distance. The remainder (not shown) had homozygous genotypes. Marker genotypes of the six lines are similar but for the two markers highlighted in yellow. Markers showing a heterozygous genotype represent the true heterozygous genotype, not heterogeneity at the markers since only a single plant was sampled for DNA.

Table S6 Line effects on trait values are significant. Data for each trait listed below were fit to a linear mixed model, with results indicating differences between lines. Line was treated as a fixed effect while block and replicate were treated as random effects. Separately analyzing lines by generation or general height class, or adding a model term for generation and height class, did not affect conclusions. The models were used to estimate trait means and confidence intervals (Figure 3B, S1).

Table S7 Analysis for significant effects using a mixed model indicates that line, location, and line×location are all significant. See methods for model; sample size N = 121.

Acknowledgments

We wish to thank E. Omar Lozano for technical support, Dr. Heriberto Cerutti for assistance in designing the RNAi gene construction, and Dr. John Rajewski for assistance with coordinating field plantings and harvesting. We thank UNL Center for Biotechnology Plant Transformation Core Facility, Shirley Sato and Natalia Nersesian for assistance in the development of the sorghum transformants.

Author Contributions

Conceived and designed the experiments: RDLRS ID SAM. Performed the experiments: RDLRS MRS GW DONL HK YW. Analyzed the data: RDLRS MRS GW DONL HK YW. Wrote the paper: SAM.

References

1. Gressel J (2008) Genetic Glass Ceilings: Transgenics for Crop Biodiversity. John Hopkins Univ. Press.
2. Cowling W (2013) Sustainable plant breeding. Plant Breeding 132:1–9.
3. Tsaftaris AS, Polidoros AN, Kapazoglou A, Tani E, Kovačević NM (2008) Epigenetics and Plant Breeding. Plant Breeding Reviews, Volume 30. John Wiley & Sons, Inc. Hoboken, NJ, USA. doi: 10.1002/97804703801 30.ch2
4. Groszmann M, Greaves IK, Albert N, Fujimoto R, Helliwell CA, et al. (2011) Epigenetics in plants-vernalisation and hybrid vigour. Biochim Biophys Acta 1809:427–37.
5. Shivaprasad PV, Dunn RM, Santos BA, Bassett A, Baulcombe DC (2012) Extraordinary transgressive phenotypes of hybrid tomato are influenced by epigenetics and small silencing RNAs. EMBO J 31:257–66.
6. Schmitz RJ, Amasino RM (2007) Vernalization: a model for investigating epigenetics and eukaryotic gene regulation in plants. Biochim Biophys Acta 1769:269–75.
7. Heo JB, Sung S (2011) Encoding memory of winter by noncoding RNAs. Epigenetics 6:544–7.
8. Cheptou PO, Donohue K (2013) Epigenetics as a new avenue for the role of inbreeding depression in evolutionary ecology. Heredity (Edinb) 110:205–6.
9. Dooner HK, Weil DF (2007) Give-and-take: interactions between DNA transposons and their host plant genomes. Curr Opin Genet Dev 17:486–492.
10. Smith AM, Hansey CN, Kaeppler SM (2012) TCUP: A novel hAT transposon active in maize tissue culture. Front Plant Sci 3:6.
11. Abdelnoor RV, Yule R, Elo A, Christensen AC, Meyer-Gauen G, et al. (2003) Substoichiometric shifting in the plant mitochondrial genome is influenced by a gene homologous to MutS. Proc Natl Acad Sci USA 100:5968–5973.
12. Xu YZ, Arrieta-Montiel MP, Virdi KS, de Paula WB, Widhalm JR, et al. (2011) MSH1 is a nucleoid protein that alters mitochondrial and plastid properties and plant response to high light. Plant Cell 239:3428–41.
13. Shedge V, Davila J, Arrieta-Montiel MP, Mohammed S, Mackenzie SA (2010) Extensive rearrangement of the Arabidopsis mitochondrial genome elicits cellular conditions for thermotolerance. Plant Physiol 152:1960–1970.
14. Xu YZ, Santamaria Rde L, Virdi KS, Arrieta-Montiel MP, Razvi F, et al. (2012) The chloroplast triggers developmental reprogramming when MUTS HOMOLOG1 is suppressed in plants. Plant Physiol 159:710–20.
15. Pinheiro J, Bates D, DebRoy S, Sarkar D, the R Development Core Team (2013) Nlme: Linear and Nonlinear Mixed Effects Models. R package version 3.1-109.
16. Schloss SJ, Mitchell E, White M, Kukatla R, Bowers E, et al. (2002) Characterization of RFLP probe sequence for gene discovery and SSR development in Sorghum bicolor (L.) Moench. Theor Appl Genet 105, 912–920.
17. Li M, Yuyama N, Luo L, Hirata M, Cai H (2009) In silico mapping of 1758 new SSR markers developed from public genomic sequences for sorghum. Mol Breed 24: 41–47.
18. Miller FR (1984) Registration of RTx430 sorghum parental line. Crop Sci 24:1224.
19. El-Soda M, Malosetti M, Zwaan BJ, Koornneef M, Aarts MG (2014) Genotype×environment interaction QTL mapping in plants: lessons from Arabidopsis. Trends Plant Sci 19(6):390–398.
20. Multani DS, Briggs SP, Chamberlin MA, Blakeslee JJ, Murphy AS, et al. (2003) Loss of an MDR transporter in compact stalks of maize br2 and sorghum dw3 mutants. Science 302:81–4.

21. Colomé-Tatché M, Cortijo S, Wardenaar R, Morgado L, Lahouze B, et al. (2012) Features of the Arabidopsis recombination landscape resulting from the combined loss of sequence variation and DNA methylation. Proc Natl Acad Sci USA 109:16240–5.
22. Migicovsky Z, Kovalchuk I (2013) Changes to DNA methylation and homologous recombination frequency in the progeny of stressed plants. Biochem Cell Biol 91:1–5.
23. Brown PJ, Rooney WL, Franks C, Kresovich S (2008) Efficient mapping of plant height quantitative trait loci in a sorghum association population with introgressed dwarfing genes. Genetics 180:629–637.
24. Zhang YY, Fischer M, Colot V, Bossdorf O (2013) Epigenetic variation creates potential for evolution of plant phenotypic plasticity. New Phytol 197:314–22.
25. Johannes F, Porcher E, Teixeira FK, Saliba-Colombani V, Simon M, et al. (2009) Assessing the impact of transgenerational epigenetic variation on complex traits. PLoS Genet 5(6):e1000530.
26. Reinders J, Wulff BB, Mirouze M, Marí-Ordóñez A, Dapp M, et al. (2009) Compromised stability of DNA methylation and transposon immobilization in mosaic Arabidopsis epigenomes. Genes Dev 23:939–50.
27. Hauben M, Haesendonckx B, Standaert E, Van Der Kelen K, Azmi A, et al. (2009) Energy use efficiency is characterized by an epigenetic component that can be directed through artificial selection to increase yield. Proc Natl Acad Sci USA 106:20109–14.
28. Stroud H, Ding B, Simon SA, Feng S, Bellizzi M, et al. (2013) Plants regenerated from tissue culture contain stable epigenome changes in rice. Elife 2:e00354.

14

Molecular Genetic Diversity of Major Indian Rice Cultivars over Decadal Periods

Gangaprasad Choudhary[◑], **Nagireddy Ranjitkumar**[◑], **Malathi Surapaneni**[◑], **Dondapati Annekitty Deborah**, **Abhilash Vipparla**, **Ghanta Anuradha**, **Ebrahimali Abubacker Siddiq**, **Lakshminarayana Reddy Vemireddy***

Institute of Biotechnology, Acharya NG Ranga Agricultural University, Rajendranagar, Hyderabad, Andhra Pradesh, India

Abstract

Genetic diversity in representative sets of high yielding varieties of rice released in India between 1970 and 2010 was studied at molecular level employing hypervariable microsatellite markers. Of 64 rice SSR primer pairs studied, 52 showed polymorphism, when screened in 100 rice genotypes. A total of 184 alleles was identified averaging 3.63 alleles per locus. Cluster analysis clearly grouped the 100 genotypes into their respective decadal periods i.e., 1970s, 1980s, 1990s and 2000s. The trend of diversity over the decadal periods estimated based on the number of alleles (*Na*), allelic richness (*Rs*), Nei's genetic diversity index (*He*), observed heterozygosity (*Ho*) and polymorphism information content (PIC) revealed increase of diversity over the periods in year of releasewise and longevitywise classification of rice varieties. Analysis of molecular variance (AMOVA) suggested more variation in within the decadal periods than among the decades. Pairwise comparison of population differentiation (*Fst*) among decadal periods showed significant difference between all the pairs except a few. Analysis of trends of appearing and disappearing alleles over decadal periods showed an increase in the appearance of alleles and decrease in disappearance in both the categories of varieties. It was obvious from the present findings, that genetic diversity was progressively on the rise in the varieties released during the decadal periods, between 1970s and 2000s.

Editor: Rajagopal Subramanyam, University of Hyderabad, India

Funding: The authors acknowledge ANGRAU (Acharya NG Ranga Agricultural University) for providing the financial and lab facilities under Rashtriya Krishi Vikas Yojana (RKVY). The funders had no role in study design, data collection and analysis, decision to publish, or preparation of the manuscript.

Competing Interests: The authors have declared that no competing interests exist.

* E-mail: vlnreddi@gmail.com

◑ These authors contributed equally to this work.

Introduction

Ever since the domestication of crop plants, man has been improving them giving selection emphasis to traits that suit his agro-ecological and socioeconomic needs. In rice, like many crops, selection preference has been for improvement of yield enhancing traits like compact panicle with more grains/panicle, large seed size, non-shattering habit etc. The selection process continued for centuries result in cultivars far different from the wild/weedy progenitor species in their habit and potential. Since the advent of the short statured high yielding varieties in the mid-sixties, the selection priority of breeders has been for higher stability or performance, need based crop duration, tolerance to various stresses and consumer preferred grain quality. Excessive breeding emphasis in this direction given during the last 50 years knowingly or unknowingly has led to some sort of genetic uniformity among the currently cultivated high yielding varieties. The improved germplasm is being excessively depended on for needed variability for progressive improvement of the crop with no precise knowledge of extent of exploitable variability beyond simply inherited traits. Also it is not clear to what extent breeding strategies *in vogue* have been facilitating to broaden or narrow the genetic diversity in the breeding nurseries of rice.

Precise knowledge of exploitable genetic diversity and genetic relatedness among the constituents of the improved gene pool is vital for meticulously planning and executing target directed breeding especially when genetic relatedness among the improved parental sources appears to become increasingly narrow. Unlike morpho-physiological traits used earlier to estimate genetic variability/relatedness, molecular markers have become quite handy in precisely understanding the extent of genetic divergence among varieties being chosen these days as parental sources in breeding programs. Many recent scientific papers tracing the trend of genetic diversity in crop varieties released over the years reveal the diversity levels to often fluctuate strongly from one time period to the successive periods [1–3] and sometimes show conflicting results [4,5]. By and large no clear pattern has emerged from such studies in the past, as both increasing and decreasing trends in diversity have been observed [6–8]. Diversity loss in rice during the last few centuries is obvious from the rapid decline of rice varieties from 400,000 before colonialism to 30,000 by mid-19th Century. This number has further come down during the era of the Green revolution as few high yielding varieties replaced thousands of native varieties [9]. Estimation of genetic diversity in varieties released in different crops during the last century reveals no clear trends [10]. In rice, both decreased [11] as well as increased trend of diversity [12] has been reported.

Since the introduction of semidwarf high yielding varieties in the mid 1960s, simultaneously launch of the All India Coordinated Rice Improvement Programmes (AICRIP) and active involvement of International Rice Research Institute (IRRI), Philippines facilitated rapid exchange of germplasm among breeders, institutes and between countries culminating in the release of around 1000 high yielding varieties in the country. Despite such large number of varieties developed using diverse germplasm, molecular marker based diversity analysis has shown the genetic base of Indian rice gene pool to be surprisingly narrow [13,14]. Moreover, with regard to trends of genetic diversity in major Indian rice cultivars, however, little work has been done Recently, hypervariable microsatellite markers evenly distributed in rice genome have been demonstrated to be quite effective in estimating genetic diversity [15] Keeping the foregoing, the present study was undertaken to understand the level and trend of genetic diversity in the high yielding varieties developed in India over decadal periods between 1970s and 2000s employing hypervariable microsatellite markers.

Materials and Methods

Plant Material

The experimental material comprised of 100 rice cultivars, which included 89 high yielding varieties released between 1970 and 2010 in different states of India and 11 traditional varieties (Table 1 and Table S1). They were grouped according to the decade of their release (herefrom regarded as "year of release-wise") for general cultivation across regions *viz.*, 1970s, 1980s, 1990s and 2000s. They were as well classified into "longevitywise" as differences in genetic diversity could be observed between them, especially when a variety is popular within and across decades.

DNA Extraction

Genomic DNA was isolated from 20 day-old seedlings germinated in sterile petri dishes lined with moist filter paper using the CTAB (Cetyl Try methyl Ammonim Bromide) method described [16] with some modifications. The purity and concentration of the isolated genomic DNA samples were estimated by UV-absorption spectrophotometer (Beckman DU 650 model) as per the procedure described by Sambrook [17]. Quantification of DNA was done by analyzing the purified DNA on 0.8% agarose gel with lambda (λ) *Hind* III DNA as standard. Based on the intensity and thickness of genomic DNA bands, as compared to lambda (λ) *Hind* III DNA, the concentration and quality of DNA in individual samples were determined.

SSR Marker Analysis

In all, 64 hypervariable microsatellite markers distributed on all the 12 chromosomes of rice were picked up from our previous study [15]. These markers covering 236.8 Mb of physical distance with average distance between them being 5.92 Mb. PCR reactions were carried out in 10 µl reaction volume containing of 10× PCR buffer (10 mM Tris-HCl pH 8.3, 50 mM KCl), 1.5 mM $MgCl_2$, 0.2 mM each dNTPs, 5 pmol of each forward and reverse primer, 0.5U of Taq DNA polymerase (NEB), and 5 ng of genomic DNA. Reactions were carried out in GenAmp PCR system 9700 (Applied Biosystems, USA) thermal cycler using the following temperature profile: an initial denaturation of 5 min at 94°C followed by 35 cycles of 45 s at 94°C, 45 s at 55°C and 1 min at 72°C, then a final extension of 5 min at 72°C. Amplification products were resolved on 3% metaphor agarose gels using a horizontal gel electrophoresis unit (CBS Scientific, USA). The DNA fragments were then visualized under UV-transilluminator and documented using ALPHA IMAGER gel documentation system (Alpha Innotech, USA) which was stored for further scoring and permanent records.

Data Analysis

Only clear and unambiguous bands of SSR markers were scored. The sizes of the amplified fragments were estimated with the help of Alpha image software by Gel documentation system using 100 bp DNA ladder (NEB) as size standard. Markers were scored for the presence (1) or absence (0) of the corresponding band among the genotypes. To measure the informativeness of the markers, polymorphism information content (PIC) for each of the SSR markers was computed according to the formula: $PIC = 1 - \Sigma Pi^2 - \Sigma\Sigma Pi^2\ Pj^2$ where 'i' is the total number of alleles detected for SSR marker and 'Pi' is the frequency of the ith allele in the set of hundred genotypes investigated and $j = i+1$ [18].

Genetic diversity parameters *viz.*, number of alleles (*Na*), observed heterozygosity (*Ho*), Shannon Index (*I*) and Nei's genetic diversity index (*He*) [19] were evaluated using POP-GENE v 1.31 (http://www.ualberta.ca/~fyeh). The allelic richness (*Rs*), is a measure of the number of alleles independent of the sample size, was measured using FSTAT [20]. Significant difference in genetic diversity parameters between each pair of the decadal periods was calculated using the Wilcoxon matched

Table 1. Rice varieties used in the present study.

Decadal Period	Released Rice Varieties	Total No.
Landraces	INRC10192, Lalnakanda, Hasansona, Solumpiket, Basmati370, Acharmati, Bate Aus, Aus Boro, Dular, Aus Bako, Azucena	11
1970s	Manoharsali, Taichung Native-1, Jaya, Tella Hamsa, Rajeswari, Annapoorna, Jyothi, PR-106, N-22 (Nagina-22), Surekha (WL-13400), Prabhat (MTU-3626), Swarnadhan, Masur, WGL3200	14
1980s	Annada, Rasi, SasyaSree, PLA1100, Swarna (MTU-7029), Himalaya-2, Kalinga III, Vikas, CO-43, Keshari, Parijat, VL Dhan-16, VLDhan-206, Pathara, Prasanna, Sonasali, Pothana, Sabita, HKR-120, Mandya Vijaya, Suraksha, Satya, ASD-17, CSR-10, Kasturi, Kanak, Pusa Basmati-1, Tikkana (NLR-27999), Pinakini (NLR-9672-96), Samba Mahsuri (BPT-5204), Vanaprabha	30
1990s	Krishnaveni (MTU-2077), Haryana Basmati, IR-64, Chandana, VLDhan-221, Kavya (WGL-48684), Sneha, Swarnamukhi (NLR-145), Badami, Pusa-44, Ghanteswari, Khandagiri, Nilagiri, Himalaya-2216, Mahi Sugandha, Taraori Basmati, Basmati-386, Krishna Hamsa, Nidhi, VLDhan-61, Triguna, CSR-13, Lalithagiri, Uydyagiri	24
2000s	Bharani (NLR-30491), Srikakulam Sannalu, Somasila (NLR-33358), Sravani (NLR-33359), Cottondora Sannalu (MTU1010), PR-115, Mugad Sugandha-1, Sumathi, Vandana, Yamini (CSR 30), Pusa sugandha-3, Super, Pusa-1121, WGL-32100, MTU-1061 (Indra), Taramati, Suganda samba, Vasumati, Sharbathi, Improved Samba mahsuri and Swarna sub1	21

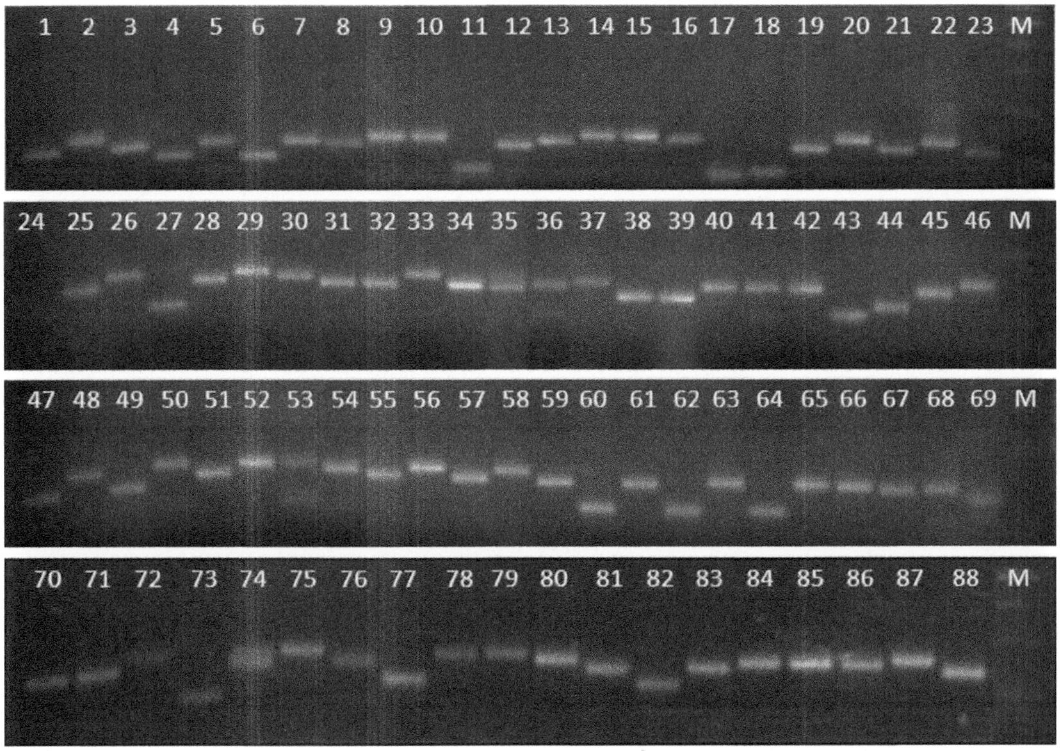

Figure 1. A representative gel picture of screening of rice varieties with RM562. For decoding of the numbers refer Table S1.

pairs test, a nonparametric alternative to the *t*-test (http://www.stattools.net/Wilcoxon_Pgm.php). In addition, number of rare alleles (number of alleles in less than 5% of the population) and high frequency or common alleles (number of alleles in more than 85% of the population) were estimated using Microsoft Excel 2007. The *F*-statistics (*Fst*) were used to analyze genetic differentiation of the varieties in all possible pairs of decades based on allelic discrepancy at each locus by the procedure of AMOVA (Analysis of Molecular Variance), besides the genetic variation within and among the populations using the software ARLEQUIN v 2.0 [21]. The appearance and disappearance of alleles and number of private alleles have been estimated using the CONVERT v 1.31software [22]. In order to overcome the problem of unequal sample size, resampling was done using an in-house script for R software [23]. The UNJ (Unweighted Neighbour Joining method) cluster analysis followed by bootstrap analysis with1000 permutations for total cultivars was carried out using DARwin 5.0.145 (http://darwin.cirad.fr/). The dendrogram based on unbiased genetic distances among decadal periods was constructed by UPGMA (Unweighted pair-group method with arithmetic average) employing POPGENE v 1.31.

Results and Discussion

Hypervariable Microsatellite Marker Analysis

Sixty four hvRM (hypervariable rice microsatellite) markers distributed evenly on all the 12 chromosomes were chosen to assess the genetic relatedness among the 100 genotypes. Fifty-two of them were found to show polymorphism (81.25%)(Table 2 and Figure 1). In all, 184 alleles were identified by amplification of the 52 polymorphic hvRMs with an average number of alleles of 3.6 per locus, with the number ranging from 3 (RM16416) to 7 (RM8207) (Table 2). This was significantly lower than the average number of alleles reported by Jain *et al* (7.8) [24], Spada *et al.* (7.2) [25] and Zhu *et al.* (4.37) [26] and higher than the Chuan-Guang and Gui-Quan [27]. Yu *et al.* [28] studied 193 rice accessions drawn from 26 countries using 101 SSR primer pairs and detected an average allele number of 6.3 per locus, which is also higher than the value reported here. Luce *et al.* [29] analyzed 419 rice accessions from the gene banks in five European countries using 16 SSR loci (different from the ones we selected) and reported an average of 9.1 alleles per locus. The higher value than the present study (3.6) could probably be as a result of larger number of accessions used by these authors. Of the 52 loci, 27 were with three alleles while 16 loci with four alleles (Figure 2). All the 52 markers revealed high PIC values the range being between 0.67 (RM16416) and 0.97 (RM14735) with more than 50% loci in the range of 0.8 to 0.9 (Figure 2). Some of the hvRM markers like RM11340, RM12548, RM13584, RM14270, RM14735, RM15580, and RM22273 showed high PIC values in our previous study as well [15]. The average PIC value estimated in the present study (0.87) is more than that of the previous studies by Giarrocco *et al.* [30] (0.69) and Jayamani *et al.* [31] (0.67) possibly because of hypervariable microsatellite markers used. Recently, it has been demonstrated that microsatellite markers of hypervariable nature would be more polymorphic than non-hypervariable markers [15] The *He* values ranged from 0.15 (RM12031) to 0.76 (RM14735) with an average of 0.59 and more than 40% of the loci were in the frequency of 0.7 to 0.8 (Figure 2). The average *He* value of the present study is slightly lower than the earlier studies [11,32] due to the inclusion of only major Indian rice cultivars, which largely belong to *indica* sub group.

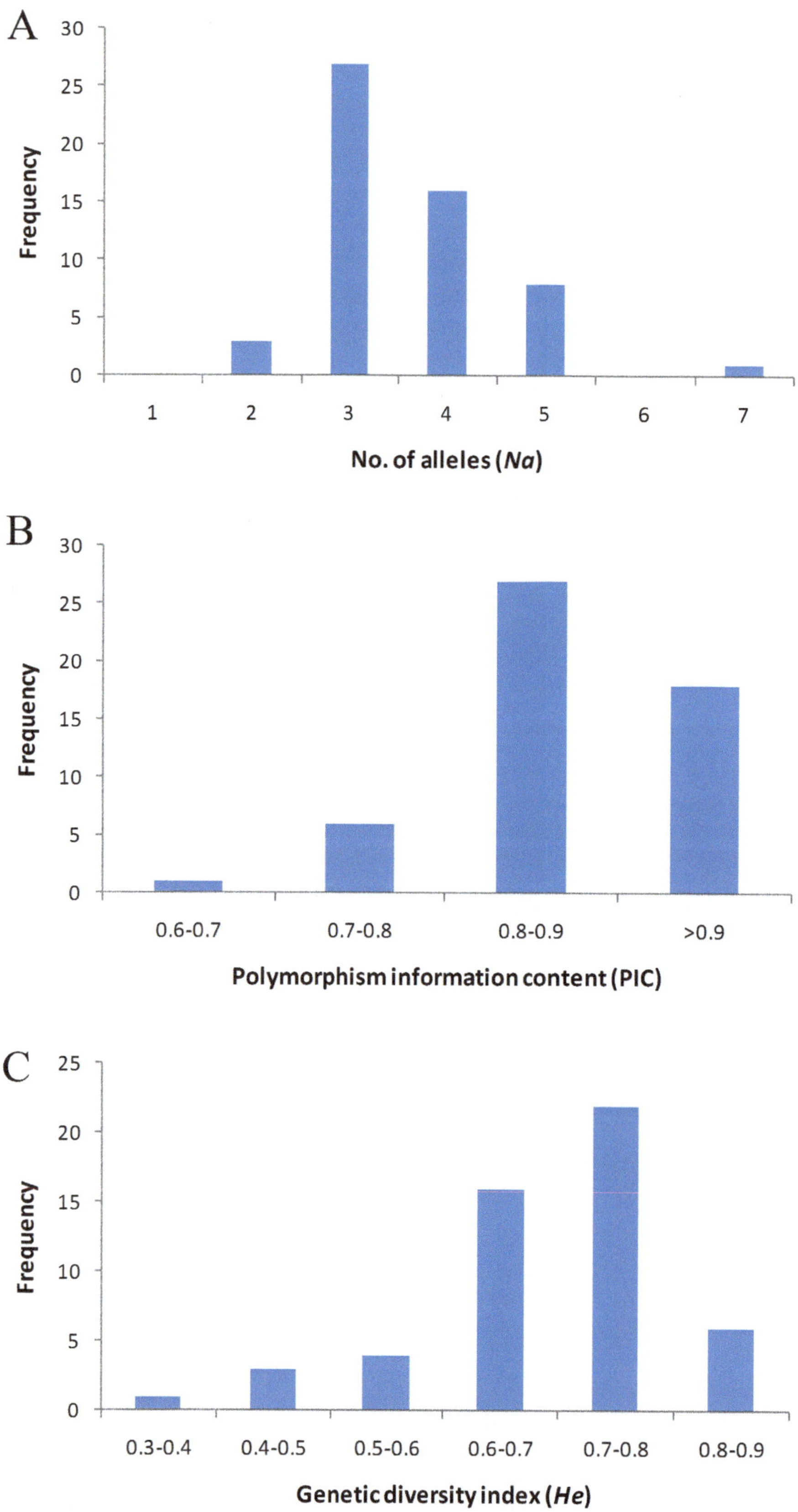

Figure 2. Distribution of number of alleles (*Na*) (A), polymorphism information content (PIC) (B) and genetic diversity (*He*) (C) estimated from 100 rice varieties using 52 hypervariable microsatellite loci.

Table 2. Molecular diversity parameters of the microsatellite markers used in the study.

Locus	Chr.	SSR motif	*Na*	*Rs*	PIC	*Ho*	*He*	*I*
RM562	1	(AAG)13	5	1.63	0.94	0.04	0.73	1.34
RM11313	1	(AAT)23	3	1.49	0.86	0.00	0.56	0.90
RM11340	1	(AT)42	4	1.47	0.89	0.07	0.50	0.89
RM11356	1	(AT)31	3	1.54	0.88	0.12	0.56	0.89
RM11597	1	(AT)42	3	1.48	0.9	0.00	0.57	0.92
RM12031	1	(AG)40	3	1.18	0.74	0.02	0.15	0.32
RM12253	1	(AAT)23	4	1.57	0.91	0.12	0.60	1.02
RM12292	1	(AT)38	3	1.62	0.92	0.14	0.66	1.09
RM12353	2	(AAT)32	4	1.40	0.86	0.16	0.56	0.91
RM12548	2	(AT)46	3	1.57	0.9	0.01	0.60	0.99
RM13131	2	(AT)32	3	1.44	0.91	0.01	0.55	0.90
RM13584	2	(AATC)5	3	1.57	0.88	0.05	0.63	1.05
RM14270	3	(AT)46	4	1.60	0.93	0.00	0.66	1.09
RM14735	3	(AT)42	5	1.60	0.98	0.06	0.76	1.52
RM14778	3	(AT)37	3	1.62	0.89	0.01	0.63	1.04
RM15004	3	(AAT)38	4	1.64	0.91	0.05	0.65	1.14
RM15580	3	(AT)50	3	1.56	0.85	0.02	0.59	0.95
RM16416	4	(AAT)12	3	1.32	0.69	0.02	0.34	0.53
RM16577	4	(AT)29	4	1.60	0.91	0.13	0.62	1.03
RM17405	4	(AAT)36	3	1.47	0.88	0.00	0.49	0.78
RM17669	4	(AT)30	3	1.58	0.9	0.03	0.58	0.95
RM5693	5	(AAT)18	4	1.59	0.78	0.59	0.61	1.08
RM5844	5	(AAT)20	5	1.68	0.94	0.01	0.68	1.24
RM5907	5	(AAT)19	3	1.50	0.86	0.01	0.48	0.84
RM18384	5	(AAG)22	3	1.42	0.9	0.26	0.40	0.80
RM18639	5	(AAT)17	5	1.66	0.78	0.81	0.68	1.26
RM19545	6	(AAT)21	4	1.58	0.91	0.00	0.58	0.94
RM20037	6	(AT)38	4	1.60	0.92	0.01	0.63	1.05
RM20710	6	(AT)46	5	1.70	0.97	0.49	0.76	1.52
RM21693	7	(AT)44	5	1.60	0.93	0.05	0.66	1.19
RM21941	7	(AAT)25	3	1.56	0.84	0.05	0.58	0.98
RM6965	7	(AAG)15	3	1.49	0.91	0.03	0.54	0.90
RM22250	8	(AAT)30	5	1.71	0.94	0.05	0.71	1.34
RM22544	8	(AAT)20	4	1.69	0.89	0.41	0.73	1.35
RM22565	8	(ACAT)15	4	1.67	0.9	0.08	0.68	1.26
RM22688	8	(AAT)28	3	1.34	0.74	0.01	0.38	0.56
RM22273	8	(AT)35	4	1.66	0.83	0.34	0.68	1.24
RM23017	8	(AAT)18	3	1.42	0.83	0.05	0.51	0.79
RM23036	8	(AGAT)15	4	1.62	0.9	0.18	0.64	1.11
RM23362	8	(AAG)19	5	1.60	0.94	0.46	0.66	1.22
RM23741	9	(AAT)28	4	1.59	0.78	0.85	0.59	0.99
RM24015	9	(AGAT)9	4	1.68	0.86	0.68	0.70	1.29
RM24044	9	(AAG)11	3	1.45	0.88	0.19	0.50	0.84
RM24260	9	(AAT)31	3	1.55	0.88	0.00	0.63	1.04
RM5708	10	(AAT)22	4	1.68	0.83	0.47	0.69	1.30
RM8207	10	(AAG)23	7	1.69	0.95	0.05	0.69	1.34
RM25262	10	(AAT)38	3	1.41	0.79	0.08	0.47	0.66
RM25969	11	(AAG)18	3	1.34	0.81	0.01	0.40	0.72
RM26190	11	(AGAT)13	3	1.47	0.92	0.01	0.59	0.96
RM26632	11	(AAAG)9	3	1.53	0.85	0.01	0.57	0.95

Table 2. Cont.

Locus	Chr.	SSR motif	*Na*	*Rs*	PIC	*Ho*	*He*	*I*
RM27840	12	(AAT)37	3	1.54	0.86	0.02	0.58	0.95
RM28279	12	(AATC)8	3	1.56	0.84	0.28	0.61	1.01
Mean			3.69	1.55	0.87	0.15	0.59	1.02
SD			0.90	0.11	0.06	0.22	0.11	0.24

Chr.- Chromosome; *Na*-Number of alleles; PIC-Polymorphism Information Content; *Rs*: Allelic richness; *Ho*- Observed heterozygosity; *He*- Nei's genetic diversity; *I*-Shannon Index; SD-Standard Deviation.

Overall Genetic Diversity in Popular Varieties of Rice

The unweighted neighbour-joining (UNJ) dendrogram constructed on the basis of genetic similarity matrix grouped the 100 genotypes into five clusters *viz.,* landraces, 1970s, 1980s, 1990s, and 2000s (Figure 3). The phylogeny tree reveals that appearing of some of the varieties released during one decade in another decade due to the presence of common parents in their pedigree. For instance, ASD17 and CSR10 belonging to the decadal period of the 1980s clustered with those of the 1970s as IR8 and Jaya of the 1970s being common in their parentage. Varieties of the decadal periods1990s and 2000s comprised largely of varieties released during their respective decades. However, varieties like Tikkana and Pinakini of the 1980s and Somasila and Srikakulam Sannalu of 2000s have been found to cluster with those of the 1990s. Interestingly, Basmati varieties irrespective of their year of release formed a separate sub-cluster. This result is in agreement with previous reports by Glaszman [33], Nagaraju *et al* [34] and Narshimulu *et al* [15]. Equally and interestingly the recently released varieties developed through marker-assisted breeding *viz.,* Swarna-sub1 and improved Samba mahsuri grouped with landrace cluster instead of the expected grouping with the 2000s decade where in their original parents *viz.,* Swarna and Samba mahsuri exist. This could be due to the presence of part of the donor genomes of FR13A, a flood resistant landrace the source for *sub-1* gene and SS1113, the source of bacterial blight resistance genes (*xa13, xa5* and *Xa21*) in improved Samba mahsuri, remaining in the improved versions of the varieties even after many backcrosses. Longevitywise clustering of the 100 genotypes too was in total agreement with the decadewise clustering. Even though some varieties were popular beyond the decade(s) of their release, they do not affect the clustering pattern of the genotypes as well as the genetic diversity of successive decadal periods. For instance, Tella Hamsa and Jyoti remain popular even as late as in the 2000s, albeit released decades back in the states of Andhra Pradesh and Kerala, respectively.

Genetic distance between each pair of decadal periods suggest that it was the highest between the landraces and the high yielding varieties of the 1990s (0.2715), while the lowest between the decades 1990s and 2000s (0.0905) in the year of releasewise group. In case of longevitywise group also genetic distance was highest between the landraces and the high yielding varieties of the 1990s (0.210) but lowest between those of the 1980s and 2000s (0.0237) (Table 3). A dendrogram constructed based on Nei's genetic distance also confirms the above findings as the genetic distance between landrace cultivars and the improved varieties released in 1990s being the highest (Figure 4).

Trends of Genetic Diversity within Decadal Period

Genetic diversity trend of varietal groups released during different decades since the release of semi-dwarf high yielding varieties (1970s to 2000s) along with sets of landraces and Basmati accessions was studied based on number of alleles *(Na)*, Nei's genetic diversity index (*He*) and polymorphism information content (PIC) (Figure 5 and Table S2). As well, the diversity within decadal periods was estimated through separate analysis of year of releasewise as well as longevitywise varieties. The diversity trends as estimated through a set of parameters are as under:

Overall increasing trend of number of alleles (*Na*) was observed from landraces to varieties of 2000s in the categories of both year of releasewise and longevitywise classified rice varieties. *Na* values were increased by 13.38% and 13.87% from landraces to 2000s while it was 12.79% and 9.83% from 1970s to the 2000s, respectively for both the categories. On an average, landraces to total modern cultivars percentage of *Na* increase has been observed to be 9.02 for year of releasewise and 10.91 for longevitywise varieties. This is in contrast to the observation of Wei *et al.* [11], who reported decreasing trend of the *Na* score of 7.8% from the 1950s to the 1990s.

Genetic diversity index (*He*) also reveals that the genetic diversity has been increased approximately 10% from landraces to 2000s and landraces to modern cultivars in the year of releasewise category while it was 16.02% and 13.39% for longevitywise category, respectively. However, from 1970s to 2000s the trend of increase was 4.71% and 8.92% for year of releasewise and longevitywise varieties, respectively. The observed trends in the present study are in agreement with Mantegazza *et al.,* [12], who have reported increase of genetic diversity of Italian rice cultivars between 1850 and 2001. The same trend has been reported also by Morin *et al.,* [35]. However, these findings vary with the report of Wei *et al.,* [11] who observed 7% loss of genetic diversity between 1950s and 1990s.

Polymorphism information content (PIC) also shows increasing trend over decadal periods, though not to the level of *Na* and *He*. The percent increase of PIC values from landraces to all modern cultivars, from landraces to 2000s and from 1970s to 2000s as estimated by year of releasewise were 2.75, 4.25 and 5.77%, respectively, while the corresponding values estimated longevitywise are 4.17, 5.88 and 4.87%. Overall, the trends as measured by the parameters of *Na*, *He* and PIC appear to slightly slow down in the decadal period of 1990s but regains its increasing trend in that of 2000s.

The genetic diversity in Indian cultivars ($He = 0.54$) as estimated by the Nei's genetic diversity index in the present study is nearly equal to that of the representative world collection of *indica* varieties ($He = 0.55$). It is, however, overwhelmingly high as compared to the representative world collection of temperate *japonica* ($He = 0.39$) [36]. Thus, the findings of the present study confirms further, the *indica* varietal group to be much more genetically diverse than *japonica* group which is in agreement with many earlier reports [32,37].

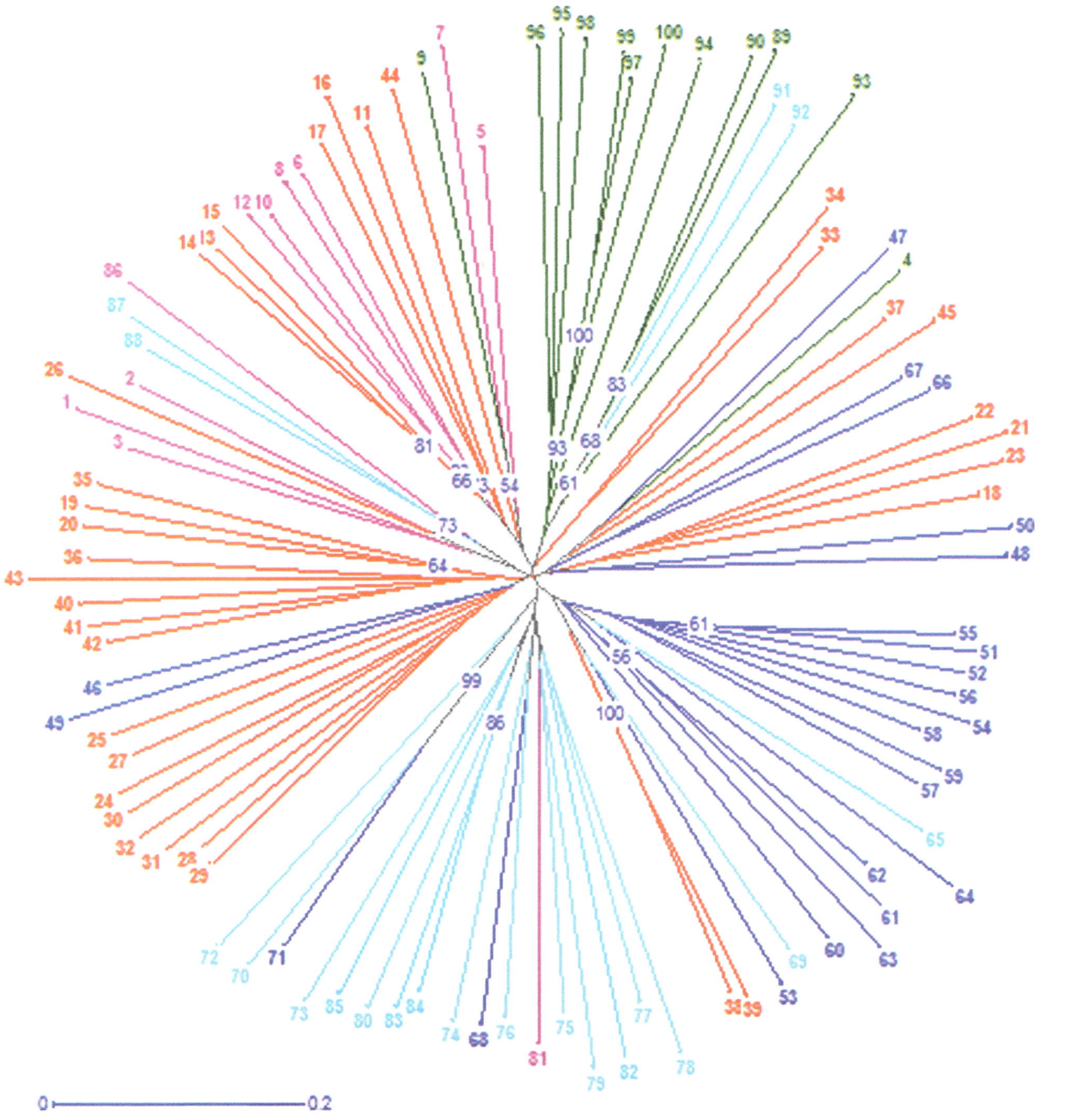

Figure 3. Genetic relationship among 100 rice varieties estimated using Unbiased Neighbour-Joining dendrogram of 52 hypervariable microsatellite loci. Landraces- green colored;1970s-pink colored;1980-red colored;1990s-dark blue colored;2000s-sky blue colored. Rice varieties were represented in numbers. For full details of the varieties refer Table S1.

Overall, the present investigation reveals a genetic diversity as estimated using different parameters such as allele number (*Na*), polymorphism information content (PIC) and genetic diversity index (*He*), to increase progressively in the advancing decadal periods from 1970s to 2000s and interestingly, it was higher as compared to even landraces. Use of genotypes, which cover longer periods always offer added advantage in temporal studies. In the present study only a few landraces which are being cultivated for their certain features by local farmers has been included to compare genetic diversity in varieties released before and after 1970s. The analysis of genetic diversity using various parameters clearly reveals the increasing trend from landraces to the decadal period 2000s. While this observation is in agreement with those of Mantegazza *et al* [12], who have reported steady increase in the levels of gene diversity in Italian rice germplasm, it was not so with the findings of Steele *et al* [38] who have reported no change in the level of diversity following the introduction of modern rice varieties. Wei *et al* [11] and Yuan *et al* [32] report, however, loss

Table 3. Nei's Unbiased measures of genetic distance among decadal periods of major Indian rice cultivars.

Decade	Landraces	1970s	1980s	1990s	2000s
Landraces	****	0.204	0.182	0.210	0.169
1970s	0.206	****	0.052	0.059	0.056
1980s	0.189	0.099	****	0.046	0.024
1990s	0.272	0.126	0.118	****	0.029
2000s	0.211	0.132	0.100	0.091	****

Lower diagonal: Year of releasewise; Upper diagonal: Longevitywise.

of more alleles in modern cultivars as compared to varieties released in the 1950s. Put together, the present findings and previous reports, it is reasonable to assume that global rice genetic diversity tend to increase over the decades by gaining as well as losing to different proportion of alleles. In the absence of phenotypic expression of alleles gained or lost in rice genome it would be difficult to conclude if they are of adaptive or agronomic value. Qualitative study of variation in gene diversity, over the decadal periods, as estimated by the total number of alleles that appeared and disappeared showed an increase in the appearance and decrease in the disappearance of alleles from landraces to 2000s in both the year of releasewise and longevitywise analysis (Figure 6). Though we presume that disappearance of certain alleles might be of deleterious nature and that appearance of new alleles to be of positive breeding value, it need not be so always. Thus, there is a need for still more precisely planned further study. As all the major cultivars of India are belonging to the *indica* subspecies of *Oryza sativa*, we could not compare with the *japonica* varieties in terms of allele appearance and disappearance. However, earlier [11] it was proved that more alleles loss has been observed in *indica* varieties than in *japonica*. This phenomenon can be attributed to the higher diversity nature of *indica* varieties compared to *japonica*.

Many reasons may be attributed to the progressively increased genetic diversity over the decadal periods between 1970s and 2000s. In the initial decades, the advent of semidwarf plant type based exotic high yielding varieties like Taichung (Native)-1 and IR8 and varieties derived from crosses involving them with traditional Indian varieties could have introduced large variability for yield, earliness, photo-insensitivity, seed non-dormancy etc. In the subsequent decades, the focus has been shifted towards the development of polygenic durable resistance to multiple pathogens, adaptation to diverse rice ecologies and consumer preference driven quality improvement. For instance, high yielding varieties of the initial decades were more susceptible to indigenous as well as introduced diseases and insect pests like bacterial blight, tungro virus, hoppers etc. warranting to exploit exotic cultivar and wild/weedy sources. Convergent breeding efforts have made most of the high yielding varieties of today multiple pest resistant. Given the challenges ahead which have all the potential to impede further progress in crop improvement, there is need to enrich the variability employing multiparent breeding and exploitation of hidden variability from progenitor species and primitive cultivars.

Trends of Genetic Diversity among Decadal Periods

Analysis of molecular variance (AMOVA) has revealed most of the variation was existed within decadal periods i.e., 92.12% and 96.66%, respectively for year of releasewise and longevitywise as against very low variation between the decadal periods i.e., 3.34% for longevitywise classified varieties and 7.88% for year of releasewise classification (Table 4). These results are in agreement with Yuan *et al.* [32]. Pairwise comparison of population differentiation (*Fst*) among decadal periods reveals significant genetic differentiation among all except between the landraces and the decadal period of 1970s as estimated by both year of releasewise and longevitywise analysis. Also there was no

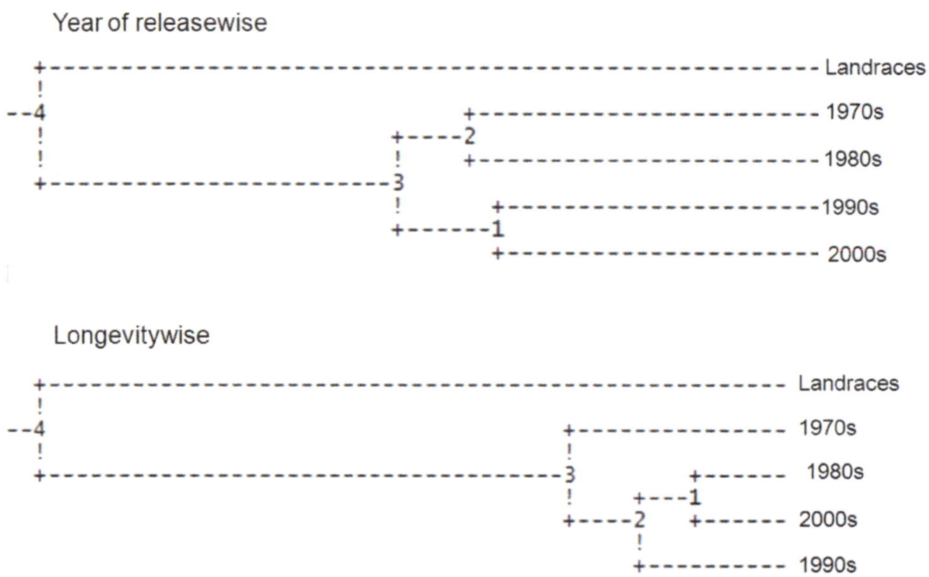

Figure 4. UPGMA dendrogram based on Nei's genetic distances using POPGENE v 1.31 showing the genetic relationship of rice genotypes among decadal periods. The genetic distances between different groups are as follows. Year of releasewise: 4 and Landraces - 10.96, 4 and 3–5.01, 3 and 2–0.97, 2 and 1970s - 4.97, 2 and 1980s- 4.97, 3 and 1–1.41, 1 and 1990s-4.52 and 1 and 2000s-4.52. Longevitywise: 4 and Landraces −9.55, 4 and 3–6.76, 3 and 1970s - 2.78, 3 and 2 -0.93, 2 and 1 - 0.66, 1 and 1980s - 1.18, 1 and 2000s - 1.18, 1 and 1990s - 1.85.

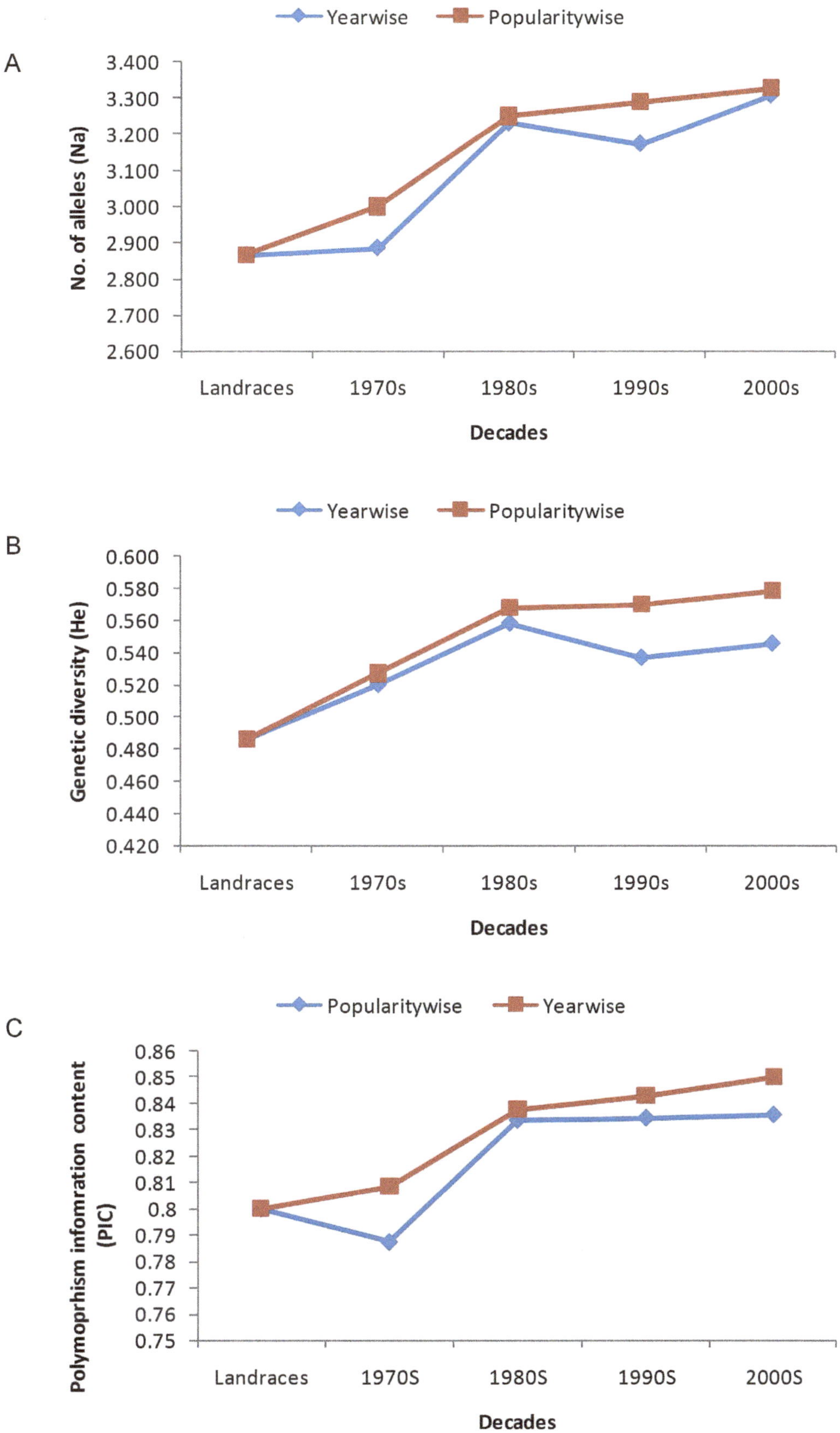

Figure 5. Changes in number of alleles (*Na*), Nei's genetic diversity (*He*) and PIC values over decadal periods.

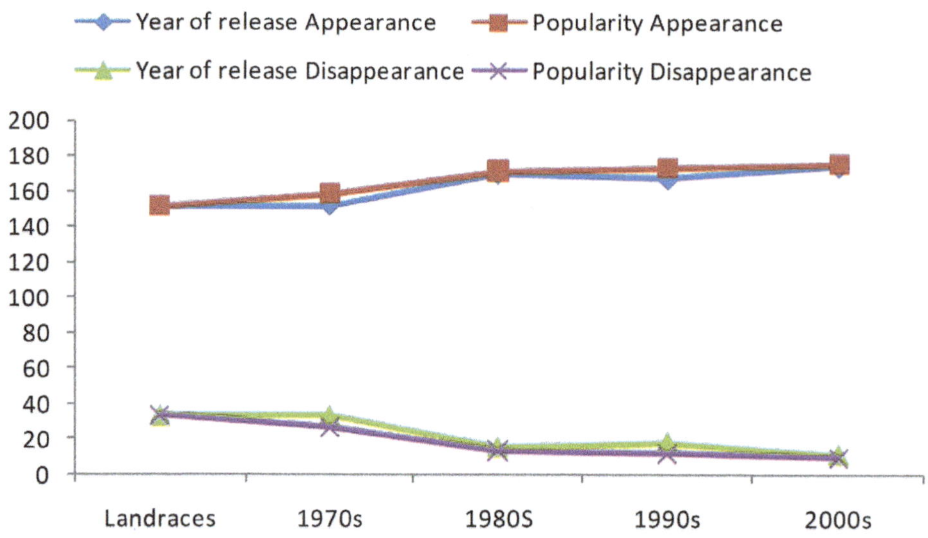

Figure 6. Appearance and disappearance of alleles over decadal periods.

significant differentiation between the 1970s and rest of the decades 1980s to 2000s in longevitywise analysis (Table 5).

Locus-by-locus AMOVA showed significant ($p<0.01$) differences in genetic variation among the decadal periods as exhibited by 14 loci *viz*., RM11356, RM12031, RM13131, RM14735, RM15004, RM15580, RM5693, RM5907, RM18384, RM22554, RM22565, RM22688, RM23741, and RM8207. Interestingly, three loci viz., RM11313, RM12353, and RM6965 have been found to explain more than 20% of the variation among the decades (Table S3).

Analysis of Population Specific Alleles

Analysis for identification of unique population specific alleles has revealed all the decadal periods found to show specific alleles with the exception of the 1970s as estimated by year of releasewise as well as longevitywise classification. As expected, the landraces exhibited as many as six (RM12031, RM15004, RM5844, RM21693, RM22250 and RM23362) and seven (RM12031, RM15004, RM5844, RM21693, RM22250, RM23362 and RM8207) specific alleles compared to the decadal periods in both year of rleasewise and longevitywise categories, respectively (Table S4). While the decadal periods 1980s (RM18384) and 1990s (RM5708) having one specific allele each, the period 2000s is with two (RM5844 and RM23741) specific alleles. Of the 10 population specific alleles identified in all, five are located in genic regions (RM5844, RM18384, RM22250, RM23362 and RM23741) while rest in non-genic regions. Seven of the 10 alleles specific to populations are comprised of tri-nucleotide repeats, especially of AAT repeat motif.

Qualitative analysis of genetic diversity revealed that the landraces were comprising of more number of population specific alleles as compared to the modern high yielding varieties suggesting loss of some of the landrace specific alleles on account of intensive human selection over the decades. The present findings are in agreement with those of earlier workers [11,32] who have also reported loss of alleles in the progressively improved modern varieties. The reason can be attributed to conscious and simultaneous selection against alleles of adaptive value characteristic to landraces resulting in less number of certain alleles and selection for alleles of agronomic value in the modern cultivars.

Effect of Genic-SSRs, Non-genic-SSRs and Gene-specific Markers on Genetic Diversity Trends

Ever since the domestication of rice, man's breeding priority has been yield enhancement and "breeding out" the wild progenitor traits like lax panicle, shattering, awns, etc., and later the focus had been shifted towards "breeding in" the traits of plant architecture and resistance to biotic and abiotic stresses and acceptable cooking quality. In that long breeding process, except the genomic regions, which govern domestication related traits, a major part of the rice

Table 4. Analysis of molecular variance (AMOVA) of major Indian rice cultivars.

Sample	Source of variation	df	Variance components	Percentage of variation
Year of releasewise	Among populations	4	0.549**	7.88
	Within populations	195	6.426***	92.12
	Total	199	0	
Popularitywise	Among populations	4	0.231**	3.34
	Within populations	287	6.690***	96.66
	Total	291	0	

df: degrees of freedom; p>0.001-*** : p>0.01-**.

Table 5. Pairwise population differentiation (*Fst*) of major Indian rice cultivars among decadal periods.

Decade	Landraces	1970s	1980s	1990s	2000s
Landraces	0.00000	0.06822	0.05477***	0.08557***	0.05925***
1970s	0.05112	0.00000	0.0138	0.02299	0.02338
1980s	0.03997*	0.03957	0.00000	0.03998***	0.01241
1990s	0.13083***	0.07904***	0.10569***	0.00000	0.02389**
2000s	0.09004***	0.07927**	0.07507***	0.0665***	0.00000

Lower diagonal: Year of releasewise; Upper diagonal: Longevitywise. p>0.001-***; p>0.01-**;p>0.1-*.

genome remains unchanged and hence large unfolded genetic diversity. To estimate genetic variability in crop improvement research, increasingly crop specific microsatellite markers are used. Since most of the microsatellites are non-coding, estimates made and inferences drawn based on them may not reflect true genomic diversity of a crop. To understand precisely the changing trends of genetic diversity in breeding material over periods, it is necessary to consider markers located both in genic and non-genic regions. In the present study, thus, microsatellite markers located in both genic and non-genic regions have been chosen to understand the level and trend of genetic diversity in varieties developed over the last four decades in comparison to the traditional varieties. The study using two marker types separately and together interestingly showed an increasing trend of genetic diversity (Figure S1, Table S5, Table S6). Analysis of gene specific marker data obtained from earlier studies also [39–41] clearly suggests an increasing trend of genetic diversity in Indian rice varieties developed over a period of four decades. In contrast, Qi *et al* [42] have reported initially declining trend of diversity in varieties developed between 1950 and 1980s and subsequently increasing trend as estimated based on both SSR markers and phenotypic traits in Chinese rice varieties.

Analysis of Genetic Diversity Based on Region, Ecology, Grain Size and Days to 50% Flowering Periodwise

When genetic diversity parameters viz., number of alleles (*Na*), number of effective alleles (*Ne*), Shannon index (*I*) and Nei's genetic diversity index (*He*) estimated according to their region of release in India (South, North and East), ecology (irrigated and rainfed), grain size (long and medium) and days to 50% flowering (early, medium early, medium and late) of a representative set of Indian rice varieties, observed variation was not correlating with their classification (Figure S2, Table S7). The reasons could be attributed to the repeated use of founder varieties such as T(N)1 and IR8 as well as their derivatives like Jaya, Rajeswari, Rasi, Prabhat, PR106 and Suraksha in the ancestry of many of the subsequently evolved varieties irrespective of their trait classification. Further, observation of pedigree records of some of the emerging new varieties released during the past two decades revealed that they were having some of the quality rice varieties such as Basmati370, Sona and BPT5204 (Samba mahsuri) as their donors but not the original founder varieties or their derivatives. The reason for this trend could be assumed as the changing consumer preferences towards quality rices. Keeping this trend in view breeders also being used quality rice varieties as the donors in their breeding programs.

Conclusions

Despite the perception that genetic variability has declined over the decades in the improved rice gene pool the present findings and previous reports suggest genetic diversity in the improved cultivars has been increasing over the decades, while losing some of the alleles which could be of unimportant and not relevant to the changing breeding objectives. However, it is important to understand the kind of alleles being gained and lost over the decades so as to plan our future breeding/selection strategies for directed improvement of the crop. The study has overall reveals that adequate diversity still exists in the advanced cultivar gene pool for steady varietal improvement. In the diversity conservation exercise being practiced at national and global levels, therefore, much emphasis is to be given to improved germplasm as that of the customarily conserved gene pool comprising wild/weedy species and landraces so that the genes of importance are conserved and profitably utilized to meet future challenges. It is, however, important to monitor periodically the level and trend of genetic diversity while pursuing aggressively rice breeding research.

Supporting Information

Figure S1 Changes in number of alleles *(Na)*, genetic diversity (*He*) over decadal periods using genic and non-genic SSRs along with gene-specific markers(previous studies).

Figure S2 Number of alleles *(Na)*, number of effective alleles (*Ne*), Shannon index (*I*) and Nei's genetic diversity (*He*) estimated in different groups of rice varieties i.e., region, ecology, grain size and days to 50% floweringwise.

Table S1 Details of the rice cultiavars used in the study.

Table S2 Summary statistics of genetic diversity parameters of major Indian rice cultivars.

Table S3 Locus by locus AMOVA among populations.

Table S4 Population specific alleles and their corresponding markers.

Table S5 List of genic and non-genic SSRs and their functions.

Table S6 Summary statistics of genetic diversity parameters of Indian rice cultivars using different microsatellite classes.

Table S7 Summary statistics for genetic diversity parameters of Indian rice varieties classified according to their region, ecology, grain size and days to 50% flowering.

Acknowledgments

We are very thankful to Dr.Mir for helping in resampling analysis using R-Software. Thanks are also due to the anonymous reviewers for their valuable comments for the improvement of the manuscript.

Author Contributions

Conceived and designed the experiments: LRV EAS. Performed the experiments: GC NR DAD AV. Analyzed the data: LRV MS. Contributed reagents/materials/analysis tools: GA. Wrote the paper: LRV EAS.

References

1. Christiansen MJ, Andersen SB, Ortiz R (2002) Diversity changes in an intensively bred wheat germplasm during the 20th century. Molecular Breeding 9: 1–11.
2. Tian QZ, Zhou RH, Jia JZ (2005) Genetic diversity trend of common wheat (*Triticum aestivum* L.) in China revealed with AFLP markers. Genetic Resources and Crop Evolution 52: 325–331.
3. White J, Law JR, MacKay I, Chalmers KJ, Smith JSC, et al. (2007) The genetic diversity of UK, US and Australian cultivars of Triticum aestivum measured by DArT markers and considered by genome. Theoretical and Applied Genetics 116: 439–453.
4. Huang X-Q, Wolf M, Ganal MW, Orford S, Koebner RMD, et al. (2007) Did Modern Plant Breeding Lead to Genetic Erosion in European Winter Wheat Varieties? Crop Science 47: 343.
5. Roussel V, Koenig J, Beckert M, Balfourier F (2004) Molecular diversity in French bread wheat accessions related to temporal trends and breeding programmes. TAG Theoretical and Applied Genetics 108: 920–930.
6. Fu Y-B, Peterson GW, Yu J-K, Gao L, Jia J, et al. (2006) Impact of plant breeding on genetic diversity of the Canadian hard red spring wheat germplasm as revealed by EST-derived SSR markers. Theoretical and Applied Genetics 112: 1239–1247.
7. Hazen SP, Zhu L, Kim H-S, Tang G, Ward RW (2002) Genetic diversity of winter wheat in Shaanxi province, China, and other common wheat germplasm pools. Genetic Resources and Crop Evolution 49: 439–448.
8. Hysing S-C, Säll T, Nybom H, Liljeroth E, Merker A, et al. (2008) Temporal diversity changes among 198 Nordic bread wheat landraces and cultivars detected by retrotransposon-based S-SAP analysis. Plant Genetic Resources 6.
9. Heal G, Walker B, Levin S, Arrow K, Dasgupta P, et al. (2004) Genetic diversity and interdependent crop choices in agriculture. Resource and Energy Economics 26: 175–184.
10. Wouw M, Hintum T, Kik C, Treuren R, Visser B (2010) Genetic diversity trends in twentieth century crop cultivars: a meta analysis. Theoretical and Applied Genetics 120: 1241–1252.
11. Wei X, Yuan X, Yu H, Wang Y, Xu Q, et al. (2009) Temporal changes in SSR allelic diversity of major rice cultivars in China. Journal of Genetics and Genomics 36: 363–370.
12. Mantegazza R, Biloni M, Grassi F, Basso B, Lu B-R, et al. (2008) Temporal Trends of Variation in Italian Rice Germplasm over the Past Two Centuries Revealed by AFLP and SSR Markers. Crop Science 48: 1832–1840.
13. Upadhyay P, Singh VK, Neeraja CN (2011) Identification of Genotype Specific Alleles and Molecular Diversity Assessment of Popular Rice (*Oryza sativa* L.) Varieties of India. International Journal of Plant Breeding and Genetics 5: 130–140.
14. Neeraja CN, Vemireddy LR, Malathi S, Siddiq EA (2009) Identification of alternate dwarfing gene sources to widely used Dee-Gee-Woo-Gen allele of sd1 gene by molecular and biochemical assays in rice (*Oryza sativa* L.). Electronic Journal of Biotechnology 12.
15. Narshimulu G, Jamaloddin M, Vemireddy LR, Anuradha G, Siddiq E (2011) Potentiality of evenly distributed hypervariable microsatellite markers in marker-assisted breeding of rice. Plant Breeding 130: 314–320.
16. Murray MG, Thompson WF (1980) Rapid isolation of high molecular weight plant DNA. Nucleic Acids Research 8: 4321–4326.
17. Sambrook J, Russell DW (2001) Molecular cloning: A laboratory manual. Cold Spring Harbour Laboratory Press, Cold Spring Harbour, New York.
18. Botstein D, White RL, Skolnick M, Davis RW (1980) Construction of a genetic linkage map in man using restriction fragment length polymorphisms. American Journal of Human Genetics 32: 314–331.
19. Nei M (1973) Analysis of gene diversity in subdivided populations. Proc Natl Acad Sci U S A 70: 3321–3323.
20. Goudet J (1995) FSTAT (Version 1.2): A Computer Program to Calculate F-Statistics. J Heredity 86: 485–486.
21. Excoffier L, Laval G, Schneider S (2005) Arlequin (version 3.0): an integrated software package for population genetics data analysis. Evol Bioinform Online 1: 47–50.
22. Glaubitz JC (2004) convert: A user-friendly program to reformat diploid genotypic data for commonly used population genetic software packages. Molecular Ecology Notes 4: 309–310.
23. Mir RR, Kumar J, Balyan HS, Gupta PK (2012) A study of genetic diversity among Indian bread wheat (*Triticum aestivum* L.) cultivars released during last 100 years. Genetic Resources and Crop Evolution 59: 717–726.
24. Jain S, Jain RK, McCouch SR (2004) Genetic analysis of Indian aromatic and quality rice (*Oryza sativa* L.) germplasm using panels of fluorescently-labeled microsatellite markers. Theoretical and Applied Genetics 109: 965–977.
25. Spada A, Mantegazza R, Biloni M, Caporali E, Sala F (2004) Italian rice varieties: historical data, molecular markers and pedigrees to reveal their genetic relationships. Plant Breeding 123: 105–111.
26. Zhu MY, Wang YY, Zhu YY, Lu BR (2004) Genetic diversity of rice landraces from Yunnan revealed by SSR analysis and its significance for conservation. J Huazhong Agri Univ 23: 187–191.
27. Chuan-Guang, Gui-Quan Z (2010) SSR analysis of genetic diversity and the temporal trends of major commercial inbred indica rice cultivars in south china in 1949–2005. Acta Agron Sin 36: 1843–1852.
28. Yu SB, Xu WJ, Vijayakumar CHM, Ali J, Fu BY, et al. (2003) Molecular diversity and multilocus organization of the parental lines used in the International Rice Molecular Breeding Program. Theoretical and Applied Genetics 108: 131–140.
29. Luce C, Noyer J, Tharreaeu D, Ahmadi N, Feyt H (2001) The use of microsatellite markers to examine the diversity of the genetic resources of rice (*Oryza sativa* L.) adapted to European conditions. Acta Horticulture 546: 221–235.
30. Giarrocco LE, Marassi MA, Salerno GL (2007) Assessment of the Genetic Diversity in Argentine Rice Cultivars with SSR Markers. Crop Science 47: 853–858.
31. Jayamani P, Negrão S, Martins M, Maçãs B, Oliveira MM (2007) Genetic Relatedness of Portuguese Rice Accessions from Diverse Origins as Assessed by Microsatellite Markers. Crop Science 47: 879–884.
32. Yuan X-P, Wei X-H, Hua L, Yu H-Y, Wang Y-P, et al. (2007) A Comparative Study of SSR Diversity in Chinese Major Rice Varieties Planted in 1950s and in the Recent Ten Years (1995–2004). Rice Science 14: 78–84.
33. Glaszmann JC (1987) Isozymes and classification of Asian rice varieties. Theoretical and Applied Genetics 74: 21–30.
34. Nagaraju J, Kathirvel M, Kumar RR, Siddiq EA, Hasnain SE (2002) Genetic analysis of traditional and evolved Basmati and non-Basmati rice varieties by using fluorescence-based ISSR-PCR and SSR markers. Proc Natl Acad Sci U S A 99: 5836–5841.
35. Morin SR, Calibo M, Garcia-Belen M, Pham J-L, Palis F (2002) Natural hazards and genetic diversity in rice. Agriculture and Human Values 19: 133–149.
36. Garris AJ, Tai TH, Coburn J, Kresovich S, McCouch S (2005) Genetic structure and diversity in *Oryza sativa* L. Genetics 169: 1631–1638.
37. Yonemaru J, Yamamoto T, Ebana K, Yamamoto E, Nagasaki H, et al. (2012) Genome-wide haplotype changes produced by artificial selection during modern rice breeding in Japan. PLoS One 7: e32982.
38. Steele KA, Gyawali S, Joshi KD, Shrestha P, Sthapit BR, et al. (2009) Has the introduction of modern rice varieties changed rice genetic diversity in a high-altitude region of Nepal? Field Crops Research 113: 24–30.
39. Ramkumar G, Sivaranjani AKP, Pandey MK, Sakthivel K, Shobha Rani N, et al. (2010) Development of a PCR-based SNP marker system for effective selection of kernel length and kernel elongation in rice. Molecular Breeding 26: 735–740.
40. Sundaram RM, Sakthivel K, Hariprasad AS, Ramesha MS, Viraktamath BC, et al. (2010) Development and validation of a PCR-based functional marker system for the major wide-compatible gene locus S5 in rice. Molecular Breeding 26: 719–727.
41. Singh A, Singh PK, Singh R, Pandit A, Mahato AK, et al. (2010) SNP haplotypes of the BADH1 gene and their association with aroma in rice (*Oryza sativa* L.). Molecular Breeding 26: 325–338.
42. Qi Y, Zhang D, Zhang H, Wang M, Sun J, et al. (2006) Genetic diversity of rice cultivars (*Oryza sativa* L.) in China and the temporal trends in recent fifty years. Chinese Science Bulletin 51: 681–688.

15

Adenoviruses Using the Cancer Marker EphA2 as a Receptor In Vitro and In Vivo by Genetic Ligand Insertion into Different Capsid Scaffolds

Michael Behr[1¤a], Johanna K. Kaufmann[1¤b], Patrick Ketzer[1], Sarah Engelhardt[1], Martin Mück-Häusl[2¤c], Pamela M. Okun[3], Gabriele Petersen[4], Frank Neipel[5], Jessica C. Hassel[6], Anja Ehrhardt[2,7], Alexander H. Enk[6], Dirk M. Nettelbeck[1]*

1 Oncolytic Adenovirus Group, German Cancer Research Center (DKFZ), Heidelberg, Germany, **2** Max von Pettenkofer-Institute, Department of Virology, Ludwig-Maximilians-University Munich, Munich, Germany, **3** Department of General Pediatrics, Division of Inherited Metabolic Diseases, Heidelberg University Hospital, Heidelberg, Germany, **4** COS, CellNetworks Deep Sequencing Core Facility, University Heidelberg, Heidelberg, Germany, **5** Institute for Clinical and Molecular Virology, Erlangen University Hospital, Erlangen, Germany, **6** Department of Dermatology, Heidelberg University Hospital, Heidelberg, Germany, **7** Institute of Virology and Microbiology, Center for Biomedical Education and Research, Department of Human Medicine, Faculty of Health, University Witten/Herdecke, Witten, Germany

Abstract

Adenoviral gene therapy and oncolysis would critically benefit from targeted cell entry by genetically modified capsids. This requires both the ablation of native adenovirus tropism and the identification of ligands that remain functional in virus context. Here, we establish cell type-specific entry of HAdV-5-based vectors by genetic ligand insertion into a chimeric fiber with shaft and knob domains of the short HAdV-41 fiber (Ad5T/41sSK). This fiber format was reported to ablate transduction *in vitro* and biodistribution to the liver *in vivo*. We show that the YSA peptide, binding to the pan-cancer marker EphA2, can be inserted into three positions of the chimeric fiber, resulting in strong transduction of EphA2-positive but not EphA2-negative cells of human melanoma biopsies and of tumor xenografts after intratumoral injection. Transduction was blocked by soluble YSA peptide and restored for EphA2-negative cells after recombinant EphA2 expression. The YSA peptide could also be inserted into three positions of a CAR binding-ablated HAdV-5 fiber enabling specific transduction; however, the Ad5T/41sSK format was superior *in vivo*. In conclusion, we establish an adenovirus capsid facilitating functional insertion of targeting peptides and a novel adenovirus using the tumor marker EphA2 as receptor with high potential for cancer gene therapy and viral oncolysis.

Editor: Ilya Ulasov, Swedish Medical Center, United States of America

Funding: This work was supported by the Helmholtz-University Young Investigator Group grant VH NG 212 and the Deutsche Krebshilfe (German Cancer Aid) grant 10-7946. The funders had no role in study design, data collection and analysis, decision to publish, or preparation of the manuscript.

Competing Interests: The authors have declared that no competing interests exist.

* E-mail: d.nettelbeck@dkfz.de

¤a Current address: Dr. von Froreich Bioscientia Medical Laboratory, Hamburg, Germany
¤b Current address: Harvey Cushing Neuro-Oncology Laboratories, Department of Neurosurgery, Brigham & Women's Hospital and Harvard Medical School, Boston, Massachusetts, United States of America
¤c Current address: Institute of Virology, Helmholtz Center Munich, Munich, Germany

Introduction

Recombinant adenoviruses (Ads) are in pre-clinical and clinical development as oncolytic agents and vectors for vaccination and gene therapy [1,2].uch Ad-based drugs can considerably benefit from or even depend on a cell type-specific mode of action in order to limit side effects, while retaining therapeutic potency. Cell type-specificity of Ad-based vectors and oncolytics has been well established at the post-entry level by transcriptional targeting and microRNA regulation [3,4].In contrast, targeting of Ad cell entry, which would better avoid side effects and virus sequestration, remains a challenge. Genetic strategies for Ad entry targeting facilitate simple clinical grade production (no chemical modifications or separate adapters are required) and ensure that progeny oncolytic Ads produced in the patients' tumors retain the improved properties. However, strategies for genetic entry targeting of Ads are hampered by the rigid Ad capsid structure, the inability of antibody molecules, as preferred targeting moieties, to fold properly after fusion to cytosolically expressed Ad capsid proteins, and the complex interactions of host factors modulating Ad cell entry in patients (see below).

Entry targeting of Ads requires two steps: ablation of native tropism for healthy cells and re-targeting of a receptor specifically expressed on target cells by insertion of a corresponding ligand into the Ad capsid. Cell entry of native Ads is initiated by binding of the capsid protein fiber to the attachment receptor, which is Coxsackie-Ad receptor (CAR) for the most widely used serotype HAdV-5 [5].Virus internalization is then triggered by binding of the penton base to cellular integrins [6].For therapeutic applications of Ads not intended to target hepatic tissue, liver de-targeting is of key importance to avoid toxic side effects. However, ablating CAR binding by mutating the fiber did not reduce liver transduction after systemic virus injection [7].Additional deletion of integrin binding by mutating the penton base resulted in reduced liver transduction in some but not all studies [7].Moreover, penton base mutations might be counterproductive in the context of Ad targeting, as virus entry and post-entry steps of retargeted viruses can be compromised [8].Liver transduction by HAdV-5, irrespective of their interaction with CAR and integrins, was attributed to blood coagulation factors bound to hexon and fiber [9–11].Which receptor mediates transduction of hepatocytes by HAdV-5 *in vivo* is still being discussed [12,13].

Our approach for entry targeting of HAdV-5-derived viruses is to replace the fiber shaft and knob domains with the corresponding domains of the HAdV-41 short fiber (Ad5T/41sSK) and to insert peptide ligands into this chimeric capsid. HAdV-41 binds CAR via a second long fiber, while no cell-binding activity has been attributed to the short fiber. Correspondingly, strongly reduced transduction *in vitro* and liver transduction *in vivo* have been demonstrated by several groups for HAdV-5-based vectors containing short fibers of HAdV-41 or of the closely related HAdV-40 [14–19].We have previously demonstrated that infectivity of Ads with chimeric Ad5T/41sSK fiber (then termed F5/41s) can be restored by genetic peptide ligand insertion using the integrin binding RGD4C-peptide as a model peptide [14].In fact, we identified several functional insertion sites, thus establishing the chimeric Ad5T/41sSK fiber as a flexible fiber scaffold for ligand insertion: the HI and EG loops on the side of the knob and for the IJ loop on its top, resulting in superior transduction efficiency compared with C-terminal fusions. However, as integrins are ubiquitously expressed, the RGD4C peptide was not suitable to demonstrate the potential of the Ad5T/41sSK format for cell type-specific cell entry and transduction. Therefore, the aim of the present study was to establish cell entry targeting by the Ad5T/41sSK strategy using a cell-selective peptide ligand and to compare this strategy with a HAdV-5 fiber-based targeting approach.

The YSA peptide, a 12-mer identified by phage display, selectively binds to the receptor tyrosine kinase EphA2, but not to related kinases [20].We focused our study on this peptide ligand because in contrast to several other tested peptides it retained cell-binding activity in the context of the Ad fiber. Importantly, EphA2 is gaining increasing attention as target for cancer therapy because it is (i) upregulated on most solid tumors and on tumor endothelium, (ii) better accessible on tumors that often lack cell-associated ligands, (iii) functionally associated with tumor progression, and (iv) was recently reported to be a cancer stem cell marker [21,22].Several EphA2-targeted therapeutic modalities have shown proof of concept in pre-clinical studies, including kinase inhibitors, antibodies, immunotoxins, engineered T cells, soluble receptors, and vaccines [22–24].

Here, we investigated Ad entry targeting *in vitro* and *in vivo* by genetically inserting the EphA2-binding YSA peptide into different receptor-blind Ad fiber scaffolds. Specifically, we explored sites for functional YSA peptide insertion into the knob domains of the short HAdV-41 fiber and of the HAdV-5 fiber. In addition to virus production by combined fiber transfection/virus superinfection as we have done before [14],we investigated direct engineering of fiber genes in the virus genomes, which is of advantage or required for ease of virus manufacturing and for viral oncolysis, respectively. Selectivity and efficiency of Ad cell entry mediated by the YSA peptide was investigated in cell culture, human metastases biopsies, and animal xenograft models comparing three fiber formats: (i) the chimeric Ad5T/41sSK fiber, (ii) a long-shafted chimeric fiber containing the HAdV-5 fiber tail and shaft domains and the short HAdV-41 fiber knob, and (iii) a long-shafted but CAR-binding ablated HAdV-5 fiber.

Results

Specific transduction of EphA2-positive cells by Ads with YSA peptide inserted into chimeric fibers containing the knob of the HAdV-41 short fiber

We investigated entry targeting of Ads by genetic insertion of a targeting peptide into chimeric fibers with HAdV-41 knob as a detargeted scaffold. To this end, we inserted the 12-mer EphA2-binding peptide YSA [20] flanked by short linkers into the HI, IJ or EG loops of this knob domain. To explore the relevance of shaft length on YSA-mediated Ad transduction, we combined these YSA-containing knobs with the short HAdV-41 fiber shaft (Ad5T/41sSK viruses, Fig. 1A) or the long HAdV-5 fiber shaft (Ad5TS/41sK viruses, Fig. 1B). In a third set of fibers, we incorporated the long HAdV-5 fiber shaft with a mutated heparin sulfate proteoglycan (HSPG)-binding motif (Ad5TS*/41sK viruses, Fig. 1C). This mutation was reported to confer improved detargeting [17,25,26].After plasmid transfection, all constructs were expressed and possessed trimerization capacity, but trimerization was clearly less efficient for the long-shafted constructs, especially those containing the peptide in the HI loop (Fig. 2A). Using a combined transfection/superinfection protocol (see Materials and Methods), we were able to produce high titer pseudotyped LacZ reporter Ad vectors with all fiber formats. Incorporation of fiber molecules into viral particles was efficient for the short-shafted fiber and, despite reduced trimerization capacity, for the long-shafted IJ-YSA fibers (Fig. 2B). Long-shafted EG-YSA and HI-YSA fibers showed reduced or lacked fiber incorporation into virus particles, respectively.

Next, we analyzed EphA2 expression in a panel of pancreatic cancer cells, melanoma cells, endothelial cells, and a hepatic cell line. We detected strong EphA2 expression for pancreatic cancer cells, endothelial cells, and for 4 of 6 melanoma cell cultures (Fig. S1). EphA2 was not detected for the two melanoma cell lines SK-MEL-28 and Mel624 and the hepatic cell line HepG2. Transduction experiments with EphA2-positive cell lines PANC-1, MIA PaCa-2, Capan-1 (pancreatic cancer) and C8161 (melanoma) yielded results that were similar for the respective cell lines: All three viruses pseudotyped with the short-shafted YSA fibers showed transduction, which was substantially stronger than for the control viruses without YSA peptide (Fig. 2C; 10-fold to 35-fold increase in β-Gal activity). For viruses with unmodified HAdV-5 shaft, the insertion of YSA into the HI, IJ and EG loops showed weak, strong, and intermediate transduction efficiency, respectively. The long-shafted IJ-YSA virus was even superior to the short-shafted viruses ($p<0.05$ for all cell lines.). For all viruses with the mutated long shaft, we observed inefficient transduction (shown in PANC-1 and C8161 cells). In EphA2-negative SK-MEL-28 cells, the EG-YSA virus and especially the IJ-YSA virus with unmodified long shaft showed transduction significantly higher than the control viruses. We conclude that short-shafted Ad5T/41sSK viruses with YSA peptide inserted into the EG, HI or IJ loop selectively transduce EphA2-positive tumor cells, while the long shafted Ad5TS/41sK-IJ-YSA virus showed even stronger, but less selective transduction.

Next, we investigated how the transduction efficiency of pseudotyped YSA viruses compares with viruses containing the established, integrin-binding RGD peptide in tumor and endothelial cells. Most previous studies on genetic peptide ligand insertion used RGD-containing peptides and demonstrated improved, but not targeted cell entry [27–29].Indeed, HAdV-5 viruses with RGD peptide in the HI loop are being investigated in clinical studies, because of their improved cell entry efficiency [30].We found that in the context of the Ad5T/41sSK chimeric

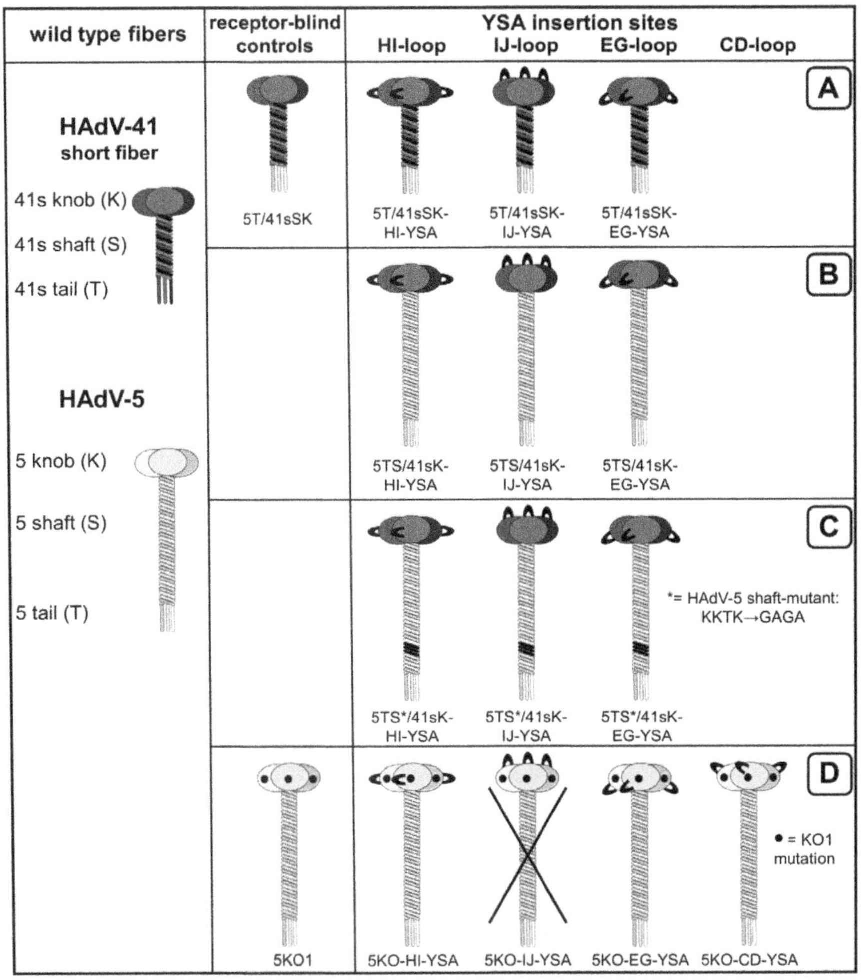

Figure 1. Schematic outline of fiber constructs with YSA peptide insertion investigated in this study. (**A**) Short-shafted chimeric fibers with the HAdV-5 tail fused to the shaft and knob of the HAdV-41 short fiber. (**B,C**) Long-shafted chimeric fibers containing the HAdV-5 tail and wild-type (B) or mutant (C) shaft fused to the HAdV-41 short fiber knob. (**D**) CAR binding-ablated HAdV-5 fibers. YSA peptide insertion sites are indicated by loops. Insertion of the linker sequence in different positions of the HAdV-5 fiber knob IJ loop resulted in loss of trimerization capacity (data not shown) and viruses were not produced for this insertion site.

fiber YSA peptide-mediated transduction of EphA2-positive cells was similar or even superior to transduction mediated by the RGD4C peptide inserted in the HI loop (Fig. 3A and B), the most effective insertion site for this peptide [14].In EphA2-negative SK-MEL-28 cells, only the RGD virus resulted in transduction significantly higher than the control virus without peptide. Finally, we could show that transduction of EphA2-positive cells by Ad5T/41sSK-HI-YSA and Ad5T/41sSK-IJ-YSA but not Ad5T/41sSK-HI-RGD is effectively blocked by soluble YSA peptide, while a control peptide with randomized sequence did not block transduction (Fig. 3B). Overall, these results demonstrate potent YSA-mediated transduction specifically of EphA2-positive cells by Ads with YSA peptide inserted into the chimeric Ad5T/41sSK fiber.

Specific transduction of EphA2-positive cells by Ads with YSA peptide inserted into the CAR binding-ablated HAdV-5 fiber

For specific applications, e.g. those requiring tumor-specific gene transfer but not liver de-targeting (see discussion), Ads with a HAdV-5-based fiber might be sufficient or advantageous compared with more extensively modified chimeric fiber-containing viruses. For this reason and to compare our chimeric fiber format with the native one, we investigated which insertion sites enable genetic incorporation of the YSA peptide into the CAR binding-ablated HAdV-5 fiber knob KO1 (S408E and P409A mutations, Ref [31], Fig. 1D). We found that the YSA peptide can be inserted into the CD, EG and HI loops of the fiber retaining trimerization capacity (somewhat less efficient for the CD loop) and incorporation capability into virus particles generated by the transfection/superinfection protocol (Fig. 4A and B). We also explored the IJ loop as a possible position for peptide ligand insertion considering the favorable location on the top of the knob domain. However, we observed loss of fiber trimerization after inserting a linker at positions G560/H561 or I564/N565 (not shown). Pseudotyped Ads with the YSA peptide inserted into the EG, HI or CD loop of the KO1 knob mediated efficient transduction of EphA2-positive cells, which was 8-fold to 230-fold superior to the control virus Ad5KO1 without peptide insertion (Fig. 4C). The Ad5KO-HI-YSA virus was more efficient than the Ad5KO-EG-YSA and Ad5KO-CD-YSA viruses in all cell lines tested as well as superior

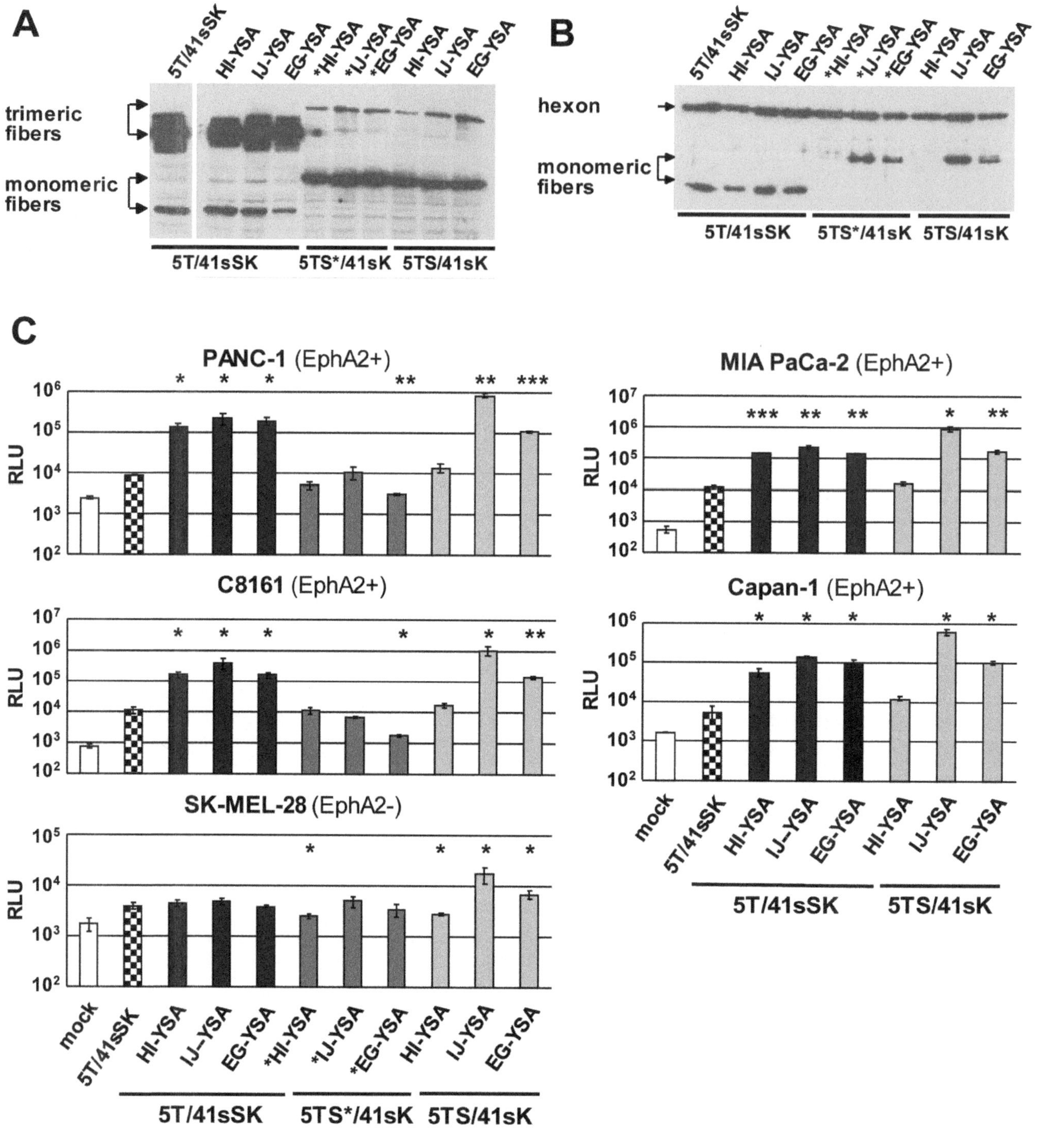

Figure 2. Long- and short-shafted chimeric fibers with genetic insertion of the YSA peptide into the HAdV-41 short fiber knob: trimerization, incorporation into virus particles, and viral transduction. (**A**) Immunoblot of unboiled cell lysates after transient transfection of fiber expression plasmids into 293T cells. (**B**) Immunoblot of purified virus particles of pseudotyped LacZ reporter viruses generated by the transfection/superinfection protocol. (**C**) Transduction of EphA2-positive and EphA2-negative cells with pseudotyped LacZ reporter viruses. SK-MEL-28 cells were transduced in quadruplicates, all other cell lines were transduced in triplicates. Columns and error bars show mean values and standard deviations of β-Gal activity, respectively. RLU, relative luminescence units; * $p<0.05$, ** $p<0.01$, *** $p<0.001$ versus 5T41sSK in the respective cell line.

to Ad5T/41sSK-EG-YSA. Furthermore, the Ad5KO-HI-YSA virus showed similar or superior transduction efficiency compared with the Ad5WT virus, containing the wild-type HAdV-5 fiber. In EphA2-negative cells, none of the viruses with YSA peptide inserted into the KO1 fiber increased transduction efficiency compared with the parental virus (Fig. 4C), demonstrating a lack of off-target transduction. In conclusion, the transduction efficiency of Ads with YSA peptide insertion into the HAdV-5 fiber depends on the insertion site with the HI insertion showing the best transduction efficiency and specificity.

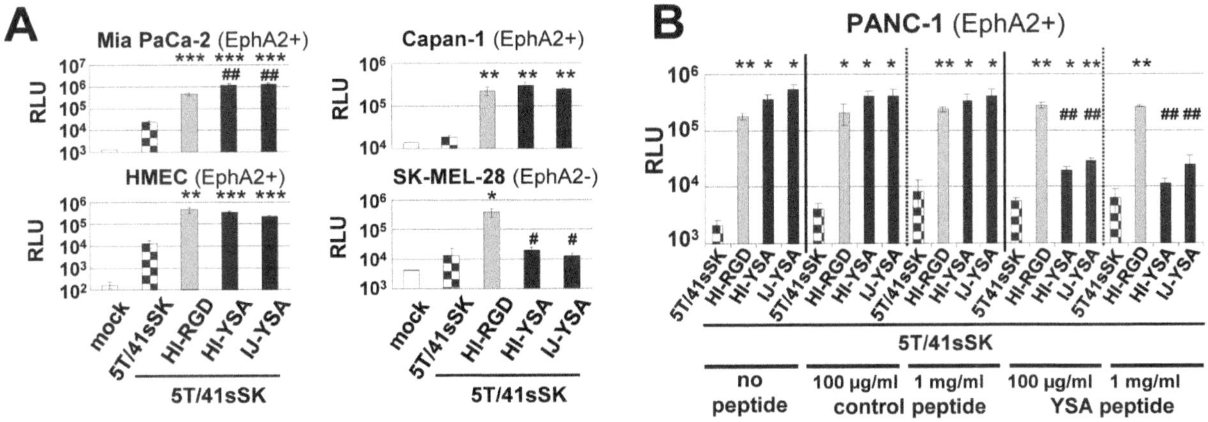

Figure 3. Efficiency and selectivity of transduction by Ad vectors containing short-shafted chimeric fibers with genetic insertion of the YSA peptide. (**A**) Transduction of EphA2-positive and EphA2-negative cells by pseudotyped Ad vectors containing short-shafted chimeric fibers with genetic insertion of the YSA peptide in comparison to matching vectors with genetic insertion of the RGD4C peptide. Cells were transduced in quadruplicates. (**B**) Transduction of EphA2-positive PANC-1 cells after competition with soluble YSA peptide or with control peptide of randomized sequence. Cells were transduced with pseudotyped Ads in triplicates. Columns and error bars show mean values and standard deviations of β-Gal activity, respectively. RLU, relative luminescence units; */# $p<0.05$, **/## $p<0.01$ *** $p<0.001$ versus 5T/41sSK (*) or 5T/41sSK-HI-RGD (#).

EphA2-mediated transduction of tumor and endothelial cells by genomically fiber-modified Ads

We next developed EphA2-targeted viruses with modified fibers encoded by their genome rather than pseudotyped by transfection. Genomically modified viruses would simplify manufacturing procedures and facilitate large scale production of viruses. Also, genomic fiber insertion is required in the context of viral oncolysis. However, it is not clear how genomic fiber replacement affects the production of infectious viruses. In fact, a previous study reported growth defects and/or defective particles for Ad vectors with the native fiber gene replaced by a gene encoding the entire short HAdV-41 fiber with peptide insertions [32].Using BAC recombineering technology, we cloned six genomes of first generation GFP/Luc reporter viruses. In three genomes the HAdV-5 fiber was replaced by a YSA-containing fiber representing each of the fiber formats (Ad5T/41sSK-IJ-YSA, Ad5TS/41sK-IJ-YSA and Ad5KO1-HI-YSA). The other three genomes represented the control viruses Ad5T/41sSK, Ad5wt, and Ad5KO1. All viruses could be produced at high titer and showed efficient fiber incorporation (Fig. 5A). Results of transduction experiments with these genetically modified viruses reproduced those obtained with pseudotyped viruses produced by the transfection/superinfection protocol, demonstrating that genomic fiber insertion was successful (Fig. 5B–D): In EphA2-positive pancreatic cancer cells, melanoma cells and endothelial cells, we again observed a dramatic increase in transduction efficiency for YSA-containing viruses compared with receptor-blind control viruses (15-fold to 236-fold). Insertion of the YSA peptide into the KO1 fiber resulted in the highest increase in transduction efficiency. The IJ-YSA fiber chimeric virus with long shaft was again somewhat stronger than the matching virus with a short shaft. Of note, YSA viruses resulted in stronger transduction even compared with the virus containing the wild-type HAdV-5 fiber in EphA2-positive cells. This was the case not only for cells weakly transduced with viruses containing the HAdV-5 fiber, i.e. C8161 cells (36-fold to 220-fold), but also in the EphA2- and CAR-positive [33,34] cells MIA PaCa-2, PANC-1 and HUVEC. In EphA2-negative SK-MEL-28 and HepG2 cells, Ad5T/41sSK-IJ-YSA and Ad5KO1-HI-YSA showed no significant increase in transduction efficiency compared with the respective controls without peptide insertion, whereas Ad5TS/41sK-IJ-YSA again showed somewhat increased transduction. Visualization of GFP expression confirmed that Ads with YSA peptide resulted in transduction of markedly increased percentages of EphA2-positive cells compared with control viruses without peptide and also compared with Ad5wt (Fig. S2). In EphA2-negative cells, transduction efficiency as determined by this GFP-based read-out was strongly reduced for Ad5T/41sSK-IJ-YSA, Ad5KO1-HI-YSA, Ad5T/41sSK and Ad5KO1 in comparison to Ad5wt, but less so for Ad5TS/41sK-IJ-YSA. These results confirm that Ad5T/41sSK-IJ-YSA and Ad5KO1-HI-YSA viruses feature the best entry targeting capacity for EphA2-positive cells. Of note, transduction of SK-MEL-28 cells by YSA viruses but not by control viruses without peptide insertion could be restored by overexpression of recombinant EphA2 (Fig. 6, Fig. S3), demonstrating that EphA2 mediates cell entry of YSA peptide-containing viruses.

YSA peptide-mediated adenoviral transduction of pancreatic cancer and melanoma xenografts *in vivo*

Production of high titer preparations of genomically capsid-modified viruses allowed us to perform animal studies to explore YSA peptide-mediated adenoviral transduction *in vivo*. To this end, we used NOD-SCID mice with subcutaneous xenograft tumors of PANC-1 pancreatic cancer cells or C8161 melanoma cells. EphA2 expression *in vivo* was verified in both tumor models (Fig. S4). In a first experiment, we injected animals bearing PANC-1 or C8161 tumors intratumorally (i.t.) with Ad5T/41sSK-IJ-YSA or Ad5T/41sSK (Fig. 7A). In both tumor models we observed significantly stronger transduction with Ad5T/41sSK-IJ-YSA (2.9-fold for PANC-1 and 5.9-fold for C8161), demonstrating that YSA-peptide-mediated transduction is functional *in vivo*. In a second experiment we injected C8161 tumors i.t. with the YSA-containing Ads representing each of the fiber formats (Ad5T/41sSK-IJ-YSA, Ad5TS/41sK-IJ-YSA and Ad5KO1-HI-YSA) or with control viruses (Fig. 7B). Increased transduction for Ad5T/41sSK-IJ-YSA versus Ad5T/41sSK was confirmed. Ad5TS/41sK-IJ-YSA showed somewhat higher transduction than Ad5T/41sSK-IJ-YSA. These results are in accord with the *in vitro* transduction results. However, in contrast to the *in vitro* data Ad5KO1 showed similar transduction efficiency than Ad5wt after

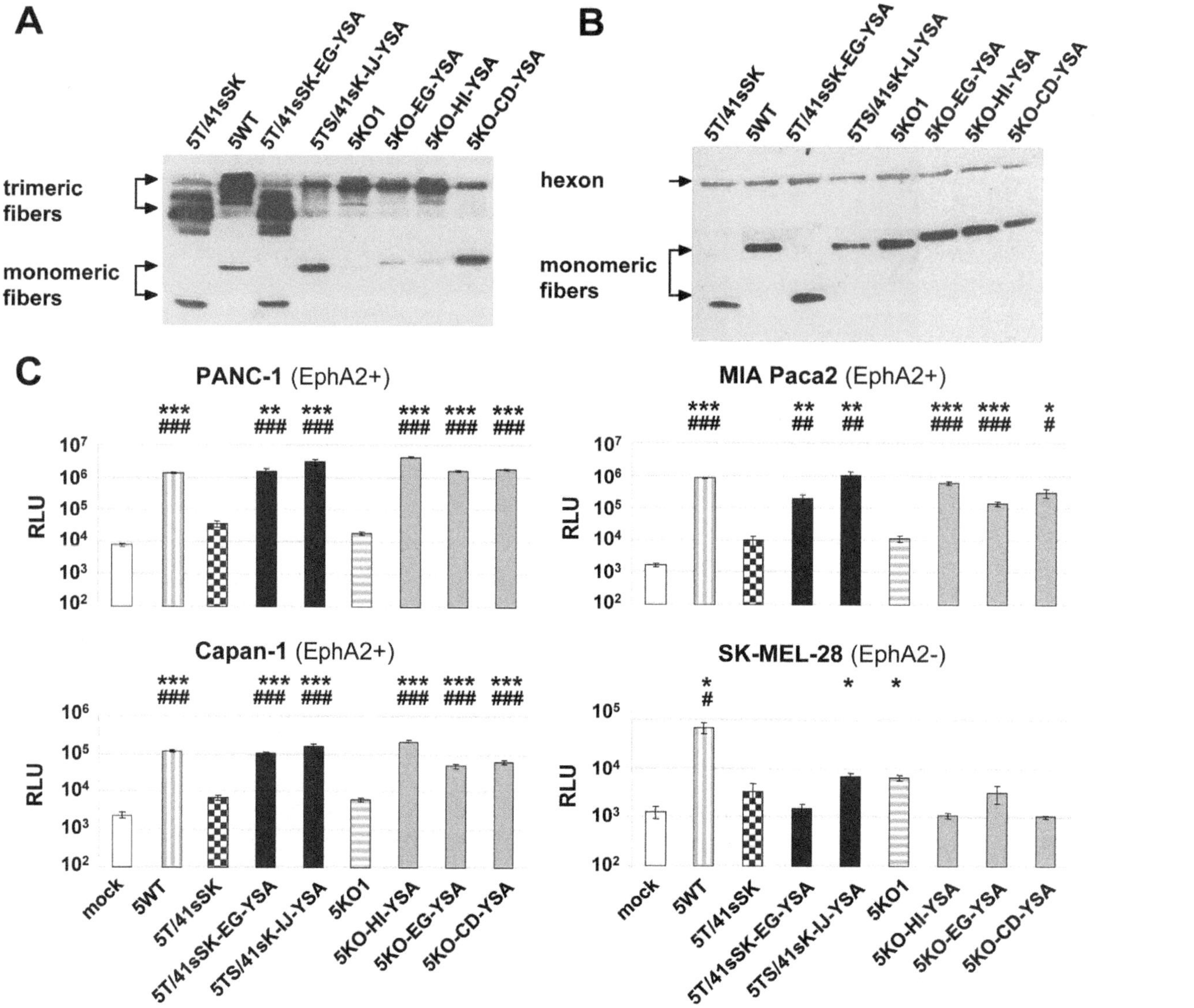

Figure 4. Fibers with genetic insertion of the YSA peptide into the CAR binding-ablated HAdV-5 knob: trimerization, incorporation into virus particles, and transduction. (**A**) Immunoblot of unboiled cell lysates after transient transfection of fiber expression plasmids into 293T cells. (**B**) Immunoblot of purified virus particles of pseudotyped LacZ reporter viruses. (**C**) Transduction of EphA2-positive and EphA2-negative cells with pseudotyped LacZ reporter viruses. Cells were transduced in quadruplicates, columns and error bars show mean values and standard deviations of β-Gal activity, respectively. RLU, relative luminescence units; */#p<0.05, **/##p<0.01, ***/###p<0.001 versus 5T/41sSK (*) or 5KO1 (#).

i.t. injection into C8161 xenografts revealing loss of de-targeting. Accordingly, the increase in transduction by YSA peptide insertion was minor (1.6-fold), but reached significance. Therefore, de- and re-targeting of Ads for specific entry via EphA2 after i.t. injection *in vivo* was best implemented with the Ad5T/41sSK fiber format. Overall, these results show that peptide-mediated entry targeting using the Ad5T/41sSK fiber format is functional *in vivo* after i.t. virus injection.

YSA peptide-mediated adenoviral transduction in biopsies of human melanoma metastases

We next investigated whether EphA2-targeted Ads result in increased transduction not only in monolayer tumor cell cultures and tumor cell xenografts, but also in freshly biopsied tumor material from patients. These represent clinically more relevant substrates with respect to tumor cell physiology and presence of tumor microenvironment. Biopsies of melanoma metastases were cut into living tissue slices immediately after surgery and were subsequently transduced with genomically capsid-modified, YSA-containing Ads representing each of the fiber formats or with control viruses. EphA2 expression was detected in living tissue slices of all biopsies obtained from 5 patients. However, expression varied between patients and was weaker than for monolayer cultures (Fig. S5A). This was expected as biopsies contain a mixture of cells of which only a fraction are tumor cells. We observed strongest transduction efficiency as determined by GFP expression for viruses with YSA peptide and for the control virus with native HAdV-5 fiber, confirming YSA peptide-mediated transduction (Fig. S5B). As GFP signals in 2D photographs do not quantitatively represent transduction efficiency, in part due to the wrinkled shape of the slices after 3 day culture, we quantified luciferase expression for these slices (Fig. 8A) and for biopsies from

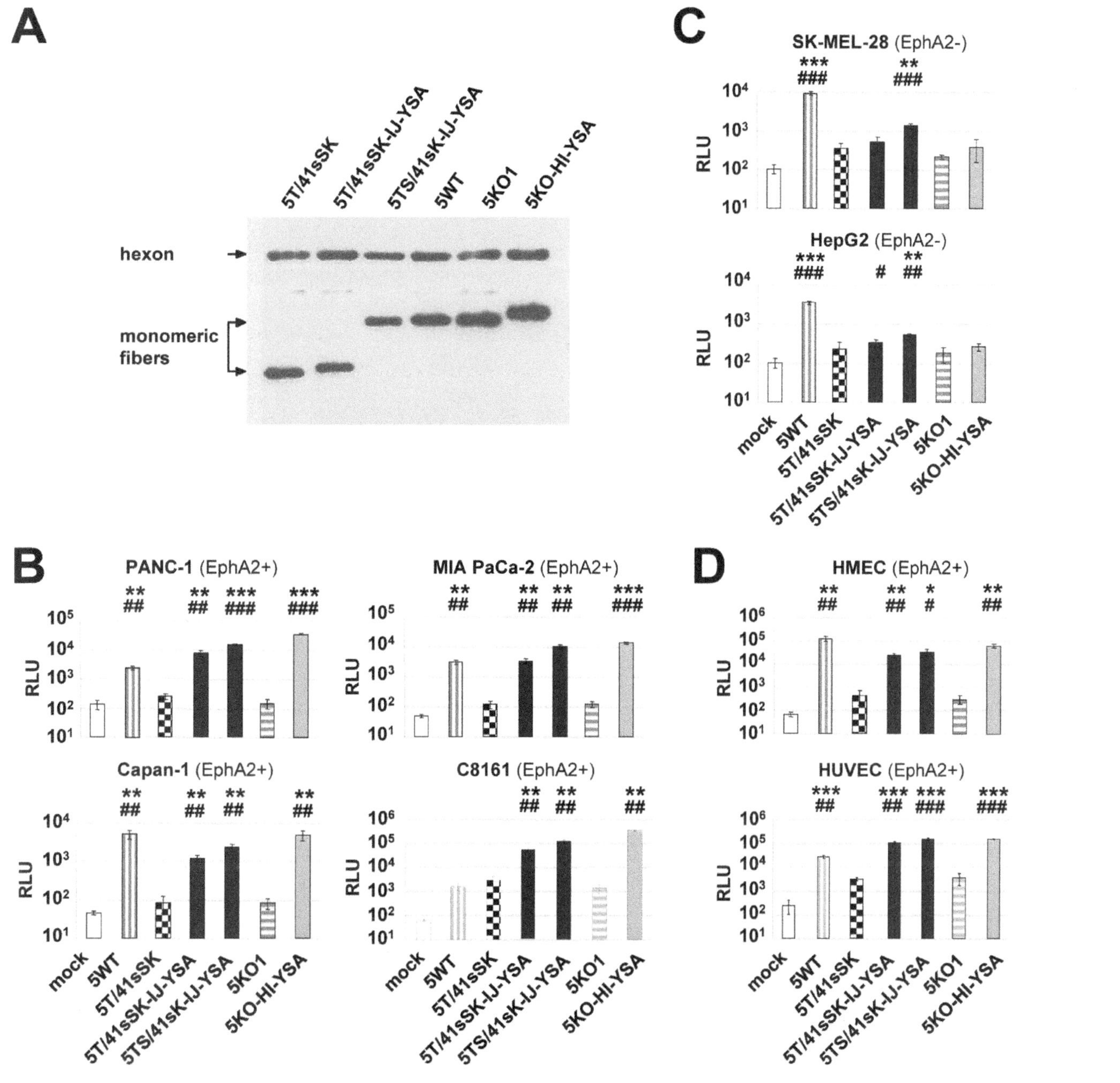

Figure 5. EphA2-targeted Ads with genomically inserted fibers containing the YSA peptide: incorporation of fibers into virus particles and transduction. (**A**) Immunoblot of purified virus particles generated by genomic insertion of recombinant fibers. (**B–D**) Transduction of EphA2-positive cancer cells (B), EphA2-negative cells (C) and EphA2-positive endothelial cells (D) with genomically fiber-modified Luc/GFP reporter viruses. C8161 cells were transduced in triplicates, all other cell lines were transduced in quadruplicates. Columns and error bars show mean values and standard deviations of Luc expression, respectively. RLU, relative luminescence units; */#$p<0.05$, **/##$p<0.01$, ***/###$p<0.001$ versus 5T/41sSK (*) or 5KO1 (#).

four additional patients (Fig. 8B-E, one biopsy yielded only enough slices for transduction with Ad5T/41sSK-IJ-YSA and Ad5T/41sSK). Note that luciferase readings are high because we performed high titer transductions to enable monitoring of GFP expression. We observed a markedly increased transduction for Ad5T/41sSK-IJ-YSA compared with Ad5T/41sSK in 5 of 5 biopsies (significance was reached in 3 biopsies with 4.3- to 576-fold differences, not in the remaining biopsies due to limited material). The strongest increase in transduction was observed for the biopsy that showed strongest EphA2 expression. Ad5TS/41sK-IJ-YSA showed higher transduction than Ad5T/41sSK in 4 of 4 biopsies, but was lower than for Ad5T/41sSK-IJ-YSA in 3 of 4 biopsies, which was in contrast to results in monolayer cultures. Finally, transduction of melanoma living tissue slices by Ad5KO1-HI-YSA was strongly increased compared with Ad5KO1 in 4 of 4 biopsies (2.1-fold to 17.1-fold, significant in 2 biopsies). These results confirm re-targeted adenoviral transduction mediated by the inserted YSA peptide for Ad5T/41sSK-IJ-YSA and Ad5KO1-HI-YSA also in clinically relevant tumor biopsy material.

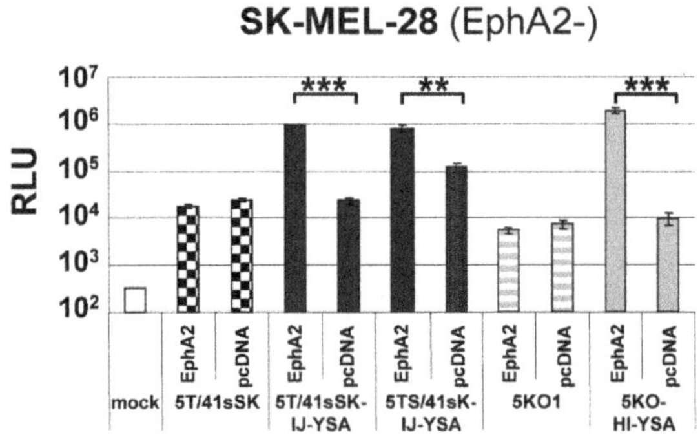

Figure 6. Transduction of EphA2-negative cells expressing recombinant EphA2 with YSA peptide-containing Ads. SK-MEL-28 cells transfected with EphA2 expression plasmid (pcDNA-EphA2) or control plasmid (pcDNA) were transduced with genomically fiber-modified Luc/GFP reporter viruses. Cells were transduced in quadruplicates, columns and error bars show mean values and standard deviations of Luc expression, respectively. RLU, relative luminescence units; ** $p<0.01$, ***$p<0.001$ for indicated comparisons. For detection of EphA2 expression of transfected cells see Fig. S3.

Discussion

In this study we establish cell type-specific Ad cell entry by functional peptide ligand insertion into a de-targeted chimeric fiber scaffold containing the short HAdV-41 fiber shaft and knob (Ad5T/41sSK). Moreover, we describe Ad vectors entry-targeted to the highly relevant pan-cancer surface marker EphA2 by genomic insertion of the gene encoding the chimeric fiber or a CAR binding-ablated HAdV-5 fiber with YSA peptide ligand. Transduction of EphA2-positive cells *in vitro* was dramatically increased by peptide ligand insertion for both fiber formats (up to 236-fold), underlining the potency of the YSA peptide for re-targeting of viral cell binding and entry. We previously identified the EG, HI and IJ loops of Ad5T/41sSK as sites for functional insertion of the RGD4C peptide, which binds to widely expressed integrins [14]. Here we confirmed the feasibility of these insertion sites using the EphA2-binding YSA peptide. In contrast to the RGD peptide, which showed superior activity in the HI loop, we observed similar activity in each of the three loops for the YSA peptide (Fig. 2), revealing that insertion sites affect peptide-receptor interactions differently dependent on the peptide. With the YSA peptide, we could demonstrate for the first time that the chimeric Ad5T/41sSK fiber facilitates cell type-specific Ad transduction using a panel of EphA2-positive and EphA2-negative cells (Figs. 2 and 5). Peptide competition and recombinant receptor expression studies clearly prove that transduction is peptide-specific and mediated by EphA2 (Figs. 3 and 6). These results establish a rationale for future studies that should explore whether further ligand peptides are functional in the Ad5T/41sSK fiber scaffold, enabling virus entry targeting via other surface markers.

We demonstrate that the introduction of the modified fibers into the virus genome, thereby replacing the native fiber, is feasible (Fig. 5) in addition to pseudotyping via the two step transfection/superinfection protocol, as also used in our previous study ([14], Figs. 2–4). We observed efficient fiber trimerization, virus production and fiber incorporation into virus particles for the genomically modified viruses with 5T/41sSK fibers (and KO1 fibers, see below). A previous study showed for genomically fiber-modified viruses expression, but no incorporation of HAdV-41 short fibers with peptide inserted into the AB, CD, or HI loops, or G region when a second HAdV-7 fiber was present [32]. Of note, both the specific insertion positions and the inserted peptides differed from those in our study. Furthermore, the fiber tail domain was derived from HAdV-41, which we previously reported to result in reduced fiber trimerization and incorporation when compared with the native HAdV-5 tail used by us [14]. Still, in the Schoggins study the short HAdV-41 fiber without peptide insertion was preferentially incorporated into viral particles in presence of the HAdV-7 fiber indicating that the peptide insertions affected fiber incorporation [32]. We conclude that the EG, HI and IJ loops of the HAdV-41 short fiber knob are recommended for insertion of ligand peptides, establishing a panel of fiber

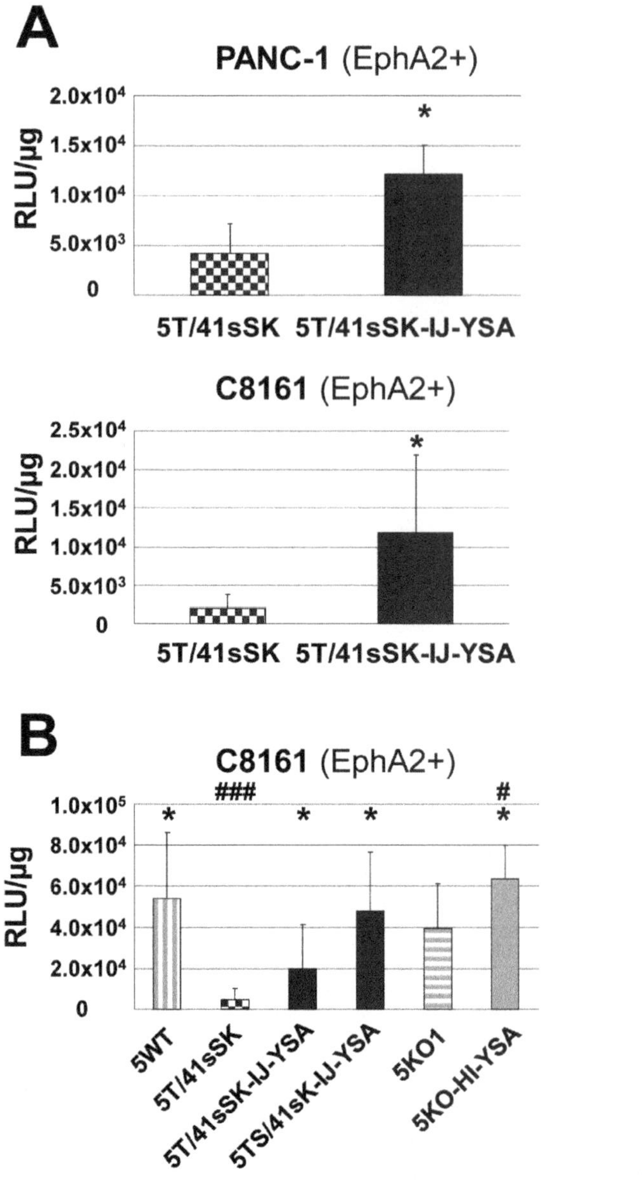

Figure 7. Transduction of EphA2-positive tumor xenografts *in vivo*. (**A,B**) 2×10^{10} vp of genomically fiber-modified, EphA2-targeted Luc/GFP reporter and control viruses were injected intratumorally into NOD-SCID mice carrying tumor xenografts (n = 6 to 9 tumors per group). Luciferase reporter gene expression in tumors was quantified 3 days after virus injection. Columns and error bars show mean values and standard deviations, respectively. RLU, relative luminescence units. * $p<0.05$ versus 5T/41sSK, $^{\#}p<0.05$ and $^{\#\#\#}p<0.001$ versus 5KO1.

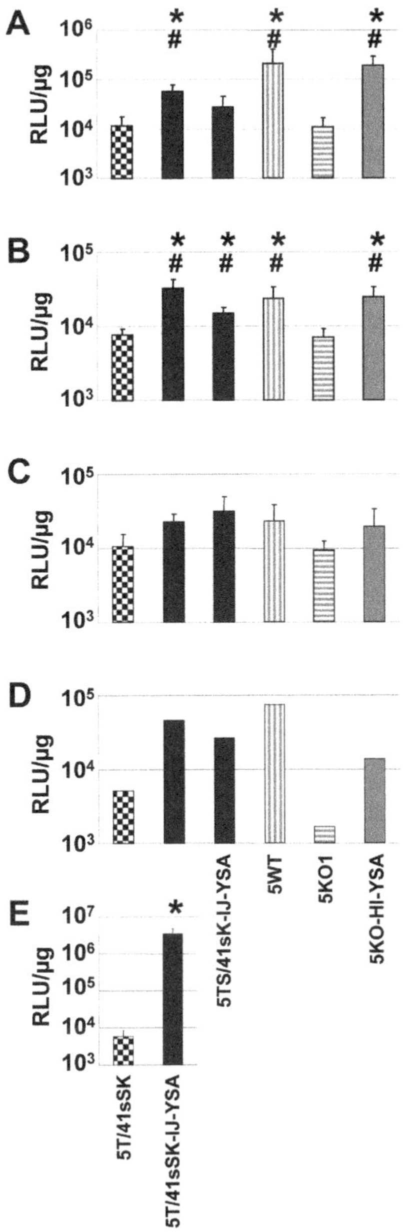

Figure 8. Transduction of living tissue slices of human melanoma metastases *ex vivo*. (A–E) Living tissue slices of melanoma metastases from 5 different patients were transduced with 10^{10} vp/slice of genomically fiber-modified, EphA2-targeted Luc/GFP reporter and control viruses. The number of transduced slices per virus was dependent on the size of the biopsy and was n = 4 (A,B), n = 3 (C), n = 2 (D) and n = 5 (E). Luciferase reporter gene expression was quantified 3 days post-transduction. Columns (A–E) and error bars (A,B,C,E) show mean values and standard deviations, respectively. RLU, relative luminescence units. $^{*/\#}p<0.05$ versus 5T/41sSK (*) or 5KO1 (#).

scaffolds for this purpose. However, fiber trimerization and incorporation into virus particles need to be evaluated individually for each ligand. The Schoggins paper also reports maturation and cell entry defects for virus particles containing the HAdV-41 short fibers with insertions in the G region. Note that these viruses differ from our viruses also by deletion of the penton RGD motif, which might contribute to virus cell entry defects. Furthermore, maturation defects were dependent on the sequence of the inserted peptide. Although we did not explicitly investigate virus particle maturation, we were successful in producing high titer viruses that show efficient fiber incorporation and transduction similar or even superior to matching viruses containing the HAdV-5 fiber (Fig. 5). These results argue that the Ad5T/41sSK fiber scaffold described here and in our previous study [14] is compatible with genomic fiber modification, facilitating the development of entry-targeted Ad vectors at high quantity and quality and of entry-targeted oncolytic Ads *per se*.

Our study further demonstrates that entry targeting is not enhanced by replacement of the short HAdV-41-derived shaft in the chimeric Ad5T/41sSK fiber with the long HAdV-5 shaft, a strategy we pursued hypothesizing that ligand-receptor interactions would be improved. In fact, we observed reduced fiber trimerization for chimeric viruses with long shaft (5TS/41sSK and 5TS*/41sSK constructs, Fig. 2), indicating that trimerization is hampered when fusing the HAdV-41 short fiber knob with the HAdV-5 shaft. As only trimeric fibers can be incorporated into viral particles, is it somewhat surprising that Ad5TS*/41sSK and Ad5TS/41sSK fibers with YSA peptide inserted into the IJ loop show normal fiber content in purified viruses (Figs. 2 and 5). This result shows that even low-level trimerization can be sufficient to ensure fiber incorporation. However, incorporation of fibers into viral particles was reduced or lost after YSA peptide insertion into the EG or HI loops, respectively (Fig. 2). Trimerization and incorporation deficiency of the Ad5TS/41sSK-HI-YSA fiber is in contrast to results for the HAdV-5 fiber, which is susceptible to YSA peptide insertion into the HI loop (Figs. 4, 5 and [53,54]). This difference can be explained by the different sequence, length and intramolecular environment of the HI loops in the HAdV-41 short fiber versus the HAdV-5 fiber. However, note that short shafted HAdV-41 fibers accepted YSA insertions into the HI loop without loss of trimerization and incorporation (Figs. 2 and 5), because of the above mentioned critical role of the shaft domain for fiber trimerization and incorporation into virus particles. Ad5TS/41sK viruses with YSA peptide inserted into the EG loop and especially into the IJ loop resulted in significant transduction of EphA2-negative cells. This property might be independent of the inserted ligand, as Nakamura et al. previously reported increased Ad transduction *in vitro* and *in vivo* for the HAdV-40 short fiber knob when fused to a long fiber shaft [16]. Possible explanations are direct cell binding via the shaft domain [26] or shaft length-dependent modification of cell binding properties of the knob. The latter was shown for CAR-binding knobs, which mediate strongly reduced adenoviral transduction when fused to a short shaft [16]. Finally, YSA-mediated viral transduction was lost irrespective of the peptide insertion site, when a HAdV-5 fiber shaft containing a mutation of the putative HSPG binding motif (KKTK to GAGA) was used (Fig. 2), which is in accord with previous studies [14,35–38] and is hypothesized to result from reduced shaft flexibility and/or defective post-entry virus trafficking.

The chimeric Ad5T/41sSK fiber represents a promising novel scaffold for peptide ligand insertion, especially considering its reported reduced biodistribution to the liver after i.v. injection [14,15,17–19]. However, its suitability for applications in the context of oncolytic Ads, which depend on full replication potency, remains to be demonstrated. For specific applications requiring tumor-specific activity but not liver-detargeting, we considered the CAR binding-ablated HAdV-5 fiber as candidate scaffold suitable for ligand insertion. We show in this study that functional YSA peptide insertion is feasible in the CD, FG and HI loops with the HI loop giving the best results (Fig. 4). However, the CD and FG insertion sites might still be suitable for other peptide ligands, since our data for the chimeric Ad5T/41sSK fiber show that insertion site preferences depend on the inserted peptide. Several previous studies showed functional peptide ligand insertion into the HI loop or fusion to the C-terminus of the HAdV-5 fiber [39]. Moreover, Ads with RGD peptide insertion into the HI loop, featuring enhanced infectivity, are being investigated in clinical gene therapy and virotherapy studies [30]. Internal insertion sites other than the HI loop have been rarely studied, but should be considered in the future. Transduction by Ad5KO-HI-YSA was slightly more effective and selective compared with Ad5T/41sSK-IJ-YSA *in vitro* (Figs. 4 and 5). Thus, both virus formats warrant further investigation in gene therapy and virotherapy applications.

In xenograft experiments, we found that i.t. virus injection resulted in strongly reduced tumor transduction for Ad5T/41sSK when compared with Ad5WT, while this was not the case for Ad5KO1. Of note, this was observed also for xenografts of CAR-negative C8161 cells (Fig. 7). These results point at CAR-independent tumor transduction *in vivo* mediated by HAdV-5 fibers, a conclusion also supported by a previous study reporting that CAR binding ablation does not de-target HAdV-5 vectors after i.t. injection [37]. Thus, we show that Ad5T/41sSK possesses superior de-targeting features after i.t. application, while CAR binding-ablated HAdV-5 fibers require further optimization for efficient de-targeting. Importantly, YSA peptide insertion into the Ad5T/41sSK fiber resulted in a significant increase in the transduction of EphA2-positive pancreatic cancer and melanoma xenografts, indicating peptide-mediated viral entry targeting *in vivo* (Fig. 7). Ad5KO-HI-YSA viruses showed a minor but significant increase in tumor transduction compared with Ad5KO1 (Fig. 7). Peptide-mediated increase in adenoviral transduction after i.t. injection was frequently shown for Ads with RGD peptide genetically inserted into the HAdV-5 HI loop [28,40].However, RGD-containing viruses also mediate increased transduction of healthy tissues, as the targeted integrins are widely expressed [29].Inconsistent results were reported for i.t. injection of viruses with other genetically inserted peptide ligands that mediate targeted transduction *in vitro*. Some studies reported lack of peptide mediated transduction after i.t. injection in the context of CAR-binding[41] or CAR binding-ablated HAdV-5 [37].Other studies reported increased transduction (3-fold) or selective oncolytic activity after i.t. injection of Ads with genetically inserted peptide in the HAdV-5 fiber binding or not binding CAR, respectively [42,43].

Our i.t. injection data show for the YSA peptide that the affinity was sufficient to mediate targeted viral cell entry *in vivo*. Future studies will need to investigate whether this strategy of genetic ligand peptide insertion into the Ad5T/41sSK fiber mediates targeted transduction after systemic adenovirus injection. Previous reports showed that peptide insertion into the HAdV-5 HI loop, even when mediating effective transduction *in vitro*, did not necessarily target virus transduction after systemic application [37,44].Thus, entry-targeted Ads or their delivery mode might require further improvement in order to overcome additional barriers to tumor homing after systemic injection. We believe that attempts to reduce the interaction with host factors that neutralize or sequester virus particles, such as antibodies and blood coagulation factors [12],or strategies to overcome anatomical barriers, such as vessel walls and tumor matrix [45–47] are warranted. Strategies that have been reported but remain to be investigated in combination with entry-targeted viruses include additional capsid modifications [48],virus shielding [49],increased tumor blood vessel permeabilization [46],carrier cells [50],matrix degradation [51],and cell junction openers [52].

EphA2 attracts increasing attention as a target for cancer therapy [22] and therefore represents a promising target for tumor-delivery of therapeutic viruses. This is because EphA2 is overexpressed in many tumors, including tumor stem cells, and functionally associated with tumor progression [22]. Importantly, a high affinity peptide ligand for EphA2 is available, the YSA peptide. Our study demonstrates functional genetic insertion of the YSA peptide into the Ad capsid and shows entry targeting *in vitro* and *in vivo* with best results for the chimeric Ad5T/41sSK fiber format. Two previous studies by van Geer *et al.* support the feasibility of the YSA peptide for entry targeting of Ad transduction *in vitro* [53,54]. In these studies, the YSA peptide was inserted into the HI loop of the HAdV-5 fiber either still able to bind CAR [54] or de-targeted from both CAR and integrins [53]. The latter virus, however, was strongly attenuated compared with a matching virus containing the wild-type HAdV-5 fiber. In contrast, Ad5KO1-HI-YSA reported in our study possesses similar or even superior activity to Ad5WT in the same EphA2-positive cell lines used by van Geer and colleagues (Figs. 4 and 5), indicating that either the KO1 backbone is superior or a secondary penton base-integrin interaction supports EphA2-mediated Ad cell entry. Van Geer *et al.* reported a lack of EphA2-mediated adenoviral tumor transduction *in vivo* when injecting their double-ablated YSA Ads i.v. or i.p. into mice with xenografts of pancreatic cancer cells [53]. However, they did not perform i.t. injections, which in our study were successful to demonstrate peptide-mediated entry targeting *in vivo*.

In conclusion, we developed a capsid-modified Ad vector, Ad5T/41sSK-IJ-YSA, targeted to cancer cells *in vitro* and *in vivo* by using the pan-cancer marker EphA2 as receptor for virus entry. Ad5T/41sSK-IJ-YSA offers interesting opportunities for applications in gene therapy and virotherapy of several cancers reported to overexpress EphA2. Moreover, the chimeric fiber scaffold of Ad5T/41sSK might be exploited for targeting other tumor markers by insertion of corresponding ligands.

Materials and Methods

Ethics statement

All animal experimental procedures were approved by the responsible Animal Protection Officer at the German Cancer Research Center and by the regional authorities (Regierungspräsidium Karlsruhe), according to the German Animal Protection Law. Biopsies of human melanoma metastases were obtained from the Department of Dermatology, University Hospital Heidelberg, Germany, with written informed consent from the donor according to the Declaration of Helsinki Principles and with approval of the local Ethics Committee (Ethikkommission Medizinische Fakultät Heidelberg, Votum S-337/2008).

Cell culture

Human cell lines 293 (Q-Biogene, Montreal, Canada), 293T (ATCC), 633 [55], A375M [56], C8161 [57], Capan-1 (ATCC), HepG2 (ATCC), HMEC-1 (Life Technologies, Darmstadt, Germany), Mel624 [58], Mel888 [58], MiaPaCa-2 (ATCC), PANC-1 (ATCC), HUVEC (PromoCell, Heidelberg, Germany) and low passage human melanoma cells purified from human melanoma metastasis biopsies, PMelL, were cultivated using standard cell culture conditions as described before [14].

Fiber expression plasmids

Fiber expression plasmids were cloned based on pDV67 (kindly provided by Glen Nemerow, La Jolla, CA [55];). Plasmids encoding chimeric fibers 5T/41sSK and 5TS*/41sK with cloning site for ligand insertion were previously described [14]. 5TS/41sK constructs were obtained by re-mutation. The sequence encoding the YSAYPDSVPMMS-peptide (Ref [20]) was inserted by oligonucleotide cloning (see Materials and Methods S1 for listing of oligonucleotides). The pKO1 expression plasmid encoding a CAR binding-ablated HAdV-5 fiber (S408E and P409A mutations, ref [31]) was generated by site-directed mutagenesis. A GGSGG linker containing a BamHI-site was inserted into pKO1 by site-directed mutagenesis between (i) G450/T451 in the CD-loop; (ii) K468/S469 in the EG-loop; (iii) T542/G543 in the HI-loop; (iv) G560/H561 in the IJ-loop or (v) I564/N565, also in the IJ-loop. The YSA-peptide was inserted by oligonucleotide cloning as described above. All fiber expression plasmids were sequenced.

Transient transfection assays

For analysis of fiber trimerization, 5.0×10^5 293T cells per well were seeded in 3 ml of medium containing 2% FBS in 6-well plates. Cells were transfected the next day with 5 µg of fiber expression plasmid per well, using Lipofectamine (Life Technologies, Darmstadt, Germany) according to the manufacturer's instructions. 72 h after transfection, cells were washed in PBS and lysed in 100 µl of 100 mM Tris by gently rocking for 1 h at 4°C. 3 µl of 5 M NaCl were added, cell debris was pelleted and supernatants were analyzed by immunoblot. For transient EphA2 expression, 2.5×10^4 SK-MEL-28 cells per well were seeded in 48-well plates. The next day cells were transfected in quadruplicates with 1.5 µg per well of EphA2 expression plasmid pcDNA4-EphA2mychis [59] or of control plasmid without insert using Lipofectamine. After 24 h cells were transduced with Ads and processed for reporter gene analysis. In parallel, cell lysates of transfected cells were analyzed for EphA2 expression by immunoblotting.

Immunoblotting

30 µg boiled or unboiled lysates from cell cultures, living tissue slices, or homogenized tumors or 5.0×10^9 boiled viral particles were separated by SDS-PAGE and transferred to a nitrocellulose membrane (Schleicher und Schuell, Dassel, Germany). Membranes were probed with a mouse monoclonal antibody binding to the HAdV-5 fiber tail (clone 4D2; Abcam, Cambridge, UK) or with a mouse monoclonal antibody specific for EphA2 (clone D7; Millipore, Darmstadt, Germany). As loading control, membranes were probed with a mouse monoclonal antibody specific for β-actin (clone AC-74; Sigma, Deisenhofen, Germany) or with rabbit monoclonal antibody binding to HAdV-5 hexon (ab24240; Abcam). Bound antibodies were detected with an anti-mouse-HRP (#7076S; Cell Signaling, Danvers, MA, USA) or with an anti-rabbit-HRP (#7074; Cell Signaling) secondary antibody followed by enhanced chemiluminescence (Thermo Scientific, Schwerte, Germany).

Generation of LacZ reporter Ads pseudotyped with modified fibers by a two step transfection/superinfection protocol

Genomically fiberless LacZ reporter viruses were phenotypically equipped (pseudotyped) with a fiber by transient transfection of 293T cells with the respective fiber expression plasmid followed by superinfection with Ad5.βGal.ΔF (kindly provided by AdVec Inc., Ontario, Canada and Glen Nemerow, La Jolla, CA) as described by Hesse *et al.* [14] Viruses were harvested 48 h post-infection and purified by CsCl equilibrium density gradient ultracentrifugation. Genomic integrity was verified by PCR, restriction digests, and sequencing. Medium virus yields were between 2000 and 8900 vp/cell for viruses with HAdV-41 short fiber knob and between 900 and 1400 vp/cell for viruses with HAdV-5 KO1 mutant fiber knob.

Generation of genomically fiber modified reporter Ads expressing GFP and luciferase by BAC-recombineering

For a detailed protocol of BAC recombineering see Refs [60] and [61], which our protocol was adapted from. First, the genome of an E1/E3-deleted GFP and luciferase reporter virus (QBiogene, Irvine, CA, USA) was provided with the BAC-backbone pAd5-FRT (kindly provided by Zsolt Ruzsics, Munich, Germany [62],). The latter was amplified by PCR using primers with a 5′-overhang of 30 nucleotides corresponding to both ends of the viral genome to allow homologous recombination in SW102 bacteria (kindly provided by Stephen P. Creekmore, NCI, Frederick, MD). Second, also by homologous recombination, the fiber gene coding sequence was replaced by the sequence of a GalK/Amp selection cassette which was amplified from plasmid pT GalK-Amp (kindly provided by Stephen P. Creekmore) using primers with a 5′-overhang of 30 nucleotides corresponding to the flanking position in the Ad genome were the fiber gene should be removed. Third, again by homologous recombination, the cassette was exchanged by the fiber sequence of interest amplified from the cloned expression plasmids. Details about the cloning procedures can be obtained upon request. Ad genomes were released from BACs by Pac I digest and were transfected into 293 cells using Lipofectamine. Viral particles were amplified on 293 cells. Viruses were harvested 48 h post-infection and purified by CsCl equilibrium density gradient ultracentrifugation. Genomic integrity was verified by PCR, restriction digests, and sequencing. Medium virus yields were between 9700 and 16300 vp/cell for all viruses.

Virus quantification

Physical particle concentrations (vp/ml) of virus preparations were determined by OD_{260} determination and were used for transduction studies, which is standard procedure when comparing viruses with different capsids. Infectious virus titers described as 50% tissue culture infectious dose ($TCID_{50}$) were additionally determined for viruses with genomic modification to ensure the quality of virus preparations. Similar vp-to-$TCID_{50}$ ratios confirmed that the quality of virus preparations was not reduced by insertion of peptide ligand into the chimeric or KO1 fibers (Ad5wt: 13; Ad5T/41sSK: 58; Ad5T/41sSK-IJ-YSA: 57; Ad5TS/41sK-IJ-YSA: 53; Ad5KO1: 653; Ad5KO1-HI-YSA: 439).

Adenoviral gene transfer assays

Cells were seeded in 48-well plates at a concentration of 2.5×10^4 cells per well in 0.5 ml of medium containing 2% FBS. The following day, cells were pre-cooled by incubating the plates on ice for 20 min prior to transduction. Medium was removed from cell layers and the cells were transduced in triplicates or quadruplicates with Ads in 150 µl pre-cooled media containing 2% FBS. Transductions were performed using vp titers, as viruses compared in individual experiments possessed different capsid modifications that affect biological titers. For viruses generated by the two step transfection/superinfection protocol, 1×10^5 vp/cell were used. For the genomically modified viruses 100 vp/cell (PANC-1, C8161, HMEC) or 200 vp/cell (Capan-1, MIA Paca-2, SK-MEL-28, HUVEC, HepG2) were used. After incubation on ice for 75 min, viruses were removed, cells were washed and cultivated in media containing 2% FBS. Cell-virus incubations were performed on ice to reduce non-specific viral uptake into cells under experimental conditions were high virus titers were used.

Cells were harvested for reporter assay at 2 days post infection. Experiments were repeated at least once with the same virus preparation and were confirmed with an independent virus preparation. For peptide competition experiments with soluble 12mer YSA peptide or control peptide with randomized sequence (both PSL, Heidelberg, Germany) cells were preincubated for 20 min prior to transduction on ice with viruses in peptide-containing media.

Reporter gene assays

Quantification of β-Gal activity was done using the Tropix kit (Life Technologies) according to the manufacturer's instructions. 50 µl lysate from each well was measured for 5 s using a Bertold luminometer. For quantification of Luc activity, cells were lysed in 150 µl cell culture lysis buffer (Promega, Mannheim, Germany) and frozen once. Living tissue slices and tumor homogenate were lysed in cell culture lysis buffer followed by three cycles of freezing and thawing. After centrifugation, the protein concentration in the supernatant was determined using the Bio-Rad DC™ Protein Assay kit (München, Germany). Luciferase activity was quantified using 50 µl lysate and 5 sec measurement in a Berthold luminometer using luciferin substrate solution [63].

In vivo gene transfer

5×10^6 C8161 or PANC-1 cells were implanted s.c. into both flanks of 6- to 8-week-old female NOD/SCID mice (in-house breeding, German Cancer Research Center). On an average volume of 200 mm^3, 2×10^{10} vp in 100 µl PBS were injected i.t. into both tumors. After 72 h, mice were sacrificed by cervical dislocation and tumors were extracted, snap frozen, and homogenized with a Micro Dismembrator.

Living tissue slices

Tissues were sliced as previously described [64]. Slices were transduced with 1.0×10^{10} vp/slice for 75 min and kept at 37°C in a humidified atmosphere of 5% CO_2 for 72 h.

Statistical analysis

Differences between indicated groups were analyzed using the Student's *t* test. P values of <0.05 were considered statistically significant.

Supporting Information

Figure S1 EphA2 expression in a panel of pancreatic cancer, melanoma, hepatoma, and endothelial cells. Detection of EphA2 expression in cell cultures by immunoblot. β-actin was used as loading control. Longer exposure is shown for upper right panel. Pancreatic cancer cell lines: PANC-1, Capan-1, and MIA Paca-2; melanoma cells: C8161, Mel888, PMelL, A375M, SK-MEL-28, and Mel624; endothelial cells: HUVEC (primary), and HMEC (immortalized); hepatoma cell line: HepG2.

Figure S2 EphA2-targeted Ads with genomically inserted fibers containing the YSA peptide: transduction efficiency. Transduction of EphA2-positive PANC-1 and C8161 cells and of EphA2-negative SK-MEL-28 cells with genomically fiber-modified Luc/GFP reporter viruses at 1000 vp/cell. GFP expression was visualized 48 h post-transduction (4x magnification).

Figure S3 Transduction of EphA2-negative cells expressing recombinant EphA2 with YSA peptide-containing Ads: Detection of EphA2 expression. EphA2 expression in SK-MEL-28 cells of the experiment depicted in Fig. 6, as detected by immunoblot. Cells were transfected with EphA2 expression plasmid (pcDNA-EphA2) or control plasmid (pcDNA). Lysate of C8161 cells was used as positive control. β-actin served as loading control.

Figure S4 Transduction of EphA2-positive tumor xenografts *in vivo*: Detection of EphA2 expression. Detection of EphA2 expression in tumor xenografts of the experiments shown in the indicated subfigures of Fig. 7 by immunoblot. Corresponding cell cultures and β-actin served as control.

Figure S5 Transduction of living tissue slices of human melanoma metastases *ex vivo*. Living tissue slices were transduced with 10^{10} vp/slice of genomically fiber-modified, EphA2-targeted Luc/GFP reporter and control viruses. (**A**) Detection of EphA2 expression in living tissue slices of the experiments shown in Fig. 8 by immunoblot. Lysate of C8161 cells was used as positive control and β-actin served as loading control. (**B**) Lower panels show GFP-expression of representative slices from the experiment presented in Fig. 8A at 3 days post-transduction (4x magnification). Upper panels show bright field pictures of the same slices.

Acknowledgments

We thank S.P. Creekmore (Frederick, MD) for providing SW102 bacteria and pT GalK-Amp plasmid, C. Cziepluch (Heidelberg, Germany) for providing pancreatic cancer cell lines, E. Bautz (Heidelberg, Germany) for providing HepG2 cells, G. Fey (Erlangen, Germany) for providing 293T cells, I. Fidler (Houston, TX) for providing A375M cells, C. Garlichs (Erlangen, Germany) for providing HUVEC cells, G. Nemerow (La Jolla, CA) for providing 633 cells, and pDV67, G. Nemerow and AdVec Inc. (Hamilton, Ontario, Canada) for providing Ad5.βGal.ΔF, Zsolt Ruzsics (Munich, Germany) for providing pAd5-FRT, J. Schlom (Bethesda, MD) for providing Mel624 and Mel888 cells, and D. Welch (Birmingham, AL) for providing C8161 cells. We acknowledge H. Eskerski (Heidelberg, Germany) for technical support.

Author Contributions

Conceived and designed the experiments: MB JK PK DN. Performed the experiments: MB JK PK SE MH. Analyzed the data: MB JK PK MH AE DN. Contributed reagents/materials/analysis tools: PO GP FN JH AE AHE. Wrote the paper: MB JK DN.

References

1. Lasaro MO, Ertl HC (2009) New insights on adenovirus as vaccine vectors. Mol Ther 17: 1333–1339.
2. Yamamoto M, Curiel DT (2010) Current issues and future directions of oncolytic adenoviruses. Mol Ther 18: 243–250.
3. Dorer DE, Nettelbeck DM (2009) Targeting cancer by transcriptional control in cancer gene therapy and viral oncolysis. Adv Drug Deliv Rev 61: 554–571.
4. Sakurai F, Katayama K, Mizuguchi H (2011) MicroRNA-regulated transgene expression systems for gene therapy and virotherapy. Front Biosci 16: 2389–2401.
5. Bergelson JM, Cunningham JA, Droguett G, Kurt-Jones EA, Krithivas A, et al. (1997) Isolation of a common receptor for Coxsackie B viruses and adenoviruses 2 and 5. Science 275: 1320–1323.

6. Wickham TJ, Mathias P, Cheresh DA, Nemerow GR (1993) Integrins alpha v beta 3 and alpha v beta 5 promote adenovirus internalization but not virus attachment. Cell 73: 309–319.
7. Nicklin SA, Wu E, Nemerow GR, Baker AH (2005) The influence of adenovirus fiber structure and function on vector development for gene therapy. Mol Ther 12: 384–393.
8. Shayakhmetov DM, Eberly AM, Li ZY, Lieber A (2005) Deletion of penton RGD motifs affects the efficiency of both the internalization and the endosome escape of viral particles containing adenovirus serotype 5 or 35 fiber knobs. J Virol 79: 1053–1061.
9. Parker AL, Waddington SN, Nicol CG, Shayakhmetov DM, Buckley SM, et al. (2006) Multiple vitamin K-dependent coagulation zymogens promote adenovirus-mediated gene delivery to hepatocytes. Blood 108: 2554–2561.
10. Shayakhmetov DM, Gaggar A, Ni S, Li ZY, Lieber A (2005) Adenovirus binding to blood factors results in liver cell infection and hepatotoxicity. J Virol 79: 7478–7491.
11. Waddington SN, McVey JH, Bhella D, Parker AL, Barker K, et al. (2008) Adenovirus serotype 5 hexon mediates liver gene transfer. Cell 132: 397–409.
12. Duffy MR, Bradshaw AC, Parker AL, McVey JH, Baker AH (2011) A cluster of basic amino acids in the factor X serine protease mediates surface attachment of adenovirus/FX complexes. J Virol 85: 10914–10919.
13. Xu Z, Qiu Q, Tian J, Smith JS, Conenello GM, et al. (2013) Coagulation factor X shields adenovirus type 5 from attack by natural antibodies and complement. Nat Med 19: 452–457.
14. Hesse A, Kosmides D, Kontermann RE, Nettelbeck DM (2007) Tropism modification of adenovirus vectors by peptide ligand insertion into various positions of the adenovirus serotype 41 short-fiber knob domain. J Virol 81: 2688–2699.
15. Kashentseva EA, Douglas JT, Zinn KR, Curiel DT, Dmitriev IP (2009) Targeting of adenovirus serotype 5 pseudotyped with short fiber from serotype 41 to c-erbB2-positive cells using bispecific single-chain diabody. J Mol Biol 388: 443–461.
16. Nakamura T, Sato K, Hamada H (2003) Reduction of natural adenovirus tropism to the liver by both ablation of fiber-coxsackievirus and adenovirus receptor interaction and use of replaceable short fiber. J Virol 77: 2512–2521.
17. Nicol CG, Graham D, Miller WH, White SJ, Smith TA, et al. (2004) Effect of adenovirus serotype 5 fiber and penton modifications on in vivo tropism in rats. Mol Ther 10: 344–354.
18. Schoggins JW, Gall JG, Falck-Pedersen E (2003) Subgroup B and F fiber chimeras eliminate normal adenovirus type 5 vector transduction in vitro and in vivo. J Virol 77: 1039–1048.
19. Schoggins JW, Nociari M, Philpott N, Falck-Pedersen E (2005) Influence of fiber detargeting on adenovirus-mediated innate and adaptive immune activation. J Virol 79: 11627–11637.
20. Koolpe M, Dail M, Pasquale EB (2002) An ephrin mimetic peptide that selectively targets the EphA2 receptor. J Biol Chem 277: 46974–46979.
21. Binda E, Visioli A, Giani F, Lamorte G, Copetti M, et al. (2012) The EphA2 receptor drives self-renewal and tumorigenicity in stem-like tumor-propagating cells from human glioblastomas. Cancer Cell 22: 765–780.
22. Wykosky J, Debinski W (2008) The EphA2 receptor and ephrinA1 ligand in solid tumors: function and therapeutic targeting. Mol Cancer Res 6: 1795–1806.
23. Chow KK, Naik S, Kakarla S, Brawley VS, Shaffer DR, et al. (2013) T cells redirected to EphA2 for the immunotherapy of glioblastoma. Mol Ther 21: 629–637.
24. Tandon M, Vemula SV, Mittal SK (2011) Emerging strategies for EphA2 receptor targeting for cancer therapeutics. Expert Opin Ther Targets 15: 31–51.
25. Smith TA, Idamakanti N, Marshall-Neff J, Rollence ML, Wright P, et al. (2003) Receptor interactions involved in adenoviral-mediated gene delivery after systemic administration in non-human primates. Hum Gene Ther 14: 1595–1604.
26. Smith TA, Idamakanti N, Rollence ML, Marshall-Neff J, Kim J, et al. (2003) Adenovirus serotype 5 fiber shaft influences in vivo gene transfer in mice. Hum Gene Ther 14: 777–787.
27. Dmitriev I, Krasnykh V, Miller CR, Wang M, Kashentseva E, et al. (1998) An adenovirus vector with genetically modified fibers demonstrates expanded tropism via utilization of a coxsackievirus and adenovirus receptor-independent cell entry mechanism. J Virol 72: 9706–9713.
28. Kanerva A, Wang M, Bauerschmitz GJ, Lam JT, Desmond RA, et al. (2002) Gene transfer to ovarian cancer versus normal tissues with fiber-modified adenoviruses. Mol Ther 5: 695–704.
29. Reynolds P, Dmitriev I, Curiel D (1999) Insertion of an RGD motif into the HI loop of adenovirus fiber protein alters the distribution of transgene expression of the systemically administered vector. Gene Ther 6: 1336–1339.
30. Russell SJ, Peng KW, Bell JC (2012) Oncolytic virotherapy. Nat Biotechnol 30: 658–670.
31. Jakubczak JL, Rollence ML, Stewart DA, Jafari JD, Von Seggern DJ, et al. (2001) Adenovirus type 5 viral particles pseudotyped with mutagenized fiber proteins show diminished infectivity of coxsackie B-adenovirus receptor-bearing cells. J Virol 75: 2972–2981.
32. Schoggins JW, Falck-Pedersen E (2009) Serotype 5 Adenovirus fiber (F7F41S) chimeric vectors incur packaging deficiencies when targeting peptides are inserted into Ad41 short fiber. Virology 395: 10–20.
33. Carson SD, Hobbs JT, Tracy SM, Chapman NM (1999) Expression of the coxsackievirus and adenovirus receptor in cultured human umbilical vein endothelial cells: regulation in response to cell density. J Virol 73: 7077–7079.
34. Toyoda E, Doi R, Kami K, Mori T, Ito D, et al. (2008) Adenovirus vectors with chimeric type 5 and 35 fiber proteins exhibit enhanced transfection of human pancreatic cancer cells. Int J Oncol 33: 1141–1147.
35. Bayo-Puxan N, Cascallo M, Gros A, Huch M, Fillat C, et al. (2006) Role of the putative heparan sulfate glycosaminoglycan-binding site of the adenovirus type 5 fiber shaft on liver detargeting and knob-mediated retargeting. J Gen Virol 87: 2487–2495.
36. Kritz AB, Nicol CG, Dishart KL, Nelson R, Holbeck S, et al. (2007) Adenovirus 5 fibers mutated at the putative HSPG-binding site show restricted retargeting with targeting peptides in the HI loop. Mol Ther 15: 741–749.
37. Rittner K, Schreiber V, Erbs P, Lusky M (2007) Targeting of adenovirus vectors carrying a tumor cell-specific peptide: in vitro and in vivo studies. Cancer Gene Ther 14: 509–518.
38. Wu E, Pache L, Von Seggern DJ, Mullen TM, Mikyas Y, et al. (2003) Flexibility of the adenovirus fiber is required for efficient receptor interaction. J Virol 77: 7225–7235.
39. Glasgow JN, Everts M, Curiel DT (2006) Transductional targeting of adenovirus vectors for gene therapy. Cancer Gene Ther 13: 830–844.
40. Mizuguchi H, Hayakawa T (2002) Enhanced antitumor effect and reduced vector dissemination with fiber-modified adenovirus vectors expressing herpes simplex virus thymidine kinase. Cancer Gene Ther 9: 236–242.
41. MacLeod SH, Elgadi MM, Bossi G, Sankar U, Pisio A, et al. (2012) HER3 targeting of adenovirus by fiber modification increases infection of breast cancer cells in vitro, but not following intratumoral injection in mice. Cancer Gene Ther 19: 888–898.
42. Miura Y, Yamasaki S, Davydova J, Brown E, Aoki K, et al. (2013) Infectivity-selective oncolytic adenovirus developed by high-throughput screening of adenovirus-formatted library. Mol Ther 21: 139–148.
43. Nishimoto T, Yoshida K, Miura Y, Kobayashi A, Hara H, et al. (2009) Oncolytic virus therapy for pancreatic cancer using the adenovirus library displaying random peptides on the fiber knob. Gene Ther 16: 669–680.
44. Wu P, Kudrolli TA, Chowdhury WH, Liu MM, Rodriguez R, et al. (2010) Adenovirus targeting to prostate-specific membrane antigen through virus-displayed, semirandom peptide library screening. Cancer Res 70: 9549–9553.
45. Li ZY, Ni S, Yang X, Kiviat N, Lieber A (2004) Xenograft models for liver metastasis: Relationship between tumor morphology and adenovirus vector transduction. Mol Ther 9: 650–657.
46. Seki T, Carroll F, Illingworth S, Green N, Cawood R, et al. (2011) Tumour necrosis factor-alpha increases extravasation of virus particles into tumour tissue by activating the Rho A/Rho kinase pathway. J Control Release 156: 381–389.
47. Shayakhmetov DM, Li ZY, Ni S, Lieber A (2002) Targeting of adenovirus vectors to tumor cells does not enable efficient transduction of breast cancer metastases. Cancer Res 62: 1063–1068.
48. Short JJ, Rivera AA, Wu H, Walter MR, Yamamoto M, et al. (2010) Substitution of adenovirus serotype 3 hexon onto a serotype 5 oncolytic adenovirus reduces factor X binding, decreases liver tropism, and improves antitumor efficacy. Mol Cancer Ther 9: 2536–2544.
49. Kim J, Kim PH, Kim SW, Yun CO (2012) Enhancing the therapeutic efficacy of adenovirus in combination with biomaterials. Biomaterials 33: 1838–1850.
50. Willmon C, Harrington K, Kottke T, Prestwich R, Melcher A, et al. (2009) Cell carriers for oncolytic viruses: Fed Ex for cancer therapy. Mol Ther 17: 1667–1676.
51. Kim MH, Bodenstine TM, Sumerel LA, Rivera AA, Baker AH, et al. (2006) Tissue inhibitor of metalloproteinases-2 improves antitumor efficacy of a replicating adenovirus in vivo. Cancer Biol Ther 5: 1647–1653.
52. Beyer I, van Rensburg R, Strauss R, Li Z, Wang H, et al. (2011) Epithelial junction opener JO-1 improves monoclonal antibody therapy of cancer. Cancer Res 71: 7080–7090.
53. van Geer MA, Bakker CT, Koizumi N, Mizuguchi H, Wesseling JG, et al. (2009) Ephrin A2 receptor targeting does not increase adenoviral pancreatic cancer transduction in vivo. World J Gastroenterol 15: 2754–2762.
54. Van Geer MA, Brevoord D, Kuhlmann KF, Bakker CT, Mizuguchi H, et al. (2010) A fiber modified adenovirus vector that targets to the EphrinA2 receptor reveals enhanced gene transfer to ex vivo pancreatic cancer. Int J Oncol 36: 233–244.
55. Von Seggern DJ, Huang S, Fleck SK, Stevenson SC, Nemerow GR (2000) Adenovirus vector pseudotyping in fiber-expressing cell lines: improved transduction of Epstein-Barr virus-transformed B cells. J Virol 74: 354–362.
56. Hendrix MJ, Seftor EA, Seftor RE, Fidler IJ (1987) A simple quantitative assay for studying the invasive potential of high and low human metastatic variants. Cancer Lett 38: 137–147.
57. Welch DR, Bisi JE, Miller BE, Conaway D, Seftor EA, et al. (1991) Characterization of a highly invasive and spontaneously metastatic human malignant melanoma cell line. In J Cancer 47: 227–237.
58. Weber J, Salgaller M, Samid D, Johnson B, Herlyn M, et al. (1994). Expression of the MAGE-1 Tumor Antigen Is Up-Regulated by the Demethylating Agent 5-Aza-2′-Deoxycytidine. Cancer Res 54 1766–1771.
59. Hahn AS, Kaufmann JK, Wies E, Naschberger E, Panteleev-Ivlev J, et al. (2012) The ephrin receptor tyrosine kinase A2 is a cellular receptor for Kaposi's sarcoma-associated herpesvirus. Nat Med 18: 961–966.

60. Warming S, Costantino N, Court DL, Jenkins NA, Copeland NG (2005) Simple and highly efficient BAC recombineering using galK selection. Nucleic Acids Res 33: e36.
61. Ruzsics Z, Wagner M, Osterlehner A, Cook J, Koszinowski U, et al. (2006) Transposon-assisted cloning and traceless mutagenesis of adenoviruses: Development of a novel vector based on species D. J Virol 80: 8100–8113.
62. Wodrich H, Henaff D, Jammart B, Segura-Morales C, Seelmeir S, et al. (2010) A capsid-encoded PPxY-motif facilitates adenovirus entry. PLoS Pathog 6: e1000808.
63. Ketzer P, Haas SF, Engelhardt S, Hartig JS, Nettelbeck DM (2012) Synthetic riboswitches for external regulation of genes transferred by replication-deficient and oncolytic adenoviruses. Nucleic Acids Res 40: e167.
64. Kaufmann JK, Bossow S, Grossardt C, Sawall S, Kupsch J, et al. (2013) Chemovirotherapy of malignant melanoma with a targeted and armed oncolytic measles virus. J Invest Dermatol 133: 1034–1042.

16

Allelic Combinations of Soybean Maturity Loci *E1, E2, E3* and *E4* Result in Diversity of Maturity and Adaptation to Different Latitudes

Bingjun Jiang[1]ᶿ, Haiyang Nan[2,3]ᶿ, Youfei Gao[1]ᶿ, Lili Tang[2], Yanlei Yue[1], Sijia Lu[2,3], Liming Ma[1], Dong Cao[2], Shi Sun[1], Jialin Wang[2], Cunxiang Wu[1], Xiaohui Yuan[2], Wensheng Hou[1], Fanjiang Kong[2]*, Tianfu Han[1]*, Baohui Liu[2]*

1 The National Key Facility for Crop Gene Resources and Genetic Improvement and MOA Key Lab of Soybean Biology (Beijing), Institute of Crop Sciences, The Chinese Academy of Agricultural Sciences, Beijing, China, **2** The Key Laboratory of Soybean Molecular Design Breeding, Northeast Institute of Geography and Agroecology, Chinese Academy of Sciences, Harbin, China, **3** University of Chinese Academy of Sciences, Beijing, China

Abstract

Soybean cultivars are extremely diverse in time to flowering and maturation as a result of various photoperiod sensitivities. The underlying molecular genetic mechanism is not fully clear, however, four maturity loci *E1, E2, E3* and *E4* have been molecularly identified. In this report, cultivars were selected with various photoperiod sensitivities from different ecological zones, which covered almost all maturity groups (MG) from MG 000 to MG VIII and MG X adapted from latitude N 18° to N 53°. They were planted in the field under natural daylength condition (ND) in Beijing, China or in pots under different photoperiod treatments. Maturity-related traits were then investigated. The four *E* maturity loci were genotyped at the molecular level. Our results suggested that these four *E* genes have different impacts on maturity and their allelic variations and combinations determine the diversification of soybean maturity and adaptation to different latitudes. The genetic mechanisms underlying photoperiod sensitivity and adaptation in wild soybean seemed unique from those in cultivated soybean. The allelic combinations and functional molecular markers for the four *E* loci will significantly assist molecular breeding towards high productivity.

Editor: Fan Chen, Institute of Genetics and Developmental Biology, Chinese Academy of Sciences, China

Funding: This work was supported by the National Natural Science Foundation of China (31201232, 31071445, 31171579, 31201222 and 31371643); Open Foundation of Key Laboratory of Soybean Molecular Design Breeding, Chinese Academy of Sciences; "Hundred Talents" Program of Chinese Academy of Sciences; Strategic Action Plan for Science and Technology Innovation of Chinese Academy of Sciences (XDA08030100); Heilongjiang Natural Science Foundation of China (ZD201001, JC201313); China Agriculture Research System (CARS-04) and Chinese Academy of Agricultural Sciences (CAAS) Innovation Project. The funders had no role in study design, data collection and analysis, decision to publish, or preparation of the manuscript.

Competing Interests: The authors have declared that no competing interests exist.

* Email: kongfj@iga.ac.cn (FK); hantianfu@caas.cn (TH); liubh@iga.ac.cn (BL)

ᶿ These authors contributed equally to this work.

Introduction

Soybean (*Glycine max* (L.) Merrill) is a short-day crop with high protein and oil contents. Many cultivars were bred with different maturity to adapt to various ecological environments. For the convenience of breeding layout, 13 MGs from MG000 to MGX were classified in North America [1–3]. Chinese soybean researchers divided cultivars into different maturity groups as well [4–5]. Soybean production has increased in America in response to recent increases in global demand and maturity is the key factor determining soybean productivity. Therefore, it is rather important to understand the mechanism of soybean maturity diversity and adaptation.

Flowering and maturity were highly controlled by major genes in soybean. Up to now, nine maturity loci have been identified as *E1–E8* and *J* [1], [6–12]. These loci have different roles under different photoperiods. Wang et al [13] found that long daylength condition (LD) might strengthen while short daylength condition (SD) might weaken these maturity loci. More results and progress of maturity genes was reviewed by Xia et al [14]. Furthermore, four loci were identified at molecular level. *E1* gene was identified as a transcription factor which functions as a flowering repressor with a putative nuclear localization signal and a B3-related domain [15]. *E2* is an orthologue of Arabidopsis flowering gene *GIGANTEA* [16]. *E3* and *E4* are phytochrome genes *GmPhyA3* [17] and *GmPhyA2* [18], respectively. In addition, two homologs of soybean *Flowering Locus T* (*FT*) genes, *GmFT2A* and *GmFT5A* were identified and coordinately regulate flowering [19]. Four identified maturity genes *E1*, *E2*, *E3* and *E4* delay flowering and maturity under LD through down regulating *GmFT2A* and *GmFT5A* [15], [16], [19]. As for other loci, more studies should be done before learning their molecular identities.

Although these four known loci *E1*, *E2*, *E3* and *E4* provide an important key to learn the mechanism of flowering and maturity,

Table 1. Cultivars from North America and their respective maturity group.

Code	PI number	Variety	MG	Code	PI number	Variety	MG
CAAE001	PI548594	Maple Presto	000	**CAAE021**	PI534646	Flyer	IV
CAAE002	PI567787	OAC Vision	000	**CAAE022**	PI598222	TN4-94	IV
CAAE003	PI548648	Canatto	00	**CAAE023**	PI564849	Nathan	V
CAAE005	PI592523	Glacier	00	**CAAE024**	PI572239	Holladay	V
CAAE006	PI629004	MN0201	0	**CAAE025**	PI633609	Lonoke	V
CAAE007	PI596541	Traill	0	**CAAE026**	PI561400	Rhodes	V
CAAE008	PI612764	MN0901	0	**CAAE027**	PI633610	Desha	VI
CAAE009	PI599300	Surge	NA	**CAAE028**	PI592756	Dillon	VI
CAAE010	PI548641	Haroson	I	**CAAE029**	PI617045	NC-Roy	VI
CAAE011	PI614833	NE1900	I	**CAAE030**	PI599333	Musen	VI
CAAE012	PI608438	Titan	I	**CAAE031**	PI531068	Stonewall	VII
CAAE013	PI561858	Holt	II	**CAAE032**	PI595645	Benning	VII
CAAE014	PI567786	OAC Talbot	II	**CAAE033**	PI617041	Santee	VII
CAAE015	PI533655	Burlison	II	**CAAE034**	PI555453	Hagood	VII
CAAE016	PI595926	Athow	III	**CAAE035**	PI603953	Motte	VIII
CAAE017	PI548634	Zane	III	**CAAE036**	PI548970	Foster	VIII
CAAE018	PI593258	Macon	III	**CAAE037**	PI548663	Dowling	VIII
CAAE019	PI578057	Saline	III	**CAAE038**		Jupiter	X
CAAE020	PI614155	NS93-4118	IV				

NA, not available.

Table 2. Cultivars and accessions from China and far-east Russia and their adaption latitudes.

Code	Variety	Latitude	Code	Variety	Latitude
CAAE039	Dengke 2	N 47°–53°	**CAAE064**	Fengchengzaochadou	N 25°–30°
CAAE040	Huajiang 4	N 48°–50°	**CAAE065**	Jin 6606	N 41°–43°
CAAE041	Heihe 27	N 46°–48°	**CAAE066**	Jinzhou 8–14	N 40°–43°
CAAE042	Heihe 3	N 46°–50°	**CAAE071**	Bahong 1	N 39°–42°
CAAE043	Heihe 43	N 46°–48°	**CAAE072**	Mianyanghuangwofeng	N 30°
CAAE044	Suinong 14	N 44°–48°	**CAAE073**	Lüpidou	N 25°–28°
CAAE045	Hefeng 25	N 39°–43°	**CAAE074**	Liuyuezao	N 25°–28°
CAAE047	Jilin 3	N 44°–46°	**CAAE075**	Ruijinxiaohuangdou	N 24°–27°
CAAE049	Jiunong 21	N 42°–46°	**CAAE076**	Edou 2	N 30°–35°
CAAE050	Jilin 30	N 41°–43°	**CAAE077**	Yulindahuangdou	N 23°
CAAE052	Jindou 19	N 35°–40°	**CAAE081**	Qiudou 1	N 25°–30°
CAAE053	Tiefeng 31	N 35°–40°	**CAAE082**	Jiangledaqingdou	N 27°
CAAE054	Jidou 12	N 35°–38°	**CAAE084**	Guixia 1	N 22°–26°
CAAE055	Qihuang 28	N 34°–37°	**CAAE085**	Nandou 12	N 29°–32°
CAAE056	Zhonghuang 13	N 30°–40°	**CAAE086**	Zigongdongdou	N 29°
CAAE057	Xudou 9	N 32°–35°	**CAAE087***	Heiheyesheng	N 50°
CAAE058	Xudou 1	N 32°–35°	**CAAE088***	Bayanyesheng	N 46°
CAAE059	Fengshouhuang	N 35°–38°	**CAAE089***	Baiyangdianyesheng	N 39°
CAAE061	Dandou 2	N 40°–42°	**CAAE090***	Guangxiyesheng	N 23°
CAAE062	Yuejin 4	N 35°–38°	**CAAE091**	Mohe 1	N 53°
CAAE063	Jinda 814	NA	**CAAE092**	Ziweicika 4/75	N 50°

CAAE092 is a cultivar from far east Russia; *indicates wild soybeans: CAAE087, CAAE088, CAAE089 and CAAE090.

we mainly got knowledge based on few cultivars but not on a population level. Population-level knowledge will provide another different view of these four loci's role on maturity and adaptation. Therefore, in this study we selected a set of soybean cultivars which cover 12 maturity groups from MG000 to MGVIII and MGX plus some cultivars with wide range of latitude from N 18° to N 53°. These cultivars were subjected to different photoperiod treatments. Traits of beginning bloom (R1), physiological maturity (R7) and full maturity (R8) were investigated [20]. Maturity loci *E1*, *E2*, *E3* and *E4* were genotyped in the population. Further association analysis was done. The results showed that allelic combinations of these four *E* genes significantly determine the ecological-economical adaption of cultivars although they have different impacts on maturity. In addition, the genetic mechanisms underlying photoperiod sensitivity and adaptation in wild soybean seemed unique from those in cultivated soybean.

Materials and Methods

Soybean cultivars were selected from North America (Table 1), China and Russia (Table 2) [21–23]. Four wild soybean accessions were also included for genotyping only (Table 2), which are CAAE087 (Heiheyesheng) collected in Heihe (N 50°22′, E 127°53′), Heilongjiang, China; CAAE088 (Bayanyesheng) in Bayan (N 46°08′, E 127°39′), Heilongjiang, China; CAAE089 (Baiyangdianyesheng) in Baoding (N 38°51′, E 115°30′), Hebei, China; and CAAE090 (Guangxiyesheng) in Nanning (N 22°48′, E 108°19′), Guangxi, China. They covered 12 maturity groups from MG 000 to MG VIII and MG X and ranged from N 18° to N 53° indicated in Table 1 and Table 2.

Due to seed availability, 59 cultivars were selected for both field and pot experiments, 12 cultivars only in the field experiment, and 4 cultivars only in the pot experiment. For the field experiment, seeds were sowed on May 14th, 2012 in Beijing (N 39°97′, E 116°34′) and maturity-related traits of R1, R7 and R8 were recorded regularly [20]. For the pot experiment, seeds were sowed in 10-liter pots on May 16th 2012 and grown under ND in Beijing. After emergence, the seedlings were thinned until each pot contained five uniform plants. These uniform plants were grown until the unifoliate expanded then treated with different photoperiods (LD, 16 h light/8 h dark; SD, 12 h light/12 h dark; and ND). For SD, the plants were transferred to dark room to shorten the daylength. For LD, incandescent bulbs (50 μmol m^{-2} s^{-2} at the top of plants) with automatic timer controls were used to extend the daylength. Additional details of plant growth and treatments were the same as reported by Wu et al [24]. The days to first flowering of each plant was recorded. Both experiments finished on Oct 15th, 2012. Photoperiod sensitivity (PS) was thus calculated as the following function, where DFF_{LD} is the days to first flowering (R1) from the expansion of the first pair of unifoliates (V1) under LD while DFF_{SD} under SD [25].

$$\mathrm{PS}(\%) = (DFF_{LD} - DFF_{SD})/DFF_{LD} \times 100\%$$

Genomic DNA was isolated from soybean unifoliate leaves using TianGen New Plant Genomic DNA Isolation Kit (DP320). Maturity loci *E1*, *E2*, *E3* and *E4* were genotyped using functional allele specific molecular markers [26].

Table 3. The days from VE to R1, R7 and R8 of soybean Cultivars under ND condition at Beijing.

Code	Phenotype			Code	Phenotype		
	VE-R1	VE-R7	VE-R8		VE-R1	VE-R7	VE-R8
CAAE001	20.7±1.7	61.0±3.8	70.9±1.8	**CAAE038**	122.3±1.0	NA	NA
CAAE002	20.0±1.8	63.9±3.4	71.5±1.9	**CAAE039**	20.4±0.8	61.5±4.8	71.7±2.0
CAAE003	20.1±0.4	57.2±1.0	68.0±0.0	**CAAE040**	19.0±0.0	67.0±6.7	76.7±3.4
CAAE005	20.3±1.8	68.4±1.6	73.7±1.0	**CAAE041**	20.1±0.4	69.7±0.5	76.7±1.3
CAAE006	19.9±1.3	69.3±3.0	77.0±2.2	**CAAE042**	20.4±1.5	71.3±1.8	78.3±2.1
CAAE007	20.3±2.0	70.1±2.2	77±2.2	**CAAE043**	19.1±0.5	71.0±2.2	81.3±6.5
CAAE008	22.4±1.6	93.4±10.6	108.5±15	**CAAE044**	21.1±0.7	71.3±2.4	80.7±6.3
CAAE009	21.9±1.2	86.9±2.1	100.7±3.6	**CAAE045**	20.2±0.4	71.3±1.8	78.3±2.1
CAAE010	22.5±1.4	85.3±2.0	90.9±1.6	**CAAE047**	23.1±2.3	83.9±1.3	94.4±4.2
CAAE011	25.8±1.9	81.1±11.1	103.9±5.9	**CAAE049**	23.4±1.2	83.9±9.9	101.4±7.0
CAAE012	21.9±1.4	89.7±1.6	101.3±2.6	**CAAE050**	24.9±1.1	98.6±5.2	109.9±7.3
CAAE013	22.7±1.4	94.2±1.8	109.9±4.7	**CAAE052**	24.5±0.8	99.9±2.8	114.2±2.4
CAAE014	21.6±0.5	92.5±3.0	100.3±3.1	**CAAE053**	23.1±0.3	113.6±3.0	121.7±2.2
CAAE015	25.2±1.9	102.1±7.3	113.9±5.3	**CAAE054**	37.4±0.6	119.8±6.6	127.3±7.0
CAAE016	24.9±1.0	106.3±7.0	118.5±3.5	**CAAE055**	42.7±2.3	127.0±8.0	136.8±5.1
CAAE017	26.9±1.8	117.1±2.3	125.5±1.5	**CAAE056**	36.4±2.0	113.7±6.3	121.1±2.7
CAAE018	24.7±0.6	117.9±1.8	124.9±3.3	**CAAE057**	31.3±1.0	111.4±2.5	119.9±1.4
CAAE019	28.6±1.2	123.4±2.7	131.6±3.3	**CAAE058**	24.2±1.0	95.9±2.8	106.2±5.2
CAAE020	24.2±1.2	116.4±2.8	125.7±1.9	**CAAE059**	25.5±0.5	111.2±3.1	118.7±2.6
CAAE021	26.3±1.3	123.4±3.3	133.4±2.9	**CAAE063**	41.9±2.1	116.0±2.0	122.0±0.0
CAAE022	32.5±4.4	125.1±4.1	134.6±1.4	**CAAE064**	41.3±1.0	101.9±3.7	111.3±4.7
CAAE023	56.1±3.9	131.9±7.8	139.3±2.4	**CAAE065**	27.1±1.2	103.9±9.1	115.7±5.1
CAAE024	50.6±2.6	132.9±6.1	141.0±3.9	**CAAE066**	42.5±1.5	116.0±3.2	126.8±6.3
CAAE025	53.7±1.1	140.6±4.5	142±0.0[†]	**CAAE071**	49.1±4.0	132.6±6.4	140.0±5.6
CAAE026	54.9±1.9	141.5±4.4	143.2±1.8	**CAAE072**	91.3±2.7	141.0±1.4	NA
CAAE027	61.8±4.6	142.5±4.9	142.0±0.0[†]	**CAAE073**	54.2±1.1	140.0±4.7	144.0±2.8[†]
CAAE028	63.3±5.5	141.8±4.3	144.0±1.7[†]	**CAAE074**	69.0±0.0	125.4±4.6	133.5±4.6
CAAE029	59.0±5.2	142.0±3.1	145.0±0.0[†]	**CAAE075**	94.0±2.1	NA	NA
CAAE030	75.1±2.0	141.9±3.0	145.0±0.0[†]	**CAAE076**	65.5±5.5	143.8±1.6	NA
CAAE031	77.7±3.3	142.8±1.1	NA	**CAAE077**	71.8±9.8	NA	NA
CAAE032	70.8±3.2	NA	NA	**CAAE081**	NA	NA	NA
CAAE033	70.4±3.1	145.0±0.0	NA	**CAAE082**	96.8±1.8	NA	NA
CAAE034	77.7±1.6	NA	NA	**CAAE084**	98.3±2.9	NA	NA
CAAE035	75.9±4.4	145.0±0.0	NA	**CAAE085**	95.1±2.6	NA	NA

Table 3. Cont.

Code	Phenotype			Code	Phenotype		
	VE-R1	VE-R7	VE-R8		VE-R1	VE-R7	VE-R8
CAAE036	74.3±4.5	NA	NA	CAAE086	NA	NA	NA
CAAE037	69.6±4.6	146.1±0.9	NA				

†Partial plants (<50%) matured. NA, not available.

Ethics Statement

No specific permissions were required for domestic research of the collections of wild soybean accessions in Heihe (N 50°22′, E 127°53′), Bayan (N 46°08′, E 127°39′), Baoding (N 38°51′, E 115°30′) and Nanning (N 22°48′, E 108°19′), China. All the field studies did not involve endangered or protected species.

Results

Soybean cultivars have diverse flowering and maturity dates

Seventy-one cultivars were planted under ND in the field at Beijing. These cultivars showed rich diversity in maturity (Table 3). Two cultivars CAAE081 and CAAE086 failed to flower until the experiment ended. Nine cultivars CAAE032, CAAE034, CAAE036, CAAE038, CAAE075, CAAE077, CAAE082, CAAE084 and CAAE085, flowered but did not reach R7. Six cultivars CAAE031, CAAE033, CAAE035, CAAE037, CAAE072 and CAAE076 did not reach R8 although they flowered and podded. For other cultivars that regularly flowered and matured, the days to R1 from emergence (VE) ranged from 19.0 (CAAE040) to 75.1 (CAAE030), the days to R7 ranged from 57.2 (CAAE003) to 142.5 (CAAE027), and the days to R8 ranged from 68.0 (CAAE003) to 145.0 (CAAE029 and CAAE030). Thus, the range of time to R1, R7 and R8 showed the maturity diversity of these cultivars. The cultivars that reached R1 less than 50 days after emergence could mature before the frost in Beijing, those that reached R1 in greater than 50 days but less than 70 days could mature partially, and the ones that reached R1 after 70 days could hardly mature (Table 3). In Figure 1 and Figure 2, maturity-related traits of R1, R7 and R8 generally increased from early MG to late MG and from high latitude to low latitude.

Soybean cultivars react variously to photoperiod treatments

Sixty-three cultivars were planted in pots. They were treated with different photoperiods after V1. They began flowering in 18.4 to 32.8 days after emergence under SD, in 20.0 to 122.3 days under ND, and in 20.5 to 113.7 under LD, while under LD three cultivars CAAE031 (PI531068, MGVII), CAAE084 (Qiudou 1), and CAAE086 (Zigongdongdou) did not flower (Table 4). The PS was calculated according to the equation [25], which ranged between 10% and 80%. For the three cultivars that did not flower, it could be set at 100%. Thus, CAAE035 (PI603953, MGVIII), CAAE072 (Mianyanghuangwofeng), CAAE085 (Nandou 12), CAAE075 (Ruijinxiaohuangdou) and the former three cultivars are most sensitive to photoperiod (PS>75%). CAAE001 (PI548594, MG000), CAAE002 (PI567787, MG000), CAAE005 (PI592523, MG00), CAAE007 (PI596541, MG0), CAAE042 (Heihe 3) and CAAE091 (Mohe 1) could be classified as photoperiod insensitive because their photoperiod sensitivities are lower than 10%. These data suggested that these soybean cultivars diversify significantly in photoperiod sensitivity. Moreover, as shown in Figure 2, the photoperiod sensitivity of these cultivars generally increased from early MG to late MG and from high to low latitude.

Genotyping soybean cultivars of maturity loci *E1*, *E2*, *E3* and *E4*

Eighty-five cultivars were genotyped at four maturity loci *E1*, *E2*, *E3* and *E4* (Table 4). Thirty-eight cultivars are from North America, which cover 12 maturity groups from MG000 to MGVIII and MGX. Other cultivars are from China except

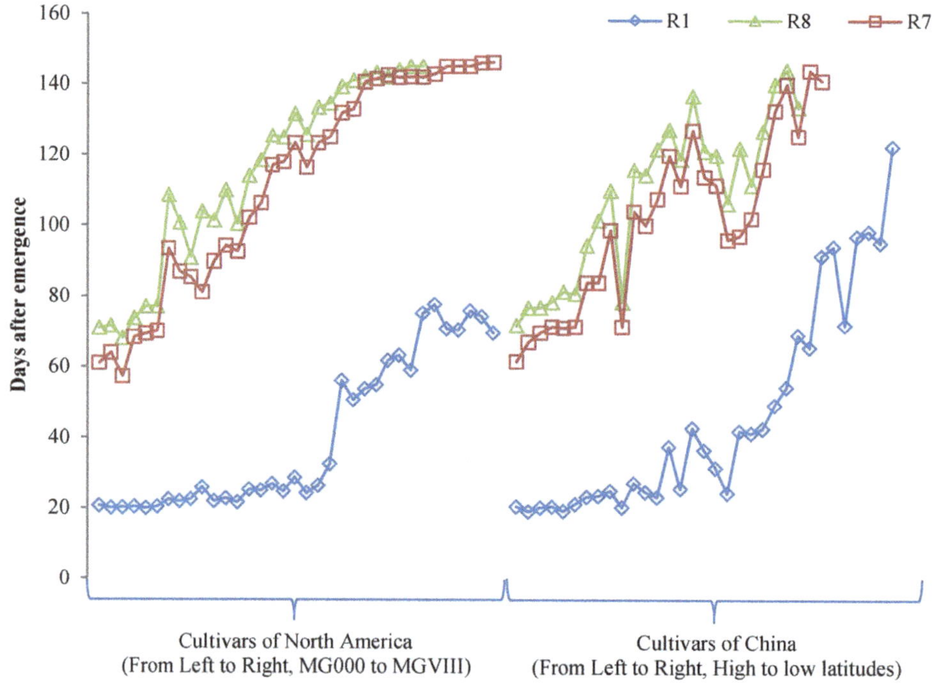

Figure 1. Maturity-related traits of soybean R1, R7 and R8 in the field at Beijing. Left, cultivars of North America, which are approximately sorted by maturity group. Right, cultivars of China, which are sorted roughly by adaption latitude.

CAAE092, which is from Russia (Tables 1 and 2). There are ten genotypes in total in this population (Table 5). The genotypes of *E1/E2/E3/E4* and *E1/e2/E3/E4* are the majority types, which were identified in 28 and 19 cultivars respectively (Table 5). Three genotypes of *e1-as/E2/e3/E4*, *e1-as/e2/e3/e4* and *e1/e2/e3/E4* were identified only in one variety each (Table 5). For the *E1* locus, allele *e1* and *el-as* are always detected in early-maturing cultivars from MG000 to MGIV or from high latitudes adapted cultivars. Moreover, all of the four wild soybeans were *E1/E2/E3/E4* type.

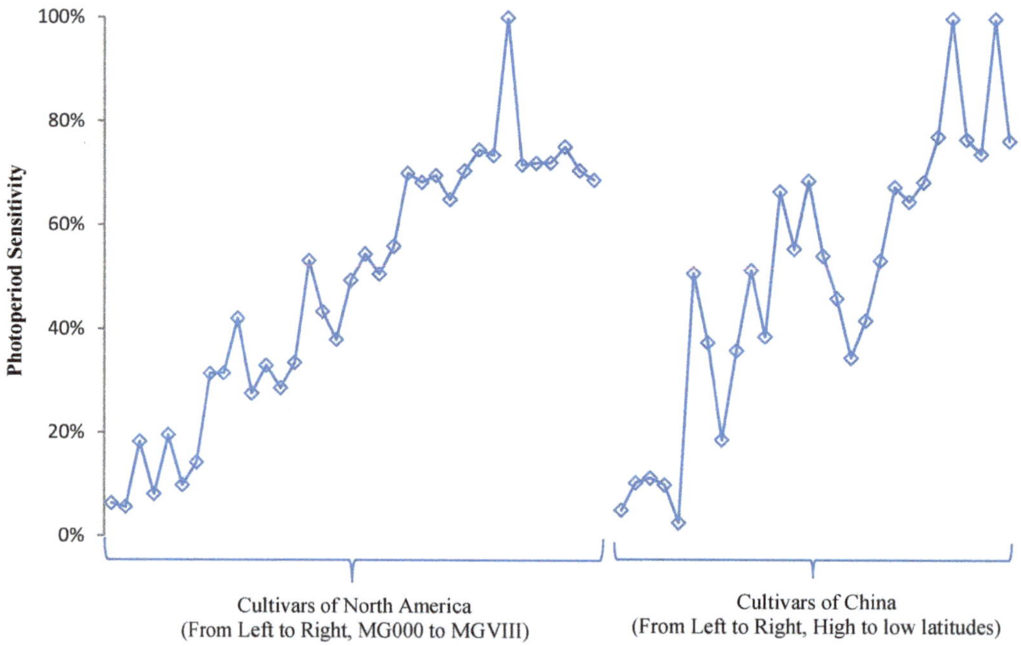

Figure 2. Photoperiod sensitivity of soybean cultivars. Left, cultivars of North America, which are approximately sorted by maturity group. Right: cultivars of China, which are sorted roughly by adaption latitude.

Table 4. The days to R1 from V1 under different photoperiod treatments and the resultant PS at Beijing.

Cultivar	Days from V1 to R1			PS (%)	Cultivar	Days from V1 to R1			PS (%)
	SD	ND	LD			SD	ND	LD	
CAAE001	21.4±1.1	20.4±0.7	22.9±2.4	6.3	**CAAE035**	26.9±1.3	83.2±1.1	108.2±8.9	75.2
CAAE002	22.6±1.1	20.9±1.4	23.9±2.2	5.6	**CAAE036**	26.7±2.4	83.9±1.4	90.6±0.5	70.6
CAAE003	21.4±1.5	24.7±1.7	26.2±3.3	18.2	**CAAE037**	27.6±0.7	73.9±3.1	88.4±0.5	68.8
CAAE005	21.4±2.0	20.0±0.6	23.2±0.9	8.1	**CAAE050**	23.1±0.9	30.2±1.4	37.1±5.0	37.6
CAAE006	21.7±2.1	22.0±2.5	27.0±5.4	19.5	**CAAE053**	23.1±1.7	28.9±2.2	37.8±6.5	38.8
CAAE007	20.6±1.3	20.1±0.6	22.9±1.7	9.8	**CAAE052**	22.9±0.8	31.9±1.9	47.2±3.8	51.5
CAAE008	24.1±3.2	25.5±0.9	28.1±4.1	14.2	**CAAE054**	24.1±1.1	44.1±1.8	54.3±3.0	55.6
CAAE009	22.6±2.7	29.8±1.9	32.9±3.4	31.4	**CAAE057**	27.0±0.8	45.0±2.7	50.2±1.9	46.2
CAAE010	22.8±1.7	28.5±2.1	33.2±1.6	31.4	**CAAE056**	28.0±1.7	49.5±1.4	61.3±3.2	54.3
CAAE011	24.8±1.6	32.5±1.5	42.7±3.4	42.0	**CAAE061**	26.7±3.1	50.3±4.6	54.3±7.1	50.9
CAAE012	25.1±3.0	27.9±2.3	34.6±3.1	27.5	**CAAE076**	27.3±0.8	73.3±1.5	86.9±0.9	68.5
CAAE013	22.2±2.3	28.6±2.1	33.1±2.5	33.0	**CAAE064**	32.8±2.0	45.9±2.4	56.5±2.8	41.9
CAAE014	24.8±1.9	28.7±2.6	34.7±4.7	28.5	**CAAE059**	24.1±1.4	44.8±1.6	77.4±2.0	68.8
CAAE015	26.7±2.1	31.2±2.6	40.1±3.7	33.5	**CAAE084**	30.8±1.9	106.4±0.7	NA	100.0
CAAE016	22.0±1.9	31.2±3.2	46.9±5.2	53.1	**CAAE045**	24.3±2.1	27.6±1.5	30.0±2.6	18.9
CAAE017	26.2±2.1	35.6±6.0	46.3±0.5	43.4	**CAAE041**	21.2±1.7	22.8±0.9	23.6±1.3	10.2
CAAE018	25.1±0.4	31.5±0.8	40.5±1.6	38.0	**CAAE042**	23.0±1.7	24.3±1.4	23.7±1.7	2.8
CAAE019	25.3±2.1	39.4±1.0	49.9±3.1	49.3	**CAAE040**	20.2±1.5	20.4±0.5	22.8±1.6	11.5
CAAE020	21.3±1.6	34.1±6.0	46.7±4.2	54.4	**CAAE065**	26.4±1.4	34.1±2.9	41.3±2.7	36.0
CAAE021	23.7±2.7	36.7±3.1	47.9±3.6	50.5	**CAAE066**	26.5±0.5	48.6±1.3	56.9±1.9	53.4
CAAE022	23.2±2.2	45.2±2.5	52.5±5.0	55.9	**CAAE074**	32.2±1.0	62.9±0.3	91.5±0.9	64.8
CAAE024	25.0±0.8	60.2±1.2	83.3±0.6	70.0	**CAAE073**	26.6±1.2	63.8±5.1	82.0±0.7	67.6
CAAE025	27.9±1.4	67.1±1.4	87.9±1.1	68.3	**CAAE072**	28.2±1.8	92.7±3.3	124.4±0.8	77.3
CAAE026	26.4±1.3	67.2±0.8	86.9±0.8	69.6	**CAAE091**	20.4±2.8	22.9±0.7	21.6±1.2	5.2
CAAE027	29.5±1.0	67.0±0.5	84.0±0.0	64.9	**CAAE085**	29.8±1.7	98.6±1.3	126.8±1.7	76.5
CAAE028	26.3±2.0	68.1±1.3	89.0±2.3	70.5	**CAAE075**	28.9±1.4	98.3±1.4	124.5±1.2	76.8
CAAE029	24.1±0.8	82.1±3.3	94.7±0.5	74.5	**CAAE058**	23.9±1.4	30.9±1.9	36.6±7.0	34.7
CAAE030	25.0±0.8	83.8±1.4	94.1±0.6	73.4	**CAAE077**	29.6±2.8	96.4±0.5	113.7±1.0	74.0
CAAE031	24.7±0.8	83.6±0.5	NA	100.0	**CAAE062**	27.6±0.9	55.3±2.3	83.0±1.4	66.7
CAAE032	25.1±0.7	72.6±1.3	88.6±1.1	71.7	**CAAE092**	18.4±2.0	20.9±1.6	20.5±1.8	10.6
CAAE033	24.9±0.4	73.9±2.2	88.9±1.9	72.0	**CAAE086**	32.8±0.8	122.3±0.8	NA	100.0
CAAE034	24.9±0.6	84.6±1.1	89.0±1.7	72.0					

NA, not available.

Table 5. Genotype of soybean cultivars at four maturity loci *E1*, *E2*, *E3* and *E4*.

Genotype				Num	Variety
E1	E2	E3	E4		
E1	*E2*	*E3*	*E4*	28	CAAE023 (MGV), CAAE024 (MGV), CAAE025 (MGV), CAAE026 (MGV), CAAE027 (MGVI), CAAE028 (MGVI), CAAE029 (MGVI), CAAE030 (MGVI), CAAE031 (MGVII), CAAE032 (MGVII), CAAE033 (MGVII), CAAE034 (MGVII), CAAE035 (MG VIII), CAAE036 (MG VIII), CAAE037 (MG VIII), CAAE038 (MGX), CAAE071, CAAE072, CAAE075, CAAE081, CAAE082, CAAE084, CAAE085, CAAE086, CAAE087, CAAE088, CAAE089, CAAE090
E1	*e2*	*E3*	*E4*	19	CAAE047, CAAE049, CAAE050, CAAE052, CAAE054, CAAE055, CAAE056, CAAE057, CAAE058, CAAE061, CAAE062, CAAE063, CAAE064, CAAE065, CAAE066, CAAE073, CAAE074, CAAE076, CAAE077
E1	*e2*	*e3*	*E4*	3	CAAE044, CAAE045, CAAE059
e1-as	*E2*	*E3*	*E4*	8	CAAE016 (MGIII), CAAE017 (MGIII), CAAE018 (MGIII), CAAE019 (MGIII), CAAE020 (MGIV), CAAE021 (MGIV), CAAE022 (MGIV), CAAE053
e1-as	*E2*	*e3*	*E4*	1	CAAE011 (MGI)
e1-as	*e2*	*E3*	*E4*	8	CAAE003 (MG00), CAAE008 (MG0), CAAE009, CAAE010 (MGI), CAAE012 (MGI), CAAE013 (MGII), CAAE014 (MGII), CAAE015 (MGII)
e1-as	*e2*	*e3*	*E4*	7	CAAE005 (MG00), CAAE006 (MG0), CAAE007 (MG0), CAAE040, CAAE041, CAAE042, CAAE043
e1-as	*e2*	*e3*	*e4*	1	CAAE039
e1	*e2*	*e3*	*E4*	1	CAAE092
e1	*e2*	*e3*	*e4*	3	CAAE001 (MG000), CAAE002 (MG000), CAAE091

Maturity loci *E1, E2, E3* and *E4* have different impacts on maturity and photoperiod response

In general, recessive alleles *e1*, *el-as*, *e2*, *e3* and *e4* promoted flowering and maturity but with different impacts (Figure 3). The allele *e4* was detected in only four cultivars (CAAE039 with the genotype *e1-as/e2/e3/e4*, and CAAE001, CAAE002 and CAAE091 with the genotype of *e1/e2/e3/e4*) in the population. The four cultivars with recessive *e4* alleles were adapted to high latitude and showed photoperiod insensitivity suggesting the importance of the *e4* allele for high latitude adaptation. The cultivars with the allele *e1*, *e1-as* or *e2* exhibited a narrower range of the days from VE to R1 (VE-R1) than that of the days from VE to R7 (VE-R7) and that of the days from R1 to R7 (R1–R7) (Figure 3). In contrast, the cultivars with the allele *e3* or *e4* showed a consistently narrow range although some outliers existed (Figure 3). Moreover, these four recessive alleles promoted flowering under different photoperiod conditions (Figure 4), and the cultivars with more recessive alleles of *e1*, *e1-as*, *e2*, *e3* and *e4* had a lower PS during photoperiod treatments (Figure 5).

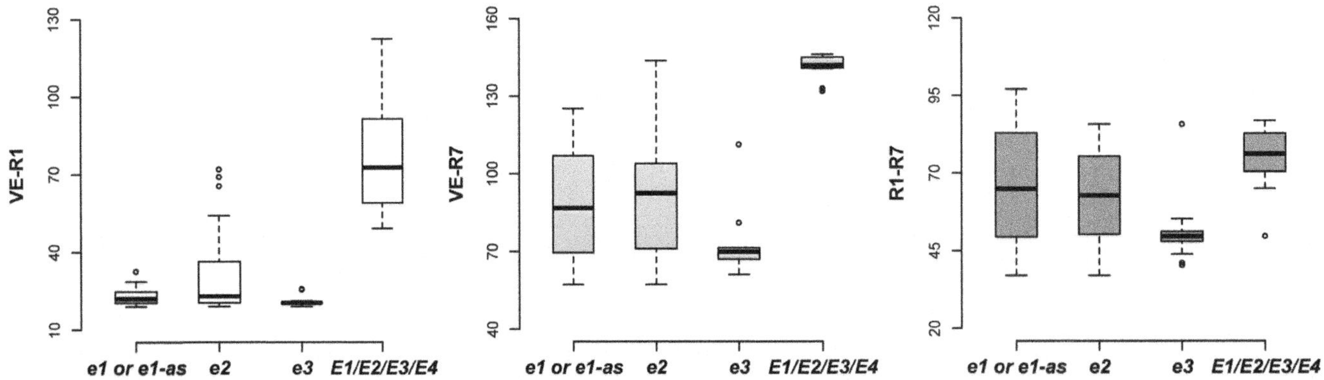

Figure 3. Quartile box plots showing days between the stages of VE, R1 and R7. Circles show outliers.

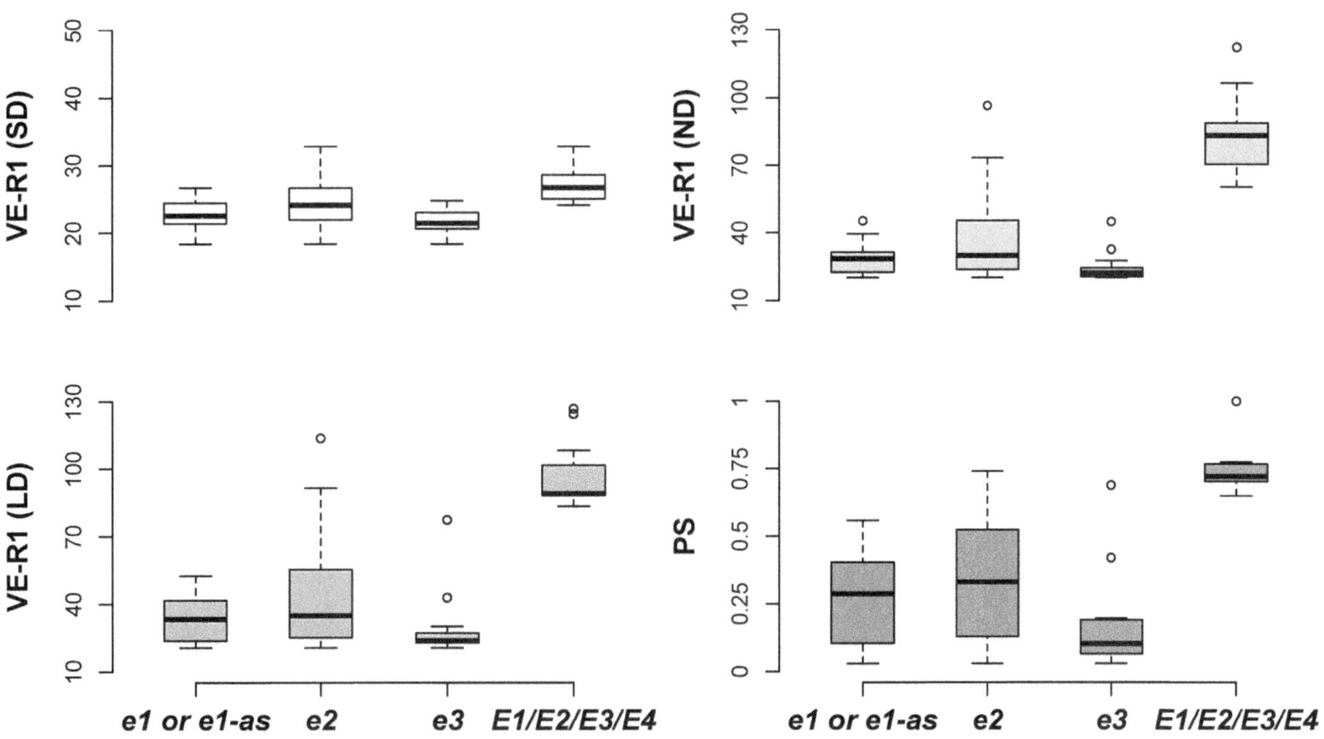

Figure 4. Quartile box plots showing days between the stages of VE and R1 under different photoperiod conditions and PS. Circles show outliers.

Discussion

Soybean cultivars from different maturity groups show diversity in flowering, maturity and photoperiod sensitivity

The tested soybean cultivars were selected from North America, China and Russia. Some of them covered from MG000 to MGVIII and MGX, almost all of the total 13 MG [2] and the others were collected from N 18° to N 53° to cover the wide range of latitude, which represents the main soybean producing area in China. In addition, some wild soybean accessions were also included. Thus, the population of lines used here should exhibit the diversity of maturity not only in phenotype but also in genotype. The field experiment under ND provided strong evidence. Some cultivars failed to flower (R1), some could not reach pod yellowing (R7) and some could not reach full maturity (R8) (Table 3). Even for those cultivars that flowered and matured, the days to first flowering, physiological maturity and full maturity varied significantly (Table 3). What's more, in the experiment of photoperiod treatments, the population also showed diversity of photoperiod sensitivity as expected. The days to first flowering varied between different photoperiod treatments and between different cultivars. The photoperiod sensitivity also varied from 10% to 80% (Table 4). Six cultivars were classified as photoperiod insensitive: CAAE001 (PI548594, MG000, *e1/e2/e3/e4*), CAAE002 (PI567787, MG000, *e1/e2/e3/e4*), CAAE005 (PI592523, MG00, *e1-as/e2/e3/E4*), CAAE007 (PI596541, MG0, *e1-as/e2/e3/E4*), CAAE042 (Heihe 3, *e1-as/e2/e3/E4*) and CAAE091 (Mohe 1, *e1/e2/e3/e4*). These results were consistent with previous report that soybean photoperiod insensitivity was at least conditioned by three genetic mechanisms according to allelic combinations of *E1*, *E2* and *E4*: *e3/e4*; *e1/e3 or e1/e4* and *e1-as/e3/E4*. In the genetic mechanism of *e1-as/e3/E4*, novel unidentified gene/genes participated in photoperiod insensitivity [26], [27].

E genes have different impacts on flowering and maturation

Soybean, as a short-day crop, has many cultivars with diversified maturity structure. *E1*, *E2* and *E3* are involved with different impacts. In the field experiment under ND, the recessive allele *e1*, *el-as* and *e2* significantly narrowed the variation of VE-R1 more than that of VE-R7 and R1–R7 (Figure 3), suggesting that *E1* and *E2* genes have significant impact on pre-flowering development other than post-flowering responses. However, the loci *E3* and *E4* might function not only in pre-flowering development but also in post-flowering development, indicated by the narrow variation in VE-R1, VE-R7 and R1–R7 of cultivars with *e3* and e4 alleles (Figure 3). This result is consistent with Xu et al [26]. that *E3* and *E4* respond not only to pre-flowering but also to post-flowering by increasing pod filling duration, number of nods and pod numbers by up-regulating the expression of growth habit gene *Dt1*. This result implies the significance of *E3* and *E4* loci for molecular genetic breeding to increase soybean productivity. Moreover, the outliers suggested that the disfunction of *E3* and *E4* might be interrupted by other genes. Similarly, the loci *E1* and *E3* are related with photoperiod sensitivity (Figure 4).

For the genotype of *E1/E2/E3/E4*, the photoperiod sensitivity was mostly above 70%. Compared with genotypes *E1/E2/E3/E4*, *e1-as/E2/E3/E4*, *e1-as/e2/E3/E4* and *e1-as/e2/e3/E4* (Figure 5), the photoperiod sensitivity decreased with the numbers of recessive

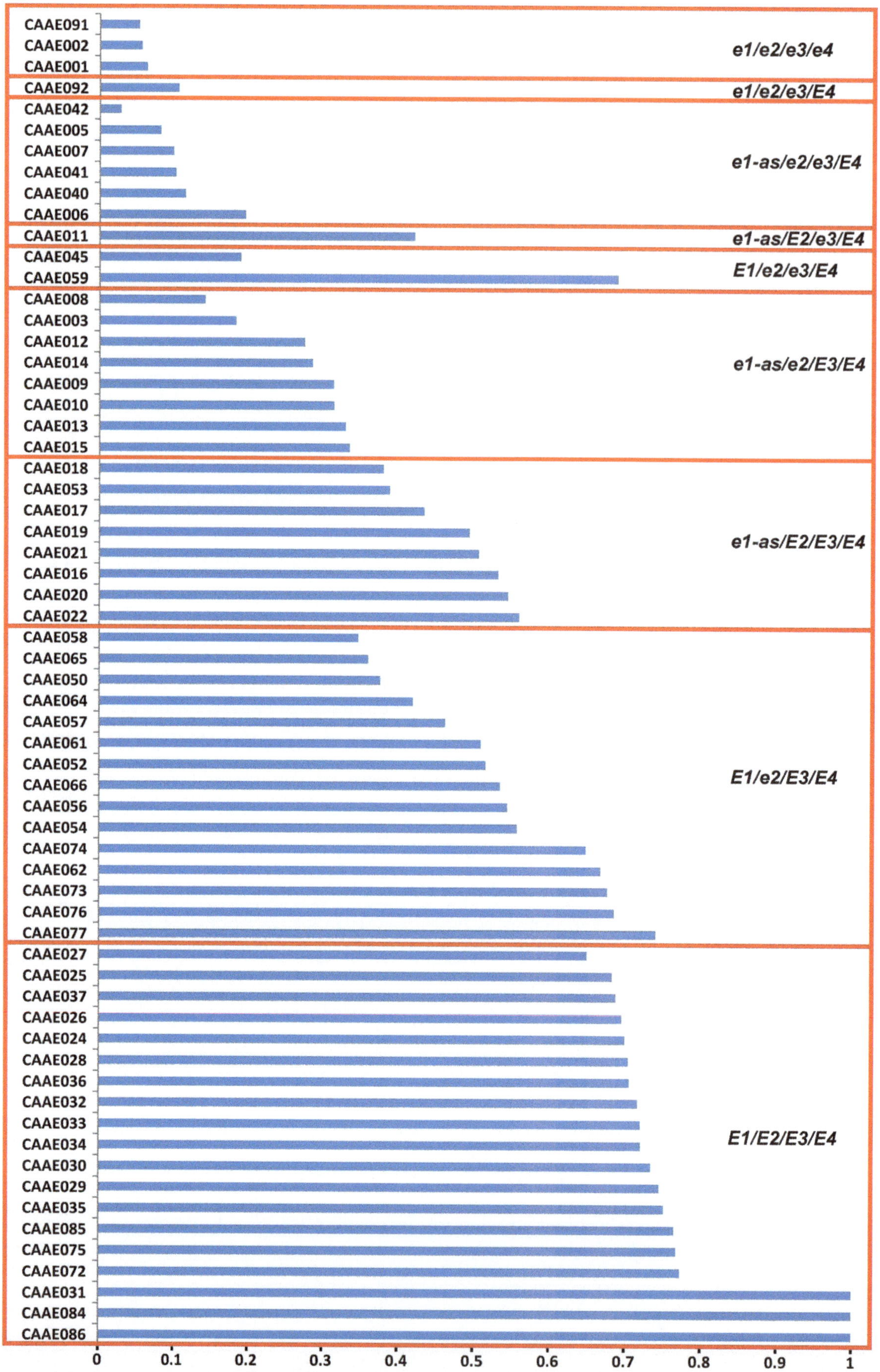

Figure 5. Photoperiod sensitivity grouped by *E* genotypes.

alleles. While comparing *E1/e2/E3/E4* and *e1-as/E2/E3/E4* with *E1E2E3E4*, it was suggested that *E1* plays a more important role than *E2* because *e1* decreased the photoperiod sensitivity more significantly and narrowed its range. These results further proved that soybean photoperiod insensitivity was involved by four maturity loci *E1, E3, E4* and *E7* while *E2* locus was not involved [28], [29]. In these tested cultivars of China, *E1/e2/E3/E4* is much more abundant. It ranged from N 18° -N 42° while *E1/E2/E3/E4* was distributed south of N 39° except for the four wild soybeans (CAAE087, Heiheyesheng; CAAE088, Bayanyesheng; CAAE089, Baiyangdianyesheng; and CAAE090, Guangxiyesheng). For *e1* or *el-as* alleles, the associated cultivars where mostly located above N 45° except for CAAE053 (Tiefeng 31, N 35°–40°). Thus, to improve the ecological adaptability of cultivars, the *E1* gene must function less because of its most strong impact on delaying maturity while other *E* genes might become important in adaptation. Unlike cultivated soybean, two wild soybean accessions, CAAE087 (Heiheyesheng) and CAAE088 (Bayanyesheng) adapted north of N 46° in China where cultivars generally had less photoperiod sensitivity where genotyped as *E1/E2/E3/E4*. It is greatly important for wild soybean to adapt and survive during season alteration. The genetic mechanisms underlying photoperiod insensitivity and adaptation in wild soybean were therefore unique from those in cultivated soybean.

Allelic combinations of *E* genes determine maturity groups

In the population analyzed here, MGV to MGX have the same genotype at *E1/E2/E3/E4*. It suggests that in these maturity groups, other unknown maturity genes should be involved in the determination of Mature Group. MGIII to MGIV are mainly genotyped as *e1-as/E2/E3/E4*, and MGII is *e1-as/e2/E3/E4*. For MGII to MGX, each group has one genotype. MGI has the genotypes of *e1-as/e2/E3/E4* and *e1-as/E2/e3/E4*; MG0 and MG00 both have the genotypes of *e1-as/e2/E3/E4* and *e1-as/e2/e3/E4*; and MG000 has *e1/e2/e3/e4*. From MG000 to MGI, each maturity group has two genotypes, which means that photoperiod is the first key factor in these region to determine maturity group. However, the more recessive alleles at *E* genes, the earlier cultivars mature. Due to the limited number of cultivars used in each MG, the maturity genotypes for each group may be underestimated and additional genotypes for each MG may be identified with a larger sample of cultivars. Although it is not possible to enumerate all genotypes of a given maturity group, this limited sample of cultivars showed that allelic combinations of *E* genes determine maturity groups in general.

Conclusions

The *E* genes (*E1*, *E2*, *E3* and *E4*) have different roles in maturity and photoperiod sensitivity and their allelic combinations determine maturity group and adaptation to different latitude.

Acknowledgments

We thank Randall L. Nelson (Soybean/Maize Germplasm, Pathology and Genetics Research Unit, USDA and Department of Crop Sciences, University of Illinois, Urbana, Illinois, USA), Lijuan Qiu and Zhangxiong Liu (Institute of Crop Sciences, Chinese Agricultural Academy of Sciences, China) for providing some soybean germplasm. We also appreciate Dr. Elroy R. Cober (Eastern Cereal and Oilseed Research Centre, Agriculture and Agri-Food Canada, Canada) and Dr. Jun Abe (Research Faculty of Agriculture, Hokkaido University) for proofreading of the manuscript.

Author Contributions

Conceived and designed the experiments: BJ FK TH BL. Performed the experiments: BJ HN YG LT YY SL LM JW SS DC. Analyzed the data: BJ HN YG CW XY WH. Contributed reagents/materials/analysis tools: FK TH BL. Contributed to the writing of the manuscript: BJ FK.

References

1. Cober ER, Voldeng HD (2001) A new soybean maturity and photoperiod-sensitivity locus linked to *E1* and *T*. Crop Sci. 41: 698–701.
2. Hartwig E (1970) Growth and reproductive characteristics of soybeans [*Glycine max* (L) Merr.] grown under short-day conditions. Trop. Sci. 12: 47–53.
3. Zhang L, Kyei-Boahen S, Zhang J, Zhang M, Freeland T, et al. (2007) Modifications of optimum adaptation zones for soybean maturity groups in the USA. Crop Manage. http://dx. doi. org/10.1094/ CM-2007-0927-01-RS.
4. Hao G, Chen X, Pu M (1992) Maturity groups of soybean cultivars in China. Acta Agr. Sin. 18: 275–281.
5. Wang G (1981) The research about the ecological division of soybean cultivars in China. Sci. Agr. Sin.: 39–46.
6. Bernard R (1971) Two major genes for time of flowering and maturity in soybeans. Crop sci. 11: 242–244.
7. Bonato ER, Vello NA (1999) *E6*, a dominant gene conditioning early flowering and maturity in soybeans. Genet. Mol. Biol. 22: 229–232.
8. Buzzell R (1971) Inheritance of a soybean flowering response to fluorescent-daylength conditions. Can. J. Genet. Cytol. 13: 703–707.
9. Buzzell R, Voldeng HD (1980) Inheritance of insensitivity to long daylength. Soyb. Genet. Newsl. 7: 26–29.
10. Cober ER, Molnar SJ, Charette M, Voldeng HD (2010) A new locus for early maturity in soybean. Crop sci. 50: 524–527.
11. McBlain B, Bernard R (1987) A new gene affecting the time of flowering and maturity in soybeans. J. Hered. 78: 160–162.
12. Ray JD, Hinson K, Mankono J, Malo MF (1995) Genetic control of a long-juvenile trait in soybean. Crop sci. 35: 1001–1006.
13. Wang Y, Wu C, Zhang X, Wang Y, Han TF (2008) Effects of soybean major maturity genes under different photoperiods. Acta Agr. Sin. 34: 1160–1168.
14. Xia Z, Zhai H, Liu B, Kong F, Yuan X, et al. (2012) Molecular identification of genes controlling flowering time, maturity, and photoperiod response in soybean. Plant Syst. Evol.: 1217–1227.
15. Xia Z, Watanabe S, Yamada T, Tsubokura Y, Nakashima H, et al. (2012) Positional cloning and characterization reveal the molecular basis for soybean maturity locus *E1* that regulates photoperiodic flowering. Proc. Natl. Acad. Sci. 109: E2155–E2164.
16. Watanabe S, Xia Z, Hideshima R, Tsubokura Y, Sato S, et al. (2011) A map-based cloning strategy employing a residual heterozygous line reveals that the *GIGANTEA* gene is involved in soybean maturity and flowering. Genetics 188: 395–407.
17. Watanabe S, Hideshima R, Xia Z, Tsubokura Y, Sato S, et al. (2009) Map-based cloning of the gene associated with the soybean maturity locus *E3*. Genetics 182: 1251–1262.
18. Liu B, Kanazawa A, Matsumura H, Takahashi R, Harada K, et al. (2008) Genetic redundancy in soybean photoresponses associated with duplication of the *phytochrome A* gene. Genetics 180: 995–1007.
19. Kong F, Liu B, Xia Z, Sato S, Kim BM, et al. (2010) Two coordinately regulated homologs of *FLOWERING LOCUS T* are involved in the control of photoperiodic flowering in soybean. Plant physiol. 154: 1220–1231.
20. Fehr W, Caviness C, Burmood D, Pennington J (1971) Stage of development descriptions for soybeans, *Glycine max* (L.) Merrill. Crop sci. 11: 929–931.
21. Jilin Academy of Agricultural Sciences (1985) Soybean cultivars in China. 1 ed. Agriculture Press, Beijing.
22. Soybean Institute, Jilin Academy of Agricultural Sciences (1993) Soybean cultivars in China (1978–1992). 1 ed. Agr. Press, Beijing.
23. Li WD, Zhang MC (2006) Huang-Huai-Hai summer-sowed soybeans and their parameters. 1 ed. China Agr. Sci. Technol. Press, Beijing.
24. Wu C, Ma Q, Yam KM, Cheung MY, Xu Y, et al. (2006) In situ expression of the *GmNMH7* gene is photoperiod-dependent in a unique soybean (*Glycine max* [L.] Merr.) flowering reversion system. Planta 223: 725–735.
25. Fei Z, Wu C, Sun H, Hou W, Zhang B, et al. (2009) Identification of photothermal responses in soybean by integrating photoperiod treatments with planting-date experiments. Acta Agr. Sin.: 1525–1531.
26. Xu M, Xu Z, Liu B, Kong F, Tsubokura Y, et al. (2013) Genetic variation in four maturity genes affects photoperiod insensitivity and *PHYA*-regulated post-flowering responses of soybean. BMC Plant Biol. 13: 91.
27. Tsubokura Y, Matsumura H, Xu M, Liu B, Nakashima H, et al. (2013) Genetic variation in soybean at the maturity locus *E4* is involved in adaptation to long days at high latitudes. Agronomy 3: 117–134.
28. Cober ER, Tanner J, Voldeng HD (1996) Genetic control of photoperiod response in early-maturing, near-isogenic soybean lines. Crop sci. 36: 601–605.
29. Cober ER, Tanner J, Voldeng HD (1996) Soybean photoperiod-sensitivity loci respond differentially to light quality. Crop sci. 36: 606–610.

17

Genetic Diversity, Population Structure, and Resistance to *Phytophthora capsici* of a Worldwide Collection of Eggplant Germplasm

Rachel P. Naegele[1], Samantha Boyle[1], Lina M. Quesada-Ocampo[2], Mary K. Hausbeck[1]*

1 Department of Plant and Microbial Sciences, Michigan State University, East Lansing, Michigan, United States of America, **2** Department of Plant Pathology, North Carolina State University, Raleigh, North Carolina, United States of America

Abstract

Eggplant (*Solanum melongena* L.) is an important solanaceous crop with high phenotypic diversity and moderate genotypic diversity. Ninety-nine genotypes of eggplant germplasm (species (*S. melongena, S. incanum, S. linnaeanum* and *S. gilo*), landraces and heirloom cultivars) from 32 countries and five continents were evaluated for genetic diversity, population structure, fruit shape, and disease resistance to Phytophthora fruit rot. Fruits from each line were measured for fruit shape and evaluated for resistance to two *Phytophthora capsici* isolates seven days post inoculation. Only one accession (PI 413784) was completely resistant to both isolates evaluated. Partial resistance to Phytophthora fruit rot was found in accessions from all four eggplant species evaluated in this study. Genetic diversity and population structure were assessed using 22 polymorphic simple sequence repeats (SSRs). The polymorphism information content (PIC) for the population was moderate (0.49) in the population. Genetic analyses using the program STRUCTURE indicated the existence of four genetic clusters within the eggplant collection. Population structure was detected when eggplant lines were grouped by species, continent of origin, country of origin, fruit shape and disease resistance.

Editor: Boris Alexander Vinatzer, Virginia Tech, United States of America

Funding: Funding was provided by MSU GREEEN grant GR12-006. The funders had no role in study design, data collection and analysis, decision to publish, or preparation of the manuscript.

Competing Interests: The authors have declared that no competing interests exist.

* E-mail: hausbec1@anr.msu.edu

Introduction

Cultivated eggplant, *Solanum melongena* L., is a high-value vegetable commodity in Europe and Asia. China and India are the major producers with 27.7 and 11.9 million tons per year, respectively [2011 FAO]. Eggplant is the third most important solanaceous crop worldwide after potato and tomato, and fourth most important in the U.S. [2011 FAO]. In the U.S., eggplants are a minor crop grown for specialty markets with an approximate production of 62 thousand tons annually [2011 FAO].

Unlike most other cultivated solanaceous crops (tomatoes, peppers, and potatoes), eggplants are an Old World species. Domestication of the cultivated eggplant is thought to have occurred in Asia as early as 59 B.C. [1–3]. Since that time, it has been transported and cultivated around the world [4,5]. Primary and secondary centers of diversity for eggplant are Asia and the Mediterranean basin in Europe, respectively. Studies have indicated that the progenitors of domesticated eggplant (*S. melongena*) originated in Africa and were derived from the closely related *Solanum incanum* (part of the eggplant complex) [6] and *Solanum linnaeanum* [5]. Both *S. incanum* and *S. linnaeanum* can form partially fertile hybrids with *S. melongena* making them potential sources for desirable traits such as abiotic and biotic disease resistance [7–9]. Wild relatives have traditionally been a good source of disease resistance for cultivated species that exhibit lower genetic diversity [10–15]. Domesticated heirloom varieties and landrace accessions may also harbor resistance, and are often more similar to modern cultivated varieties than wild species, making them a good source for desirable traits [16–17].

Multiple *Phytophthora* species are capable of causing disease symptoms on eggplants. Infected eggplants can display root and/or fruit symptoms. One causal agent of Phytophthora fruit rot is *Phytophthora capsici* L., an oomycete that affects multiple solanaceous species including eggplant, pepper, and tomato [18–22]. In the field, chemical management is expensive and provides limited protection against *Phytophthora capsici*-induced fruit rot in eggplant, which is the most common *Phytophthora*-induced symptom in eggplants [23]. Cultivated eggplants have some level of root rot resistance to moderately virulent isolates of *P. capsici* [20]. No cultivars, to date, have displayed any type of fruit resistance. Host resistance, an important part of a successful, sustainable management program, is not available for management of Phytophthora fruit rot in eggplants and currently, no known lines or cultivars are resistant to *P. capsici*. Partial fruit rot resistance to *P. capsici* has been identified in other solanaceous species such as peppers and tomatoes ([24], Granke et al. unpublished), but to our knowledge this has not been evaluated in eggplant.

In addition to disease resistance, fruit shape is an important attribute for each cultivar and many studies have been performed to identify the genetic basis of fruit shape in the Solanaceae [25–30]. Size, shape and color vary greatly between eggplant market classes, and it will be important to maintain this phenotypic diversity when incorporating disease resistance [4,31–32]. This

Table 1. Eggplant germplasm used for the study of morphological and molecular variation.

Species	ID	Accession	Plant ID	Country	Source
S. melongena	101	C-S-16	-	Spain	UT Valencia
S. melongena	102	Grif 1276	46B	Thailand	USDA-GRIN
S. melongena	103	Grif 14182	New Orleans Market	U.S.	USDA-GRIN
S. melongena	104	Grif 14186	Hastings purple thornless	U.S.	USDA-GRIN
S. melongena	105	H15	-	Spain	UT Valencia
S. melongena	106	IVIA-371	-	Spain	UT Valencia
S. melongena	107	MM 108 bis	-	France	AVRDC
S. melongena	108	MM 114	Berengena larga negra	Spain	AVRDC
S. melongena	109	MM 1171	Large Santa Olalla	Spain	AVRDC
S. melongena	110	MM 1363	-	Costa Rica	AVRDC
S. melongena	111	MM 1364	-	Costa Rica	AVRDC
S. melongena	112	MM 1365	-	Guatemala	AVRDC
S. melongena	113	MM 141	Violette d'Avignon	France	AVRDC
S. melongena	114	MM 1750	Listada di Gandia	Spain	AVRDC
S. melongena	115	MM 346	Berengena redonda	Spain	AVRDC
S. melongena	116	MM 39	Noire de Chateaurenard	France	AVRDC
S. melongena	117	MM 522	Waimanolo long B1	U.S.	AVRDC
S. melongena	118	MM 56	Violette de Toulouse	France	AVRDC
S. melongena	119	MM 61	Zebrina	Spain	AVRDC
S. melongena	120	MM 64	Ronde de Valence	France	AVRDC
S. melongena	121	MM 69	Monstrueuse de New York	U.S.	AVRDC
S. melongena	122	MM 91	Black Beauty	U.S.	AVRDC
S. melongena	123	PI 102727	No. 202	Uzbekistan	USDA-GRIN
S. melongena	124	PI 105346	Lao Lai Hei Chieh	China	USDA-GRIN
S. melongena	125	PI 115505	Giant of Benares	India	USDA-GRIN
S. melongena	126	PI 140446	5917	Iran	USDA-GRIN
S. melongena	127	PI 140456	7015	Iran	USDA-GRIN
S. melongena	128	PI 141968	No. 1	China	USDA-GRIN
S. melongena	129	PI 143410	Badenjan	Iran	USDA-GRIN
S. melongena	130	PI 169641	1448	Turkey	USDA-GRIN
S. melongena	131	PI 169650	2259	Turkey	USDA-GRIN
S. melongena	132	PI 171851	6753	Turkey	USDA-GRIN
S. melongena	133	PI 175914	9043	Turkey	USDA-GRIN
S. melongena	134	PI 179500	9877	Iraq	USDA-GRIN
S. melongena	135	PI 179997	10598	India	USDA-GRIN
S. melongena	136	PI 181896	Aleppo 3	Syria	USDA-GRIN
S. melongena	137	PI 181963	Homs 21	Syria	USDA-GRIN
S. melongena	138	PI 193599	Long Violet	Ethiopia	USDA-GRIN
S. melongena	140	PI 199516	M 19	Greece	USDA-GRIN
S. melongena	141	PI 200881	-	Afghanistan	USDA-GRIN
S. melongena	142	PI 204731	-	Turkey	USDA-GRIN
S. melongena	143	PI 213193	M-57/29	Greece	USDA-GRIN
S. melongena	144	PI 217962	Banjal Bemba	Pakistan	USDA-GRIN
S. melongena	145	PI 223844	-	Philippines	USDA-GRIN
S. melongena	146	PI 230333	Kairyo-onaga	Japan	USDA-GRIN
S. melongena	147	PI 230334	Kitta Horyo	Japan	USDA-GRIN
S. melongena	148	PI 230335	Taiwan-naga	Japan	USDA-GRIN
S. melongena	149	PI 232078	Kopek	South Africa	USDA-GRIN
S. melongena	150	PI 232079	Mofale	South Africa	USDA-GRIN
S. melongena	151	PI 233916	-	El Salvador	USDA-GRIN

Table 1. Cont.

Species	ID	Accession	Plant ID	Country	Source
S. melongena	152	PI 234632	Early Round Purple	South Africa	USDA-GRIN
S. melongena	153	PI 241506	Badanjan	Iran	USDA-GRIN
S. melongena	154	PI 249570	Makhua Proh	Thailand	USDA-GRIN
S. melongena	155	PI 256077	No. 1	Afghanistan	USDA-GRIN
S. melongena	156	PI 263727	Rosita	Puerto Rico	USDA-GRIN
S. melongena	157	PI 267104	Cylinder A-132	Soviet	USDA-GRIN
S. melongena	158	PI 269600	423	Pakistan	USDA-GRIN
S. melongena	159	PI 276104	Motale	South Africa	USDA-GRIN
S. melongena	160	PI 286099	No. 62-46-2	U.S.	USDA-GRIN
S. melongena	161	PI 286100	No. 62-48-2	U.S.	USDA-GRIN
S. melongena	162	PI 290467	Lungi de Impant	Hungary	USDA-GRIN
S. melongena	163	PI 290469	Cu-e-da-juan	Hungary	USDA-GRIN
S. melongena	164	PI 304839	G2562	Brazil	USDA-GRIN
S. melongena	165	PI 320501	24	Canada	USDA-GRIN
S. melongena	166	PI 320504	28	Canada	USDA-GRIN
S. melongena	167	PI 320509	35	Canada	USDA-GRIN
S. melongena	168	PI 349612	Terongglatik	Indonesia	USDA-GRIN
S. melongena	169	PI 351129	Kurume Long	Japan	USDA-GRIN
S. melongena	170	PI 358232	Dolg	Macedonia	USDA-GRIN
S. melongena	171	PI 358242	Morska Pata	Macedonia	USDA-GRIN
S. melongena	172	PI 358244	Renski dolg	Macedonia	USDA-GRIN
S. melongena	173	PI 368822	Sredno Dolg	Macedonia	USDA-GRIN
S. linnaeanum	174	PI 388846	WL-74	Italy	USDA-GRIN
S. linnaeanum	175	PI 388847	WL-85	Italy	USDA-GRIN
S. melongena	176	PI 391646	Liu-ye-ch'ieh	China	USDA-GRIN
S. melongena	178	PI 413782	22–73	Cote D'Ivoire	USDA-GRIN
S. melongena	179	PI 413783	3–73	Burkina Faso	USDA-GRIN
S. melongena	180	PI 413784	13–73	Burkina Faso	USDA-GRIN
S. melongena	181	PI 419198	Tsu Yang	China	USDA-GRIN
S. linnaeanum	182	PI 420415	52	Colombia	USDA-GRIN
S. gilo	183	PI 441908	BGH 5008	Brazil	USDA-GRIN
S. melongena	184	PI 452122	Lunga Violetta di Romagna	Italy	USDA-GRIN
S. melongena	185	PI 452123	Tonda di Manfredonia	Italy	USDA-GRIN
S. melongena	186	PI 462370	Neznyj 36	Soviet	USDA-GRIN
S. melongena	187	PI 470273	-	Indonesia	USDA-GRIN
S. melongena	189	PI 478390	O 81	China	USDA-GRIN
S. melongena	190	PI 491192	Kemer	Turkey	USDA-GRIN
S. incanum	191	PI 500922	Chipusni	Zambia	USDA-GRIN
S. melongena	192	PI 560903	Six Leaves	China	USDA-GRIN
S. melongena	193	PI 561139	37	Kazakhstan	USDA-GRIN
S. melongena	194	PI 561140	36	Kazakhstan	USDA-GRIN
S. melongena	195	PI 593748	56A	Thailand	USDA-GRIN
S. melongena	196	PI 593806	171	Thailand	USDA-GRIN
S. melongena	198	PI 593885	314	Thailand	USDA-GRIN
S. melongena	199	PI 595220	Gator	United States	USDA-GRIN
S. melongena	200	PI 600912	Little fingers	U.S.	USDA-GRIN
S. melongena	201	PI 606714	Pompano market	U.S.	USDA-GRIN
S. melongena	202	PI 639121	Puerto Rican beauty	Puerto Rico	USDA-GRIN
S. melongena	203	PI 639122	Blackee	U.S.	USDA-GRIN

Figure 1. Eggplant Phytophthora fruit rot disease rating scale shown on various eggplant genotypes: 0 = no visible symptoms, a rating of 1 = <25% symptomatic area, 25%>2<50%, 50%>3<75%, and a rating of 4≥75% symptomatic area of the fruit.

phenotypic diversity does not always translate to high levels of genetic diversity [4,33]. Modern varieties of eggplant often have lower genetic diversity, and new traits are often bred into commercial varieties from landraces or wild relatives with higher genetic diversity [34]. The characterization of genetic diversity is important for maintenance and utilization of germplasm resources (wild, landrace, heirloom, breeding lines and cultivars), and the development of core collections [3,35]. Genetic bottlenecks (domestication, selection of lines by market class, etc.) have limited the variability existing within cultivated lines [3].

Population structure analysis has recently been gaining popularity as a method to understand and visualize spatial and temporal differences between subpopulations [19,36–40]. Information on population structure can provide insight about connections between phenotypic variation and the distribution of genetic diversity. Population structure should also be taken into account when testing and incorporating desirable traits. If population structure is present in materials being evaluated for association mapping, spurious associations may be found between a particular genotype and the trait of interest [31,41]. Many studies have looked at the genetic diversity of eggplants within specific countries or regions, and a recent study compared genetic differentiation and structure in three centers of diversity; however, few studies have looked at diversity and population structure in a global collection of eggplants [2,4,31,34,42–46].

We evaluated the fruit shape index, Phytophthora fruit rot resistance, genetic diversity and population structure of a diverse collection of eggplant germplasm using 22 simple sequence repeats (SSRs). Our objectives were to evaluate a worldwide collection of eggplants for population structure and genetic diversity, and to determine if the population structure is associated with fruit shape or resistance to *Phytophthora capsici*. These results provide an initial basis for understanding the worldwide population structure of eggplants for breeding and conservation, and its relationship with disease resistance and fruit shape.

Materials and Methods

Permissions

No specific permits were required for the described field studies, which took place in an experimental field plot at Michigan State University (MSU). This field plot was used by the authors of this paper affiliated to the aforementioned institution (RN, SB, MK, LQ) for field trials to evaluate eggplant germplasm and cultivated species.

Plant Material

Ninety-four accessions of eggplants (*S. melongena*), three accessions of *S. linnaeanum*, one accession of *S. gilo*, and one accession of *S. incanum* were obtained from the United States Department of Agriculture Germplasm Resource Information Network (ars-grin.usda.gov), Universidad de Technologia de Valencia, and the INRA (French National Institute for Agricultural Research) (Table 1). Accessions represented 32 countries on five different continents, and included primary and secondary centers of diversity. Seeds were sown into 72-cell trays containing a soilless peat mixture (Suremix Michigan Grower Products, Inc. Galesburg, MI) in a polyethylene greenhouse (MSU Horticulture Teaching and Research Center, Holt, MI). Eight weeks after sowing, seedlings were transplanted into a field at the MSU Plant Pathology Research Center (East Lansing, MI). Individual accessions were planted into single plots. Each individual line was established in 3 m long plot and 12 lines were planted per row. Within rows, plants were spaced 0.45 m apart. Rows were spaced 2.4 m apart, covered with black plastic mulch, and grown according to local practices. Immature eggplant fruits of marketable size were hand harvested throughout the growing season and brought to the lab for inoculation and evaluation.

Isolates

Two virulent *P. capsici* isolates, previously evaluated on pepper, tomato and eggplant, were selected from the long-term collection of Dr. Mary K. Hausbeck at MSU [18–20]. Isolates were characterized by host of origin, mefenoxam sensitivity [insensitive (I) or sensitive (S)] and mating type (A1 or A2). Isolate 12889 (pepper, I, A1) and isolate OP97 (cucumber, S, A1) were maintained on unclarified V8 agar at 25°C under constant light. Prior to inoculations, isolates were activated by inoculating and recovering each isolate from an individual pepper fruit to ensure virulence.

Inoculation and Evaluation

For inoculation, a single 6 mm-diameter plug from an actively growing *P. capsici* isolate on V8 agar was placed, mycelium side down onto a non-wounded eggplant fruit surface-disinfested in 10% bleach for 5 min and rinsed with distilled water. Control eggplants were inoculated with a single 6 mm-diameter sterile plug of agar. Plugs were covered with a sterile microcentrifuge tube and affixed into place with petroleum jelly. Eggplants were placed into a humidity chamber consisting of an aluminum pan with a ring of moistened paper towel around the edge, covered with plastic wrap, sealed with tape and kept under constant light at room

Table 2. Fruit shape parameters and mean disease ratings for each isolate.

		Mean[a]		Fruit[b]			
Species	Accession	12889	OP97	Ratio	Length	Width	Cluster[c]
S. melongena	C-S-16	S	S	5.59	4.12	22.98	1
S. melongena	Grif 1276	MS	R	1.14	4.30	4.92	2
S. melongena	Grif 14182	S	S	2.59	6.11	14.97	3
S. melongena	Grif 14186	S	S	1.75	7.83	13.48	1
S. melongena	H15	R	S	1.93	6.17	11.81	2
S. melongena	IVIA-371	S	S	2.08	7.26	14.98	4
S. melongena	MM 108 bis	S	S	5.23	4.21	21.52	1
S. melongena	MM 114	S	S	6.57	3.44	22.23	2
S. melongena	MM 1171	S	S	2.81	5.43	15.14	1
S. melongena	MM 1363	S	S	5.05	5.12	25.77	2
S. melongena	MM 1364	S	S	3.02	6.02	17.78	5
S. melongena	MM 1365	S	MS	1.87	7.88	14.55	2
S. melongena	MM 141	S	S	4.67	5.14	23.84	1
S. melongena	MM 1750	S	S	2.49	7.11	17.52	5
S. melongena	MM 346	S	S	1.31	8.73	11.35	2
S. melongena	MM 39	S	S	5.37	4.29	22.89	1
S. melongena	MM 522	S	S	8.03	3.27	26.14	1
S. melongena	MM 56	MS	S	2.42	6.83	16.33	2
S. melongena	MM 61	S	S	1.97	5.88	11.53	2
S. melongena	MM 64	S	S	1.16	8.65	9.99	2
S. melongena	MM 69	S	S	1.34	8.78	11.43	4
S. melongena	MM 91	S	S	1.92	7.49	14.19	4
S. melongena	PI 102727	S	S	2.44	6.10	14.76	4
S. melongena	PI 105346	S	S	1.17	8.94	10.33	4
S. melongena	PI 115505	S	S	1.72	6.57	11.27	2
S. melongena	PI 140446	S	S	1.77	7.14	12.55	2
S. melongena	PI 140456	S	S	3.48	6.31	21.85	5
S. melongena	PI 141968	S	S	4.46	4.52	19.91	2
S. melongena	PI 143410	S	S	1.35	7.94	10.68	5
S. melongena	PI 169641	S	S	3.78	5.19	19.34	4
S. melongena	PI 169650	S	S	4.66	4.22	19.20	4
S. melongena	PI 171851	S	S	4.31	4.14	17.78	4
S. melongena	PI 175914	S	S	2.92	4.83	14.08	5
S. melongena	PI 179500	S	S	3.64	4.41	16.01	2
S. melongena	PI 179997	S	S	3.34	4.75	15.88	5
S. melongena	PI 181896	MS	S	1.91	6.60	12.54	5
S. melongena	PI 181963	S	S	3.99	4.04	15.95	5
S. melongena	PI 193599	S	MS	1.84	6.45	11.64	5
S. melongena	PI 199516	S	S	1.74	8.76	14.23	2
S. melongena	PI 200881	S	S	3.82	5.85	22.10	1
S. melongena	PI 204731	S	S	2.73	7.55	18.06	5
S. melongena	PI 213193	S	S	1.07	9.21	9.75	5
S. melongena	PI 217962	S	S	3.26	4.75	15.36	4
S. melongena	PI 223844	S	S	2.89	5.27	14.82	2
S. melongena	PI 230333	S	S	7.15	3.62	25.64	4
S. melongena	PI 230334	S	S	6.84	3.10	21.22	5
S. melongena	PI 230335	S	S	7.43	3.52	26.16	1
S. melongena	PI 232078	S	S	4.01	4.57	18.15	5
S. melongena	PI 232079	S	S	2.36	5.61	13.14	1

Table 2. Cont.

		Mean[a]		Fruit[b]			
Species	**Accession**	**12889**	**OP97**	**Ratio**	**Length**	**Width**	**Cluster[c]**
S. melongena	PI 233916	MS	S	2.37	5.6	13.16	5
S. melongena	PI 234632	S	S	0.93	8.72	7.90	5
S. melongena	PI 241506	MS	S	2.73	6.08	16.37	2
S. melongena	PI 249570	S	S	1.53	5.79	9.96	5
S. melongena	PI 256077	S	MS	3.07	5.27	16.05	4
S. melongena	PI 263727	S	S	1.95	8.01	12.82	4
S. melongena	PI 267104	S	S	3.93	4.86	18.811	5
S. melongena	PI 269600	S	S	1.71	7.13	11.56	4
S. melongena	PI 276104	S	S	2.22	7.03	15.56	2
S. melongena	PI 286099	S	S	5.81	4.52	25.27	5
S. melongena	PI 286100	S	S	6.45	4.73	28.12	5
S. melongena	PI 290467	S	S	3.34	6.28	20.96	3
S. melongena	PI 290469	S	S	2.16	6.99	14.89	4
S. melongena	PI 304839	S	S	2.85	6.17	17.51	4
S. melongena	PI 320501	S	S	2.18	7.23	15.48	1
S. melongena	PI 320504	S	S	4.58	6.58	29.07	1
S. melongena	PI 320509	S	S	2.46	6.93	17.07	4
S. melongena	PI 349612	S	S	1.44	5.09	7.38	4
S. melongena	PI 351129	S	S	5.61	4.70	26.48	4
S. melongena	PI 358232	S	S	4.59	4.90	22.35	4
S. melongena	PI 358242	S	S	2.16	7.18	14.53	4
S. melongena	PI 358244	S	S	5.36	4.70	24.73	4
S. melongena	PI 368822	S	S	3.18	5.81	18.22	1
S. linnaeanum	PI 388846	MR	S	1.03	2.38	2.46	3
S. linnaeanum	PI 388847	MR	S	1.07	1.89	2.01	3
S. melongena	PI 391646	S	S	5.33	8.20	25.77	1
S. melongena	PI 413782	S	R	0.79	1.48	1.16	3
S. melongena	PI 413783	MR	R	0.46	4.37	2.02	3
S. melongena	PI 413784	R	R	0.69	5.85	4.05	3
S. melongena	PI 419198	S	MS	5.63	4.38	24.25	1
S. linnaeanum	PI 420415	S	S	1.06	2.07	2.14	3
S. gilo	PI 441908	R	MR	0.83	5.15	4.27	3
S. melongena	PI 452122	S	S	5.79	4.08	23.75	1
S. melongena	PI 452123	S	S	1.29	9.60	12.16	1
S. melongena	PI 462370	S	S	1.15	10.94	12.02	5
S. melongena	PI 470273	S	S	3.30	4.88	15.86	3
S. melongena	PI 478390	S	S	0.75	9.54	7.08	4
S. melongena	PI 491192	S	S	4.95	4.46	22.06	1
S. incanum	PI 500922	MS	R	1.03	2.56	2.63	3
S. melongena	PI 560903	S	S	0.95	8.70	8.15	1
S. melongena	PI 561139	S	S	2.94	5.58	16.15	1
S. melongena	PI 561140	S	S	3.48	4.65	16.2063	4
S. melongena	PI 593748	S	S	2.65	5.77	15.18	1
S. melongena	PI 593806	S	S	3.79	4.20	15.86	1
S. melongena	PI 593885	S	S	1.12	6.00	6.57	2
S. melongena	PI 595220	S	S	2.70	4.56	12.54	5
S. melongena	PI 600912	S	S	4.58	3.35	15.3	2
S. melongena	PI 606714	S	S	2.40	5.60	13.26	3

Table 2. Cont.

Species	Accession	Mean[a]		Fruit[b]			Cluster[c]
		12889	OP97	Ratio	Length	Width	
S. melongena	PI 639121	S	S	1.96	7.30	14.02	1
S. melongena	PI 639122	S	S	1.60	6.70	10.77	5

[a]Mean disease rating across all experimental replicates for each isolate, 12889 and OP97.
[b]Mean fruit parameters for ratio (fruit length:fruit width), length (cm), and width (cm).
[c]STRUCTURE genetic cluster assignment based on 22 SSR markers.

temperature (25°C). Three fruits (replicates) from each eggplant line were evaluated per isolate. The experiment was performed three times (runs). An experiment replicate included three fruits for each isolate of every eggplant accession evaluated in a completely randomized design (CRD) blocked by isolate. One line (PI 500922) was repeated only one time for a total of two experimental replicates of three fruit per isolate due to poor fruit set. Two control fruits were inoculated with a sterile plug of V8 agar for each line.

Eggplant fruits were evaluated for disease severity seven days after inoculation. Fruits were evaluated on the following progressive scale based on the percentage of symptomatic fruit surface to account for differences in fruit size: 0 = no visible symptoms (resistant (R)), 1 = <25% of the fruit was symptomatic (moderately resistant (MR)), 2 = 25% to <50% (moderately susceptible (MS)), 3 = 50% to <75% (susceptible (S)), 4 = ≥75% symptomatic area (susceptible (S)) (Fig. 1). Visible mycelia growth was assessed as 0 = absent, 1 = present. *Phytophthora* isolations were performed on 10% of symptomatic fruits by peeling back the external layer of the fruit and plating three small portions of fruit tissue at the disease margin onto V8 agar plates amended with benomyl, ampicillin, PCNB, and mefenoxam to confirm the causal agent of the symptoms [47]. *Phytophthora capsici* was identified using morphological characteristics according to Waterhouse [48] and isolate mefenoxam sensitivity was confirmed by transferring the recovered isolates to V8 plates amended with 100 ppm mefenoxam according to Lamour and Hausbeck [47].

Ten immature fruits of marketable size collected from each line were measured for maximum length (cm) and maximum width (cm) using a hand caliper. Fruit shape was calculated as the ratio of maximum length to midpoint width for each line. Fruit shape ratios were rounded to the nearest whole number. Values between 0 and 1 were considered round, 2–3 were considered oval, 4–5 were semi-elongate and >5 were considered elongate.

Phenotype Statistical Analyses

Mean values for disease ratings for each accession were estimated using the PROC MEANS function of SAS software v9.3 (SAS Institute, Cary, NC). Significant differences between disease values (ratings) for eggplant accessions and isolates were estimated using the PROC MIXED function of SAS software. Significant differences were detected between experiment runs and each run was analyzed separately using Fisher's LSD test ($P = 0.05$). Accession by isolate interactions were calculated using the ANOVA slice option of PROC MIXED when $P \leq 0.05$. Lines with a consistent disease mean value of ≥2 in each run of the experiment were considered susceptible, with a consistent mean value <2 were termed moderately susceptible, lines with a

Figure 2. Fruit size and shape differences between eggplants. *Solanum incanum* (left) and *S. linnaeanum* (right) fruit (A), *S. melongena* fruit (B) and *S. linnaeanum* and *S. melongena* fruit varying in shape, size and color (C). U.S. quarter used for size reference.

Table 3. *Solanum* spp. fruit shape, width and length variation between countries of origin.

	Fruit					
Category[a]	**Shape**[b]		**Width (cm)**[c]		**Length (cm)**[d]	
China	3.0	cd	7.4	A	15.9	de
France	3.8	b	5.8	Cd	18.9	b
Iran	2.3	de	6.9	Ab	15.4	de
Italy	3.5	bc	6.8	Abc	18.0	bcd
Japan	6.8	a	3.7	E	24.9	a
Macedonia	3.8	b	5.6	Cd	20.0	b
S. Africa	2.4	de	6.5	Abc	13.7	ef
Spain	3.1	c	6.0	Bc	15.9	d
Thailand	2.2	d	5.6	Cd	11.8	f
Turkey	3.9	b	5.1	D	18.4	bc
USA	3.6	bc	5.7	Cd	16.9	cd

[a]Categories with less than five individuals representing a country were not included in analyses.
[b]Mean fruit shape calculated as the ratio of fruit length to fruit width.
[c]Mean fruit width at midpoint measured in cm.
[d]Mean fruit length from peduncle to blossom end measured in cm.

consistent mean value <1 were moderately resistant, and lines with a mean value $=0$ were resistant. Significant differences for pathogen growth were estimated using the PROC GLIMMIX function of SAS at $P=0.05$.

Fruit shape significant differences between lines and countries were calculated using the PROC mixed function of SAS software v9.3. Countries represented by less than four accessions were excluded from analyses. Unequal sample sizes among countries

Table 4. Polymorphic primers evaluated against 99 eggplant lines.

SSR	**Forward sequence**	**Reverse sequence**	**Allele**[a]	**Genetic Diversity**	**PIC**[b]	**Source**
BM61461	CTCATTACCACTTCATACAAAACAG	TGCAGTAGGTGTTGCTACGG	6	0.18	0.18	SolCAP
GPMS203	CACCAACACATCTTTTTCAACC	ATAATAGTGGTTGCGGCGAC	4	0.24	0.23	SolCAP
CB164833	CGGGCAGGTGCTATTATAAAAC	CGGCCGAGGTACAAGCC	3	0.57	0.49	SolCAP
T0633	GATGGGCTATGCTTGCTGTT	ACATCCCCAATGTTGTTGTG	2	0.03	0.03	SolCAP
CA516334	ACCCACCTTCATCAACAACC	ATTTGTGGCTTTTCGAAACG	6	0.61	0.55	SolCAP
GPMS178	GATTTTTGACATGTCACATTCATG	AACGTTGAAAAATAAAGTAAGCAAG	5	0.73	0.69	SolCAP
GP1102	GAACCCTTCATTCCTGTATGT	TTTGCCCGCATTATGTAAATC	2	0.45	0.35	SolCAP
C2_At5g34850	AGTGAAGTGGCTACATCCAAAATCTC	GAACAAAACATGCCCTACTGTAGGAA	7	0.60	0.51	SolCAP
C2_At1g69210	AGCTCTATTCATTTAAAACTAGTCCTCAT	TCTTTTCTTGTATTGGCGGCTAAATTC	2	0.50	0.37	SolCAP
AF348141	CCTTACGGGGAAAACCTAGC	CCATACGGACGTTGTCCTCT	5	0.68	0.62	NCBI
CAMS362	CCCCTTCTGACCTTGATTGA	TATGCCCCTCCTGTGATAGC	4	0.48	0.42	Minamiyama 2007
GO496268.1	CGTTGCCTGTTTACCAACCT	CCTTCTTCTGCACTTCCACA	2	0.48	0.37	NCBI
C2_At5g13200	TATGGGTCCGCCTGCAGTTCCAAC	AAGTTTTCCCCATGCCGCTTCTGT	3	0.11	0.10	SolCAP
C2_At1g32410	TGTTAGTGTCTGGAGGGATTGTATTG	AGATTCGGTGTAGAGACTGGAAGTATC	4	0.64	0.57	SolCAP
CSM7F	CGACGATCACCTTGATAACG	CCTTAAATGCAGAGTTTCCAAAG	2	0.50	0.37	Hurtado 2012
CSM27	TGTTTGGAGGTGAGGGAAAG	TCCAACTCACCGGAAAAATC	3	0.57	0.50	Hurtado 2012
CSM30	CACTGTTCCTGGTTGCTGTG	TTTAGCTTTAGCCCATCTACCG	3	0.50	0.40	Hurtado 2012
CSM31	CAACCGATATGCTCAGATGC	CGGGTATGGTCATGTTTTGC	6	0.76	0.71	Hurtado 2012
CSM43	ATTTTAACCCCGGGAAAATG	ACCGCTTCTAGGTTTTGCAC	4	0.62	0.55	Hurtado 2012
CSM44	CGTCGTTGTAACCCATCATC	TTGCCAAATTCCTTGTGTTC	3	0.46	0.36	Hurtado 2012
CSM54	ATGTGCCTCCATTCTGCAAG	TGGGTGGGATGCTGAGTAAG	3	0.44	0.37	Hurtado 2012
CSM73	TTCAACATAGCCTGGACCATT	AATGCAGGGTTTGGACTTCA	4	0.63	0.56	Hurtado 2012

[a]Number of unique alleles detected in the population.
[b]Polymorphism information content for each marker.

Table 5. Genetic differentiation (pairwise Fst) estimates of SSRs for *S. melongena* grouped by continent.

	Fst[b]			
Category[a]	**Africa**	**Asia**	**Europe**	**N. America**
Asia	0.00	-		
Europe	0.04*	0.02	-	
N. America	0.04*	0.00	0.03*	-
S. America	0.03	0.11*	0.05	0.06*

[a]Categories with less than four lines were excluded from analyses and are not shown.
[b]Average values for SSRs are presented; * indicates value was outside the 2.5% and 97.5% confidence intervals at 1000 bootstraps.

were accounted using the Kenward-Rogers degrees of freedom option implemented in SAS software. Line mean values for fruit shape, length and width were calculated using the lsmeans statement of SAS software. Correlations between fruit shape parameters and disease susceptibility were estimated using the PROC CORR function of SAS. Disease susceptibility correlations were evaluated for each isolate and experimental replicate separately.

Genetic Analyses

Genomic DNA was extracted from young green leaves of eggplants using the Nucleo Spin II DNA extraction kit (Machery-Nagel Germany, CAT#740770) according to the manufacturer's instructions. DNA was normalized to 5 ng/ µl using the Nano-Drop ND 1000 spectrophotometer and NanoDrop 2.4.7c software (NanoDrop Technologies Inc., Wilmington, DE).

One hundred ninety-two primers from previously published SSR markers ([3,49], solgenomics.org) or designed (Primer 3 http://primer3.sourceforge.net/) from putative Solanaceae defense-related genes (NCBI http://ncbi.nlm.nih.gov) were tested against a subset of the eggplant collection to identify polymorphic markers. Reactions were performed in 15 µl total volume and contained 1 µl DNA, and 0.15 µl GoTaq (Promega Corporation Madison, WI), 0.9 µl 25 µM $MgCl_2$, 0.3 µl dNTPs, and 0.6 µl each of forward and reverse primers (Integrated DNA Technologies, Inc., Coralville, IA), with 8.45 µl ddH_2O. PCR reactions were performed in a programmable thermal cycler (Eppendorf, Westbury, NY) using the program: initial denaturation, 94°C (3 min) followed by 35 cycles at 94°C (30 s), 60°C (30 s) and 72°C (1 min), with a final extension step of 10 min at 72°C. PCR products were analyzed by electrophoresis in 4% (wt/vol) agarose gel in 1× Tris-borate-EDTA buffer, stained with ethidium bromide (5 ug/ml) for visualization and compared to a 100-bp ladder (Invitrogen Life Technologies, Burlington, ON, Canada) to determine amplicon sizes. SSR markers identified as polymorphic in the population were used for genetic diversity, population structure and trait associations.

Table 6. Genetic diversity estimates for SSRs for *S. melongena* grouped by continent and country of origin.

	Diversity estimates[b]		
Category[a]	**AlleleNo**	**G_D**	**PIC**
Africa	2.71	0.51	0.43
Asia	3.10	0.48	0.41
Europe	2.81	0.48	0.41
N. America	3.10	0.50	0.44
S. America	2.67	0.46	0.39
China	2.18	0.39	0.32
France	2.14	0.38	0.31
Iran	2.23	0.42	0.35
Japan	1.91	0.35	0.28
Macedonia	2.18	0.40	0.33
South Africa	2.14	0.41	0.34
Spain	2.50	0.42	0.36
Thailand	2.36	0.42	0.36
Turkey	2.36	0.40	0.34
USA	2.77	0.48	0.42

[a]Categories with less than four lines were excluded from analyses and are not shown.
[b]Mean values are presented for the average number of alleles (AlleleNo), genetic diversity (G_D) and the polymorphism information content (PIC).

Table 7. Genetic differentiation (pairwise Fst) estimates of SSRs for *S. melongena* grouped by country.

	Fst[b]								
Category[a]	**China**	**France**	**Iran**	**Japan**	**Macedonia**	**S. Africa**	**Spain**	**Thailand**	**Turkey**
France	0.00	-							
Iran	0.00	0.04	-						
Japan	0.01	0.04	0.05	-					
Macedonia	0.06	0.13*	0.09*	0.08	-				
S. Africa	0.04	0.08	0.01	0.07	0.05	-			
Spain	0.06	0.04	0.04	0.13*	0.13*	0.09*	-		
Thailand	0.05	0.04	0.07*	0.15*	0.10*	0.04	0.01	-	
Turkey	0.04	0.09	0.06	0.15*	0.03	0.05	0.09	0.04	-
USA	0.05	0.10*	0.04	0.17*	0.10*	0.05	0.04*	0.07*	0.05

[a]Categories with less than four lines were excluded from analyses and are not shown.
[b]Average values for SSRs are presented; * indicates value was outside the 2.5% and 97.5% confidence intervals at 1000 bootstraps.

Genetic Diversity and Population Structure

Genetic diversity was estimated using Powermarker v3.25 [50] and significance at each locus was determined with 1000 permutations using the Exact test; overall genetic diversity was estimated using the Mantel test as implemented in Powermarker. Genetic distance matrix values were calculated using Euclidean distance with the unweighted pair group method with arithmetic mean (UPGMA) and visualized in MEGA5 [50,51].

Population structure of the germplasm was analyzed using STRUCTURE v2.3.4 [52]. Following preliminary analyses, burnin length, MCMC chain replication and lambda were selected to be 200,000, 500,000 and 1.52, respectively. Population number (k) was determined empirically by comparing posterior distribution likelihoods independently among 3 independent runs of $K = 1$ to 20 as described by Evanno et al. [53]. Data included 22 polymorphic SSRs and were analyzed using the admixture model and correlated allele frequencies without previous population information [52,54]. Wright's subpopulation fixation index (Fst), the proportion of the total genetic variance within a subpopulation, significance between populations was determined using 1000 bootstrap replicates as implemented in Powermarker [55].

Visualization of the resulting Q (proportion of membership based on a 0 to 1 scale) of each accession into predefined categories (country, continent, species, disease susceptibility and fruit shape) was generated using the Population Sorting Tool (PST) in R [19,56] (J.J. Morrice, unpublished). Individuals with $Q \geq 0.6$ membership in a single subpopulation were labeled as such. Individuals with $Q<0.6$ membership in a single subpopulation were considered admixed. Significance of population structure predefined categories was estimated using the population differentiation test implemented in Powermarker. Significance at each locus and overall was determined using 1000 permutations. Countries represented by less than four individuals were excluded from analyses. Significance of pairwise Fst differentiation was based on 2.5% and 97.5% confidence intervals ($P = 0.05$) based off of 1000 bootstrap replications.

Results

Phytophthora capsici Disease Resistance and Fruit Shape

Significant differences between experimental runs indicated the effect of environmental variability during fruit growth and development on fruit disease resistance was high. In each repetition of the experiment, there were significant differences among plant accessions ($P<0.0001$). No significant differences in disease severity were found between isolates in any replicate of the experiment ($P = 0.32$, $P = 0.43$, and $P = 0.43$). The interaction between accessions and isolate was significant for each run (Run 1: $P = 0.0008$; Run 2: $P<0.0001$; and Run 3 $P<0.0001$) of the experiment. Differences in pathogen growth (absence/presence) and the interaction between pathogen growth and accession were not significant in any replicate (approximately $P = 1.0$ for each). The majority of the accessions were susceptible at 7 days post inoculation to isolates OP97 (89%) and 12889 (87%), respectively (Table 2). Symptoms included brown discoloration of the fruit and water soaking, with occasional external mycelial growth (Fig 1). Eggplant accession PI 413784 was the only line completely Rto both isolates tested. Susceptibility to one isolate did not always result in susceptibility to the other isolate. Lines PI 413782 and Grif 1276 were R (rating $= 0$) to isolate OP97 and S or MS to isolate 12889. *S. melongena* lines, MM1365, PI 193599, PI 263727 and PI 419198 were MS (rating <2) to isolate OP97. Eggplant lines H15 and PI 441908 were R to isolate 12889 and S and MS to isolate OP97, respectively. Two of the *S. linnaeanum* lines, PI

Table 8. Genetic differentiation (pairwise Fst) estimates of SSRs for eggplant germplasm grouped by disease resistance.

	Fst[b]		
Category[a]	**R/MR**	**MS**	**S**
R/MR	-	0.19	0.52*
MS	0.01	-	0.15
S	0.00	0.03*	-

[a]12889 appears below the diagonal and OP97 values are above the diagonal; MS = moderately susceptible, R/MR = resistant/moderately resistant, S = susceptible.
[b]Average values for SSRs are presented; * indicates value was outside the 95% confidence interval at 1000 bootstraps.

388847 and PI 388846, and the *S. incanum* accession PI 500922 were R or MR to isolate 12889. Lines PI 181896, PI 233916, MM 56, and PI 413783 were MS to isolate 12889. *Phytophthora capsici* isolates were successfully recovered from diseased fruits and mefenoxam sensitivity was confirmed for each isolate (*data not shown*).

Fruit shape and size varied considerably in the population (Fig 2, Table 2). *S. melongena* accessions had fruit shape ratios ranging from 1 (round) to 8 (elongate). The wild species evaluated (*S. linnaeanum, S. gilo*, and *S. incanum*) had a fruit shape ratio of approximately 1 (round) with fruit ≤3 cm (Table 2). *Solanum melongena* line PI 413783 had the lowest fruit shape ratio (0.46) and line MM 522 had the highest fruit shape ratio (8).

When evaluated by country, *S. melongena* fruits from Japan had the highest length:width ratio indicating fruits were slender and elongated. Fruits from Thailand had the lowest fruit shape ratio, indicating fruits were more round. Fruit length and width also varied greatly between countries. Fruits from China were the widest and Japan the narrowest. Fruits from Japan were also the longest and fruits from Thailand were the shortest (Table 3).

Diversity of SSR Loci in the Eggplant Germplasm Collection

The 192 primers evaluated yielded 22 polymorphic markers that were used for characterizing and evaluating genetic diversity of the eggplant collection (Table 4). A total of 83 alleles were detected among the 22 SSRs, ranging from 2 to 7 alleles per locus with an average allele diversity of 3.8 alleles per locus. The mean genetic diversity index of the collection was 0.49 ranging from 0.03 (T0633) to 0.76 (CSM31) (Table 4). The mean polymorphism information content (PIC) value was 0.42 and individual markers ranged from 0.03 to 0.71 for the population. The highest PIC value was 0.35 in PI 290467 and the lowest PIC value was 0.085 in Grif 1276. Genetic diversity was equally distributed within continents (0.46–0.51), and pairwise Fsts indicated low to moderate genetic differentiation between continents (0.00–0.11) (Table 5). Genetic diversity within countries was similar (0.35–0.48) (Table 6), and pairwise Fst values suggested low (0.00) to great (0.17) genetic differentiation among countries (Table 7). Pairwise Fsts for disease resistance to isolates 12889 and OP97 showed little (0.00) to very great (0.52) genetic differentiation between categories (Table 8). MR/R phenotypes only had high and significant ($P = 0.05$) genetic differentiation with S phenotypes for isolate OP97. Significant genetic differentiation was also detected between the MS and the S category for isolate 12889. No significant genetic differentiation was found between the R/MR category and the MS categories for either *Phytophthora* isolate (Table 8–9). Genetic diversity of fruit shape categories was moderate to high (0.43–0.52) (*data not shown*). Pairwise Fst differentiation between fruit shape categories was low (0.00) to moderate (0.1) (Table 9). Individuals with an elongate fruit shape were significantly differentiated from those with a round or oval fruit shape. Significant differentiation was also detected between round shaped individuals and semi-elongated individuals (Table 9).

Population Structure Analysis

Population structure of the 99 accessions was estimated using the STRUCTURE software and the 22 polymorphic SSRs. Accessions were grouped into four genetic clusters (Ln = - 3381.8). *S. linnaeanum, S. gilo* and *S. incanum* accessions were placed into genetic Cluster 3, while *S. melongena* individuals were distributed through each of the clusters (Table 2). Seventy-eight individuals could be assigned to a single cluster based on membership, while the remaining 21 individuals could not be assigned and were classified as admixed. When compared with the UPGMA tree, STRUCTURE-inferred clusters largely overlapped with the grouping of branches based on genetic distance (Fig 3). Relationships between the inferred clusters according to the UPGMA tree indicated that Cluster 3 was more differentiated from Clusters 1, 2 and 4, and that Clusters 1 and 4 and Clusters 2 and 4 were less differentiated from each other. Pairwise Fsts between clusters were significant and ranged from 0.08 to 0.17, indicating 8–17% of the

Table 9. Genetic differentiation (pairwise Fst) estimates of SSRs for *S. melongena* germplasm grouped by fruit shape.

	Fst[b]		
Category[a]	**Elongate**	**Oval**	**Round**
Oval	0.10*		
Round	0.06*	0.00	
Semi-Elongate	0.04	0.02	0.03*

[a]Fruit shape category based on the ratio of mean length:mean width for each line.
[b]Average values for SSRs are presented; * indicates value was outside the 2.5% and 97.5% confidence interval at 1000 bootstraps.

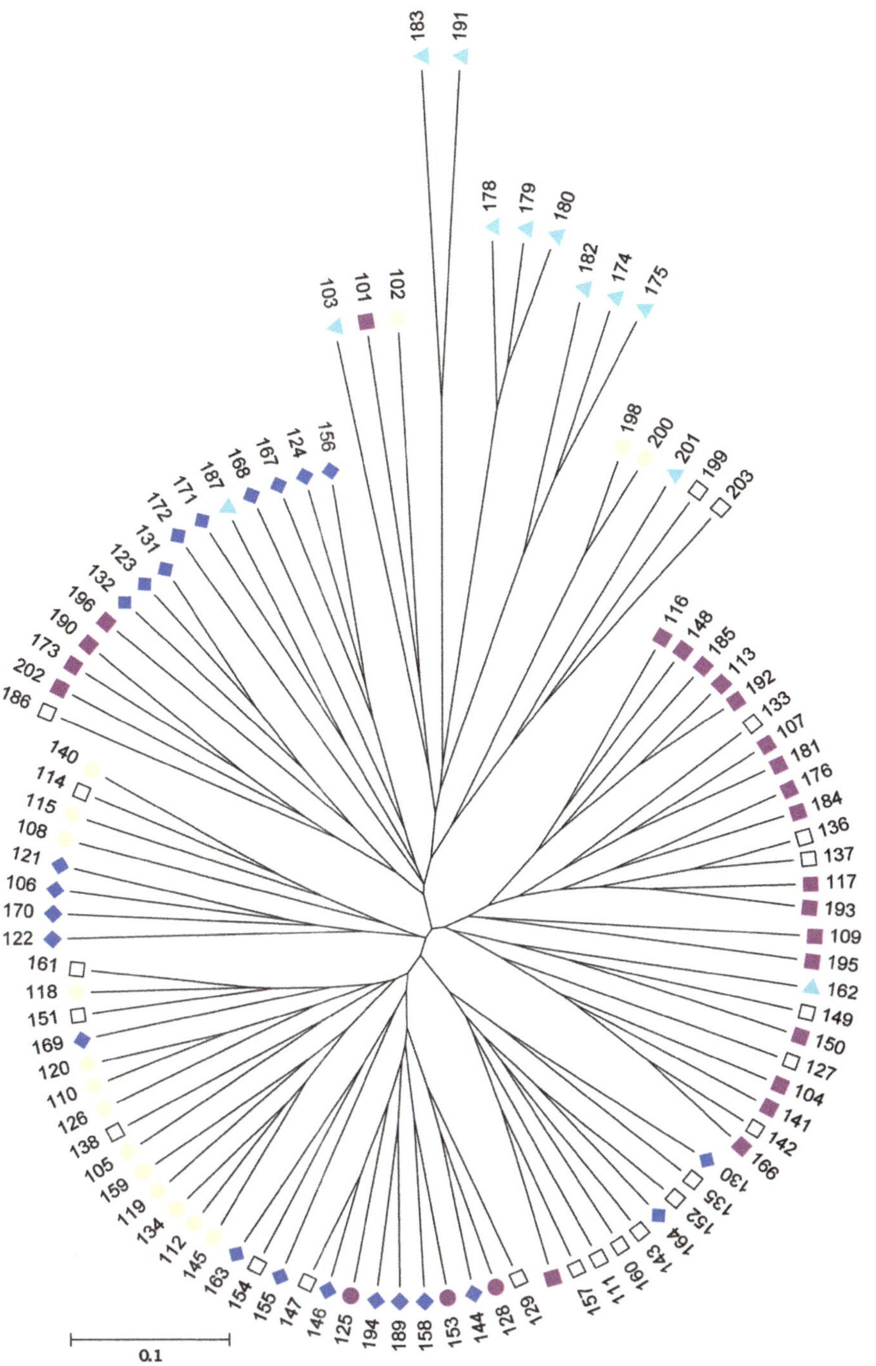

Figure 3. UPGMA genetic distance matrix differences between eggplant lines. Lines are colored based on their STRUCTURE inferred subpopulations. Cluster 1 individuals are denoted by purple squares, Cluster 2 individuals are black-outlined light yellow circles, Cluster 3 individuals are sky blue counterclockwise triangles, Cluster 4 individuals are steel blue diamonds, and admixed individuals are open squares.

variation was explained by genetic differences between clusters. Cluster 1 had moderate differentiation from Clusters 2, 3 and 4. Cluster 2 had great differentiation from Cluster 3, and Cluster 4 had moderate differentiation from Clusters 2 and 3 following the guidelines suggested by Hartl and Clark [55]. Population structure was detected when individuals were grouped by continent of origin (Fig.4A), country of origin (Fig. 4B), species, fruit shape (Fig. 5) and disease resistance to isolates 12889 and OP97 (Fig. 6), as some clusters were more frequent than others in each grouping (Fig. 4–6). Cluster 3 individuals were not represented in Asia, and Cluster 4 individuals were not represented in Africa (Fig 4). For both isolates, individuals from Cluster 4 were not represented in the MR/R categories for either isolate, had low representation in the MS category, and were highly represented in the S category. Cluster 3 individuals were highly represented in both the R/MR and S categories, but not the MS for both isolates (Fig. 6). When grouped by fruit shape (round, oval, semi-elongate and elongate), Cluster 1 was under represented in the oval and elongate fruit

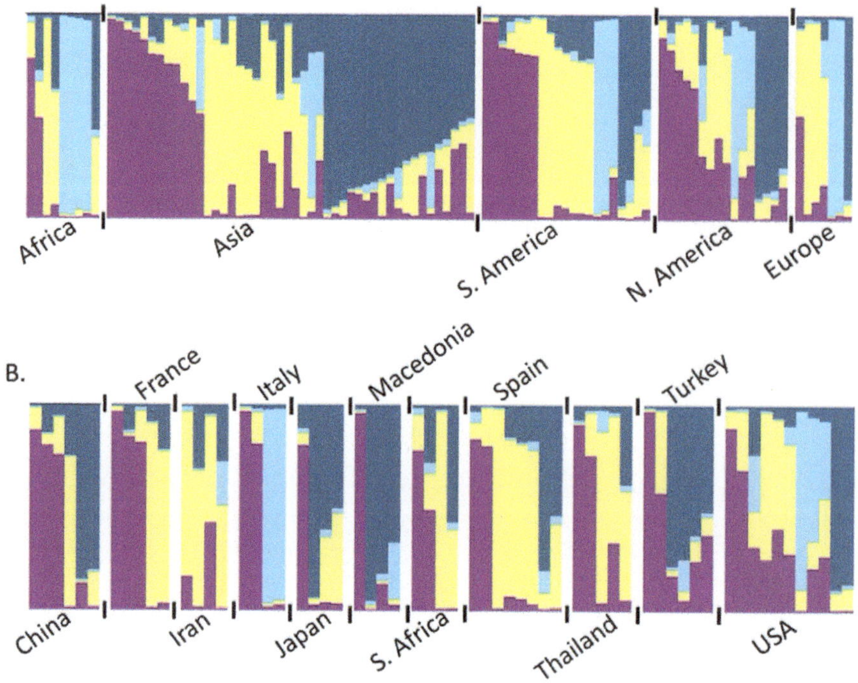

Figure 4. Population structure grouped by country (A) and continent (B) of origin for eggplant (*S. melongena*) germplasm. Cluster 1 (purple), Cluster 2 (light yellow), Cluster 3 (sky blue) and Cluster 4 (dark blue). A white space and black tick marks separate subgroups of individuals. (A) Population structure grouped by country of origin for the *S. melongena* germplasm. Only countries represented by four or more individuals were included (A).

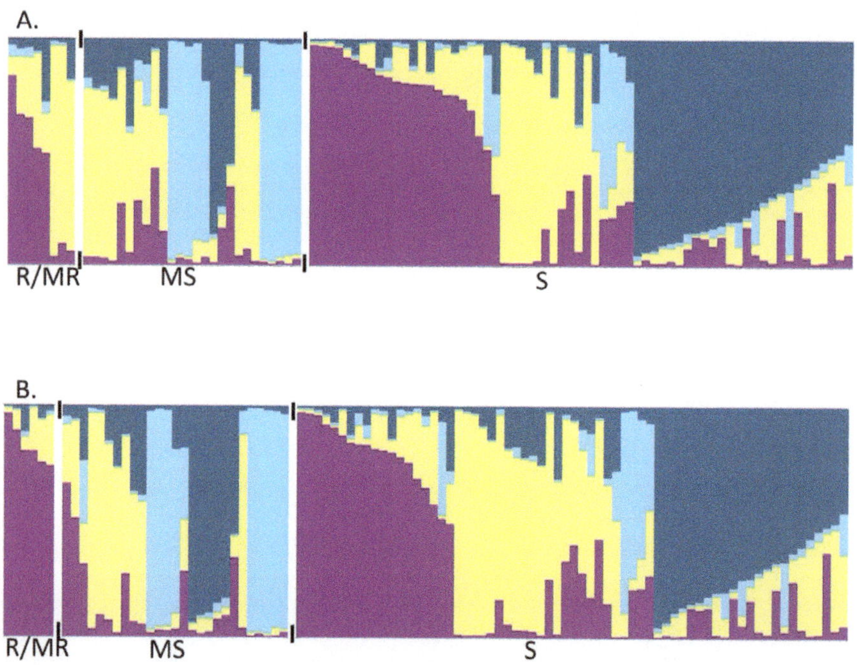

Figure 5. Population structure grouped by disease resistance to isolate 12889 (A) and OP97 (B). Individuals were grouped into a resistant and moderately resistant category (R/MR), a moderately susceptible category (MS), and a susceptible category (S) based on their mean disease ratings. Cluster 1 (purple), Cluster 2 (light yellow), Cluster 3 (sky blue) and Cluster 4 (dark blue). A white space and black tick marks separate subgroups of individuals.

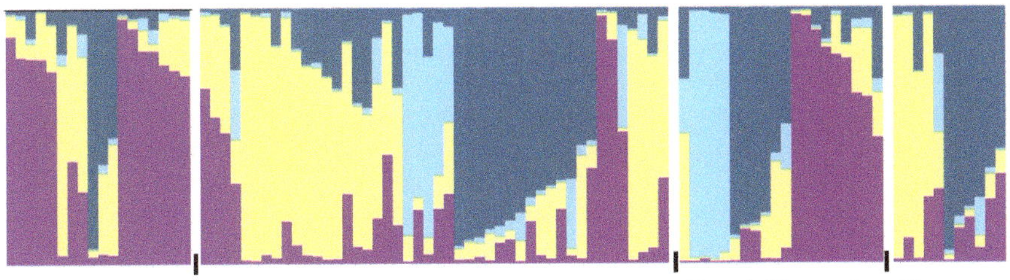

Figure 6. Population structure grouped by *S. melongena* fruit shape. Cluster 1 (purple), Cluster 2 (light yellow), Cluster 3 (sky blue) and Cluster 4 (dark blue). A white space and black tick marks separate subgroups of individuals.

shape categories. Clusters 2 and 4 both had low representation in the round category. Cluster 3 was not represented in round or elongated individuals and had minor representation in the oval shape category.

Discussion

This study investigated Phytophthora fruit rot resistance, fruit shape, population structure and genetic diversity in a worldwide collection of eggplant. The overall estimate of genetic diversity of the collection was moderate (0.5) in our study, similar to a recent report on eggplant diversity, though this is likely to be an underestimation due to limited sampling [3]. Bayesian clustering, using SSR markers, identified four genetic clusters in the eggplant collection. Most individuals belonged predominantly to one of the four clusters, while 20% were admixed according to the inferred clustering. Admixture, an indicator of migration or interbreeding between genetic clusters, was moderate in our population. Inferred genetic clusters did not directly correspond with the predefined categories of continent, country, fruit shape or Phytophthora fruit rot resistance, though some clusters did appear more frequently in one category compared to another.

On eggplant, fruit rot is the most common symptom of *P. capsici* seen in the field [23]. Symptoms start as small water soaked lesions, turning brown and eventually covering the whole fruit. Advanced symptoms can include complete rotting of the fruit and visible mycelia on the external surface of the fruit [23]. Isolate-specific interactions and partial fruit rot resistance have been identified in other solanaceous species (tomatoes and peppers) suggesting a multigenic host response, but no studies have looked at Phytophthora fruit rot in eggplant [24] (Granke et al. unpublished). In our study, the eggplant accessions evaluated demonstrated partial and isolate-specific resistance to Phytophthora fruit rot. Most lines evaluated were completely susceptible to both isolates (~90%). Several eggplant accessions displayed isolate-specific resistance; these individuals were placed into genetic Clusters 2 and 3, and were from S. America, Asia, Africa and Europe. Two of these geographic regions are known centers of eggplant diversity, and likely harbor additional sources of resistance [2–5]. Only one of the 99 lines evaluated, a Cluster 3 landrace eggplant collected in Burkina Faso in the early 1900s, had complete resistance to both isolates evaluated. This accession also showed high levels of genetic similarity to the wild eggplant relatives evaluated, *S. incanum, S. gilo*, and *S. linnaeanum*. While further evaluation with more isolates is necessary, PI 413784 appears to be a promising source of host resistance to Phytophthora fruit rot in eggplant.

When categorized by disease resistance (S, MS, MR, R) for each isolate, there was significant genetic differentiation among eggplant genotypes infected with isolates OP97 or 12889. Individuals that were R and MR to isolate OP97 were significantly differentiated from individuals that were S. Only S individuals were significantly differentiated from the MS individuals when inoculated with isolate 12889. These results emphasize the importance of utilizing different *P. capsici* isolates when breeding for resistance. The three wild relatives, *S. linnaeanum, S. gilo*, and *S. incanum*, showed partial or isolate-specific resistance to the two isolates evaluated in this study.

When grouped by species, *S. linnaeanum, S. gilo*, and *S. incanum* individuals evaluated were predominantly in genetic Cluster 3. *Solanum incanum* has long been part of the eggplant complex and ancestral individuals are thought to be one of the progenitors of modern eggplant [43,44]. *Solanum linnaeanum* is a related species and has only recently been included as a possible progenitor of the modern eggplant with limited crossability [5,8]. Genetic Cluster 3 individuals were also detected in the *S. melongena* category, supporting gene movement between *S. melongena* and its wild relatives, *S. incanum* and *S. linnaeanum*. These *S. melongena* individuals may have been misclassified, but are more likely the result of introgression since the wild species were small fruited and prickly.

Cultivated eggplant, similar to pepper and tomato, is a phenotypically diverse species with varying levels of genotypic diversity [42,44,46,57]. *Solanum melongena* fruit shape, size and color is a byproduct of domestication, selection, and breeding for specific market classes. Phenotypic evaluation of eggplant fruit shape varied greatly among the *S. melongena* accessions evaluated, while the wild species, *S. incanum, S. gilo, and S. linnaeanum*, had no variation in fruit shape. Maintaining market class variation may be difficult when incorporating traits like fruit rot resistance, which was most often observed in accessions with a limited size range.

Significant differences in this population were observed in fruit shape, length and width among eggplant lines when grouped by country of origin, representing different market classes, in this study. Since eggplant has market classes particular to geographic areas, it was expected that population structure categorized by fruit shapes and country of origin would correspond with the inferred genetic clusters. Significant differentiation was seen between *S. melongena* individuals with elongated fruit shapes and those with round and oval fruit shapes. Individuals with a round fruit shape were also significantly differentiated from semi-elongate fruit shape individuals. These results are consistent with limited breeding among market classes. However, inferred population structure did not correspond with the fruit categories and this may be the result of limited sampling in each geographic location. While only genetic Cluster 3 was not represented in the round or elongate shape category, all other clusters were represented by at least one individual in each category.

When grouped by country and continent, significant population structure and moderate genetic diversity was evident among the categories evaluated. The highest levels of genetic diversity were seen within the continents of Africa and N. America. The highest level of genetic diversity for countries was in the USA. The increased genetic diversity in Africa is likely due to the prevalence and intercrossing of wild and related species, as Africa is the center of origin for several eggplant species. The increased genetic diversity in N. America and the USA may be the result of breeding programs integrating wild relatives and varieties from around the world. The diversity could also be from the movement of Asian and European varieties into the U.S., which may be marketed under different names. Overall differentiation among countries was similar to the differentiation among continents, and future core collections should include individuals from areas with high genetic diversity and genetic differentiation. In particular, genotypes from China were not significantly differentiated from any other country, while genotypes from Thailand, Japan, Spain, Macedonia and the U.S. were frequently significantly differentiated from other countries. Similarly, Asia was not significantly differentiated from populations from Europe, Africa and N. America, while N. America, Europe and Africa were all significantly differentiated from each other. Asia, as a center of diversity and domestication, and in particular genotypes from China, may be more akin to the ancestral population from which these other pools were derived, making them more similar and less differentiated from other eggplants.

Cultivated eggplant, compared to other solanaceous species, is an understudied crop with worldwide importance. This study provides an overview of the population structure, genetic diversity and Phytophthora fruit rot resistance of a geographically diverse set of eggplant. The estimates of genetic diversity and the four genetic clusters found in this study are likely to be lower than actual genetic diversity and structure of eggplant due to limited sampling and molecular markers. A previous study using a subset of SSRs in a smaller collection of eggplant was able to identify more allelic variation at each locus [3]. While population structure was significant for disease resistance, fruit shape, continent and country, the genetic clusters did not completely correspond with these predefined categories in our study, which may be due, in part, to unequal samples in each category. Future studies involving eggplant diversity, disease resistance and other agronomic traits should aim to include individuals from around the world for maximum diversity, and will need to consider the effect of population structure on marker-trait associations.

Acknowledgments

The authors would like to acknowledge the technical assistance of J. Mitchell, A. Tomlinson and J. Kurjan. We would also like to thank Dr. L.L. Granke for critical evaluation of this manuscript, and we would like to thank Dr. J. Prohens and the entities listed in Table 1 for providing eggplant seeds.

Author Contributions

Conceived and designed the experiments: RN. Performed the experiments: RN SB. Analyzed the data: RN SB. Contributed reagents/materials/analysis tools: MK. Wrote the paper: RN LMQ-O MKH.

References

1. Wang J-X, Gao T-G, Knapp S (2008) Ancient Chinese Literature Reveals Pathways of Eggplant Domestication. Ann Bot 102: 891–897.
2. Meyer RS, Karol KG, Little DP, Nee MH, Litt A (2012) Phylogeographic relationships among Asian eggplants and new perspectives on eggplant domestication. Mol Phylogenet Evol 63: 685–701.
3. Hurtado M, Vilanova S, Plazas M, Gramazio P, Fonseka HH, et al. (2012) Diversity and Relationships of Eggplants from Three Geographically Distant Secondary Centers of Diversity. PLoS ONE 7: e41748.
4. Prohens J, Blanca JM, Nuez F (2005) Morphological and Molecular Variation in a Collection of Eggplants from a Secondary Center of Diversity: Implications for Conservation and Breeding. J Am Soc Hortic Sci 130: 54–63.
5. Weese TL, Bohs L (2010) Eggplant origins: Out of Africa, into the Orient. Taxon 59: 49–56.
6. Daunay MC, Lester RN, Gebhardt CG, Hennart JW, Jahn M, et al. (2001) Genetic resources of eggplant (*Solanum melongena*) and allied species: A new challenge for molecular geneticists and eggplant breeders. Solanaceae V: Advances in taxonomy and utilization. Njimegen: Nijmegen University Press. pp. 251–274.
7. Daunay MC, Lester RN, Laterrot H (1991) The use of wild species for the genetic improvement of Brinjal eggplant (*Solanum melongena*) and tomato (*Lycopersicon esculentum*). Solanaceae III: Taxonomy, chemistry and evolution. Kew: Kew Royal Botanic Gardens. pp. 389–412.
8. Hasan S, Lester RN (1992) Crossability relationship of Solanum melongena L. with its nearest wild relatives. Acta Hortic: 123–133.
9. Daunay M-C, Dalmon A, Lester R (1999) Management of a collection of Solanum species for eggplant (*Solanum melongena*) breeding purposes. in: Nee M, Symon DE, Lester RN, Jessop JP (eds). Solanaceae IV: Advances in biology and utilization. Kew: Royal Botanic Gardens. pp. 369–383.
10. Jones D, Jones GA, Hagen T, Creech E (1985) Wild species of Nicotiana as a new source of tobacco resistance to the tobacco hornworm Manduca sexta. Entomol Exp Appl 38: 157–164.
11. Spooner DM, Bamberg JB (1994) Potato genetic resources: Sources of resistance and systematics. Am Potato J 71: 325–337.
12. Bubici G, Cirulli M (2008) Screening and selection of eggplant and wild related species for resistance to *Leveillula taurica*. Euphytica 164: 339–345.
13. Benetka V, Novotná K, Štochlová P (2012) Wild populations as a source of germplasm for black poplar (Populus nigra L.) breeding programmes. Tree Genet Genomes 8: 1073–1084.
14. Diamanti J, Capocasa F, Balducci F, Battino M, Hancock J, et al. (2012) Increasing Strawberry Fruit Sensorial and Nutritional Quality Using Wild and Cultivated Germplasm. PLoS ONE 7: e46470.
15. Laserna-Ruiz I, De-Los-Mozos-Pascual M, Santana-Méridas O, Sánchez-Vioque R, Rodríguez-Conde MF (2012) Screening and selection of lentil (*Lens Miller*) germplasm resistant to seed bruchids (*Bruchus spp.*). Euphytica 188: 153–162.
16. Yun SJ, Gyenis L, Hayes PM, Matus I, Smith KP, et al. (2005) Quantitative Trait Loci for Multiple Disease Resistance in Wild Barley. Crop Sci 45: 2563.
17. Muñoz-Falcón JE, Prohens J, Vilanova S, Nuez F (2009) Diversity in commercial varieties and landraces of black eggplants and implications for broadening the breeders' gene pool. Ann Appl Biol 154: 453–465.
18. Foster JM, Hausbeck MK (2010) Resistance of Pepper to Phytophthora Crown, Root, and Fruit Rot Is Affected by Isolate Virulence. Plant Dis 94: 24–30.
19. Quesada-Ocampo LM, Hausbeck MK (2010) Resistance in tomato and wild relatives to crown and root rot caused by *Phytophthora capsici*. Phytopathology 100: 619–627.
20. Foster JM, Naegele RP, Hausbeck M (2013) Evaluation of eggplant rootstocks and pepper varieties for potential resistance to isolates of *Phytophthora capsici* from Michigan and New York. Plant Dis 97: 1037–1041.
21. Quesada-Ocampo LM, Granke LL, Mercier MR, Olsen J, Hausbeck MK (2011) Investigating the genetic structure of *Phytophthora capsici* populations. Phytopathology 101: 1061–1073.
22. Granke LL, Quesada-Ocampo L, Lamour K, Hausbeck MK (2012) Advances in research on *Phytophthora capsici* on vegetable crops in the United States. Plant Dis 96: 1588–1600.
23. Gevens AJ, Roberts P (2011) Vegetable Diseases Caused by *Phytophthora capsici* in Florida. Available: http://edis.ifas.ufl.edu/vh045. Accessed 2013 Feb 19.
24. Naegele R, Hill TA, Ashrafi H, Reyes Chin-Wo S, Van Deynze A, et al. (2014) QTL mapping of fruit rot resistance to the plant pathogen *Phytophthora capsici* Leonian in a recombinant inbred line Capsicum annuum L. population. Phytopathology 104(5):479–83.
25. Nunome T, Ishiguro K, Yoshida T, Hirai M (2001) Mapping of Fruit Shape and Color Development Traits in Eggplant (*Solanum melongena* L.) Based on RAPD and AFLP Markers. Breed Sci 51: 19–26.
26. Zygier S, Chaim AB, Efrati A, Kaluzky G, Borovsky Y, et al. (2005) QTLs mapping for fruit size and shape in chromosomes 2 and 4 in pepper and a comparison of the pepper QTL map with that of tomato. Theor Appl Genet 111: 437–445.
27. Tsaballa A, Pasentsis K, Darzentas N, Tsaftaris AS (2011) Multiple evidence for the role of an Ovate-like gene in determining fruit shape in pepper. BMC Plant Biol 11: 46.
28. Rodríguez GR, Muños S, Anderson C, Sim S-C, Michel A, et al. (2011) Distribution of *SUN*, *OVATE*, *LC*, and *FAS* in the Tomato Germplasm and the Relationship to Fruit Shape Diversity. Plant Physiol 156: 275–285.
29. Qiao J, Chen Y, Wang L, Liu F, Zhang Y, et al. (2012) QTL analysis for fruit shape in eggplant based on genetic linkage map. Acta Hortic Sin 39: 1115–1122.

30. Xu J, Ranc N, Muños S, Rolland S, Bouchet J-P, et al. (2013) Phenotypic diversity and association mapping for fruit quality traits in cultivated tomato and related species. Theor Appl Genet 126: 567–581.
31. Polignano G, Uggenti P, Bisignano V, Gatta C (2010) Genetic divergence analysis in eggplant (*Solanum melongena* L.) and allied species. Genet Resour Crop Evol 57: 171–181.
32. Daunay M-C (2008) Eggplant. In: Handbook of crop breeding, Vegetables II: Fabaceae, Liliaceae, Umbelliferae, and Solanaceae.
33. Mace ES, Lester RN, Gebhardt CG (1999) AFLP analysis of genetic relationships among the cultivated eggplant, *Solanum melongena* L., and wild relatives (Solanaceae). Theor Appl Genet 99: 626–633.
34. Muñoz-Falcón JE, Vilanova S, Plazas M, Prohens J (2011) Diversity, relationships, and genetic fingerprinting of the Listada de Gandía eggplant landrace using genomic SSRs and EST-SSRs. Sci Hortic 129: 238–246.
35. Chen X, Min D, Yasir TA, Hu Y-G (2012) Genetic Diversity, Population Structure and Linkage Disequilibrium in Elite Chinese Winter Wheat Investigated with SSR Markers. PLoS ONE 7: e44510.
36. Cao K, Wang L, Zhu G, Fang W, Chen C, et al. (2012) Genetic diversity, linkage disequilibrium, and association mapping analyses of peach (*Prunus persica*) landraces in China. Tree Genet Genomes 8: 975–990.
37. Courtois B, Frouin J, Greco R, Bruschi G, Droc G, et al. (2012) Genetic Diversity and Population Structure in a European Collection of Rice. Crop Sci 52: 1663.
38. Fricano A, Bakaher N, Corvo MD, Piffanelli P, Donini P, et al. (2012) Molecular diversity, population structure, and linkage disequilibrium in a worldwide collection of tobacco (*Nicotiana tabacum* L.) germplasm. BMC Genet 13: 18.
39. Iqbal MJ, Mamidi S, Ahsan R, Kianian SF, Coyne CJ, et al. (2012) Population structure and linkage disequilibrium in *Lupinus albus* L. germplasm and its implication for association mapping. Theor Appl Genet 125: 517–530.
40. Soto-Cerda BJ, Maureira-Butler I, Muñoz G, Rupayan A, Cloutier S (2011) SSR-based population structure, molecular diversity and linkage disequilibrium analysis of a collection of flax (*Linum usitatissimum* L.) varying for mucilage seed-coat content. Mol Breed 30: 875–888.
41. Wang ML, Sukumaran S, Barkley NA, Chen Z, Chen CY, et al. (2011) Population structure and marker–trait association analysis of the US peanut (*Arachis hypogaea* L.) mini-core collection. Theor Appl Genet 123: 1307–1317.
42. Karihaloo JL, Brauner S, Gottlieb LD (1995) Random amplified polymorphic DNA variation in the eggplant, *Solanum melongena* L. (Solanaceae). Theor Appl Genet 90: 767–770. doi:10.1007/BF00222010.
43. Furini A, Wunder J (2004) Analysis of eggplant (*Solanum melongena*)-related germplasm: morphological and AFLP data contribute to phylogenetic interpretations and germplasm utilization. Theor Appl Genet 108: 197–208.
44. Behera TK, Sharma P, Singh BK, Kumar G, Kumar R, et al. (2006) Assessment of genetic diversity and species relationships in eggplant (*Solanum melongena* L.) using STMS markers. Sci Hortic 107: 352–357.
45. Ali Z, Xu ZL, Zhang DY, He XL, Bahadur S, et al. (2011) Molecular diversity analysis of eggplant (*Solanum melongena*) genetic resources. Genet Mol Res 10: 1141–1155.
46. Li H, Chen H, Zhuang T, Chen J (2010) Analysis of genetic variation in eggplant and related Solanum species using sequence-related amplified polymorphism markers. Sci Hortic 125: 19–24.
47. Lamour KH, Hausbeck MK (2000) Mefenoxam insensitivity and the sexual stage of *Phytophthora capsici* in Michigan cucurbit fields. Phytopathology 90: 396–400.
48. Waterhouse GM (1963) Key to the species of *Phytophthora* de Bary. Kew, Surrey, UK: Commonwealth Mycological Society.
49. Minamiyama Y, Tsuro M, Kubo T, Hirai M (2007) QTL Analysis for Resistance to *Phytophthora capsici* in Pepper Using a High Density SSR-based Map. Breed Sci 57: 129–134.
50. Liu K, Muse SV (2005) PowerMarker: an integrated analysis environment for genetic marker analysis. Bioinformatics 21: 2128–2129.
51. Tamura K, Peterson D, Peterson N, Stecher G, Nei M, et al. (2011) MEGA5: Molecular Evolutionary Genetics Analysis Using Maximum Likelihood, Evolutionary Distance, and Maximum Parsimony Methods. Mol Biol Evol 28: 2731–2739.
52. Pritchard JK, Stephens M, Donnelly P (2000) Inference of Population Structure Using Multilocus Genotype Data. Genetics 155: 945–959.
53. Evanno G, Regnaut S, Goudet J (2005) Detecting the number of clusters of individuals using the software structure: a simulation study. Mol Ecol 14: 2611–2620. doi:10.1111/j.1365-294X.2005.02553.x.
54. Falush D, Stephens M, Pritchard JK (2003) Inference of Population Structure Using Multilocus Genotype Data: Linked Loci and Correlated Allele Frequencies. Genetics 164: 1567–1587.
55. Hartl DL, Clark AG (2007) Principles of population genetics. 4th ed. Sunderland, Massachusetts: Sinauer Associates, Inc. Publishers.
56. R Development Core Team (2012) R: A language and environment for statistical computing. Vienna,Austria: R Foundation for Statistical Computin-g.Available: http://www.R project.org/.
57. Vilanova S, Manzur JP, Prohens J (2012) Development and characterization of genomic simple sequence repeat markers in eggplant and their application to the study of diversity and relationships in a collection of different cultivar types and origins. Mol Breed 30: 647–660.

Analysis of Genetic Diversity and Population Structure of Rice Germplasm from North-Eastern Region of India and Development of a Core Germplasm Set

Debjani Roy Choudhury[1]ᶲ, Nivedita Singh[1]ᶲ, Amit Kumar Singh[1], Sundeep Kumar[1], Kalyani Srinivasan[2], R. K. Tyagi[2], Altaf Ahmad[3], N. K. Singh[4], Rakesh Singh[1]*

1 Division of Genomic Resources, National Bureau of Plant Genetic Resources, New Delhi, 110 012, India, **2** Division of Germplasm Conservation, National Bureau of Plant Genetic Resources, New Delhi, 110 012, India, **3** Department of Botany, Faculty of Science, Jamia Hamdard (Hamdard University), New Delhi, 110062, India, **4** National Research Centre on Plant Biotechnology, Indian Agricultural Research Institute, New Delhi, 110 012, India

Abstract

The North-Eastern region (NER) of India, comprising of Arunachal Pradesh, Assam, Manipur, Meghalaya, Mizoram, Nagaland and Tripura, is a hot spot for genetic diversity and the most probable origin of rice. North-east rice collections are known to possess various agronomically important traits like biotic and abiotic stress tolerance, unique grain and cooking quality. The genetic diversity and associated population structure of 6,984 rice accessions, originating from NER, were assessed using 36 genome wide unlinked single nucleotide polymorphism (SNP) markers distributed across the 12 rice chromosomes. All of the 36 SNP loci were polymorphic and bi-allelic, contained five types of base substitutions and together produced nine types of alleles. The polymorphic information content (PIC) ranged from 0.004 for Tripura to 0.375 for Manipur and major allele frequency ranged from 0.50 for Assam to 0.99 for Tripura. Heterozygosity ranged from 0.002 in Nagaland to 0.42 in Mizoram and gene diversity ranged from 0.006 in Arunachal Pradesh to 0.50 in Manipur. The genetic relatedness among the rice accessions was evaluated using an unrooted phylogenetic tree analysis, which grouped all accessions into three major clusters. For determining population structure, populations K = 1 to K = 20 were tested and population K = 3 was present in all the states, with the exception of Meghalaya and Manipur where, K = 5 and K = 4 populations were present, respectively. Principal Coordinate Analysis (PCoA) showed that accessions were distributed according to their population structure. AMOVA analysis showed that, maximum diversity was partitioned at the individual accession level (73% for Nagaland, 58% for Arunachal Pradesh and 57% for Tripura). Using POWERCORE software, a core set of 701 accessions was obtained, which accounted for approximately 10% of the total NE India collections, representing 99.9% of the allelic diversity. The rice core set developed will be a valuable resource for future genomic studies and crop improvement strategies.

Editor: Swarup Kumar Parida, National Institute of Plant Genome Research (NIPGR), India

Funding: Financial support granted by Indian Council of Agricultural Research, New Delhi, India, under NPTC project, is gratefully acknowledged. The funders had no role in study design, data collection and analysis, decision to publish, or preparation of the manuscript.

Competing Interests: The authors have declared that no competing interests exist.

* Email: rsingh@nbpgr.ernet.in

ᶲ These authors contributed equally to this work.

Introduction

Plant genetic resources are of paramount importance for the future and to ensure the food and nutritional security of an increasing population. A large number of genetic materials have been conserved in Genebank, but their use is being limited due to an unmanageable number of accessions and the continuous expansion of accessions numbers. Core germplasm development has been proposed for better management and use of collections available in Genebank [1]. This requires the development of a core set of accession to more precisely characterize, explore, and conserve Genebank resources, monitor the genetic drift during preservation, and identify gaps in genetic diversity [2,3]. The idea of the core set was proposed by Frankel and Brown in 1984 [2]. As proposed, a core set is a small set of accessions (usually 10 % of the population) chosen to represent the genetic spectrum of an entire collection [4].

The sampling percentage of a core set has long been under debate [5]; 20%–30% of the sampling percentage was suggested by Yonezawa *et al.* (1995) to adequately sample a given collection [6]. Mini core sets representing ~1% of total collections and have also been used to characterize very large collections [7–10]. In fact, a mini core subset from USDA rice gene bank has been developed using 26 phenotypic trait and 70 molecular markers [11]. Still the perfect ratio and fixed size for all core set does not exist, since different crops or different constructing goal needs different sampling percentages [5]. Several methods have been used to develop core set. The stratified random sampling method, in contrast to the simple random sampling method, has been successfully applied to the development of numerous core set, e.g., fodder crops, potatoes, etc. [12–16]. One common approach for constructing a core set is grouping whole collections into major ecotypes and then selecting representatives from each ecotype [5,17–18]. However, a core set formed for the purpose of capturing accessions with rare or extreme values of the desired trait(s) (e.g. high resistance to pest or high yield and yield contributing traits) should be evaluated separately with the intention of representing the (pattern of) genetic diversity in the collection. Indeed, the pattern of genetic diversity will have genetic variations among all accessions that have been accumulated as a result of natural processes, species' characteristics and historical events [19]. Until now, there is no universally accepted method for constructing a core set as many factors affect representativeness of core set, such as sampling percentage, data type, number of traits observed, genetic diversity of germplasm, grouping method and sampling method [18,20–21].

The rice genetic resource is an important source for rice breeding and makes a valuable contribution to global wealth and food security [22]. India is considered to be the origin of rice. Based on the phylo-geographical and archeological evidence it is suggested that rice was domesticated about 10,000 years ago from its wild ancestor *O. rufipogon* in the region South of Himalayan mountain range, likely in the present day eastern and North-Eastern (NE) India, extending eastward to Nepal, Myanmar and Thailand to Southern China [23–25]. The NE India is a large geographical area comprising of seven states namely Arunachal Pradesh, Assam, Manipur, Meghalaya, Mizoram, Nagaland and Tripura states, is home to a large number of indigenous rice varieties [26]. The geography of the NE states of India is unique, having snow capped peaks of the Himalayas, the ecological hot spots of the NE foothills and the Brahmaputra valley. High rainfall, humidity, varied topography and altitude, heavy natural selection pressures, environmental stresses have made the region rich both in floristic and crop diversities [27].

A large collection of the rice genetic resource from NE India is currently conserved at National Genebank (NGB), National Bureau of Plant Genetic Resources (NBPGR), New Delhi. Since these germplasm contains unique traits which can be exploited for future crop improvement programme but has not been properly characterized at molecular level. Therefore, in the present study this collection was analyzed (i) to study the genetic diversity and population structure of all 6,984 rice germplasm accessions using 36 SNP markers, which were developed and used earlier for diversity and population structure analysis of rice varieties [28] and (ii) to develop a core set representing the maximum diversity of unique NE India rice genetic resources. Taken together, the described core set can be used to effectively for genomic studies, in rice crop improvement and conservation programs in genebank.

Materials and Methods

Plant materials

A total of 6,984 accessions of NE India (Arunachal Pradesh, Assam, Manipur, Meghalaya, Mizoram, Nagaland and Tripura,) were drawn from NGB, NBPGR, New Delhi. The NE India germplasm collection also included 24 accessions with EC ID from IRRI, Philippines and two accessions from Ghana which were growing in farmer's field since their introduction in 1991. The purpose of their inclusion was to analyse their clustering in relation to Indian germplasm collection. The details of each accession along with passport information were obtained from the online database of NGB, (www.nbpgr.ernet.in) and presented in Table S1.

DNA extraction from rice seed

Seeds of each genotype (10–12 seeds) were dehusked and used for DNA isolation using QIAGEN DNeasy plant mini kit. Kernels were ground into fine powder using tissue lyser (Tissue lyser II Retsch, Germany) with a tissue lyser adapter set (QIAGENq). DNA was extracted following the procedures described by manufacturers.

Genotyping of rice accessions using SNP markers

Genomic DNA of all the 6,984 accessions was diluted to prepare working stocks of 10 ng/µl. Sequenom Mass ARRAY multiplex assays were designed for 36 SNPs (iPLEX gold chemistry), representing conserved single-copy rice genes [29], taking three genes per rice chromosome [28]. These SNP markers are located on the short arm, centromeric region and long arm of the twelve rice chromosome. The 36-plex assay(s) were designed and validated by Sequenom Corporation (San Diego). The 30-mer pre-amplification primers and variable length genotyping primers generated by the Assay Design 3.1 software were procured and used for the validation of SNPs according to the Sequenom user manual. Mass ARRAY Typer 3.4 Software was used for the visualization of SNPs and allele calling. Chromosome number, primer ID and physical position of 36 SNPs used in this study are given in Table S2.

Table 1. Estimation of gene diversity, heterozygosity, PIC and major allele frequency in NE Indian rice collection (6984)

S. No.	States	No. of accessions	Gene diversity	Heterozygosity	PIC	Major allele frequency
1	Arunachal Pradesh	663	0.006–0.49 (0.29)	0.003–0.16 (0.07)	0.006–0.37 (0.23)	0.51–0.99 (0.78)
2	Assam	2635	0.01–0.5 (0.31)	0.01–0.32 (0.10)	0.01–0.37 (0.25)	0.50–0.99 (0.76)
3	Manipur	549	0.02–0.50 (0.26)	0.004–0.26 (0.09)	0.02–0.37 (0.21)	0.50–0.98 (0.80)
4	Meghalaya	2427	0.03–0.49 (0.34)	0.01–0.29 (0.12)	0.03–0.37 (0.27)	0.50–0.98 (0.74)
5	Mizoram	107	0.01– 0.49 (0.28)	0.01–0.42 (0.16)	0.01–0.37 (0.23)	0.52–0.99 (0.79)
6	Nagaland	377	0.008–0.49 (0.31)	0.002–0.16 (0.07)	0.008–0.37 (0.25)	0.52–0.99 (0.76)
7	Tripura	226	0.004–0.49 (0.28)	0.004–0.38 (0.08)	0.004–0.37 (0.22)	0.51–0.99 (0.79)
	Total	6984				

Mean values are given in parentheses.

Table 2. State-wise Analysis of Molecular Variance (AMOVA) of NE Indian rice collections.

States	Source	df	SS	MS	Est. Var.	%
Arunachal Pradesh	Among Pops	2	1160.573	580.286	1.523	21
	Among Indiv	660	6728.744	10.195	4.305	58
	Within Indiv	663	1051.000	1.585	1.585	21
	Total	1325	8940.317		7.413	100
Assam	Among Pops	2	4963.974	2481.987	1.751	24
	Among Indiv	2632	25094.849	9.535	3.836	51
	Within Indiv	2635	4908.500	1.863	1.863	25
	Total	5269	34967.323		7.449	100
Manipur	Among Pops	2	905.987	452.994	1.422	21
	Among Indiv	546	5785.991	10.597	4.509	58
	Within Indiv	549	866.500	1.578	1.578	21
	Total	1097	7558.478		7.510	100
Meghalaya	Among Pops	4	9032.582	2258.145	2.422	30
	Among Indiv	2422	23045.557	9.515	3.745	46
	Within Indiv	2427	4914.000	2.025	2.025	35
	Total	4853	36992.138		8.192	100
Mizoram	Among Pops	2	108.378	54.189	0.747	9
	Among Indiv	104	1247.813	11.998	4.569	56
	Within Indiv	107	306.000	2.860	2.860	35
	Total	213	1662.192		8.176	100
Nagaland	Among Pops	2	40.936	20.468	0.709	8
	Among Indiv	12	168.464	14.039	6.319	75
	Within Indiv	15	21.000	1.400	1.400	17
	Total	29	230.400		8.428	100
Tripura	Among Pops	2	400.316	200.158	1.564	22
	Among Indiv	223	2084.863	9.349	3.946	57
	Within Indiv	226	329.500	1.458	1.458	21
	Total	451	2814.679		6.967	100

Statistical analyses

The major allele frequency, gene diversity, heterozygosity and Polymorphic Information Content (PIC) for each locus were calculated for SNP markers using Power Marker 3.5 [30]. In addition, Principal Component Analyses (PCA), genetic distances [31] across the genotypes and neighbor-joining (NJ) tree were calculated using Power Marker 3.5 [30]. Phylogenetic trees for all states were constructed using MEGA software version 6.0 [32]. Principle Coordinate Analysis (PCoA) and Analysis of Molecular Variance (AMOVA) were performed using software GenAlEx V6.5 [33]. SNP data were numerically coded as follows: A = 1, C = 2, G = 3, T = 4 and missing data was coded as 0 as suggested in GenAlEx V6.5 user manual [33]. The model-based program, STRUCTURE 2.3.3 [34] was used to infer the population structure. For each K, three replications were run. Each run was implemented with a burn-in period of 100,000 steps followed by 100,000 Monte Carlo Markov Chain replicates [34] derived for each K and then plotted to find the plateau of the ΔK values [35]. The "Structure harvester" program was used (http://taylor0.biology.ucla.edu) to determine the final population. POWERCORE software [36] was used for the core development. Since numbers of rice accessions from each NE states were not equal, ranging from 107 accessions from Mizoram to 2,635 accessions from Assam. Therefore, to avoid the dominance of collections from individual states within the core set, we first developed core sub-sets using the rice collections from each state. Next, all accessions of the core sub-sets were pooled together to create the final core set of NE India rice germplasm.

Results

Genetic diversity

Genetic diversity of NE India rice collections (6,984 accessions), available at NGB, was estimated using 36 SNP markers. Alleles generated with all 36 SNP markers were scored to study the genetic diversity. Since these accessions cover seven states of NE India, they were analyzed state-wise for precise estimation of the level of their genetic diversity. The accessions collected from Arunachal Pradesh (663), Assam (2635), Manipur (549), Meghalaya (2427), Mizoram (107), Nagaland (377) and Tripura (226) were analyzed for major allele frequency, heterozygosity, gene diversity and PIC (Table 1). The highest mean major allele frequency was present in Manipur accessions (0.80) and lowest was found in Meghalaya accessions (0.74). The comparison of mean heterozygosity of alleles across the seven states revealed that Mizoram accessions were highly heterozygous (0.16); whereas

Table 3. State-wise Principal Coordinate Analyses of NE Indian rice collections, with percentage of variation explained by the first 3 axes.

States	Per cent variation	1-axis	2-axis	3-axis
Arunachal Pradesh	%	26.74	9.23	7.30
	Cum %	26.74	35.97	43.27
Assam	%	20.79	8.30	6.50
	Cum %	20.79	29.09	35.58
Manipur	%	21.54	7.23	5.73
	Cum %	21.54	28.77	34.50
Meghalaya	%	22.45	11.45	5.55
	Cum %	22.23	33.54	39.05
Mizoram	%	19.10	7.96	6.62
	Cum %	19.10	27.06	33.68
Nagaland	%	24.95	11.05	7.30
	Cum %	24.95	36.01	43.31
Tripura	%	14.96	10.15	7.57
	Cum %	14.96	25.12	32.69

Cum- Cumulative.

heterozygosity in the accessions belonging to Arunachal Pradesh (0.07) and Nagaland (0.07) were recorded minimum (Table 1). The mean gene diversity across the seven states varied from 0.34 (in the collections from Meghalaya) to 0.26 in those from (Manipur) and PIC varied from 0.21 (in the collections from Manipur) to 0.27 (from Meghalaya). Out of 36 primers used for characterization of total rice collections, five SNP primers (01-608-4_C_375, 02-3029-1_C_474, 03-3478-1_C_206, 05-48-1_C_279 and 08-4218-5_C_129) exhibited the maximum PIC and gene diversity value 0.37 and 0.49, respectively (Table S2). The primer 04-19-4_C_240 was found to be the least informative with PIC and gene diversity values equaling 0.02 and 0.02, respectively, and major allele frequency of 0.99 (Table S2), SNP markers which showed low PIC values in the present study may not be considered for future studies on genetic diversity and population structure in rice. The maximum heterozygosity was observed with primer 04-1801-20_C_428 was 0.28, while the minimum heterozygosity was observed with primer 04-19-4_C_240 (0.01).

Cluster analysis was performed using the NJ method for all the collections from seven states and unrooted phylogenetic trees were constructed (Fig. S1). Rice collections of each state were grouped in to three major clusters. Genetic distance for collections of Arunachal Pradesh varied from 0.00 to 0.75, Assam from 0.00 to 0.66, Manipur from 0.00 to 0.74, Meghalaya from 0.00 to 0.77, Mizoram from 0.00 to 0.68, Nagaland from 0.00 to 0.66, and Tripura from 0.00 to 0.67.

Population structure

A model-based program, STRUCTURE, was used to determine the genetic relationship among individual rice accessions. The membership of each accession was run from K = 1 to K = 20 for all the collections from all seven states to estimate the number of populations. Structure Harvester (http://taylor0. biology.ucla.edu) was used to determine final number of populations. The number of populations in Meghalaya collections was estimated as five, in Manipur collections as four and three for collections belonging to remaining states, i.e. Arunachal Pradesh, Assam, Mizoram, Nagaland and Tripura (Fig. S2, Table S3). The collection from Arunachal Pradesh contained 265 pure and 114 admix individuals in population1, 65 pure and 16 admix individuals in population2 and 131 pure and 72 admix individuals were present in population3. The collection from Assam showed 300 pure and 329 admix in population1, 1073 pure and 568 admix in population2, 288 pure and 77 admix in population3. The Nagaland collections 61 pure and 10 admix were present in population1, 164 pure and 91 admix in population2 and 27 pure and 24 admix were present in population3. Tripura collections showed 14 pure and 1 admix in population1, 107 pure and 21 admix in population2, 54 pure and 29 admix in population3. The collections from Mizoram had 39 pure and 19 admix in population1, 4 pure and no admix in population2 and 31 pure and 14 admix in population3. Similarly, Manipur collections showed 100 pure and 63 admix in population1, 53 pure and 55 admix in population2, 22 pure and 29 admix in population3 and 100 pure and 128 admix in population4. The accessions included in the core are marked with boxes in the bar plot (Fig. S3).

Analysis of molecular variance (AMOVA)

An AMOVA study of all rice accessions, belonging to seven different NE states, was performed to analyze the distribution of genetic diversity between and within the populations. The number of populations for each state was considered the same, as revealed by the population structure. AMOVA analyses showed that 21% diversity exists among populations in the collection from Arunachal Pradesh, while the collections from Assam, Manipur, Meghalaya, Mizoram, Nagaland and Tripura showed 24, 19, 29, 9, 8 and 22% diversity, respectively (Table 2). Maximum diversity of 29% among population was observed in Meghalaya collections, while minimum diversity of 8% was observed in Nagaland collections (Fig. S4). The highest level of diversity among individuals was present in Nagaland collections (75%) and the least diversity was observed in Meghalaya collections (46%).

Principal coordinate analyses (PCoA)

PCoA revealed significant diversity in the NE India rice germplasm collections. Meghalaya collections, which exhibited

Table 4. A comparison of percent of alleles generated in total collection vs core set.

Alleles	Arunachal Pradesh: Total collection (663 acc.)	Arunachal Pradesh: Core set (60 acc.)	Assam: Total collection (2635 acc.)	Assam: Core set (259 acc.)	Manipur: Total collection (549 acc.)	Manipur: Core set (60 acc.)	Meghalaya: Total collection (2427 acc.)	Meghalaya: Core set (242 acc.)	Mizoram: Total collection (107 acc.)	Mizoram: Core set (14 acc.)	Nagaland: Total collection (377 acc.)	Nagaland: Core set (35 acc.)	Tripura: Total collection (226 acc.)	Tripura: Core set (31 acc.)
AA	20.08	21.36	19.43	20.63	17.73	18.02	20.66	20.93	18.27	19.07	21.23	20.98	19.33	20.31
CC	21.16	17.68	18.23	18.44	13.51	15.91	16.98	17.22	15.82	16.92	16.83	17.44	18.43	16.96
GG	16.72	18.44	21.10	18.93	24.67	21.07	20.07	18.90	19.55	19.07	20.53	19.86	22.19	20.01
TT	37.83	32.12	34.94	30.20	38.62	34.43	35.17	30.64	36.55	31.70	37.11	31.33	35.32	33.82
CT/TC	1.57	3.70	2.31	3.87	2.14	3.71	2.67	4.45	3.66	4.29	1.59	3.78	2.21	3.81
GA/AG	1.53	3.65	2.32	4.36	2.00	3.94	2.42	4.25	3.76	5.48	1.49	3.17	1.46	3.05
TA/AT	0.33	0.78	0.46	0.96	0.49	0.86	0.56	1.08	0.83	1.31	0.26	0.75	0.37	0.61
GT/TG	0.70	2.00	1.07	2.20	0.80	1.83	1.29	2.10	1.46	1.91	0.87	2.24	0.67	1.22
CA/AC	0.07	0.27	0.14	0.41	0.04	0.22	0.17	0.43	0.10	0.24	0.09	0.47	0.03	0.20

the maximum number of population (k = 5) showed a very distinct PCoA plot, population1 in Meghalaya collections concentrated only in quadrant 3 and 4, whereas population4 concentrated in 1 and 2 quadrant and unlike population3, population4 and population5 did not show intermixing with each other (Fig. S5d). PCoA of collections from Arunachal Pradesh showed that population2 was very distinct, forming a separate group (Fig. S5a). PCoA plot of Assam, Manipur and Nagaland collections showed that all three populations were intermixed (Fig. S5b, Fig. S5c & Fig. S5f). In Mizoram collection population2 and Tripura collection population1 were highly distinct and formed a tight group in quadrant 1 and 4, respectively (Fig. S5e, Fig. S5g). The maximum cumulative variation (%) was observed for Nagaland rice collections with a value of 43.31% (Table 3); whereas, the minimum variation was found in Tripura accessions (32.53%). Overall, <44% variation were observed by first three components of principal coordinates, which indicates that accessions from all seven states were very diverse from one another.

Development of Core Set

Of the 6,984 rice accessions studied, a core set of 701 accessions (i.e., 60 from Arunachal Pradesh, 259 from Assam, 60 from Manipur, 242 from Meghalaya, 14 from Mizoram, 35 from Nagaland, and 31 from Tripura) were selected using POWER-CORE (Table S4). Nine allele types were produced by 36 SNP markers, of which four alleles were homozygous and five were heterozygous. Out of six types of substitutions possible (2 transitions and 4 transversion), five types of substitutions were observed and C/G or G/C type substitutions were not observed in our study. Allele frequency was calculated for all collections from seven states and core set (Table 4). A comparison of allele frequency showed that there was no loss of alleles in the resulting core set and they remained 99.9% identical.

Genetic diversity of Core Set

The genetic diversity of the core set was estimated to know the extent of diversity captured from total collection. Major allele frequency; gene diversity, heterozygosity and PIC were separately estimated for core-sets of individual states (Table 4). Comparisons of all genetic parameters (major allele frequency, gene diversity, heterozygosity and PIC) revealed that values for core set (Table 5) were greater than total collections (Table 1). For example, the Arunachal Pradesh gene diversity was 0.29, which increased to 0.39 in the core set. Similarly, the PIC for the Arunachal Pradesh collection was 0.23, while its core set PIC was 0.31. Cluster analysis showed that, core set accessions were getting distributed in to four clusters. The nine accessions with EC ID which were also part of core set were represented in three out of four clusters. The genetic diversity for the core set varied from 00.00 to 0.68 (Fig. 1).

Population structure of Core Sets

STRUCTURE analysis grouped core set of 701 rice accessions into three populations (Fig. 2). Populations were studied for the number of pure and admix individuals. Population1 had 62 pure and 74 admix, population2 had 111 pure and 53 admix and population3 had 250 pure and 151 admix (Fig. 3).

AMOVA and PCoA of Core Set

AMOVA analyses for core set showed 20% variance among populations (Table 6, Fig. 4); whereas, 48% variance was present among individuals. The partitioning of molecular variance was similar to the total collection and maximum diversity among individuals. PCoA could not be useful to identify significant

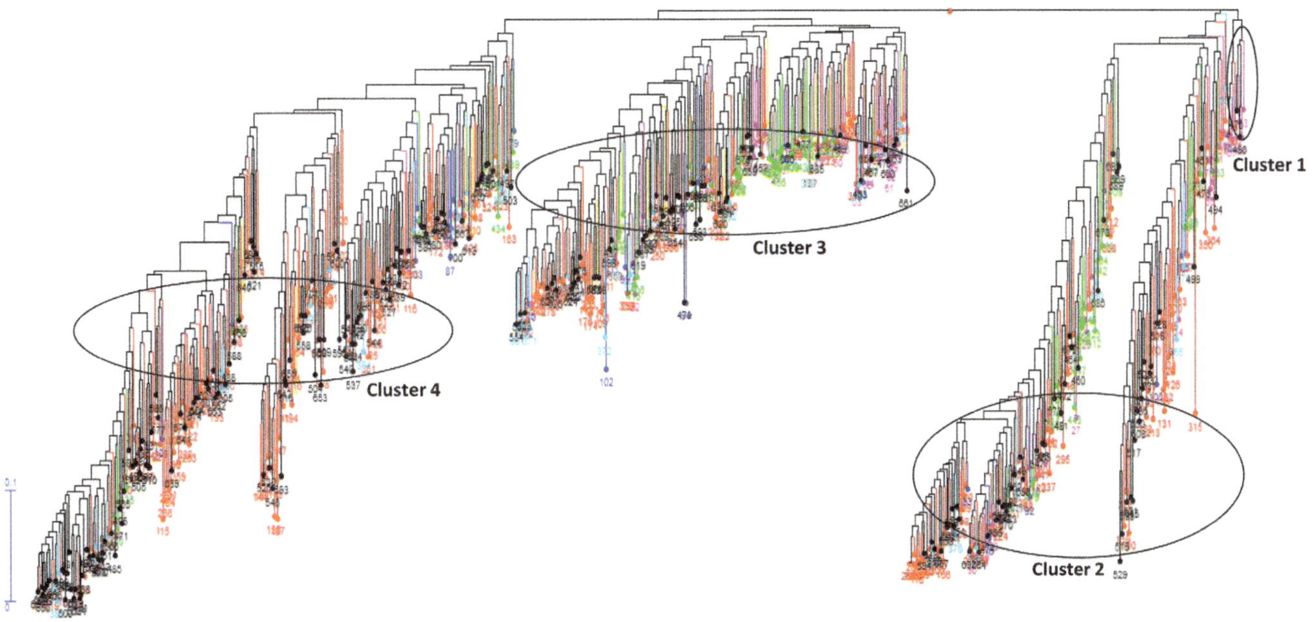

Figure 1. Phylogenetic tree of NE India rice core set (701 accessions) constructed based on SNP data. Germplasm accessions collected from the NE states are depicted with different colours Arunachal Pradesh - green, Assam - red, Manipur - purple, Meghalaya - black, Mizoram -yellow, Nagaland - light blue, Tripura - dark blue.

isolation in populations except, for population2, which was tightly clustered in quadrant 3 and 4 (Fig. 5). Maximum cumulative percent variation was explained by the first three coordinates with 43.08% variance (Table 7). A plot of percent allele frequency of total collection *versus* core set showed that nearly all alleles with similar frequency were represented in the core collection (Fig. 6) and 99.9% alleles have been retained in the collections designated as core set.

Discussion

Genome level profiling of huge germplasm collections in crop species is essential to identify small and diverse sets of accessions for their efficient use in crop improvement programmes. Despite the advancement in genomics, the Indian rice collection remained uncharacterized at the molecular level, with respect to parameters such as genetic diversity and population structure. This has been the major limiting factor in their utilization and development of improved cultivars. The present study is the first major effort to

Table 5. Estimation of gene diversity, heterozygosity, PIC and major allele frequency in NE India rice core set (701 accessions).

S. No.	States	core sample	Gene diversity	Heterozygosity	PIC	Major allele frequency
1	Arunachal Pradesh	60	0.0726–0.5000 (0.3962)	0.0377–0.2778 (0.1713)	0.0700–0.3750 (0.3124)	0.5000–0.9623 (0.6940)
2	Assam	259	0.1670–0.5000 (0.4018)	0.0843–0.3618 (0.1908)	0.1531–0.3740 (0.3172)	0.5000–0.9080 (0.69271)
3	Manipur	60	0.1372–0.5000 (0.3749)	0.0455–0.3455 (0.1745)	0.1278–0.3750 (0.2982)	0.5000–0.9259 (0.7142)
4	Meghalaya	242	0.2390–0.4995 (0.4100)	0.1013–0.3493 (0.2001)	0.2105–0.3747 (0.3231)	0.5161–0.8612 (0.6873)
5	Mizoram	14	0.0689–0.4970 (0.3506)	0.0714–0.4615 (0.2213)	0.0665–0.3735 (0.2813)	0.5385–0.9643 (0.7416)
6	Nagaland	35	0.1087–0.5000 (0.3854)	0.0385–0.3846 (0.1682)	0.1028–0.3846 (0.3049)	0.5000–0.9423 (0.7033)
7	Tripura	31	0.0351–0.5000 (0.3596)	0.0323–0.3667 (0.1524)	0.0345–0.3750 (0.2875)	0.5000–0.9821 (0.7344)

Mean values are given in parentheses.

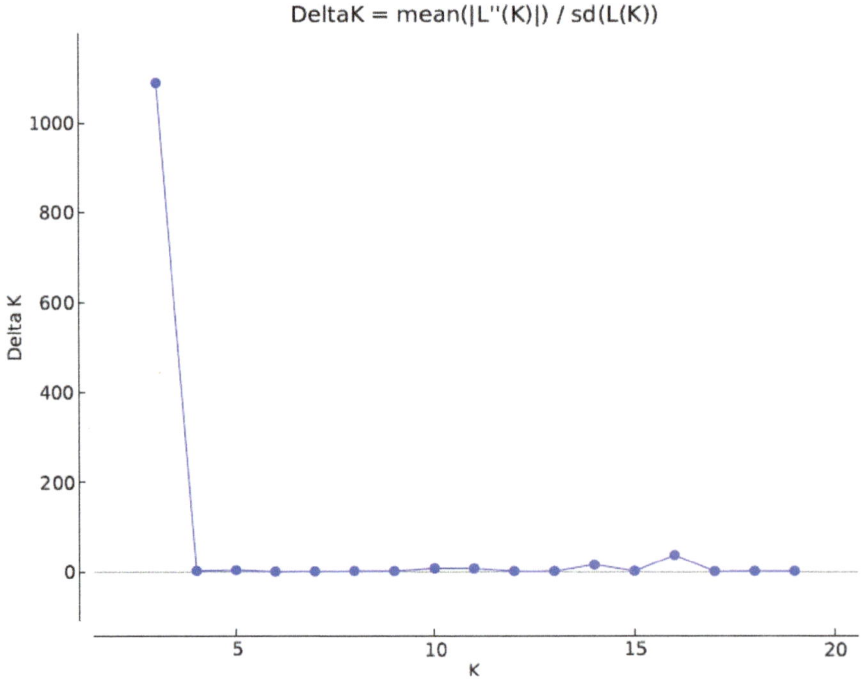

Figure 2. Estimation of populations in NE India rice core set using LnP(D) derived Δk for k from 1 to 20.

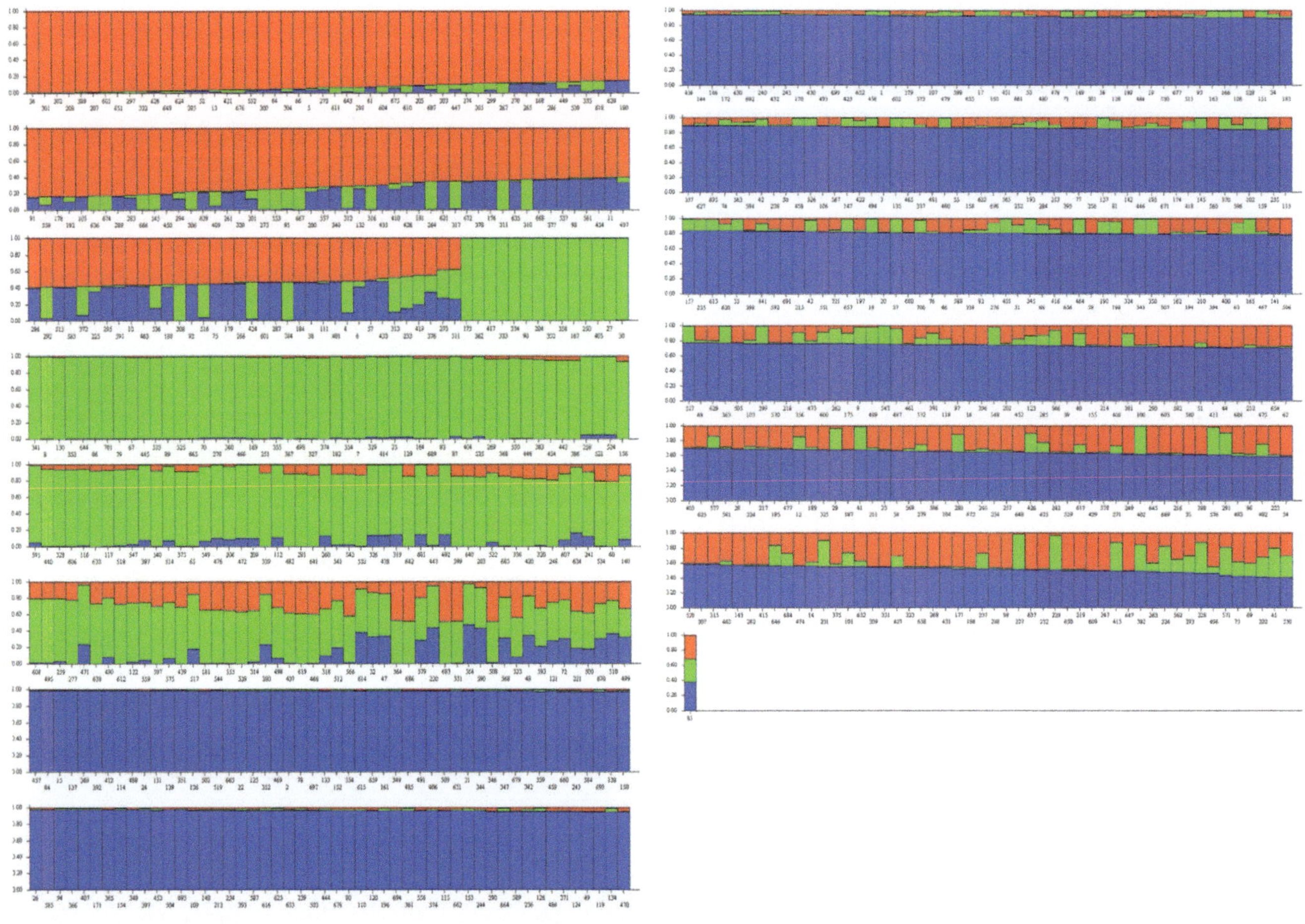

Figure 3. Model based clustering of the NE India rice core set (701 accessions).

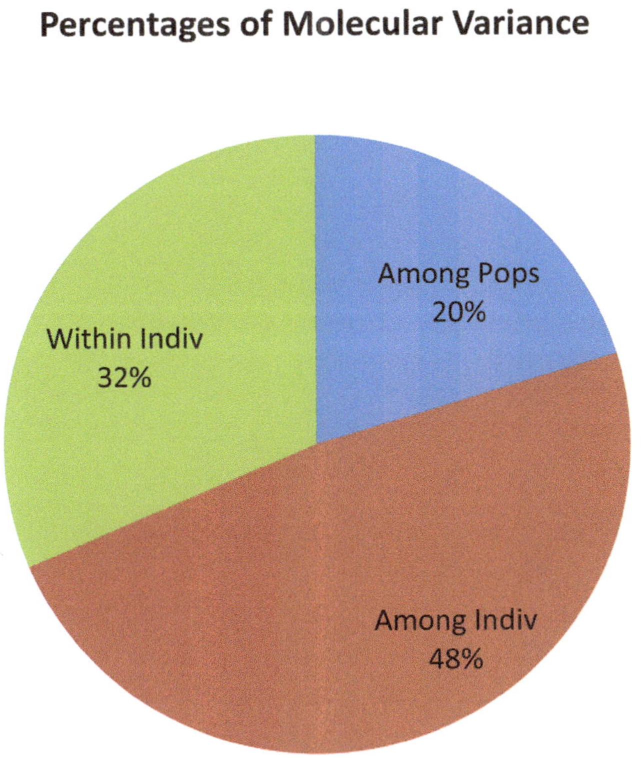

Figure 4. Analysis of Molecular Variance (AMOVA) analyses of NE India rice core set.

characterize the NE India rice collections. This study shows that highly efficient and selective SNP markers can be used to enhance genome based analysis, e.g., genome wide association mapping.

State wise analysis of the genetic diversity of the NE India collection showed interesting results. The average PIC values for these SNP markers did not vary much across the states and ranged from 0.23 (Manipur) to 0.27 (Meghalaya). The average PIC value recorded in the present study was in the range reported by researchers in previous studies. Singh *et al.*, (2013) [28] reported average PIC value 0.25 for the SNP markers in a set of 375 Indian rice varieties. The highest values for the parameters like PIC and gene diversity were found in collections from Meghalaya, which indicates the presence of a large rice genetic diversity in Meghalaya collections as compared to those from other six states. The same was also confirmed from PCoA, which displayed cluttered plot for the Meghalaya collection. The highest heterozygosity was observed in the collections from Mizoram, while Manipur had the highest major allele frequency.

Analysis of statistical genetic parameters across the collections from different states indicated a high level of variability in the NE India collection. There is relatively a high level of genetic diversity among the NE India rice collections, as reported by others using morphological [37] and molecular markers such as RAPD [38], ISSR [39] and SSR [26,40]. Comparison of our results with previous findings revealed that PIC and genetic diversity were nearly half of the SSR-based studies which were on the expected line [26,28,40]. Similar results have been reported in other species that have used SSR and SNP markers [41,42]. In the past, analysis of rice accessions from all seven states revealed three major clusters. Vairavan *et al.*, (1973) [37] reported three clusters among a set 400 accessions of NE India rice collection based on morphological analysis. Das *et al.*, (2013) [40] showed four groups were present in NE India rice varieties, while Choudhury *et al.*, (2013) [26] reported only two groups, based on SSR marker analysis.

Population structure study has been reported for the first time in NE rice collections and the number of populations estimated five for Meghalaya, four for Manipur and three for the remaining states. Admixture has been reported in all the states besides pure lines and this may be due to germplasm taken in this study which is more heterogeneous in comparison to varieties.

There have been no other studies for characterization of a large number of NE India rice accessions at the molecular level that

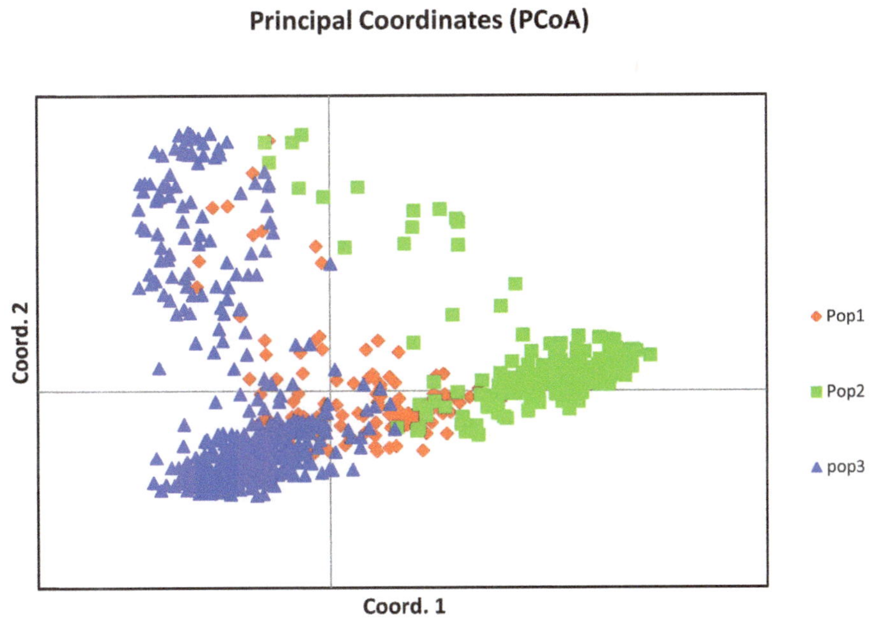

Figure 5. Principal Coordinate Analysis (PCoA) analyses of NE India rice core set.

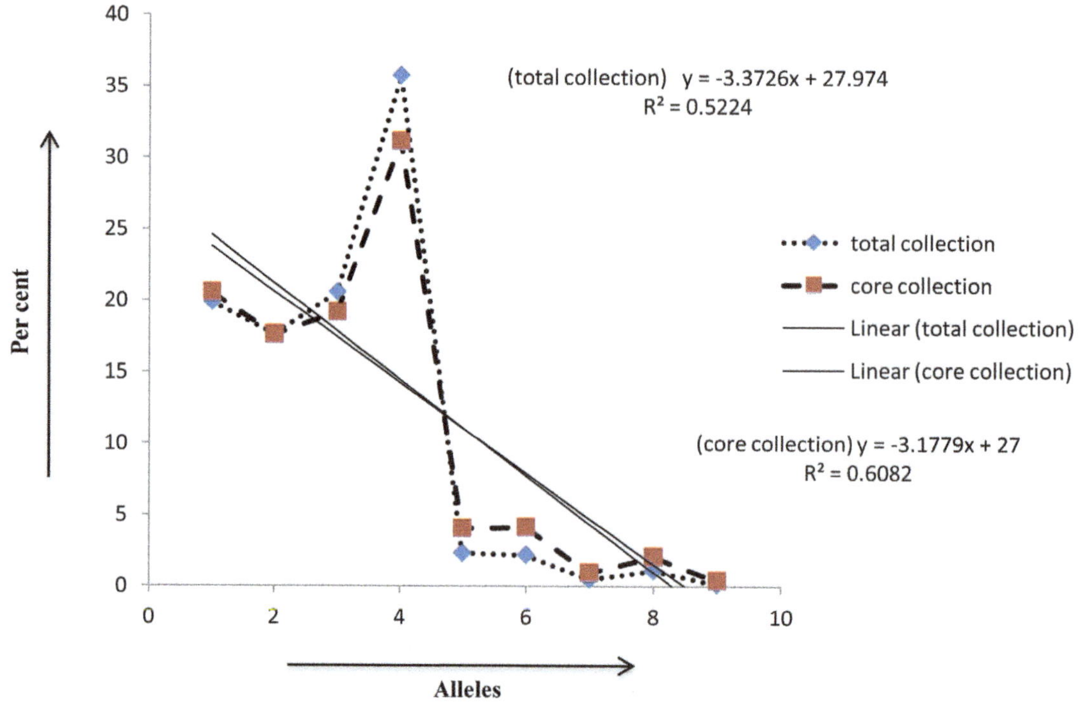

Figure 6. Comparison of allele frequency occurrence in total NE rice collections vs core set identified in the present study.

could be used for comparison of our results. AMOVA analysis showed maximum variation among individuals for accessions belonging to Nagaland (75%) and minimum for Meghalaya collection (37%). Choudhury *et al.*, (2013) [26] conducted AMOVA analysis on NE rice, which revealed 25% variation among individuals, 66% among varieties and 9% among cultivated types. The variation among individuals found in our study was higher than that reported previously, which may be attributed to the large sample size used in present study. Also partitioning of variability at population and individual levels in AMOVA among states were different because sample size taken from different states varied from 107 accessions from Mizoram to 2,635 accessions from Assam.

The genomic characterization revealed huge variability among the NE India rice collection; therefore, we decided to identify a core set for the NE India accessions to facilitate the rice breeders to effectively use the same in their crop improvement programmes and efficient conservation and management of germplasm in the genebank. In recent past, various types of molecular markers have been used to develop core sets for different annual crops. These markers include SNPs for Arabidopsis [43], SSRs for rice, wheat and common bean [44–46], AFLPs for barley [47], RFLPs for wild wheat [48] and RAPD for common bean [49]. The present study used advanced SNP markers due to their biallelic nature, co-dominance inheritance and amenability to high throughput analysis. Another crucial step for development of core set is to formulate a suitable strategy that can identify either major and minor variances or diversity, using a fully representative accession from the large collections. One of the most commonly followed approach for constructing a core set from a collection is to rely on grouping an entire collection based on growing regions and then selecting core accessions from each group followed by combining accessions from each group to form final core set [5].

In the current study, we divided the NE India rice collection into seven (on the basis of source states) subsets and later all the accessions of subsets were pooled to form a core set of NE India rice germplasm. Statistical parameters, genetic distance and cluster analysis were taken into consideration to develop this core set; these methods were followed in accordance with other studies [4,50,51,52,53,54]. We developed the NE India rice core set collection using SNP markers employing POWERCORE software (using advanced M strategy implemented through a modified heuristic algorithm). We relied on molecular data more than the passport and phenotypic data, because molecular markers can

Table 6. AMOVA of core set of NE Indian rice (701 accessions).

Source	df	SS	MS	Est. Var.	%
Among Pops	2	1649.152	824.576	1.996	20
Among Indiv.	698	8774.397	12.571	4.740	48
Within Indiv.	701	2166.500	3.091	3.091	32
Total	**1401**	**12590.049**		**9.826**	**100**

Table 7. Percentage of variation explained by the first 3 axes in Principal Coordinate Analyses of NE Indian rice core set (701 accessions).

Per cent variation	1-axis	2-axis	3-axis
%	18.41	18.04	6.63
Cum %	18.41	36.45	43.08

accurately represent the genetic diversity of the original collection and free of problems related to missing data or environmental interactions that are typically found in passport and phenotypic data. POWERCORE was used to develop a core set of 701 rice accessions (10% of total NE India rice collection). Nearly all alleles were represented in the core set; hence, there was no loss of alleles during the core set development.

In present study, we developed a core set of 10% of total collection; while there is no perfect core set constructing theory or universal methods to fit all crops, which might be due to large number of germplasm resources, absence of data for accessions, or complication and diversification of data type, and so on [50]. The core set was statistically supported with various other statistical parameters including genetic distance and cluster analysis that were taken into consideration in the final development of core set which is in accordance with previous studies [50]. The gene diversity value 0.54 and PIC value 0.48 estimated by SSR markers on rice core [22], is a higher range than the values found in our study (gene diversity value 0.39 and PIC value 0.31). This variation is obvious due to nature of marker used for development of core set. In present study, we have used SNP marker which is biallelic in nature whereas, Zhang et al., 2011, [22] have used SSR marker which is multi-allelic. A study conducted with SNP markers in tropical maize [51] reported a gene diversity value 0.29 and heterozygosity value 0.25 on core sets, and similar gene diversity was observed in present study. The gene diversity, heterozygosity, PIC values of core set were higher than those of total collection which was expected because the diversity increases with elimination of genetically similar accessions during core set development [11].

In conclusion, SNP marker based molecular characterization of NE India rice collection revealed that large variation exists among the accessions and the pattern of the genetic diversity and population structure varied across the seven states. Collections from Meghalaya were found to be the most diverse and possessed maximum number of populations. Core sets have been developed state wise as well as for the entire NE collection that can be potentially exploited by rice breeders in conventional breeding programmes and molecular breeders/researchers for genomic studies such as association mapping for gene discovery for various desirable traits and identification of suitable parents for their use in rice improvement. Further development of core set and sub-sets for NE India rice is likely to significantly reduce the task of the genebank curators as they will need to multiply and maintain small number accessions in the medium term storage as active collections to meet the requirement of the rice breeders.

Supporting Information

Figure S1 Circular phylogenetic trees of NE rice collection constructed based on SNP data using MEGA software (a) Arunachal Pradesh, (b) Assam, (c) Manipur, (d) Meghalaya, (e) Mizoram, (f) Nagaland and (g) Tripura. (Red marked accessions was selected for core by Power Core).

Figure S2 Estimation of populations in NE rice collections using LnP(D) derived Δk for k from 1 to 20 in (a) Arunachal Pradesh, (b) Assam, (c) Manipur, (d) Meghalaya, (e) Mizoram, (f) Nagaland and (g) Tripura.

Figure S3 Population structure of NE rice collection based on SNP data (a) Arunachal Pradesh, (b) Assam, (c) Manipur, (d) Meghalaya, (e) Mizoram, (f) Nagaland and (g) Tripura.

Figure S4 Analysis of Molecular Variance in NE rice collection based on SNP data (a) Arunachal Pradesh, (b) Assam, (c) Manipur, (d) Meghalaya, (e) Mizoram, (f) Nagaland and (g) Tripura.

Figure S5 Principal Coordinate Analysis in NE rice collection based on SNP data (a) Arunachal Pradesh, (b) Assam, (c) Manipur, (d) Meghalaya, (e) Mizoram, (f) Nagaland and (g) Tripura.

Table S1 List of NE rice collections studied using SNP markers.

Table S2 List of SNP primers used for genotyping of NE rice collection (6984 rice accessions) alongwith gene diversity, PIC and Major allele frequency.

Table S3 Output Table generated from Structure Harvester software for Evanno method. Yellow highlight shows the largest value in the Delta K column.

Table S4 List of rice accessions identified as NE core set based on SNP markers.

Acknowledgments

We are grateful to Dr James W. Lillard, JR., Professor and Associate Dean for Research, Morehouse School of Medicine, Atlanta, GA 30310, for manuscript editing. Thanks are due to the Director, NBPGR, New Delhi, for providing the facilities to carry out this study.

Author Contributions

Conceived and designed the experiments: RS NKS. Performed the experiments: DRC NS. Analyzed the data: RS AKS SK. Contributed reagents/materials/analysis tools: KS RKT AA. Wrote the paper: RS AKS RKT.

References

1. Frankel OH (1984) Genetic perspective of germplasm conservation. In: Arber WK. Llimensee, Peacock WL, Starlinger P, editors. Genetic manipulation: Impact on man and society. Cambridge University Press, Cambridge, England, pp. 161–170.
2. Frankel O H, Brown AHD (1984) Plant genetic resources today: A critical appraisal. In Holden JHW, Williams JT. Crop Genetic Resources: Conservation and Evaluation. London, UK: George Allen and Unwin Ltd. pp. 249–257.
3. Brown AHD (1989) The case for core collections. In: Brown AHD, Frankel OH, Marshall DR, Williams JT, editors. The Use of Plant Genetic Resources. Cambridge University Press, Cambridge, UK. pp. 136–155.
4. Ebana K, Kojima Y, Fukuoka S, Nagamine T, Kawase M (2008) Development of mini core collection of Japanese rice landrace. Breed Sci 58: 281–291.
5. Wang J, Guan Y, Wang Y, Zhu L, Wang Q (2014) A Strategy for finding the optimal scale of plant core collection based on Monte Carlo simulation. Sci. World Journal doi: 10.1155/2014/503473.
6. Yonezawa K, Nomura T, Morishima H (1995) Sampling strategies for use in stratified germplasm collections. In: Hodgkin T, Brown AHD, Hintum van Th JL, Morales EAV, editors. Core collections of plant genetic resources John Wiley and sons Chichester UK pp. 35–53.
7. Upadhyaya HD, Yadav D, Dronavalli N, Gowda CLL, Singh S (2010) Mini core germplasm collections for infusing genetic diversity in plant breeding programs. Electronic J Plant Breed 1: 1294–1309.
8. Jiang Q, Hou J, Hao C, Wang L, Ge H, et al. (2011) The wheat (*T. aestivum*) sucrose synthase 2 gene (*TaSus2*) active in endosperm development is associated with yield traits. Funct Integr Genomics 11: 49–61.
9. Upadhaya HD, Reddy KN, Sharma S, Varshney RK, Bhattacharjee R, et al. (2011) Pigeonpea composite collection and identification of germplasm for use in crop improvement programmes. Plant Genet Resour 9: 97–108.
10. Sharma M, Rathore A, Naga Mangala U, Ghosh R, Sharma S, et al. (2012) New sources of resistance to Fusarium wilt and sterility mosaic disease in a mini-core collection of pigeonpea germplasm. Eur J Plant Pathol 133: 707–714.
11. Agrama HA, Yan WG, Lee F, Fjellstrom R, Chen MH, et al. (2009) Genetic assessment of a mini-core subset developed from USDA Rice Genebank. Crop Science 49: 1336–1346.
12. Balfourier F, Charmet G, Prosperi JM, Goulard M, Monestiez P (1998) Comparison of different spatial strategies for sampling *a* core collection of natural populations *of* fodder crops. Genet Sel Evol 30:215–235.
13. Chandra S, Huaman Z, Hari Krishna S, Ortiz R (2002) Optimal sampling strategy and core collection size of Andean tetraploid potato based on isozyme data: A simulation study. Theor Appl Genet 104: 1325–1334.
14. Hu J, Zhu J, Xu HM (2000) Methods of constructing core collections by stepwise clustering with three sampling strategies based on the genotypic values of crops. Theor Appl Genet 101: 264–268.
15. Peeters JP, Martinelli JA (1989) Hierarchical cluster analysis as a tool to manage variation in germplasm collections. Theor Appl Genet 78: 42–48.
16. Franco J, Crossa J, Warburton ML, Taba S (2006) Sampling strategies for conserving maize diversity when forming core subsets using genetic markers. Crop Sci46: 854–864.
17. Reeves PA, Panella LW, Richards CM (2012) Retention of agronomically important variation in germplasm core collections: implications for allele mining. Theor Appl Genet 124: 1155–1171.
18. Upadhaya HD, Gowda CLL, Pundir RPS, Reddy VG, Singh S (2006) Development of a core subset of finger millet germplasm using geographical origin and data on 14quantitative traits. Genet Resour Crop Evol 53: 679–685.
19. Odong TL, Jansen J, van Eeuwijk FA, van Hintum TJL (2013) Quality of core collections for effective utilization of genetic resources review, discussion and interpretation. Theor Appl Genet 126: 289–305.
20. Díez CM, Imperato A, Rallo L, Barranco D, Trujillo I (2012) Worldwide core collection of olive cultivars based on simple sequence repeat and morphological markers. Crop Sci 52:211–221.
21. Rao ES, Kadirvel P, Symonds RC, Geethanjali S, Ebert AW (2012) Using SSR markers to map genetic diversity and population structure of *Solanum pimpinellifolium* for development of a core collection. Plant Genet Resour 10:38–48.
22. Zhang P, Li J, Li X, Liu X, Zhao X (2011) Population Structure and Genetic Diversity in a Rice Core Collection (*Oryza sativa L.*) investigated with SSR markers. PloS ONE 6 (12): e27565 doi: 10. 1371/journal.pone.0027565.
23. Chang TT (1976) The origin, evolution, cultivation, dissemination, and diversification of Asian and African rice. Euphytica 25: 425–441.
24. Khush GS (1997) Origin, dispersal, cultivation and variation of rice. Plant Mol Biol 35: 25–34.
25. Londo JP, Chiang YC, Hung KH, Chiang TY, Schaal BA (2006) Phylogeography of Asian wild rice, *Oryza rufipogon*, reveals multiple independent domestications of cultivated rice, *Oryza sativa*. Proc Natl Acad Sci USA 103: 9578–9583.
26. Choudhury B, Khan ML, Dayanandan S (2013) Genetic structure and diversity of Indigenous Rice varieties (*Oryza sativa*) in Eastern Himalayan region of Northeast India. Springer Plus 2: 228–237.
27. Hore DK (2005) Rice diversity collection, conservation and management in northeastern India. Genet Resour Crop Evol 52:1129–1140.
28. Singh N, Choudhury DR, Singh AK, Kumar S, Srinivasan K, et al. (2013) Comparison of SSR and SNP markers in estimation of genetic diversity and population structure of Indian rice varieties. PLoS ONE 8(12): e84136.
29. Singh NK, Dalal V, Batra K, Singh BK, Chitra G, et al. (2007) Single copy genes define a conserved order between rice and wheat for understanding differences caused by duplication, deletion, and transposition of genes. Funct Integr Genomics 7: 17–35.
30. Liu K, Muse SV (2005) Power Marker: an integrated analysis environment for genetic marker analysis. Bioinformatics 21: 2128–2129. doi:10.1093/bioinformatics/bti282.
31. Nei M, Tajima F, Tateno Y (1983) Accuracy of estimated phylogenetic trees from molecular data. II. Gene frequency data. J Mol Evol 19: 153–170.
32. Tamura K, Stecher G, Peterson D, Fillipski A, Kumar S (2013) MEGA6 Molecular Evolutionary genetics Analyses version 6.0. Mol Biol Evol 30: 2725–2729.
33. Peakall R, Smouse P (2012) GenAlEx 6.5: Genetic analysis in Excel. Population genetic software for teaching and research - an update. Bioinformatics 28: 2537–2539.
34. Pritchard JK, Stephens M, Donnelly P (2000) Inference of population structure using multilocus genotype data. Genetics 155: 945–959.
35. Evanno G, Regnaut S, Goudet J (2005) Detecting the number of clusters of individuals using the software STRUCTURE: a simulation study. Mol Ecol 14: 2611–2620.
36. Kim KW, Chung HN, Cho GT, Ma KH1, Chandrabalan D, et al (2007) PowerCore: a program applying the advanced M strategy with a heuristic search for establishing core sets. Bioinformatics 23: 2155–2162.
37. Vairavan S, Siddiq EA, Arunachalam V, Swaminathan MS (1973) A study on the nature of genetic divergence in rice from Assam and Northeast Himalayas. Theor Appl Genet 43:213–221.
38. Sarma RN, Bahar B (2005) Genetic variation of bora rice (glutinous rice) of Assam as revealed by RAPDs. Plant Genetic Resources Newsletter 144:34–38, FAO. IPGRI
39. Bhuyan N, Borah BK, Sarma RN (2007) Genetic diversity analysis in traditional lowland rice (Oryza sativa L.) of Assam using RAPD and ISSR markers. Curr Sci 93:967–972
40. Das B, Sengupta S, Parida SK, Roy B, Ghosh M, et al. (2013) Genetic diversity and population structure of rice landraces from Eastern and North Eastern States of India. BMC Genetics 14:71–85.
41. Van Inghelandt D, Melchinger AE, Lebreton C, Stich B (2010) Population structure and genetic diversity in a commercial maize breeding program assessed with SSR and SNP markers. Theor Appl Genet 120: 1289−1299.
42. Varshney RK, Baum M, Guo P, Grando S, Ceccarelli S, et al. (2010) Features of SNP and SSR diversity in a set of ICARDA barley germplasm collection. Mol Breed 26: 229−242.
43. McKhann HI, Camillery C, Berard A, Bataillon T, David JL, et al. (2004) Nested core collections maximizing genetic diversity in *Arabidopsis thaliana*. Plant J 38: 193– 202.
44. Yan WG, Agrama H, Jia M, Fjelistrom R, McClung A (2010) Geographic description of genetic diversity and relationships in the USDA rice world collection. Crop Sci 50: 2406–2417.
45. Balfourier F, Roussel V, Strelchenko P, Exbrayat-Vinson F, Sourdille P, et al. (2007) A Worldwide bread wheat core collection arrayed in a 384-well plate. Theor Appl Genet 114: 1265–1275.
46. Blair MW, Diaz LM, Buendia HF, Duque MC (2009) Genetic diversity, seed size association and population structure of a core collection of common beans (*Phaseolus vulgaris* L.). Theor Appl Genet 119: 955–972.
47. van Treuren R, Tchoudinova I, van Soest LJM, van Hintum TJL (2006) Marker- assisted acquisition and core collection formation: A case study in barley using AFLPs and pedigree data. Genet Res Crop Evol 53: 43–52.
48. Zaharieva M, Santoni S, David J (2001) Use of RFLP markers to study genetic diversity and to build a core collection of the wild wheat relative Ae-geniculata Roth (= Ae-ovarta L.). Genet Sel Evol 33S: 269–288.
49. Skroch PW, Neinhuis J, Beebe S, Tohne J, Pedraza F (1998) Comparison of Mexican common bean (*Phaseolus vulgaris* L.) core and reserve germplasm collections. Crop Sci 38: 488–496.
50. Xiao-ling L, Yong-gen L, Jin-quan L, Hai-ming X, Shahid MQ (2011) Strategies on sample size determination and qualitative and quantitative traits integration to construct core collection of rice (*Oryza sativa*). Rice Sci 18: 46–55.
51. Wen W, Franco J, Chavez-Tovar VH, Yan J, Taba S (2012) Genetic Characterization of a Core Set of a Tropical Maize Race Tuxpeño for Further Use in Maize Improvement. PLoS ONE 7(3): e32626.
52. Kuroda Y, Tomooka N, Kaga A, Vaughan D A (2009) Genetic diversity of wild soybean (*Glycine soja* Sieb. et Zucc.) and Japanese cultivated soybeans [*G. max* (L.) Merr.] based on microsatellite (SSR) analysis and the selection of a core collection. Genet Resour Crop Evol 56:1045–1055.

53. Jansen J, van Hintum TJL (2007) Genetic distance sampling: a novel sampling method for obtaining core collections using genetic distances with an application to cultivated lettuce. Theor Appl Genet 114:421–428

54. Soto-Cerda B J, Diederichsen A, Ragupathy R, Cloutier S (2013) Genetic characterization of a core collection of flax (*Linum usitatissimum* L.) suitable for association mapping studies and evidence of divergent selection between fiber and linseed types. BMC Plant Biol 13: 78–91.

19

A Strategy for Genome-Wide Identification of Gene Based Polymorphisms in Rice Reveals Non-Synonymous Variation and Functional Genotypic Markers

Subodh K. Srivastava[1], Pawel Wolinski[2], Andy Pereira[1]*

1 Crop, Soil and Environmental Sciences, University of Arkansas, Fayetteville, Arkansas, United States of America, **2** Arkansas High Performance Computing Center, University of Arkansas, Fayetteville, Arkansas, United States of America

Abstract

The genetic diversity of plants has traditionally been employed to improve crop plants to suit human needs, and in the future feed the increasing population and protect crops from environmental stresses and climate change. Genome-wide sequencing is a reality and can be used to make association to crop traits to be utilized by high-throughput marker based selection methods. This study describes a strategy of using next generation sequencing (NGS) data from the rice genome to make comparisons to the high-quality reference genome, identify functional polymorphisms within genes that might result in function changes and be used to study correlations to traits and employed in genetic mapping. We analyzed the NGS data of Oryza sativa ssp indica cv. G4 covering 241 Mb with ~20X coverage and compared to the reference genome of Oryza sativa ssp. japonica to describe the genome-wide distribution of gene-based single nucleotide polymorphisms (SNPs). The analysis shows that the 63% covered genome consists of 1.6 million SNPs with 6.9 SNPs/Kb, and including 80,146 insertions and 92,655 deletions (INDELs) genome-wide. There are a total of 1,139,801 intergenic SNPs, 295,136 SNPs in intronic/non-coding regions, 195,098 in coding regions, 23,242 SNPs at the five-prime (5′) UTR regions and 22,686 SNPs at the three-prime (3′) UTR region. SNP variation was found in 40,761 gene loci, which include 75,262 synonymous and 119,836 non-synonymous changes, and functional reading frame changes through 3,886 inducing STOP-codon (isSNP) and 729 preventing STOP-codon (psSNP) variation. There are quickly evolving 194 high SNP hotspot genes (>100 SNPs/gene), and 1,513 out of 2,458 transcription factors displaying 2,294 non-synonymous SNPs that can be a major source of phenotypic diversity within the species. All data is searchable at https://plantstress-pereira.uark.edu/oryza2. We envision that this strategy will be useful for the identification of genes for crop traits and molecular breeding of rice cultivars.

Editor: Jauhar Ali, International Rice Research Institute, Philippines

Funding: This work is supported by AP National Science Foundation, Plant Genome Research Program award DBI-0922747, http://www.nsf.gov/awardsearch/showAward?AWD_ID=0922747. The funders had no role in study design, data collection and analysis, decision to publish, or preparation of the manuscript.

Competing Interests: The authors have declared that no competing interests exist.

* Email: apereira@uark.edu

Introduction

The world's geometrically growing population has passed 7 billion, and it is estimated that rice production must increase 24% by 2025 to meet the global population demand [1]. The rice genome, considered as a global heritage, has the information and resources to continue to feed half the population of the world. The diversity of rice across the world is an excellent resource to understand the genetic events of human selection for traits of utility, and to discover new alleles for traits using plant genomics as it represents a modest size of 389 Mb for a cereal genome [2]. The genomes of individuals from the same species vary in sequence as a result of the evolutionary process and contribute to some of the specific features in individual genotypes. Therefore, sequence polymorphisms have become a prime target of interest following genome sequencing projects, with analysis and utilization of the identified variation the next goal. Single nucleotide polymorphisms (SNPs) are increasingly becoming the marker system of choice and their analyses are a part of almost all aspects of applications of genomics.

SNP discovery is valuable for crop improvement in two fundamental ways. First, it reveals DNA variation among varieties, thus providing the tools for selection in breeding programs [3]. Secondly, it provides the highest level of accuracy in mapping or anchoring of all forms of the phenotype, including biochemical, metabolic, physiological, and phenotypic performance [4]. However, for many crop plants there are an exceptionally low number of validated SNP markers available, although they are needed in large numbers for studies on genetic variation, linkage mapping, population structure analysis, association studies, map-based gene isolation and plant breeding purposes [5]. Therefore, correct identification of SNPs and their potential effects or functional annotation are important for crop improvement that will enable the selection of useful alleles of candidate genes in designing breeding experiments. Genome-wide comprehensive identification of polymorphisms among individuals within a species is crucial to studying the genetic basis of phenotypic differences and for elucidating the evolutionary history of the species [6]. Large-scale polymorphism surveys have been reported for human [7], mouse [8] and *Arabidopsis thaliana* [9]. Natural phenotypic differences are now amenable to genetic dissection, right up to the identification of causal DNA polymorphisms and functional nucleotide polymorphisms accounting for plant intraspecific developmental diversity. These variations often affect gene structure and may contribute to interspecific phenotypic traits and adaptation [10]. The high quality sequence of rice genome *Oryza sativa* ssp. *japonica* cv. Nipponbare (*Os japonica*) and *Oryza sativa* ssp. *indica* cv. 9311 (*Os indica*) contains a wealth of information that can explain the large amount of morphological, physiological and ecological variation observed in many varieties cultivated for food, since there is a high level of variability present in diverged rice genotypes [11–13]. The rice genome has also been surveyed for polymorphisms to develop several alternative models that can explain contemporary patterns of polymorphism in rice, and establish the introgression patterns of shared SNPs that reveal the breeding history and relationships among the varieties [1,14]. Rice has been extensively analyzed for SNP variation within the species, and to the reference genome for various purposes [12], including many studies related to genome-wide SNP analysis between the two model subspecies of rice. Some genotypes have been analyzed by deep sequencing, focused on non-synonymous coding SNPs (nsSNPs), and very few have been evaluated extensively for potential functions. At present, genome-wide gene based SNP analysis is still lacking, results of which could directly be utilized in crop improvement [1,12,15–17].

Rapid process in genomic information at whole genome sequence (WGS) level allow us to evaluate and analyze base by base structural variations for their similarity, and the differences reveal potential functional information. In recent years, natural variation has gained significant interest at the genome-wide level and the correlation of these variations to their agronomic traits has been the prime focus [14,18]. Development of functional markers for candidate genes has been utilized in many crops including rice. The progress of new technologies enables us to analyze multiple high quality genomes. The availability of the high-quality rice reference genome and Next Generation Sequence (NGS) information for some of the important genotypes allow us to examine the tremendous diversity of the rice gene pool at a very fine scale [19,20]. The analysis shown here is basically focused on genome-wide analysis of rice sequences to identify gene based non-synonymous SNPs at different levels of comparison to present a group of SNPs responsible for or that might play an important role at the functional level. Genome-wide SNP identification and correlation with agronomic traits could be helpful for trait characterization [15,20]. We believe that although significant analysis has been carried out for identification of SNPs between these species, there is still missing functional information of identified SNPs. The present study will focus on two rice subspecies cultivars, Guangluai-4 and Nipponbare, of the two sequenced genomes *Oryza sativa* ssp. *indica* (*Os indica*) and *Oryza sativa* ssp. *japonica* (*Os japonica*) respectively for identification and functional annotation of SNPs between them. This analysis can lead to the development of functional markers focusing on gene based analysis for trait based SNP analysis and utilization for crop improvement.

Next generation sequencing technology coupled with the growing number of genome sequences, more particularly from within the species, open the opportunities to redesign genotyping strategies based on focused genic information for more effective genetic mapping and genome analysis. Recently, genome-wide variation patterns in rice have been obtained from millions of high-quality SNPs and identified thousands of genes with significantly lower diversity in cultivated compared to wild rice, and a few of these variants have been found associated with important biological features [8,20]. The development of NGS and analysis technology empower us to analyze and decode specific information in form of genetic variation (SNPs, InDels). The occurrence and virtual identification of enormous numbers of differences in individual nucleotides between individuals, enables every SNP to be a potentially useful marker [17]. Recently, sequence-based polymorphisms have been given much attention to explore the gene-based variation between high quality genomes and their related genotypes [15,21]. This sequence based marker technology allows the development of molecular markers to target genomic regions of the sequenced genome using the NGS which facilitates isolation of co-dominants molecular markers for the targeted genomics region of any animal and plant species. This will also particularly facilitate the development of high-density molecular maps, essential for gene based cloning using genetic map positions, and identification of linked molecular markers for selecting desirable genotype in plant breeding programs.

DNA polymorphisms have been identified both in animals and plants. In animals, more than 300 diseases and traits have been studied using GWASs [22]. Among plants, in rice over 600 genes have been cloned using various functional genomics methods. Many of these genes control agriculturally useful traits such as yield, grain quality, resistances to biotic and abiotic stresses and nutrient use efficiency thus have potential utility in crop improvement [23–25]. For example, an SNP in the 5′ regulatory region of the *qSH1* gene was shown to cause loss of seed shattering owing to the absence of an abscission layer formation in rice. Another example is the *GIF1* (*grain incomplete filling 1*) gene that shows a restricted expression pattern during grain-filling compared to the wild rice allele, is probably a result of accumulated mutation in the gene's regulatory sequence through domestication [17]. Many of the gene specific SNPs can be important for trait variation, which led us to conduct genome-wide analysis of gene based SNPs to decode the genic variation underlying functions.

Materials and Methods

Data retrieval and analysis

The 73 bp reads of FASTQ data of *O. sativa* ssp. *indica* cv. Guangluai-4 (*Os indica*) with 20× genome coverage by next generation sequencing (SRA study: ERP000235) were downloaded from NCBI. The total 54,309,982 read count (~108 million reads) were mapped to the rice *O. sativa japonica* cv. Nipponbare (*Os japonica*) reference genome information (version 7) from http://rice.plantbiology.msu.edu/ using SHORE (http://1001genomes.org/).

SHORE & SHOREmap pipeline

The sequence SHORE analysis pipeline [26,27], is a mapping and analysis pipeline for short DNA sequences retrieved from databases. The SHORE, short read analysis pipeline involves five different 'OPTIONS' for processing/analysis. A) SHORE preprocess: prepares index files, local repeat and GC content files from references. B) SHORE import: prepares short read data for processing. C) SHORE Mapflowcell: Short read mapping. D) SHORE merge: Merge and filter of alignment files and E) SHORE consensus: creates consensus sequence from alignment and computes homozygous SNP file with other results. It is designed for projects whose analysis strategy involves mapping of reads to a reference sequence. We used Burrows-Wheeler Aligner (BWA) to align short reads to the rice reference genome (MSU v7)

Table 1. Dataset and statistics of genotypes used in this study.

Sl.No.	Genome	Sub-species	Cultivar	covered	SNP	SNP/kb	data used
1	*Orzya sativa*	*Japonica*	Nipponbare	100%	Ref		MSU
2	*Orzya sativa*	*Indica*	Guangluai-4	63%	1674360	6.9	NCBI

with option -n4 -g3 -c12 and a default of 5 bases Cutoff for base masking using Sanger calibrated qualities, and Cutoff for base masking using chastity values [26,28]. The Chastity value was introduced by Illumina's GA Pipeline and is used to differentiate between clusters, which are interfered by other clusters and those that are not. It is defined as the highest base intensity of a sequenced base, divided by the sum of the highest plus the second highest intensity of the specific base. Therefore the Chastity score can reach values from 0.5 to 1, or in percent from 50 to 100. The prediction of valid SNP calls have been made on the basis of SHORE statistics explained in http://sourceforge.net/apps/mediawiki/shore/index.php?title=Shore_consensus. All position SNP calls follow a concordance of ≥80% and support of at least three non-repetitive reads [26].

SHORE enables more than 50% of the overlapping reads with weighted and gapped alignments. We used the results of the alignments for discovery of putative SNPs, where each variation supported be at least 3 reads was regarded positive so as to eliminate false positive calls. SHORE calculates a quality score based on information from several features related to the quality of sequence reads and the alignment. The alignment results were subjected to quality filtering, and a cutoff of more than 25 out of 40 quality score was selected and used for functional SNP analysis. SHORE analysis includes heterozygous SNP calls (decision tree approach) for all positions with at least 25% of the bases different to the majority call. It has been reported that residual heterozygosity is present in the original individual used in the IRGSP sequencing, and a small subset of allelic sites within individuals have been identified [29] but due to statistical sampling and sequencing bias the accuracy of these SNPs is affected by heterozygous SNPs as well when sequencing heterozygous samples. Since these SNPs are not very significant for marker development, we did not follow this analysis further using these calls, and the final SNPs on annotated genic and non genic regions have been evaluated only with real homozygous SNPs [21,30]. This quality SNP data was used for functional annotation with help of the SHOREmap pipeline and analyzed further by a suite of Linux scripts [27].

SNP analysis on rice gene families and TF database

The rice genes with annotated SNPs were grouped in gene families according to their gene locus IDs in the MSU database (http://rice.plantbiology.msu.edu) and extracted with functional SNP annotation.

We could retrieve 2,726 gene loci from 32 rice gene families as described in the MSU rice genome database http://rice.plantbiology.msu.edu/annotation_community_families.shtml. In addition, SNPs were identified in transcription factors derived from the rice transcription factor database (DRTF) that currently contains 2,384 putative transcription factors (TF) gene models in *O. sativa spp. japonica*, distributed in 63 families [29,31].

Visualization of whole genome SNPs

The visualization of genome-wide SNPs was done with the CVit tool [32]. The SNP types were categorized based on the GFF (General Feature file) from the MSU rice genome annotation (version 7) and SNPs identified between Guangluai-4 and Nipponbare of the two sequenced genomes *Oryza sativa* (*Os*) *indica* and *Os japonica* respectively using SHOREmap [27].

The results of SHORE genome-wide base-by-base analysis of two reference rice genomes were used to develop the SNP2GENE database. We used MySQL as the underlying database to create the table structure, store the results and to allow for efficient searches. The front-end web interface to the database was written

Table 2. Distribution of SNPs and InDels on rice chromosomes.

		InDel Polymorphisms		SNP Polymorphisms					
Chr. No.	**Total SNPs**	**Insertions**	**Deletions**	**Intergenic/noncoding**	**3′ UTR**	**5′ UTR**	**CDS**	**Synonymous**	**Non-synonymous**
1	207659	9689	11551	37325	3315	3488	22401	9024	13377
2	174855	8136	9504	30976	2654	2598	18779	7217	11562
3	160882	7727	8897	28632	2524	2583	16287	6222	10065
4	125653	5513	6447	23141	1732	2016	16142	6302	9840
5	124752	5852	6579	20269	1791	1800	13617	5302	8315
6	138429	5939	7031	24814	1611	1572	16234	6222	10012
7	139709	6048	7006	23905	1996	1841	16690	6492	10198
8	120056	5223	6106	20683	1593	1726	14161	5432	8729
9	109797	4761	5482	20000	1214	1363	12955	4882	8073
10	123407	5039	5875	20766	1434	1388	15048	5723	9325
11	134853	5637	6202	24649	1421	1431	18649	7043	11606
12	114308	4730	5396	19351	1401	1436	14135	5401	8734
Total	1674360	80146	92655	294511	22686	23242	195098	75262	119836

Figure 1. Distribution of gene based SNPs on 12 rice chromosomes, 55801 loci and 240914 positions. The SNPs glyph shows intergenic region; black, locus; gray, on chromosomes and range bar (left side of chromosomes) Syn-SNPs, black; Nonsyn-SNPs, red; threeprimeUTR-SNPs, blue; fiveprimeUTR-SNPs, green color.

Table 3. Distribution and types of SNPs in rice gene families.

Sl. Nos	Rice Gene Families	No. of families	No. of members	SNPs	Syn	Nonsyn
1	Ascorbate Peroxidase (APx) Family	1	8	19	2	0
2	ATP-dependent metalloprotease family	1	9	9	3	1
3	Bric-A-Brac/Tramtrack/Broad Complex (BTB) Proteins	18	145	4	2	1
4	C_2H_2 zinc finger proteins	1	179	3	1	2
5	Calmodulin and calmodulin-related calcium sensor proteins	2	37	0	0	0
6	Cellulose Synthase Gene Family	2	45	0	0	0
7	Conserved peptide uORF-containing transcripts	1	39	0	0	0
8	Core Replication Machinery Proteins	1	58	0	0	0
9	Cysteine Rice Peptides	28	532	6	1	5
10	Early Auxin-responsive Aux/IAA Gene Family	1	31	3	0	0
11	Early Auxin-responsive GH3 Gene Family		14	0	0	0
12	Endo-Beta-Mannanases	1	9	5	0	1
13	F-Box Proteins	10	686	0	0	0
14	Glutaredoxin family	1	28	0	0	0
15	Glycosyl Hydrolase Family 1 Beta-Glucosidases	1	38	8	1	2
16	HKT transporters	1	7	8	1	1
17	Indole-3-acetic-acid synthetase family	1	13	0	0	0
18	LIM domain proteins	1	6	2	0	0
19	Lipocalin Gene Family	2	3	6	2	0
20	MADS-box family	1	74	71	3	9
21	Mitogen-Activated Protein Kinases (MAPKs) and MAPK kinases (MAPKKs)	2	23	19	0	3
22	Phosphatidylethanolamine Binding Protein (PEBP) Gene Family	3	19	0	0	0
23	Protein Disulfide Isomerase Superfamily	3	23	22	0	1
24	RCI2 Homologs	1	11	0	0	0
25	Rice Kinase Interactome	1	149	8	2	1
26	SBP-Box Proteins	1	19	4	0	0
27	Serine Proteases	13	222	15	2	2
28	Small auxin-up gene family	1	58	0	0	0
29	Topoisomerase 6 homologs	1	4	17	1	1
30	Type-A Response Regulators	1	10	2	0	0
31	Wall-associated Kinases	1	144	1	0	1
32	WRKY Gene Superfamily	1	83	7	0	0

in php. We also developed a custom php module for visualization of a selected individual gene. The web interface to the SNP2GENE database is available at https://plantstress-pereira.uark.edu/oryza2.

Results and Discussion

Genome-wide natural polymorphisms

The NGS genome sequence of rice genotypes *O. sativa* ssp. *indica* cv. Guangluai-4 (G4) (*Os indica*) was analyzed with the objective to identify synonymous-SNP (sSNP) and non-synonymous-SNP (nsSNP) variation within genes. The complete genome sequence of rice *Os japonica* 12 pseudomolecules including the Syngenta pseudomolecules were used for SNP survey with *Os indica* next generation sequencing reads available at NCBI (Table 1). The analysis revealed more than 65% of the total genes with genic variation. A total of 207,659 SNPs were found on Chromosome 1, which contributes to the highest number in comparison to other chromosomes, followed by 174,855 SNPs in Chromosome 2 and 160,882 in chromosome 3, proportional to their size, as also seen for the other chromosomes (Table 2). The lowest rate of SNPs were observed in chromosome 9 (109797) followed by chromosome 12 (114308), which has been reported to contain a high number of disease resistance genes with recent gene duplications [33] (Table S1). We further grouped these SNPs into different types of polymorphisms, categorized as intergenic/noncoding, 3′-region, 5′-region, CDS, synonymous and non-synonymous and SNP2GENE database develop for this to retrieve the genome-wide information. In this analyses that covers 63% of genome between *japonica* Nipponbare and *indica* Guangluai-4, we found 40,761 gene loci having polymorphisms. Further analysis of these genes reveals 75,262 synonymous and 119,836 non-synonymous changes that may have functional significance. We found 22,686 SNPs at the three-prime (3′) UTR region and 23,242 five-prime (5′) UTR regions, and functional reading frame

Table 4. Number of genes in the given range of SNPs.

SNP range	Number of genes
1–5	15124
6–10	9070
11–15	5841
16–20	3352
21–25	2105
26–30	1067
31–35	955
36–40	670
41–45	519
46–50	882
51–55	250
56–60	206
61–65	208
66–70	154
71–75	99
76–80	86
81–85	59
86–90	67
91–95	44
96–100	32
>100	194

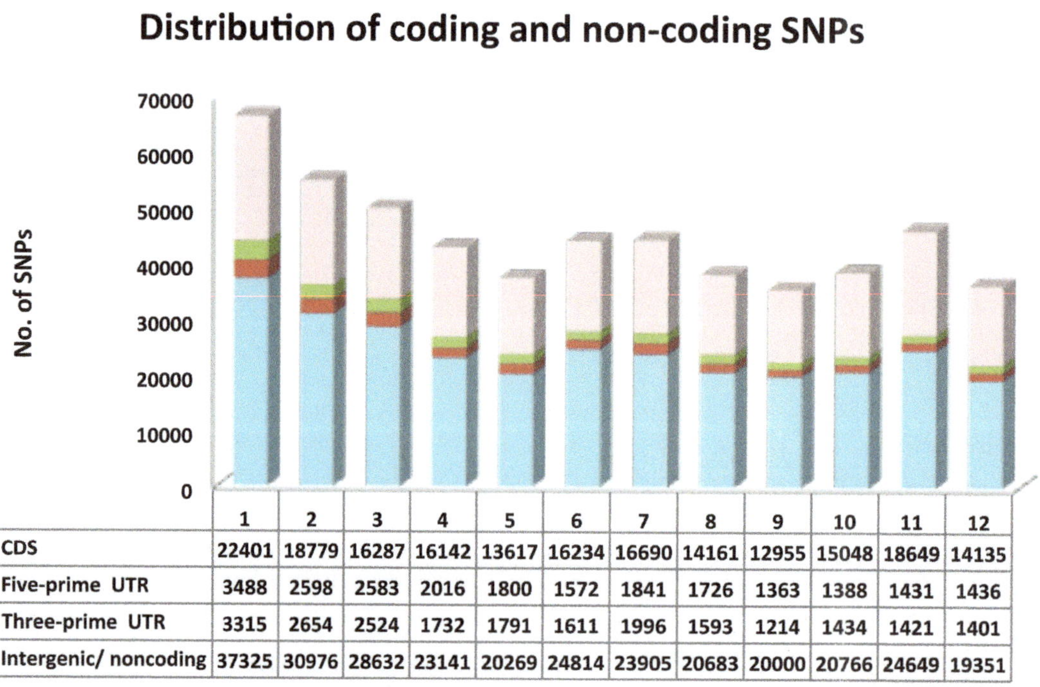

	1	2	3	4	5	6	7	8	9	10	11	12
CDS	22401	18779	16287	16142	13617	16234	16690	14161	12955	15048	18649	14135
Five-prime UTR	3488	2598	2583	2016	1800	1572	1841	1726	1363	1388	1431	1436
Three-prime UTR	3315	2654	2524	1732	1791	1611	1996	1593	1214	1434	1421	1401
Intergenic/ noncoding	37325	30976	28632	23141	20269	24814	23905	20683	20000	20766	24649	19351

Figure 2. Distributions of SNPs on coding and non-coding region of rice 1–12 chromosomes. Chromosomes show CDS, Five-prime UTRs, Three-prime UTRs and intergenic non-coding SNPs from chromosome 1–12.

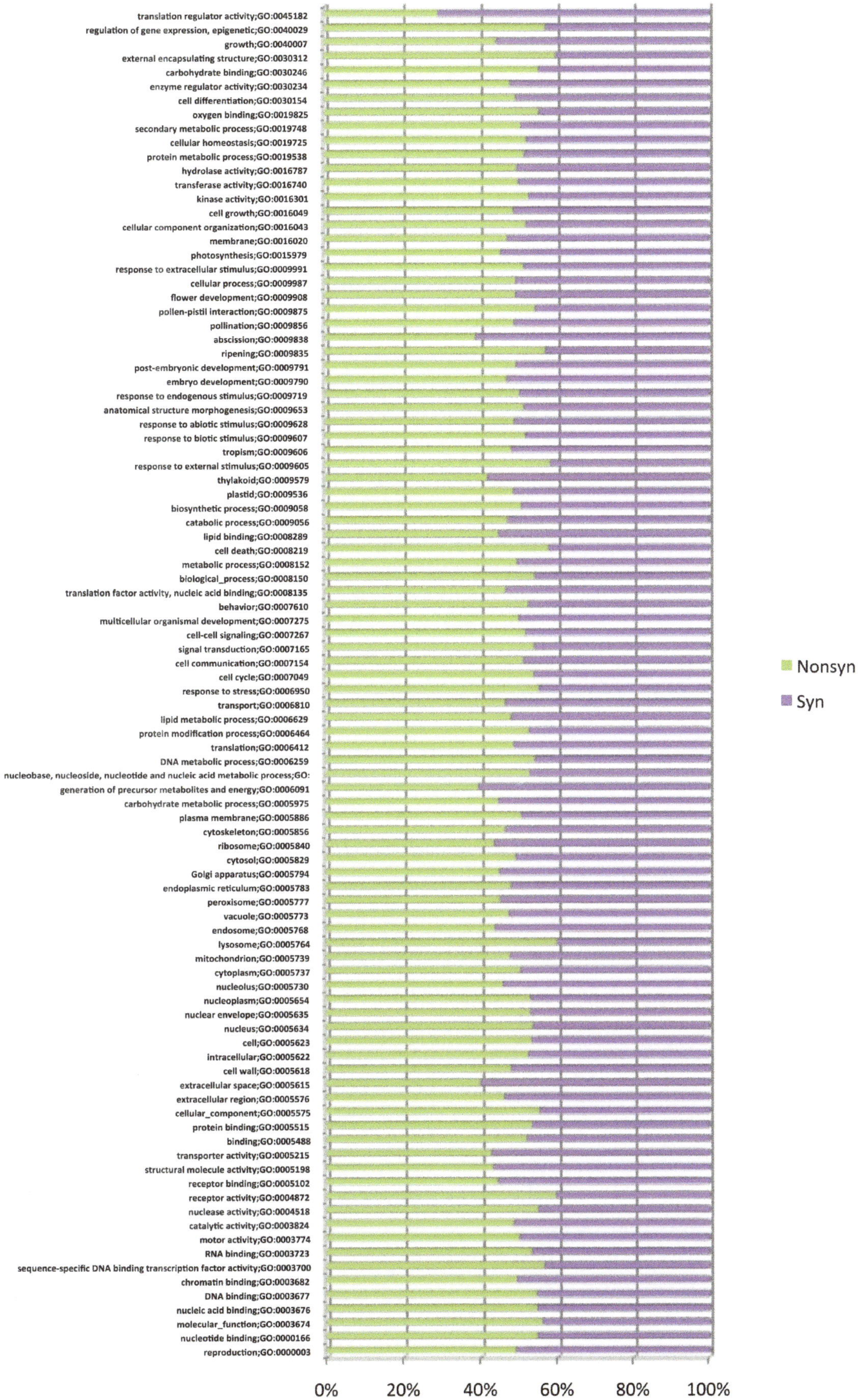
translation regulator activity;GO:0045182
regulation of gene expression, epigenetic;GO:0040029
growth;GO:0040007
external encapsulating structure;GO:0030312
carbohydrate binding;GO:0030246
enzyme regulator activity;GO:0030234
cell differentiation;GO:0030154
oxygen binding;GO:0019825
secondary metabolic process;GO:0019748
cellular homeostasis;GO:0019725
protein metabolic process;GO:0019538
hydrolase activity;GO:0016787
transferase activity;GO:0016740
kinase activity;GO:0016301
cell growth;GO:0016049
cellular component organization;GO:0016043
membrane;GO:0016020
photosynthesis;GO:0015979
response to extracellular stimulus;GO:0009991
cellular process;GO:0009987
flower development;GO:0009908
pollen-pistil interaction;GO:0009875
pollination;GO:0009856
abscission;GO:0009838
ripening;GO:0009835
post-embryonic development;GO:0009791
embryo development;GO:0009790
response to endogenous stimulus;GO:0009719
anatomical structure morphogenesis;GO:0009653
response to abiotic stimulus;GO:0009628
response to biotic stimulus;GO:0009607
tropism;GO:0009606
response to external stimulus;GO:0009605
thylakoid;GO:0009579
plastid;GO:0009536
biosynthetic process;GO:0009058
catabolic process;GO:0009056
lipid binding;GO:0008289
cell death;GO:0008219
metabolic process;GO:0008152
biological_process;GO:0008150
translation factor activity, nucleic acid binding;GO:0008135
behavior;GO:0007610
multicellular organismal development;GO:0007275
cell-cell signaling;GO:0007267
signal transduction;GO:0007165
cell communication;GO:0007154
cell cycle;GO:0007049
response to stress;GO:0006950
transport;GO:0006810
lipid metabolic process;GO:0006629
protein modification process;GO:0006464
translation;GO:0006412
DNA metabolic process;GO:0006259
nucleobase, nucleoside, nucleotide and nucleic acid metabolic process;GO:
generation of precursor metabolites and energy;GO:0006091
carbohydrate metabolic process;GO:0005975
plasma membrane;GO:0005886
cytoskeleton;GO:0005856
ribosome;GO:0005840
cytosol;GO:0005829
Golgi apparatus;GO:0005794
endoplasmic reticulum;GO:0005783
peroxisome;GO:0005777
vacuole;GO:0005773
endosome;GO:0005768
lysosome;GO:0005764
mitochondrion;GO:0005739
cytoplasm;GO:0005737
nucleolus;GO:0005730
nucleoplasm;GO:0005654
nuclear envelope;GO:0005635
nucleus;GO:0005634
cell;GO:0005623
intracellular;GO:0005622
cell wall;GO:0005618
extracellular space;GO:0005615
extracellular region;GO:0005576
cellular_component;GO:0005575
protein binding;GO:0005515
binding;GO:0005488
transporter activity;GO:0005215
structural molecule activity;GO:0005198
receptor binding;GO:0005102
receptor activity;GO:0004872
nuclease activity;GO:0004518
catalytic activity;GO:0003824
motor activity;GO:0003774
RNA binding;GO:0003723
sequence-specific DNA binding transcription factor activity;GO:0003700
chromatin binding;GO:0003682
DNA binding;GO:0003677
nucleic acid binding;GO:0003676
molecular_function;GO:0003674
nucleotide binding;GO:0000166
reproduction;GO:0000003
0%
20%
40%
60%
80%
100%
Nonsyn
Syn

Figure 3. Analysis of all rice SNPs in genes with respect to their GO-terms descriptions.

changes through 3,886 inducing STOP-codon (isSNP) and 729 preventing STOP-codon (psSNP) variation.

Gene based polymorphisms

We retrieved the total SNP information of the two rice genotypes based on 55,801 loci with 240,914 SNPs, comprising non-synonymous, synonymous, 5′-UTRs and 3′-UTRs on 12 rice chromosomes (Figure 1, Table S2 and Table S3). The functional analyses of the 55,801-gene annotation available (http//:www.plantbiology.msu.edu/) reveals 40,764 loci contribute to SNPs between these two genomes (Table S1). Since not all SNPs are important for functional analysis, the study was extended towards functional annotation of these SNPs and their effect on amino acid changes described in terms of synonymous and non-synonymous changes. These SNPs with putative functional changes were next used for gene family analysis with classifications available from annotation communities (http://rice.plantbiology.msu.edu). The locus information for the gene families were extracted and SNPs analyzed in the 2726 genes belonging to 32 different gene families and 104 sub-family groups curated by the rice community with functional and/or structural annotation submitted to the Rice Genome Annotation Project (http://rice.plantbiology.msu.edu/annotation_community_families.shtml). Our analysis shows 924 loci comprising10 different gene families are conserved, whereas other gene families have at least one variation in the family (Table 3). In total there are 239 SNP, comprising 21 synonymous and 31 non-synonymous SNPs in 2726 gene members. The largest gene family of proteins, the F-box has 10 families and 686 members that are completely conserved between these two rice sub-species, and have been reported to have conserved domains [34]. In contrast to that, 12 gene families possess synonymous and non-synonymous SNPs that have been reported to have important functions in plants.

Amongst the gene families of interest are the MADS-box family (71 SNPs) that functions in flower and fruit development [35], the protein disulfide isomerase superfamily (22 SNPs) controls diverse metabolic functions including disulfide bond formation and isomerisation during protein folding [36,37], ascorbate peroxidase (APx) family (19 SNPs) for development and response to environmental cues [38,39], mitogen-activated protein kinases (MAPKs) and MAPK kinases (MAPKKs) with 19 SNPs that play an important role in response to pathogens and disease resistance cascades, topoisomerase 6 homologs (17 SNPs) involved in overexpression of stress tolerance genes [37]and serine proteases (15 SNPs) known for proteolytic enzymes associated with several essential physiological pathways [40]. The other 15 gene families have single digit SNPs and the rest 11 are conserved (Table 3). We grouped all identified genome wide SNPs in range of SNPs to discover the highest and lowest ranging SNP loci (Table 4 and Table S4).

Synonymous polymorphisms and non-synonymous polymorphisms

The importance of 1.6 million SNPs depends upon their position and effect on gene functions, although the non-functional variation is also important as markers in the genome. In recent years many of the gene based SNPs have been reported for their role in controlling characters like grain filling, plant height, grain weight, amylose content and structure of grain in rice and maize [17,41,42]. However, it has also been found that nsSNPs and sSNPs shared similar likelihood and size of effect in association to disease in humans [22].

We found an interesting feature of SNP effect in our analysis, there were 195,446 SNPs on CDS (75,262 synonymous, 119,836 non-synonymous including 3,866 STOP codons) contributing 11% of the total SNPs effects related to change in proteins in the genomes (Figure 2), which is a significant contribution to putative phenotypic effects between these two subspecies. The highest number of non-synonymous SNP effects was found on Chr1 (13,377) followed by Chr2 (11,562). We found 3,886 SNPs which induced STOP-codons in the reading frames, and 729 SNPs preventing STOP-codon variation in the genome (Table S5 and

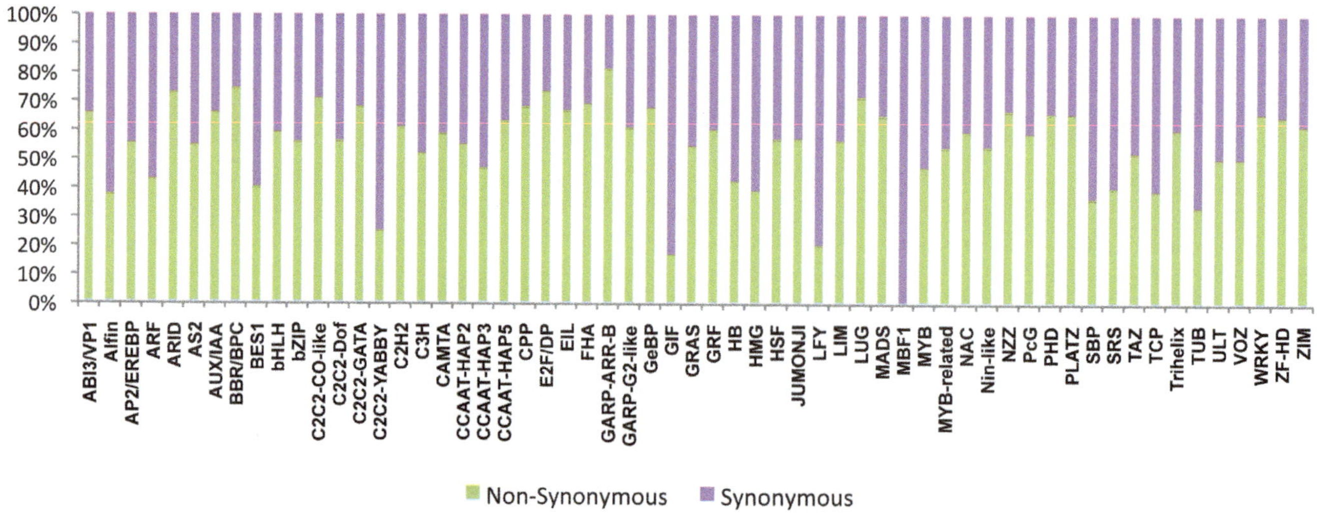

Figure 4. Synonymous and non-synonymous SNPs in transcription factor gene families. Variation in 2,384 putative transcription factors (TF) gene models in *O. sativa* spp. *japonica*, distributed in 63 families transcription factors have been predicted by DRTF.

Table 5. Distribution of polymorphism types in rice transcription factors.

TF family	Gene Number*	Total SNP	Non-Synonymous	Synonymous
ABI3/VP1	57	357	51	27
Alfin	13	103	3	5
AP2/EREBP	182	785	115	94
ARF	41	443	55	74
ARID	7	70	8	3
AS2	39	161	37	31
AUX/IAA	46	237	23	12
BBR/BPC	7	70	26	9
BES1	6	27	2	3
bHLH	184	1491	172	120
bZIP	109	1046	98	78
C2C2-CO-like	54	416	48	20
C2C2-Dof	36	166	23	18
C2C2-GATA	23	85	21	10
C2C2-YABBY	12	65	1	3
C2H2	113	984	161	104
C3H	90	677	103	96
CAMTA	8	159	24	17
CCAAT-Dr1	1	14	0	0
CCAAT-HAP2	20	224	17	14
CCAAT-HAP3	17	56	7	8
CCAAT-HAP5	18	165	46	27
CPP	16	109	17	8
E2F/DP	9	130	11	4
EIL	12	42	12	6
FHA	19	245	42	19
GARP-ARR-B	10	74	13	3
GARP-G2-like	56	534	51	33
GeBP	15	97	27	13
GIF	3	42	1	5
GRAS	58	392	89	75
GRF	18	58	12	8
HB	103	843	83	114
HMG	19	100	7	11
HRT-like	1	0	0	0
HSF	36	312	29	22
JUMONJI	17	282	37	28
LFY	1	13	1	4
LIM	13	155	45	35
LUG	11	108	15	6
MADS	83	1020	74	40
MBF1	3	7	0	2
MYB	138	1286	168	190
MYB-related	84	913	115	98
NAC	149	953	200	137
Nin-like	14	139	26	22
NZZ	1	4	2	1
PBF-2-like Whirly	1	6	0	0
PcG	34	538	78	55
PHD	79	1236	249	130

Table 5. Cont.

TF family	Gene Number*	Total SNP	Non-Synonymous	Synonymous
PLATZ	20	170	17	9
S1Fa-like	4	10	0	0
SBP	28	184	8	14
SRS	6	19	2	3
TAZ	10	131	26	24
TCP	24	82	7	11
Trihelix	23	123	21	14
TUB	21	158	12	24
ULT	2	32	2	2
VOZ	2	34	2	2
WRKY	113	1067	202	106
ZF-HD	15	113	31	17
ZIM	29	200	16	10

*The number of transcription factors have been predicted by DRTF (Gao et al., 2006).

Table S6). The highest number of induced STOP-codons was observed in Chr2 followed by Chr11 and minimum in Chr5.

The analysis was extended to identify the Ka/Ks ratios from the identified synonymous and non-synonymous changes for the kinase gene family, as an example to identify potential signatures of their biological functions, phenotypic effects, or expected evolutionary history. The family has potential roles in many traits such as pathogen defense, response to environment and almost all essential cellular functions. The predicted 1467 rice kinase genes were downloaded from http://phylomics.ucdavis.edu/kinase/. The IDs of these kinases were searched in the whole genome SNP analysis database, and 1313 genes showing SNPs were listed as kinase SNPs. We could find 839 kinases falling in the Ka/Ks (*Ka* = nonsynonymous substitutions number, and *Ks* = synonymous substitutions number) variation range from low (0.06) to high (12.0). There are 538 kinase genes that show more than 1.0 Ka/Ks value, with a potential signature for positive selection in evolution and a functional role in plants. These receptor kinases show more variability than other proteins signifying a selective role for the polymorphisms [43,44] (Table S7). Similarly some gene ontology biological processes (Figure 3) such as 'regulation of gene expression', 'response to external stimulus', 'receptor activity' show a higher level of non-synonymous changes (55–60%), signifying a positive selection for variation. Induced mutations have been reported to create a large proportion of non-sense mutations involving the introduction of novel stop-codons, and any individual mutations are therefore more likely to have a phenotypic effect, and partly explain the high mutation frequency achieved in rice [45,46]. The total 1.6 million SNPs were analyzed for the type of SNPs they carried, since this could affect the biological processes to a major extent.

High SNP genes – hotspots for variation in the rice genome

Rice has undergone significant phenotypic changes during domestication in different traits like grain size, color, shattering, seed dormancy and tillering, and geneticists have used quantitative trait locus (QTL) mapping to localize the major causative genes responsible for these traits, yielding many trait-related genes in cultivated rice [14]. These phenotypic variations are due to natural genetic variation, which is high in number as seen in the variation between the two subspecies. However, the distributions of these natural variations are not uniform throughout the genomes. It has been reported that transposon induced points (TIP) have generated approximately 14% of the genomic DNA sequence differences between subspecies *indica* and *japonica* and 10% of TIPs have been found in expressed gene regions, which represent important genetic variation [47]. These regions have genic and intergenic regions and were not distributed homogeneously throughout the genic regions.

Our analysis of SNPs based on genic regions revealed that 40,761 genes carry 535,537 SNPs. The variation was as high as 33 fold in LOC_Os04g10740, LOC_Os04g10700, LOC_Os04g03100, LOC_Os12g43630 as compared to the average of 4.5 SNP/kb within the genic region. In fact, the genic region possesses 32% of the total genome SNPs and the top 33% (13,500) of SNP carrying genes contain 72% (387,816) SNPs from the genic region. We next focused on 194 hotspot SNP genes that carried>100 SNPs/gene each, and which contribute to 5% (26,080) of genic SNPs (Table S2).

Variation in stress responsive transcription factors between sub-species

Understanding the molecular basis of natural phenotypic diversity is a major challenge in modern genetics and knowing how individual genetic polymorphisms combine to produce phenotypic change could strengthen evolutionary theory and advance applications in crop improvement [48]. Transcription factor (TF) proteins, which form gene regulatory networks (GRNs) to act in cooperative or competitive partnerships to regulate gene expression, are key components of these unique regulatory programs [49]. Therefore, TFs are key players in affecting the expression of genes and leading to qualitative phenotypic changes. The interaction between TFs and cis-regulatory DNA sequences control gene expression, constituting the essential linkages of regulatory networks. Sequence variations in TFs have been reported to control and cause phenotypic shifts in trait expression. Based on the fact that genetic interactions between transcriptions factors are a major source of phenotypic diversity within the species [48], we explored the TF locus information and analyzed the natural variation to understand their adaptive evolution. The analyzed regulatory SNPs that are located in the 3′ and 5′ flanking regions of different rice genes that might influence the activity of the transcriptional regulatory region, which comprise the cis-regulatory regions of the genes [50]. The predicted transcription factors of *Os japonica* in the DRTF currently comprise 2,384 putative transcription factor (TF) gene models, distributed in 63 families [31]. These TF genes were studied for their functional SNPs, as shown in Figure 4 and Table 5. Differences in gene expression may play a major role in speciation and phenotypic diversity.

Conclusions

Asian rice *Oryza sativa* is a cultivated, inbreeding species that feeds over half of the world's population. Genome-wide gene based analysis of sequence polymorphisms between the rice two major cultivated rice subspecies, *O. sativa* ssp *japonica* cv. Nipponbare and *O. sativa* ssp *indica* cv. Guangluai-4, revealed the distribution of SNPs and Insertion/Deletions at the gene level. The sequence polymorphisms commonly occur in both coding and non-coding regions and have significant differences in distribution between different gene families. These variations often affect differences in gene structure between the *indica* and *japonica* subspecies, and may contribute to phenotypic adaptation differences.

Supporting Information

Table S1 SNP distributions of all rice genes of 12 chromosomes.

Table S2 Distributions of 3′ UTR SNPs in rice Chromosome 1-12.

Table S3 Distributions of 5′ UTR SNPs in rice Chromosome 1-12.

Table S4 Hotspot genes containing 100 SNP and more in the locus.

Table S5 Distributions of 'Induced Stop' coding SNPs in rice Chromosomes 1-12.

Table S6 Distribution of Stop codon preventing SNPs in rice Chromosome 1.

Table S7 Distribution of Rice Kinase genes Ka/Ks ratio distribution.

Acknowledgments

We thank Nathan Weeks IT Specialist, ISU for his technical advice during the analysis. We acknowledge the help and use of the Arkansas High Performance Computing Center (AHPCC) University of Arkansas Facilities.

Author Contributions

Conceived and designed the experiments: SKS PW AP. Performed the experiments: SKS PW. Analyzed the data: SKS PW AP. Contributed reagents/materials/analysis tools: SKS PW AP. Wrote the paper: SKS PW AP.

References

1. McNally KL, Childs KL, Bohnert R, Davidson RM, Zhao K, et al. (2009) Genomewide SNP variation reveals relationships among landraces and modern varieties of rice. Proceedings of the National Academy of Sciences of the United States of America 106: 12273–12278.
2. Project IRGS (2005) The map-based sequence of the rice genome. Nature 436: 793–800.
3. Rafalski A (2002) Applications of single nucleotide polymorphisms in crop genetics. Curr Opin Plant Biol 5: 94–100.
4. McNally KL, Bruskiewich R, Mackill D, Buell CR, Leach JE, et al. (2006) Sequencing multiple and diverse rice varieties. Connecting whole-genome variation with phenotypes. Plant Physiol 141: 26–31.
5. Andersen JR, Lubberstedt T (2003) Functional markers in plants. Trends Plant Sci 8: 554–560.
6. Schacherer J, Shapiro JA, Ruderfer DM, Kruglyak L (2009) Comprehensive polymorphism survey elucidates population structure of Saccharomyces cerevisiae. Nature 458: 342–345.
7. Salisbury BA, Pungliya M, Choi JY, Jiang R, Sun XJ, et al. (2003) SNP and haplotype variation in the human genome. Mutat Res 526: 53–61.
8. Lindblad-Toh K, Winchester E, Daly MJ, Wang DG, Hirschhorn JN, et al. (2000) Large-scale discovery and genotyping of single-nucleotide polymorphisms in the mouse. Nat Genet 24: 381–386.
9. Clark RM, Schweikert G, Toomajian C, Ossowski S, Zeller G, et al. (2007) Common sequence polymorphisms shaping genetic diversity in Arabidopsis thaliana. Science 317: 338–342.
10. Alonso-Blanco C, Mendez-Vigo B, Koornneef M (2005) From phenotypic to molecular polymorphisms involved in naturally occurring variation of plant development. Int J Dev Biol 49: 717–732.
11. Huang X, Feng Q, Qian Q, Zhao Q, Wang L, et al. (2009) High-throughput genotyping by whole-genome resequencing. Genome Res 19: 1068–1076.
12. Feltus FA, Wan J, Schulze SR, Estill JC, Jiang N, et al. (2004) An SNP resource for rice genetics and breeding based on subspecies indica and japonica genome alignments. Genome Res 14: 1812–1819.
13. Han B, Xue Y (2003) Genome-wide intraspecific DNA-sequence variations in rice. Curr Opin Plant Biol 6: 134–138.
14. Xu X, Liu X, Ge S, Jensen JD, Hu F, et al. (2012) Resequencing 50 accessions of cultivated and wild rice yields markers for identifying agronomically important genes. Nat Biotechnol 30: 105–111.
15. Huang X, Wei X, Sang T, Zhao Q, Feng Q, et al. (2010) Genome-wide association studies of 14 agronomic traits in rice landraces. Nat Genet 42: 961–967.
16. Caicedo AL, Williamson SH, Hernandez RD, Boyko A, Fledel-Alon A, et al. (2007) Genome-wide patterns of nucleotide polymorphism in domesticated rice. PLoS Genet 3: 1745–1756.
17. Wang E, Wang J, Zhu X, Hao W, Wang L, et al. (2008) Control of rice grain-filling and yield by a gene with a potential signature of domestication. Nat Genet 40: 1370–1374.
18. Bevan M (2011) Genomics: endless variation most beautiful. Nature 477: 415–416.
19. Nielsen R, Paul JS, Albrechtsen A, Song YS (2011) Genotype and SNP calling from next-generation sequencing data. Nat Rev Genet 12: 443–451.
20. Arai-Kichise Y, Shiwa Y, Nagasaki H, Ebana K, Yoshikawa H, et al. (2011) Discovery of genome-wide DNA polymorphisms in a landrace cultivar of Japonica rice by whole-genome sequencing. Plant Cell Physiol 52: 274–282.
21. Sahu BB, Sumit R, Srivastava SK, Bhattacharyya MK (2012) Sequence based polymorphic (SBP) marker technology for targeted genomic regions: its application in generating a molecular map of the Arabidopsis thaliana genome. BMC Genomics 13: 20.
22. Chen R, Davydov EV, Sirota M, Butte AJ (2010) Non-Synonymous and Synonymous Coding SNPs Show Similar Likelihood and Effect Size of Human Disease Association. Plos One 5.
23. Lu L, Yan W, Xue W, Shao D, Xing Y (2012) Evolution and association analysis of Ghd7 in rice. PLoS One 7: e34021.
24. Jiang Y, Cai Z, Xie W, Long T, Yu H, et al. (2012) Rice functional genomics research: progress and implications for crop genetic improvement. Biotechnol Adv 30: 1059–1070.
25. Han B, Xue Y, Li J, Deng XW, Zhang Q (2007) Rice functional genomics research in China. Philos Trans R Soc Lond B Biol Sci 362: 1009–1021.
26. Ossowski S, Schneeberger K, Clark RM, Lanz C, Warthmann N, et al. (2008) Sequencing of natural strains of Arabidopsis thaliana with short reads. Genome Res 18: 2024–2033.
27. Schneeberger K, Ossowski S, Lanz C, Juul T, Petersen AH, et al. (2009) SHOREmap: simultaneous mapping and mutation identification by deep sequencing. Nat Methods 6: 550–551.

28. Li H, Durbin R (2009) Fast and accurate short read alignment with Burrows-Wheeler transform. Bioinformatics 25: 1754–1760.
29. Kawahara Y, de la Bastide M, Hamilton JP, Kanamori H, McCombie WR, et al. (2013) Improvement of the Oryza sativa Nipponbare reference genome using next generation sequence and optical map data. Rice (N Y) 6: 4.
30. Sumit R, Sahu BB, Xu M, Sandhu D, Bhattacharyya MK (2012) Arabidopsis nonhost resistance gene PSS1 confers immunity against an oomycete and a fungal pathogen but not a bacterial pathogen that cause diseases in soybean. BMC Plant Biol 12: 87.
31. Gao G, Zhong Y, Guo A, Zhu Q, Tang W, et al. (2006) DRTF: a database of rice transcription factors. Bioinformatics 22: 1286–1287.
32. Cannon EK, Cannon SB (2011) Chromosome visualization tool: a whole genome viewer. Int J Plant Genomics 2011: 373875.
33. IRGSP (2005) The sequence of rice chromosomes 11 and 12, rich in disease resistance genes and recent gene duplications. BMC Biol 3: 20.
34. Jain M, Nijhawan A, Arora R, Agarwal P, Ray S, et al. (2007) F-box proteins in rice. Genome-wide analysis, classification, temporal and spatial gene expression during panicle and seed development, and regulation by light and abiotic stress. Plant Physiol 143: 1467–1483.
35. Ng M, Yanofsky MF (2001) Function and evolution of the plant MADS-box gene family. Nat Rev Genet 2: 186–195.
36. Jain M, Tyagi AK, Khurana JP (2006) Overexpression of putative topoisomerase 6 genes from rice confers stress tolerance in transgenic Arabidopsis plants. FEBS J 273: 5245–5260.
37. Bent AF (2001) Plant mitogen-activated protein kinase cascades: Negative regulatory roles turn out positive. Proceedings of the National Academy of Sciences of the United States of America 98: 784–786.
38. Lazzarotto F, Teixeira FK, Rosa SB, Dunand C, Fernandes CL, et al. (2011) Ascorbate peroxidase-related (APx-R) is a new heme-containing protein functionally associated with ascorbate peroxidase but evolutionarily divergent. New Phytol 191: 234–250.
39. Dunand C, Mathe C, Lazzarotto F, Margis R, Margis-Pinheiro M (2011) Ascorbate peroxidase-related (APx-R) is not a duplicable gene. Plant Signal Behav 6: 1908–1913.
40. Tripathi LP, Sowdhamini R (2006) Cross genome comparisons of serine proteases in Arabidopsis and rice. Bmc Genomics 7.
41. Monna L, Kitazawa N, Yoshino R, Suzuki J, Masuda H, et al. (2002) Positional cloning of rice semidwarfing gene, sd-1: rice "green revolution gene" encodes a mutant enzyme involved in gibberellin synthesis. DNA Res 9: 11–17.
42. Tanaka K, Ohnishi S, Kishimoto N, Kawasaki T, Baba T (1995) Structure, organization, and chromosomal location of the gene encoding a form of rice soluble starch synthase. Plant Physiol 108: 677–683.
43. Xu S, Clark T, Zheng H, Vang S, Li R, et al. (2008) Gene conversion in the rice genome. BMC Genomics 9: 93.
44. Jiang SY, Gonzalez JM, Ramachandran S (2013) Comparative genomic and transcriptomic analysis of tandemly and segmentally duplicated genes in rice. PLoS One 8: e63551.
45. Ganal MW, Altmann T, Roder MS (2009) SNP identification in crop plants. Curr Opin Plant Biol 12: 211–217.
46. Parry MA, Madgwick PJ, Bayon C, Tearall K, Hernandez-Lopez A, et al. (2009) Mutation discovery for crop improvement. J Exp Bot 60: 2817–2825.
47. Huang X, Lu G, Zhao Q, Liu X, Han B (2008) Genome-wide analysis of transposon insertion polymorphisms reveals intraspecific variation in cultivated rice. Plant Physiol 148: 25–40.
48. Gerke J, Lorenz K, Cohen B (2009) Genetic Interactions Between Transcription Factors Cause Natural Variation in Yeast. Science 323: 498–501.
49. Nowick K, Stubbs L (2010) Lineage-specific transcription factors and the evolution of gene regulatory networks. Brief Funct Genomics 9: 65–78.
50. Mottagui-Tabar S, Faghihi MA, Mizuno Y, Engstrom PG, Lenhard B, et al. (2005) Identification of functional SNPs in the 5-prime flanking sequences of human genes. BMC Genomics 6: 18.

20

Association Mapping of Seed Oil and Protein Content in *Sesamum indicum* L. Using SSR Markers

Chun Li, Hongmei Miao, Libin Wei, Tide Zhang, Xiuhua Han, Haiyang Zhang*

Henan Sesame Research Center, Henan Academy of Agricultural Sciences, Zhengzhou, Henan, People's Republic of China

Abstract

Sesame is an important oil crop for the high oil content and quality. The seed oil and protein contents are two important traits in sesame. To identify the molecular markers associated with the seed oil and protein contents in sesame, we systematically performed the association mapping among 369 worldwide germplasm accessions under 5 environments using 112 polymorphic SSR markers. The general linear model (GLM) was applied with the criteria of log$P \geq 3.0$ and high stability under all 5 environments. Among the 369 sesame accessions, the oil content ranged from 27.89%–58.73% and the protein content ranged from 16.72%–27.79%. A significant negative correlation of the oil content with the protein content was found in the population. A total of 19 markers for oil content were detected with a R^2 value range from 4% to 29%; 24 markers for protein content were detected with a R^2 value range from 3% to 29%, of which 19 markers were associated with both traits. Moreover, partial markers were confirmed using mixed linear model (MLM) method, which suggested that the oil and protein contents are controlled mostly by major genes. Allele effect analysis showed that the allele associated with high oil content was always associated with low protein content, and vice versa. Of the 19 markers associated with oil content, 17 presented near the locations of the plant lipid pathway genes and 2 were located just next to a fatty acid elongation gene and a gene encoding Stearoyl-ACP Desaturase, respectively. The findings provided a valuable foundation for oil synthesis gene identification and molecular marker assistant selection (MAS) breeding in sesame.

Editor: Tianzhen Zhang, Nanjing Agricultural University, China

Funding: This work was supported by China National "973" Project (2011CB109304) and the earmarked fund for the China Agriculture Research System (CARS-15). The funders had no role in study design, data collection and analysis, decision to publish, or preparation of the manuscript.

Competing Interests: The authors have declared that no competing interests exist.

* Email: zhy@hnagri.org.cn

Introduction

Sesame (*Sesamum indicum* L.) is an ancient and important oilseed crop and is cultivated mainly in the tropical and subtropical regions of Asia, Africa and Southern America. The harvested area of world sesame reaches to 7.3×10^7 hm^2, and the total product per year is roughly 3.7×10^7 ton from 2001 to 2010 (FAO data). Compared with other main oil crops, e.g., soybean (18% of average oil content) [1], oilseed rape (41%) [2], sunflower (40–44%) [3] and peanut (51%) [4], sesame is one of the few oil crops with the highest oil content and quality. Sesame seeds contain 55–58% oil and almost 18% proteins. Among the fatty acid compositions in sesame seeds, oleic acid (18:1) (39.6%) and linoleic acid (18:2) (46.0%) are the two main components with the ideal ratio of almost 1:1 [5,6]. Apart from the seed yield, the content of seed storage oil and proteins is the highlight agronomic trait in sesame breeding [7].

In the past two decades, in order to clarify the high quality of sesame oil and protein, many researchers focused on exploring seed development and fatty acid and storage protein synthesis processes, as well as identifying the lipid synthesis related genes and molecular markers in sesame [8–12]. Of all the three available cDNA libraries, two libraries are constructed for seed development analysis [13,14]. However, oil and protein contents are complex quantitative traits and always are affected by genotype and environment [15]. At present, the mechanism of high oil content and quality in sesame seeds is still unclear. No loci of oil and protein content traits have ever been found in the sesame linkage maps. Even though Wei et al. [12] performed the association analysis of seed oil and protein content and fatty acid composition within 216 Chinese sesame accessions using 79 molecular primer pairs (including SSRs, SRAPs and AFLPs), only one association marker (M15E10-3) was identified under two environments. Therefore, in order to precisely detect the genes or markers associated with oil and protein content traits and to improve the sesame breeding, more efficient markers and germplasm resources with larger phenotypic variation need to be applied [16,17].

Currently, linkage analysis (QTL mapping) and association mapping are two main and common analysis tools for dissecting complex phenotypic variation. Compared with the traditional linkage analysis based on mapping populations, association mapping offers higher precision for locating QTLs and selecting molecular markers [16,18]. Till now, association mapping has been extensively used for analyzing important agronomic and quantity traits in wheat, maize, cotton, oilseed rape and other crops [18–22]. In the past several years, vast simple sequence repeat (SSR) or microsatellite markers with the high polymorphism are developed in sesame [7,23,24]. Accordingly, the association mapping is getting reliable and powerful for detecting the genes or markers associated with key traits and improving the molecular marker-assisted selection (MAS) in sesame breeding programs.

The objectives of this study are: (1) to perform the association mapping of seed oil content (OC) and protein content (PC) traits in worldwide sesame accessions using the GLM and MLM models, (2) to reflect the characteristics of sesame oil and protein contents under various environments, and (3) to determine the key SSR markers associated with seed quality. In this report, a natural population covering 369 worldwide accessions from China and other 15 countries and 112 pairs of polymorphic SSR markers were applied. The results give a foundation for investigating the seed development-related genes and seed quality in sesame.

Results

Seed oil and protein content variation in the natural population

Both the seed oil content (OC) and the protein content (PC) are often influenced by environment. To decrease the environmental effect, we collected the phenotypic data of the 369 sesame accessions under 5 environments of Pingyu and Yuanyang locations in 2011, and Pingyu, Yuanyang and Xinyang locations in 2012, The descriptive parameters under each environment were calculated (Table 1). The results showed that the OC and PC significantly varied among the 369 accessions. In total, the OC of the natural population ranged from 27.89%–58.73%, with an average of 49.59%–53.14%; meanwhile, the PC ranged from 16.72%–27.79% with an average of 20.28%–22.51%. All the datasets showed a normal or nearly normal distribution. To determine the heritability of the phenotypes, we performed the variance analysis of oil and protein contents (Table 2). Results indicated that the OC and PC traits were significantly influenced by genotype and environments (i.e., year and location). No significant interactions between variety and environmental factors (year and location) were detected. Moreover, the OC and PC traits presented the significant negative correlation under the 5 environments, as the correlation coefficient (r) between OC and PC varied from $-0.66\sim-0.72$ ($P<0.01$) in 2011 and from $-0.52\sim-0.74$ ($P<0.01$) in 2012, respectively (Data not listed).

Linkage disequilibrium

Linkage disequilibrium (LD) refers to the non-random association of alleles between the genetic loci. A total of 112 SSR markers were used for estimating the LD level among the 369 sesame germplasm accessions (Table S2). These SSR markers distributed in 33 contigs/scaffolds with a total length of 180.86 Mb, which approximately represented 67 percentage of the assembly genome size (270 Mb) and 50 percentage of the estimated genome size (360Mb). The average SSR density was 1 SSR per 1.6 Mb. To reflect the associations between the polymorphic loci of the 112 SSR markers, LD P-values were determined among the 6,216 locus pairs (i.e., 112*(112-1)/2) using Fisher's exact test and two indexes of D' and r^2 (Figure 1). The average values of D' and r^2 for the 6,216 pairs were 0.1649 and 0.0173, respectively. Of the 6216 pairs, 2584 pairs (41.57%) showed a significant linkage disequilibrium ($P<0.01$), 363 pairs (5.84%) showed a higher linkage disequilibrium of $D'>0.5$, and 33 pairs (0.5%) gave a D' value of 1.0 (i.e., complete linkage). The data indicated that linkage disequilibrium existed among the sesame accessions.

SSR marker diversity and population structure

Before analyzing the association, the polymorphism of the 112 SSR markers within the 369 germplasms was investigated (Table S2). Results showed that the number of alleles ranged from 2–5, with an average of 2.47 per locus. The PIC values of the markers

Table 1. Phenotypic variation of the seed oil and protein contents in the 369 accessions under 5 environments.

Trait	Year	Place	Min. (%)	Max. (%)	Mean (%)	Std. ev.	Skewness	Kurtosis
OC	2011	Pingyu	27.89	55.03	49.59	4.61	−2.19	1.69
	2011	Yuanyang	34.28	57.53	51.94	3.72	−2.26	2.85
	2012	Pingyu	31.95	54.80	49.82	3.43	−2.54	5.13
	2012	Yuanyang	30.77	57.88	52.23	4.22	−2.53	3.98
	2012	Xinyang	32.74	58.73	53.14	4.59	−2.32	2.67
PC	2011	Pingyu	18.74	26.49	21.65	1.30	0.90	1.28
	2011	Yuanyang	17.60	25.10	20.28	1.16	0.97	1.55
	2012	Pingyu	20.22	27.79	22.51	1.04	0.78	1.68
	2012	Yuanyang	16.72	27.44	20.94	1.25	1.05	5.27
	2012	Xinyang	17.24	26.84	20.59	1.57	0.85	1.30

Table 2. Analysis of Variance of the oil and protein contents in the population under 5 environments.

Source of variation	DF	OC		PC	
		MS	F value	MS	F value
Year	1	27.33	7.46**	211.90	412.83**
Location	2	1564.58	427.02**	551.56	1074.55**
Variety	368	62.78	17.13**	5.23	10.19**
Year×Variety	368	3.19	0.87	0.55	1.07
Location×Variety	736	2.94	0.80	0.65	1.27
Residual	367	3.66		0.51	

** The significance at $P<0.01$.
DF denotes degree of freedom; MS denotes mean square.

ranged from 0.028 (Hs373) to 0.669 (Y1994) with an average value of 0.302. The percentage of heterozygotes per marker varied from 0.27% (Hs373) to 23.12% (Hs4325), with an average of 10.14%. The data indicated that the natural population had the high heterozygosity and was suitable for association mapping analysis. Subsequently, we estimated the population structure, as admixture of subpopulation could result in LD and produce the false-positive results. The most likely number (K) of subgroups in the 369 germplasm accessions was estimated using the 112 SSR markers (Figure 2). As K values increased from 1 to 10, the value of LnP(D) elevated directly; meanwhile, the ΔK reduced straightly with a clear peak value of 643.5 at $K=2$. The results indicated that the population was roughly composed of two divergent subgroups according to the Bayesian posterior probability analysis (Figure 3).

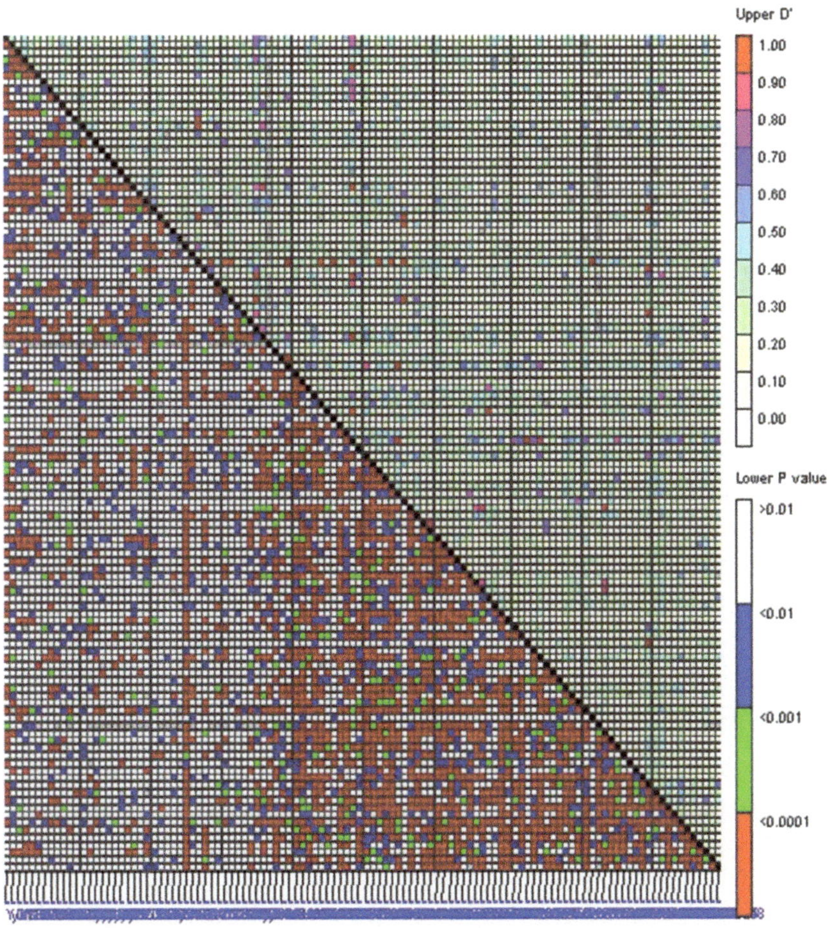

Figure 1. Disequilibrium matrix of 112 SSR polymorphic sites with both the X-axis and Y-axis. The matrix is divided into two regions by the diagonal line. The upper right region indicates the D′ value of each SSR couples, and the corresponding blocks in the lower left region indicates the significance of D′. Various value intervals of the D′ or P values are shown in different colors according to the right color columns.

Of the 369 accessions, 126 accessions were grouped into subgroup 1 (the green ones in Figure 3), 243 accessions into subgroup 2 (the red ones in Figure 3). Most (47 out of 51) of the foreign lines were located in subgroup 2. The two subgroups in the population were considered having the complex 'admixture' relationship, and no significant correlation of geographic origin with the subgroups in the Chinese lines were found.

Marker-trait associations

In this study, the association analysis was performed using general liner model (GLM) method. The stringent criterion of log$P$≥3.0 under 5 environments was used for determining the association significance between the OC and PC traits. For the OC trait, 19 markers were detected and the R^2 values ranged from 4%–29% (Table 3 & Table S2). According to the primer locations in the sesame genome, 19 markers were mapped in 11 contigs/ scaffolds. Of the 19 markers, Hs485 and Hs586 are located in scaffold00001, Hs4381, Hs02 and Hs4082 are located in C01, and Hs4061, Hs345 and Hs19563 are in C04. The distance between the markers that located in the same contig/scaffold was short than 2 Mb (Table S2). For the PC trait, 24 markers were detected with the R^2 value range of 3%–29% and mapped in 15 contigs/ scaffolds (Table 4 & Table S2). Comparison results indicated that 19 of the above 24 markers were associated with OC trait at the same time; the other 5 markers of Hs425 (in C08), Hs560 (in C11), Hs4265 (in C12), and Hs4089 and Hs1514 (in C19) were unique to PC trait. Simultaneously, in order to assay the stability, we performed association mapping using mixed liner model (MLM) method, with the criterion of log$P$≥3.0 under at least 3 environments.

For the OC trait, 9 markers were detected in MLM model and 8 markers of Hs345, Hs4381, Hs485, Hs1036, Hs4061, Hs635, Hs376 and Hs586 were confirmed using GLM and MLM methods. Especially, 4 markers of Hs345, Hs4381, Hs485 and Hs1036 had high R^2 values (≥10%) under 5 environments (Table 3 and Table S3). For the PC trait, 9 markers were found and confirmed in both models, of which 7 markers had high R^2 values (≥10%) under 5 environments (Table 4 and Table S3). These results suggested that the OC and PC traits are controlled mostly by major genes in sesame.

Marker effect on the phenotypic variation

To explore the association between the above markers and phenotypic variation and the utility potential in sesame MAS breeding program, we performed the allelic effects of the five markers associated with both traits (Table 5). For each marker, the effects estimated were in accordance with the allele character under the 5 environments. Hs345 had the largest effect on the variation of seed oil and protein content. The Hs345-1:1 and Hs345-2:2 of Hs345 showed the different variation effects on OC trait, as the average oil content in the accessions ranged from 52.05%–42.82%. In the genotypes carrying the Hs4381-1:1 allele, oil and protein contents were 43.89% and 23.25%, respectively, while the samples with the Hs485-2:2 contained 52.00% oil and 21.00% proteins. Furthermore, the specific allele indicated the negative or positive effect to a large extent on the OC or PC trait. For example, the allele effect of Hs345-2:2 on OC trait ranged from −8.13 to −12.18, whereas the effect on PC trait ranged from 2.05 to 3.82. Comparison results suggested that the alleles of all 5 markers give the opposite effects on OC and PC traits, respectively. The allele that increased the oil content certainly gave the negative effect on protein content, and vice versa. Therefore, these markers could be used for screening sesame lines with either high oil or protein content.

Comparative genome analysis

To clarify the distribution and more information of the above associated markers, we performed the comparative genome analysis of the 19 SSR markers associated with the OC trait (Table 6). All genes closest to the markers were explored. Of the 19 markers, 11 are located in gene regions and 8 are in intergenic regions. The flanking genes had various functions, such as ligase (C01.560), transcription factor (C04.26, C13.438) and kinase (C14.22). Moreover, we found that the markers of Hs4082 and Hs345 were just located next to C01.883 (ABC transporter G family member 3 gene) and C04.786 (Stearoyl-ACP Desaturase gene), respectively, which were proved involved in plant lipid biosynthesis. We also screened the upstream and downstream sequences of 500 Kb far from each marker. 17 (out of 19) markers were proved close to plant lipid pathway genes. These result further confirmed our association mapping conclusions.

Discussion

To clarify the genetic mechanisms of fatty acid and protein synthesis in sesame seeds, we herein performed the association mapping analysis of the OC and PC traits among 369 sesame accessions using 112 genic-SSR markers. These accessions were collected from 19 provinces of China and 15 other countries, and represented the genetic diversity of sesame germplasm for association mapping study. These accessions included many released Chinese and foreign cultivars. Compared with the traditional linkage analysis (QTL mapping), the association

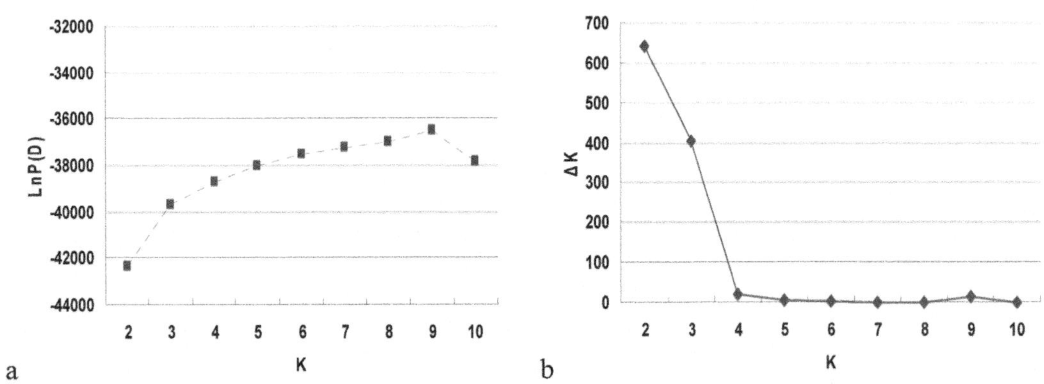

Figure 2. Estimated (a) Ln*P*(D) and (b) Δ*K* values for a given *K*.

Figure 3. Genetic composition of individuals based on Bayesian posterior probability. According to the Bayesian posterior probability, the natural population of 369 worldwide sesame lines is divided into two groups in green and red colors, respectively.

analysis based on linkage disequilibrium (LD) has been applied for the quantitative trait loci (QTLs) detection and location in many crops. Meanwhile, GLM and MLM models are applied individually for evaluating the marker association. In this article, 19 SSR markers associated with the OC trait were detected under each 5 environments in GLM model, while 24 markers were determined and associated with the protein content.

Sesame genetic diversity and the population structure

Sesame is a diploid and self-pollinated oil crop with the karyotype of 2n = 2x = 26. As all cultivars are originated from the sole cultivated species, *Sesamum indicum* L, the narrow genetic diversity presents in regional sesame resources to a large extent [24,49–51]. In addition, many reports reflect that there is no clear association between genotype and geographical origin, as many sesame accessions from the different geographic locations are clustered into the same group in the dendrogram [51–53]. Apart from the natural history of predomesticated ancestors, the diversity pattern of domestic species could be influenced by the breeding practice, germplasm collection and human activity [53–55]. In this article, in order to guarantee the broad genetic variation, we selected the 369 worldwide accessions for seed nutrition genetic analysis according to the geographical origin, molecular clustering and the morphologic diversity (Table S1). The heterozygosity ranged from 0.27% (Hs373)-23.12% (Hs4325). The result proved that the natural population could be used as the core germplasm for association mapping (Table S2). Population structure analysis showed that many sesame accessions collected from the same geographic locations were not grouped together, which further proved the unclear association between genotype and geographical origin in sesame germplasm resource [51–53].

During performing the association mapping in a population, LD patterns between the functional loci and markers should be analyzed at first [56]. We analyzed the *P*-values of linkage disequilibrium between the polymorphisms of the 112 SSR marker loci using Fisher's exact test (Figure 1). As 363 (5.84%) pairwise comparisons had the high LDs ($D'>0.5$), the linkage disequilibrium existed in 369 sesame accessions. Accordingly, we believed that the natural population is suitable for association analysis of the oil and protein contents.

Oil and protein content variation and associated SSR markers

Among large-scale sesame germplasm resources, the oil and protein contents varied significantly. Yermanos et al. [57] evaluated 721 sesame samples collected from more than 19 countries, and found that the oil content varied from 40.4–59.8% with the mean of 53.1%. The protein content ranged from 19–31% with an average of about 25% [58]. In our population, the oil content varied from 27.89%–58.73% with an average of 51.34%, while the protein content varied from 16.72%–27.79% with an average of 21.19% (Table 1). The data reflected the great variation of sesame seed compositions in germplasm [57,59,60].

Comparison analysis proved that there is a strong negative correlation of the oil content with protein content. Interestingly, the stringency relationship was also exhibited in the association analysis. As shown in the GLM analysis in Table 3 and 5, all the 19 markers associated with OC were detected for PC trait; and the alleles exhibited the opposite effects on OC and PC at the same time. These phenomena also present in other oil crops, such as oilseed rape, cotton, soybean and peanut [1,61–63]. Zhao et al. [62] found 6 QTLs with pleiotropic effects on both oil content and protein content in oilseed rape. In the cotton backcross inbred population, of 17 QTLs for oil content and 20 QTLs for protein content, 8 QTLs co-localized in the same chromosome regions and controlled oil and protein contents with opposite additive effects [63]. Hwang et al. (2014) detected 25 SNPs associated with seed oil in 13 different genomic regions through GWAS (genome wide association study), and 7 SNPs were significantly associated with both protein and oil traits. For the six of seven marker loci, a negative relationship existed between the protein effect versus that on oil [64]. Meanwhile, QTLs or markers associated with increased protein and oil contents were also found. For example, Zhao et al. (2006) found that 2 QTLs that controlled oil content were independent from protein content by conditional QTL mapping [65]; Hwang et al. (2014) found a SNP at the 4.92 Mb position on Chr 9 was associated with increased protein and oil contents [64].

In many crops, the seed oil content seems to be controlled mostly by major genes [66–69]. In this study, we detected 19 markers associated with OC trait in sesame using GLM method, and the R^2 values ranged from 4.0–29.0%. Chen et al. (2010) screened 27 QTLs related to oil content in oilseed rape and the individual explanation was high with the range of 4.20–30.20% [66]. In *Arabidopsis thaliana*, a single QTL or marker could give an explanation of 17–19% for the seed oil content [68]. In soybean, the explanation reached to 14.3–45.6% [69]. In this report, 112 polymorphic SSR markers were used for association mapping. Compared with other common molecular markers, such as SRAP and AFLP, SSR marker is more suitable for sesame diversity analysis due to the narrow genetic basis [24,49–51]. The marker distribution and density were analyzed using the sesame genome assembly data. The contigs/scaffolds carrying the 112 markers approximately covered 180 Mb of the sesame genome and occupied ~67% of the assembly size (270 Mb) and 50% of the estimated genome size (~360 Mb, in which 90 Mb was believed to be repeat sequences) (Table S2) [70,71]. Therefore, the association mapping using 112 markers is meaningful and reliable, even though some QTLs might be missed. To detect more QTLs, new SSR or SNP markers could be applied in further research.

Table 3. Association mapping of the OC trait using GLM method under the 5 environments.

Marker	2011						2012								
	Pingyu			Yuanyang			Pingyu			Yuanyang			Xinyang		
	F_{maker}	P_{marker}	R^2	F_{maker}	P_{marker}	R^2	F_{maker}	P_{marker}	R^2	F_{maker}	P_{marker}	R^2	F_{maker}	P_{marker}	R^2
Hs345	35.93	3.05E-25	0.28	26.22	4.37E-19	0.22	13.41	3.41E-10	0.13	26.44	3.11E-19	0.22	37.34	4.08E-26	0.29
Hs4381	61.72	8.30E-24	0.25	43.17	1.48E-17	0.19	40.31	1.53E-16	0.18	46.18	1.32E-18	0.20	75.46	3.73E-28	0.29
Hs1956	47.51	7.46E-19	0.22	39.98	2.83E-16	0.19	16.15	2.05E-07	0.09	24.86	8.80E-11	0.13	40.25	2.29E-16	0.19
Hs485	31.50	4.06E-18	0.20	20.63	2.29E-12	0.14	21.10	1.27E-12	0.15	21.48	7.82E-13	0.15	36.07	1.98E-20	0.23
Hs1036	41.73	6.75E-17	0.20	25.69	4.19E-11	0.13	17.87	4.24E-08	0.10	26.00	3.19E-11	0.13	43.84	1.24E-17	0.20
Hs205	17.18	3.11E-15	0.19	10.41	2.42E-09	0.12	10.37	2.68E-09	0.13	12.46	3.65E-11	0.15	16.29	1.75E-14	0.18
Hs4209	14.91	2.67E-13	0.17	8.15	2.64E-07	0.10	9.28	2.55E-08	0.12	12.37	4.41E-11	0.14	23.92	9.21E-21	0.25
Hs672	16.81	1.19E-12	0.16	11.35	1.09E-08	0.11	16.82	1.16E-12	0.16	20.27	4.38E-15	0.18	26.53	2.58E-19	0.22
Hs393	17.90	2.01E-13	0.16	10.10	9.18E-08	0.10	12.19	2.61E-09	0.12	16.74	1.32E-12	0.15	29.18	4.76E-21	0.24
Hs4061	30.93	3.93E-13	0.14	19.32	1.06E-08	0.09	21.42	1.60E-09	0.11	23.28	3.05E-10	0.11	30.72	4.66E-13	0.14
Y2129	22.99	4.73E-10	0.13	13.89	1.64E-06	0.08	15.29	4.57E-07	0.09	24.86	9.25E-11	0.13	30.65	6.69E-13	0.16
Hs618	25.22	5.54E-11	0.12	12.90	3.86E-06	0.06	7.54	6.19E-04	0.04	17.66	4.78E-08	0.09	22.91	4.26E-10	0.11
Hs377	4.14	2.01E-05	0.11	3.16	6.87E-04	0.08	3.92	4.49E-05	0.10	6.26	7.83E-09	0.15	7.09	3.63E-10	0.17
Hs635	22.15	8.38E-10	0.11	17.23	7.11E-08	0.08	10.34	4.30E-05	0.05	18.79	1.71E-08	0.09	31.73	1.98E-13	0.15
Hs02	20.98	2.38E-09	0.10	17.19	7.34E-08	0.08	10.31	4.43E-05	0.05	12.41	6.09E-06	0.06	36.20	4.55E-15	0.16
Hs376	19.25	1.15E-08	0.10	11.49	1.45E-05	0.06	12.53	5.52E-06	0.07	16.27	1.72E-07	0.08	21.26	1.89E-09	0.10
Hs1638	17.45	6.23E-08	0.09	7.29	7.96E-04	0.04	18.27	2.97E-08	0.10	19.25	1.23E-08	0.10	24.86	8.66E-11	0.13
Hs586	16.84	1.01E-07	0.08	13.48	2.25E-06	0.07	7.30	7.82E-04	0.04	10.90	2.52E-05	0.06	20.14	5.06E-09	0.10
Hs4082	8.18	3.36E-04	0.04	7.56	6.08E-04	0.04	18.30	2.69E-08	0.09	15.51	3.44E-07	0.08	20.61	3.33E-09	0.10

The markers associated with the OC trait under 5 environments are listed in the table.

Table 4. Association mapping of PC trait using GLM method under the 5 environments.

Marker	2011						2012								
	Pingyu			Yuanyang			Pingyu			Yuanyang			Xinyang		
	F_{maker}	P_{marker}	R^2	F_{maker}	P_{marker}	R^2	F_{maker}	P_{marker}	R^2	F_{maker}	P_{marker}	R^2	F_{maker}	P_{marker}	R^2
Hs345	25.46	1.41E-18	0.21	21.52	6.24E-16	0.19	13.41	3.41E-10	0.13	26.44	3.11E-19	0.22	37.34	4.08E-26	0.29
Hs4381	50.30	5.14E-20	0.21	33.86	3.25E-14	0.15	40.31	1.53E-16	0.18	46.18	1.32E-18	0.20	75.46	3.73E-28	0.29
Hs485	26.93	9.51E-16	0.18	21.07	1.32E-12	0.14	21.10	1.27E-12	0.15	21.48	7.82E-13	0.15	36.07	1.98E-20	0.23
Hs205	16.43	1.33E-14	0.18	7.90	4.52E-07	0.10	10.37	2.68E-09	0.13	12.46	3.65E-11	0.15	16.29	1.75E-14	0.18
Hs1036	31.71	2.48E-13	0.16	17.49	5.98E-08	0.09	17.87	4.24E-08	0.10	26.00	3.19E-11	0.13	43.84	1.24E-17	0.20
Hs393	16.01	4.43E-12	0.15	13.34	3.73E-10	0.12	12.19	2.61E-09	0.12	16.74	1.32E-12	0.15	29.18	4.76E-21	0.24
Hs672	17.02	8.43E-13	0.15	11.63	6.73E-09	0.11	16.82	1.16E-12	0.16	20.27	4.38E-15	0.18	26.53	2.58E-19	0.22
Hs1956	26.91	1.50E-11	0.14	22.88	4.93E-10	0.12	16.15	2.05E-07	0.09	24.86	8.80E-11	0.13	40.25	2.29E-16	0.19
Hs377	5.73	5.69E-08	0.14	4.46	6.21E-06	0.11	3.92	4.49E-05	0.10	6.26	7.83E-09	0.15	7.09	3.63E-10	0.17
Y2129	26.65	1.99E-11	0.14	16.79	1.17E-07	0.09	15.29	4.57E-07	0.09	24.86	9.25E-11	0.13	30.65	6.69E-13	0.16
Hs4209	11.18	5.03E-10	0.13	7.07	2.56E-06	0.09	9.28	2.55E-08	0.12	12.37	4.41E-11	0.14	23.92	9.21E-21	0.25
Hs4061	25.55	4.14E-11	0.12	17.49	5.59E-08	0.08	21.42	1.60E-09	0.11	23.28	3.05E-10	0.11	30.72	4.66E-13	0.14
Hs1638	20.72	3.30E-09	0.11	14.13	1.29E-06	0.08	18.27	2.97E-08	0.10	19.25	1.23E-08	0.10	24.86	8.66E-11	0.13
Hs618	22.58	5.71E-10	0.11	12.16	7.75E-06	0.06	7.54	6.19E-04	0.04	17.66	4.78E-08	0.09	22.91	4.26E-10	0.11
Hs1514	19.02	1.53E-08	0.10	15.50	3.69E-07	0.08	6.23	2.20E-03	0.04	14.59	8.48E-07	0.08	14.77	7.19E-07	0.08
Hs4089	19.68	7.67E-09	0.09	18.49	2.25E-08	0.09	6.44	1.80E-03	0.03	17.68	4.71E-08	0.09	13.84	1.61E-06	0.07
Hs425	17.07	8.21E-08	0.08	8.38	2.75E-04	0.04	7.90	4.39E-04	0.04	17.02	8.55E-08	0.08	18.86	1.61E-08	0.09
Hs560	4.53	1.96E-04	0.07	4.98	6.48E-05	0.08	4.69	1.33E-04	0.07	6.43	1.91E-06	0.10	7.36	1.97E-07	0.11
Hs635	13.54	2.12E-06	0.07	13.28	2.72E-06	0.07	10.34	4.30E-05	0.05	18.79	1.71E-08	0.09	31.73	1.98E-13	0.15
Hs02	14.67	7.44E-07	0.07	10.16	5.08E-05	0.05	10.31	4.43E-05	0.05	12.41	6.09E-06	0.06	36.20	4.55E-15	0.16
Hs4265	5.17	1.34E-04	0.06	6.58	7.07E-06	0.08	6.70	5.50E-06	0.08	8.54	1.16E-07	0.10	11.90	1.13E-10	0.14
Hs376	11.29	1.76E-05	0.06	11.04	2.22E-05	0.06	12.53	5.52E-06	0.07	16.27	1.72E-07	0.08	21.26	1.89E-09	0.10
Hs586	11.07	2.15E-05	0.06	11.38	1.60E-05	0.06	7.30	7.82E-04	0.04	10.90	2.52E-05	0.06	20.14	5.06E-09	0.10
Hs4082	12.36	6.41E-06	0.06	8.72	2.00E-04	0.04	18.30	2.69E-08	0.09	15.51	3.44E-07	0.08	20.61	3.33E-09	0.10

The markers associated with the PC trait under 5 environments are listed in the table.

Table 5. Allele effects of 5 markers simutaneously associated with OC and PC traits under 5 environments.

Marker	Allele[a]	2011				2012					
		Pingyu		Yuanyang		Pingyu		Yuanyang		Xinyang	
		OC	PC	OC	PC	OC	PC	OC	PC	OC	PC
Hs345	1:1	−0.04	−0.17	−0.59	0.25	−1.17	0.44	−1.61	0.57	−0.93	0.39
	2:2	−10.74	2.30	−8.13	2.12	−8.87	2.05	−12.18	2.96	−11.73	3.82
	1:2	−5.53	1.38	−4.59	1.67	−5.13	1.29	−5.84	2.08	−6.66	2.58
Hs4061	1:1	0.00	0.00	0.00	0.00	0.00	0.00	0.00	0.00	0.00	0.00
	2:2	8.78	−2.41	4.46	−1.95	6.37	−1.84	8.54	−2.33	7.94	−2.96
	1:2	2.58	−0.98	−0.18	−1.06	1.78	−0.80	2.58	−1.14	1.60	−0.86
Hs1956	1:1	1.12	−0.36	1.01	−0.49	0.85	−0.32	1.18	−0.39	1.67	−0.81
	2:2	−5.89	1.21	−4.13	0.79	−4.03	0.63	−5.03	0.95	−5.15	1.34
	1:2	−0.74	0.15	0.11	−0.28	−0.86	0.26	−0.60	0.49	−0.85	0.02
Hs4381	1:1	0.00	0.00	0.00	0.00	0.00	0.00	0.00	0.00	0.00	0.00
	2:2	9.75	−2.45	6.33	−1.73	6.97	−1.82	9.06	−2.17	9.80	−3.41
	1:2	5.02	−1.05	2.34	−0.49	3.16	−0.90	5.12	−0.75	4.58	−1.51
Hs485	1:1	−4.80	0.47	−3.23	0.25	−3.87	0.51	−6.88	1.62	−7.19	2.24
	2:2	2.04	−1.28	0.82	−1.21	1.71	−0.96	−0.02	0.02	0.20	−0.29
	1:2	−3.99	0.33	−3.85	−0.03	−2.59	−0.12	−4.64	1.38	−5.76	1.74

Note: [a] two of the most common alleles are listed.

Table 6. Information of candidate genes related to the markers associated with OC traits.

Marker	Related gene[a]	Annotation	Nearby lipid genes[b]	*Arabidopsis thaliana* homologous genes		Ref.
				Loci[c]	Annotation	
Hs4381	C01.460	Thioredoxin M3	C01.526	At1G55260	Lipid Transfer Protein type 5	[25]
Hs02	C01.560	E3 ubiquitin-protei	C01.548	At3G25110	Acyl-ACP Thioesterase A	[26]
		ligase RHF2A	C01.575	At3G07400	Lipid Acylhydrolase-like	
	C01.561	Unknown	C01.601	At2G30720	Acyl-CoA Thioesterase	[27]
Hs4082	C01.882	Cell cycle protein	C01.873	At1G53390	ABC Transporter	[28]
	C01.883	ABC transporter G	C01.883	At2G28070	ABC Transporter	[28]
		family member 3	C01.928	At3G44830	Diacylglycerol Acyltransferase	[29]
Hs635	C02.696	RAS-related protein Rab11A	C02.739	At1G47620	Midchain Alkane Hydroxylase	30]
Hs4209	C04.25	Transmembrane domain	C04.38	At1G31770	ABC Transporter	[31]
		containing protein	C04.56	At1G71010	Phosphatidylinositol-Phosphate Kinase	[32]
	C04.26	TCP transcription factor 12	C04.81	At1G10900	Phosphatidylinositol-Phosphate Kinase	[32]
			C04.96	At1G77660	Phosphatidylinositol-Phosphate Kinase	[32]
Hs4061	C04.681	Inositol trisphosphate 5-phosphatase 1	—			
Hs345	C04.785	Unknown	C04.767	At1G05630	Phosphoinositide 5-Phosphatase	—
	C04.786	Stearoyl-ACP desaturase	C04.786	At2G43710	Stearoyl-ACP Desaturase	[33]
Hs1956	C04.845	FAR1-related sequence	The same as Hs345			
Hs393	C13.438	bZIP transcription	C13.388	At1G10900	Phosphatidylinositol-Phosphate Kinase	[32]
		factor 40	C13.471	At5G10160	Hydroxyacyl-ACP Dehydrase	[34]
			C13.504	At1G17840	ABC Transporter	[35]
			C13.514	At4G36480	Subunit of Serine Palmitoyltransferase	[36]
Hs205	C14.21	Thylakoid membrane	C14.49	At4G33550	Lipid Transfer Protein type 3	[37]
		phosphoprotein	C14.66	At1G49430	Long-Chain Acyl-CoA Synthetase	[38]
	C14.22	Serine/threonine-protein	C14.111	At2G46210	Sphingobase-D8 Desaturase	[39]
		kinase WNK4	C14.132	At2G46090	Diacylglycerol Kinase	—
Hs672	C14.367	Cyclin-P3.1 F-box	C14.359	At2G45150	CDP-DAG Synthase	[40]
		protein	C14.413	At4G33355	Lipid Transfer Protein type 1	—
	C14.368	Unknown	C14.428	At2G44810	Acylhydrolase (DAD1-like)	—
Y2129	C15.825	Unknown	C15.92	At2G29980	Linoleate Desaturase	[41]
Hs1638	C15.840	Unknown	The same as Y2129			
Hs618	C25.41	IAA-alanine resistance	C25.69	At5G08415	Lipoate Synthase	[42]
		protein 1				
	C25.42	Beta-D-xylosidase				
Hs376	C26.474	ADP-ribosylation factor	C26.417	At1G15110	Base-Exchange-type Phosphatidylserine	[43]
					Synthase	
			C26.454	At1G71010	Phosphatidylinositol-Phosphate Kinase	—
			C26.515	At1G31812	Acyl CoA Binding Protein	[44]
Hs377	C26.475	Unknown	The same as Hs376			
Hs485	sf00001.885	Unknown	sf00001.95	At4G04930	Dihydrosphingosine Delta-4 Desaturase	[45]
	sf00001.886	Unknown				
Hs586	sf00001.754	ADP-ribosylation factor	—			
Hs1036	sf00044.12	ABC transporter I	sf00044.12	At3G20320	Acid-Binding Protein	[46]
		family member 15	sf00044.41	At2G25170	Chromatin remodeling factor	[47]

Table 6. Cont.

Marker	Related gene[a]	Annotation	Nearby lipid genes[b]	*Arabidopsis thaliana* homologous genes		Ref.
				Loci[c]	Annotation	
			sf00044.61	At4G19860	Acyl acceptor Acyltransferase	—
			sf00044.90	At1G13210	Translocase	—
			sf00044.113	At1G74320	Choline Kinase	[48]

Note: [a]Related genes are the genes containing or close to the screened markers.
[b]Nearby lipid genes refer to the genes located in the upstream and downstream of 500 Kb far from the marker.
[c]Only one of the homologous genes is listed in the table.
— refers to no known lipid genes in the location.

Candidate genes and oil components

To clarify the marker location and more genome information for OC and PC traits, we screened the genes that were close to the associated markers using the sesame genome assembly data. As a result, 36 candidate genes related to lipid pathway were identified (Table 6) [25–48]. We found that most candidate genes were involved in three pathways: (1) fatty acid and TAG (triacylglycerol) synthesis and elongation, e.g., C01.526, C01.548 and C01.928; (2) TAG degradation, e.g., C01.601; and (3) fatty acid dehydrogenation, e.g., C04.786 (Stearoyl-ACP desaturase, determining the ratio of saturated and unsaturated fatty acids) (Table S4). Therefore, the seed oil content in the sesame accessions could be regulated by three factors, i.e., oil synthesis ability, oil degradation ability and oil component ratio (e.g., 18:1 and 18:2 fatty acids). In various accessions, any alleles of the genes involved in fatty acid and TAG synthesis, TAG degradation or dehydrogenase genes could give effects on oil content. To confirm this hypothesis, further studies of seed oil synthesis should be performed in the future.

Conclusion

We systematically explored the association mapping of seed oil and protein content traits in 369 worldwide sesame accessions using the 112 SSR markers. A significant negative correlationof the oil content with the protein content existed in the population. 19 SSR markers were associated with the oil content trait with high phenotypic variation explanation, and 24 SSR markers were associated with the protein content trait using GLM method. This association results would provide an efficient platform for seed development research and MAS breeding in sesame.

Methods

Plant materials

A population of 369 core sesame germplasm resources was chosen according to the genetic diversity analysis results and phenotypic variation [51]. These core genotypes comprised 318 lines from 19 provinces of China and 51 worldwide lines from the 15 countries, which were reserved at the Sesame Germplasm Bank of Henan Sesame Research Center (HSRC), HAAS (Henan Academy of Agricultural Sciences) (Table S1). All accessions were grown with three replications at three different experimental stations of Yuanyang (113.96°E, 35.05°N), Pingyu (114.62°E, 32.97°N), and Xinyang (114.08°E, 32.13°N) in 2011 and 2012. Five or six young leaves of individual accessions were collected and reserved at −70°C for DNA extraction.

Oil and protein content analysis

After harvested, ~20 g of seeds were collected from five plants per line, and the seed OC and PC were measured on infrared determination equipment (Perten DA7250, Sweden) according to the manufactures' instructions. The standard curve for measuring the sesame oil and protein contents had been established according to the chemical analysis results of 300 sesame accessions (Unpublished data, HSRC). Three replications of each samples were assayed for phenotypic analysis. Mean value, broad-sense heritability and correlation coefficient of the phenotypic data were analyzed using the statistical analysis system software (SAS Institute Inc. 2002) [72].

SSR genotyping

The 112 polymorphic SSR pairs were selected from our SSR marker bank [24,51,73] (Table S2). DNA extraction, PCR amplification, electrophoresis and SSR genotyping analysis were performed according to the methods described by Zhang et al. [24]. The total number of polymorphic alleles at each SSR locus was calculated according to the results of all 369 lines. The polymorphic SSR alleles presented only within 4 (1%) or fewer accessions were recorded as rare alleles.

Linkage disequilibrium (LD)

As the population structure could result in the spurious associations between phenotypes and marker loci, we analyzed the extent and structure of LD within the population before selecting the appropriate association mapping strategy. To assay whether the 112 polymorphic SSR markers were segregated independently or not, LD analysis was conducted according to the dedicated procedure of the TASSEL software [49]. Both D' and r^2 were used for quantifying LD values [74,75]. Significance (P values) of D' for each SSR pairs was determined with 100,000 permutations.

Population structure and relatedness analysis

The population structure was determined using STRUCTURE 2.2 [76]. The mixture model and the independent allele frequency model were used to analyze the population dataset. Five runs of STRUCTURE were carried out for each number of populations (K) (from 1–10), and each run started with 10,000 burn-ins, followed by 100,000 iterations. While performing the STRUCTURE, we assumed that the inferred population accord with Hardy Weinberg equilibrium (HWE) and the loci are unlinked. To correct the relatedness of individuals in further analyses, the relatedness between individuals (relative kinship) was evaluated using SPAGeDi 1.2 software [77]. The matrix with the relative kinship coefficients (*K* matrix) was applied for association analysis using the Mixed Linear Model(MLM, Q+K)method.

Association mapping and marker distribution in sesame genome

Associations between the SSR markers and the oil and protein content traits were investigated using both methods of the general linear model (GLM, Q) and the mixed linear model (MLM, Q+*K*) in TASSEL 2.1 described by Bradbury et al. [49]. The mean value of the markers at $P<0.005$ was used for determining the significance of marker-trait associations.

To determine distributions of the associated markers in sesame genome, we carried out the alignment of SSR markers and transcripts with the updated sesame genome data [70,71]. In the present genome assembly, the number of N50 scaffold was 14, and the number of N90 scaffold was 64. 29,798 gene models were identified. Among related scaffolds or contigs, the sesame lipid synthesis related genes were identified according to the homologous comparison using the genes of *A. thaliana* from the Acyl Lipids pathway database [78] as queries.

Supporting Information

Table S1 Origins of 369 sesame accessions.

Table S2 Diversity statistics of 112 SSR markers in 369 sesame accessions and locations in the sesame draft genome

Table S3 Association mapping of the OC and PC traits using MLM method.

Table S4 Functions and involved pathways of the candidate genes.

Author Contributions

Conceived and designed the experiments: HZ. Performed the experiments: CL HM LW. Analyzed the data: CL HM LW. Contributed reagents/materials/analysis tools: TZ XH. Wrote the paper: CL HM LW.

References

1. Brummer EC, Graef GL, Orf J, Wilcox JR, Shoemaker RC (1997) Mapping QTL for seed protein and oil content in eight soybean populations. Crop Sci 37: 370–378.
2. Guo SX (2011) Genetic and QTL analysis of oil content and other important quality traits on *Brassica Napus* L.). PhD thesis. Sichuan Agricultural University, Rice Institute.
3. Turhan H, Citak N, Pehlivanoglu H, Mengul Z (2010) Effects of ecological and topographic conditions on oil content and fatty acid composition in sunflower. Bulgarian J Agri Sci 16:553–558.
4. Chen SL, Li YR, Cheng ZS, Liao BS, Lei Y, et al (2009) Heterosis and Genetic Analysis of Oil Content in Peanut Using Mixed Model of Major Gene and Polygene. Scientia Agricultura Sinic 42:3048–3057.
5. Anilakumar KR, Pal A, Khanum F, Bawa AS (2010) Nutritional, medicinal and industrial uses of sesame (*Sesamum indicum* L.) seeds- an overview. Agric Conspec Sci 75:159–168.
6. Bedigian D, Harlan JR (1986) Evidence for cultivation of sesame in the ancient world. Econ Bot 40: 137–154.
7. Zhang HY (2013) Sesame genetic and breeding status and development strategy in China. In: Chinese Journal of Oil Crop Sciences Editorial Office, editors. VII China National Oil Crops Conference altar Paper compilations. Wenchang, China. pp. 29–36.
8. Jin UH, Lee JW, Chung YS, Lee JH, Yi YB, et al (2001) Characterization and temporal expression of a ω-6 fatty acid desaturase cDNA from sesame (*Sesamum indicum* L.) seeds. Plant Sci 161:935–941.
9. Chun JA, Jin UH, Lee JW, Yi YB, Hyung NI, et al (2003) Isolation and characterization of a myo-inositol 1-phosphate synthase cDNA from developing sesame (*Sesamum indicum* L.) seeds: functional and differential expression, and salt–induced transcription during germination. Planta 216:874–880.
10. Leduc V, Moneret-Vautrin DA, Tzen JTC, Morisset M, Guerin L, et al (2006) Identification of oleosins as major allergens in sesame seed allergic patients. Allergy 61:349–356.
11. Magni C, Ballabio C, Restani P, Fuggetta D. Alessandri C, et al (2010) Molecular insight into IgE-mediated reactions to sesame (*Sesamum indicum* L.) seed proteins. Ann Allerg Asthma Im 105:458–464.
12. Wei W, Zhang Y, Lv H, Li D, Wang L, Zhang X (2013) Association analysis for quality traits in a diverse panel of Chinese sesame (*Sesamum indicum* L.) germplasm. J Integr Plant Biol 55:745–758.
13. Suh MC, Kim MJ, Hur CG, Bae JM, Park YI, et al (2003) Comparative analysis of expressed sequence tags from *Sesamum indicum* and *Arabidopsis thaliana* developing seeds. Plant Mol Biol 52:1107–1123.
14. Ke T, Dong C, Han M, Zhao Y, Chen H, et al (2011) Analysis of expression sequence tags from a full-length-enriched cDNA library of developing sesame seeds (*Sesamum indicum*). BMC Plant Biol 11:180.
15. Singkham N, Jogloy S, Kesmala T, Swatsitang P, Jaisil P, et al (2010) Genotypic variability and genotype by environment interactions in oil and fatty acids in high, intermediate, and low oleic acid peanut genotypes. J Agr Food chem 58:6257–6263.
16. Zhao J, Artemyeva A, Del Carpio DP, Basnet RK, Zhang N, et al (2010) Design of a *Brassica rapa* core collection for association mapping studies. Genome 53:884–898.
17. Zou J, Jiang C, Cao Z, Li R, Long Y, et al (2010) Association mapping of seed oil content in Brassica napus and comparison with quantitative trait loci identified from linkage mapping. Genome 53:908–916.
18. Abdurakhmonov IY, Saha S, Jenkins JN, Buriev ZT, Shermatov SE, et al (2009) Linkage disequilibrium based association mapping of fiber quality traits in *G. hirsutum* L. variety germplasm. Genetica 136:401–417.
19. Ecke W, Clemens R, Honsdorf N, Becker HC (2010) Extent and structure of linkage disequilibrium in canola quality winter rapeseed (*Brassica napus* L.). Theor Appl Genet 2010, 120:921–931.
20. Eleuch L, Jilal A, Grando S, Ceccarelli S, Schmising MVK, et al (2008) Genetic diversity and association analysis for salinity tolerance, heading date and plant height of barley germplasm using simple sequence repeat markers. J Integr Plant Biol 50:1004–1014.
21. Famoso AN, Zhao K, Clark RT, Tung CW, Wright MH, et al (2011) Genetic architecture of aluminum tolerance in rice (*Oryza sativa*) determined through genome-wide association analysis and QTL mapping. PLoS Genet 7:e1002221.
22. He D, Xing H, Zhao J, Li T, Tang Y, et al (2011) Population structure inferring and association analysis of fiber quality in cultivated cotton. J Northwest A & F Univ (Nat Sci Ed) 39:103–112.
23. Wei W, Qi X, Wang L, Zhang Y, Wei H, et al (2011) Characterization of the sesame (*Sesamum indicum* L.) global transcriptome using Illumina paired–end sequencing and development of EST–SSR markers. BMC genomics 12:451.
24. Zhang H, Wei L, Miao H, Zhang T, Wang C (2012) Development and validation of genic–SSR markers in sesame by RNA-seq. BMC genomics 13:316.
25. Pagnussat GC, Yu HJ, Ngo QA, Rajani S, Mayalagu S., et al (2005) Genetic and molecular identification of genes required for female gametophyte development and function in Arabidopsis. Development 132:603–614.
26. Salas JJ, Ohlrogge JB (2002) Characterization of substrate specificity of plant FatA and FatB acyl-ACP thioesterases. Arch Biochem Biophys 403:25–34.
27. Tilton G, Shockey J, Browse J (2000) Two families of acyl-CoA thioesterases in Arabidopsis. Biochem Soc Trans 28:946–947.
28. Quilichini TD, Friedmann MC, Samuels AL, Douglas CJ (2010) ATP-Binding cassette transporter G26 is required for male fertility and pollen exine formation in Arabidopsis. Plant Physiol 154:678–690.
29. Zhang M, Fan JL, Taylor DC, Ohlrogge JB (2009) DGAT1 and PDAT1 Acyltransferases have overlapping functions in Arabidopsis triacylglycerol biosynthesis and are essential for normal pollen and seed development. Plant Cell 21:3885–3901.
30. Cheng D, Nelson TC, Chen J, Walker SG, Wardwell-Swanson J, et al (2003) Identification of acyl coenzyme A:Monoacylglycerol acyltransferase 3, an intestinal specific enzyme implicated in dietary fat absorption. J Biol Chem 278:13611–13614.
31. Panikashvili D, Shi JX, Schreiber L, Aharoni A (2011) The Arabidopsis ABCG13 transporter is required for flower cuticle secretion and patterning of the petal epidermis. New Phytolt 190:113–124.
32. Xue HW, Hosaka K, Plesch G, Mueller-Roeber B (2000) Cloning of *Arabidopsis thaliana* phosphatidylinositol synthase and functional expression in the yeast pis mutant. Plant Mol Biol 42:757–764.
33. Kachroo A, Shanklin J, Whittle E, Lapchyk L, Hildebrand D, Kachroo P (2007) The Arabidopsis stearoyl-acyl carrier protein-desaturase family and the contribution of leaf isoforms to oleic acid synthesis. Plant Mol Biol 63:257–271.
34. Brown A, Affleck V, Kroon J, Slabas A (2009) Proof of function of a putative 3-hydroxyacyl-acyl carrier protein dehydratase from higher plants by mass spectrometry of product formation. FEBS lett 583:363–368.

35. Bird D, Beisson F, Brigham A, Shin J, Greer S, et al (2007). Characterization of Arabidopsis ABCG11/WBC11, an ATP binding cassette (ABC) transporter that is required for cuticular lipid secretion. Plant J 52:485–498.
36. Chen M, Han GS, Dietrich CR, Dunn TM, Cahoon EB (2006) The essential nature of sphingolipids in plants as revealed by the functional identification and characterization of the Arabidopsis LCB1 subunit of serine palmitoyltransferase. Plant Cell 18:3576–3593.
37. Xing S, Zachgo S (2008) ROXY1 and ROXY2, two Arabidopsis glutaredoxin genes, are required for anther development. Plant J 53:790–801.
38. Schnurr J, Shockey J, Browse J (2004). The acyl-CoA synthetase encoded by LACS2 is essential for normal cuticle development in Arabidopsis. Plant Cell 16:629–642.
39. Chen M, Markham JE, Cahoon EB (2012) Sphingolipid delta 8 unsaturation is important for glucosylceramide biosynthesis and low-temperature performance in Arabidopsis. Plant J 69:769–781.
40. Haselier A, Akbari H, Weth A, Baumgartner W, Frentzen M (2010) Two closely related genes of Arabidopsis encode plastidial cytidinediphosphate diacylglycerol synthases essential for photoautotrophic growth. Plant Physiol 153:1372–1384.
41. Arondel V, Lemieux B, Hwang I, Gibson S, Goodman HM, et al (1992) Map-based cloning of a gene controlling omega-3fatty-acid desaturation in Arabidopis. Science 258:1353–1355.
42. Yasuno R, Wada H (2002) The biosynthetic pathway for lipoic acid is present in plastids and mitochondria in Arabidopsis thaliana. FEBS Lett 517:110–114.
43. Yamaoka Y, Yu Y, Mizoi J, Fujiki Y, Saito K, et al (2011) PHOSPHATIDYL-SERINE SYNTHASE1 is required for microspore development in *Arabidopsis thaliana*. Plant J 67:648–661.
44. Engeseth NJ, Pacovsky RS, Newman T, Ohlrogge JB (1996) Characterization of an Acyl-CoA-binding protein from *Arabidopsis thaliana. Arch Biochem Biophys* 331:55–62.
45. Michaelson LV, Zauner S, Markham JE, Haslam RP, Desikan R, et al (2009) Functional characterization of a higher plant sphingolipid Delta4-desaturase: Defining the role of sphingosine and sphingosine-1-phosphate in Arabidopsis. Plant Physiol 149:487–498.
46. Awai K, Xu CC, Tamot B, Benning C (2006) A phosphatidic acid-binding protein of the chloroplast inner envelope membrane involved in lipid trafficking. Proc Natl Acad Sci 103:10817–10822.
47. Ogas J, Cheng JC, Sung ZR, Somerville C (1997) Cellular differentiation regulated by gibberellin in the *Arabidopsis thaliana* pickle mutant. Science 277:91–94.
48. Monks DE, Goode JH, Dewey RE (1996) Characterization of soybean choline kinase cDNAs and their expression in yeast and Escherichia coli. Plant Physiol 11: 1197–1205.
49. Laurentin HE, Karlovsky P (2006) Genetic relationship and diversity in a sesame (*Sesamum indicum* L.) germplasm collection using amplified fragment length polymorphism (AFLP). BMC genetics 7:10.
50. Zhang P, Zhang HY, Guo WZ, Zheng YZ, Wei LB, et al (2007) Genetic diversity analysis of *Sesamum indicum* L. germplasms using SRAP and EST-SSR markers. Acta Agro Sin 33:1696–1702.
51. Yue WD, Wei L, Zhang T, Li C, Miao H, et al (2012) Genetic diversity and population structure of germplasm resources in sesame (*Sesamum indicum* L.) by SSR markers. Acta Agro Sin 38:2286–2296.
52. Kim DH, Zur G, Danin-Poleg Y, Lee SW, Shim KB, et al (2002) Genetic relationships of sesame germplasm collection as revealed by inter-simple sequence repeats. Plant Breed 121:259–262.
53. Kumar H, Kaur G, Banga S (2012) Molecular characterization and assessment of genetic diversity in sesame (*Sesamum indicum* L.) germplasm collection using ISSR markers. J Crop Improve 26:540–557.
54. Stankiewicz M, Gadamski G, Gawronski SW (2001) Genetic variation and phylogenetic relationships of triazine resistant and triazine susceptible biotypes of *Solanum nigrum*-analysis using RAPD markers. Weed Res 41:287–300.
55. Garris AJ, Tai TH, Coburn J, Kresovich S, McCOUCH S (2005) Genetic structure and diversity in *Oryza sativa* L. Genetics 169:1631–1638.
56. Ersoz ES, Yu J, Buckler ES (2007) Applications of linkage disequilibrium and association mapping in crop plants. *In*: Varshney RK, Tuberosa R, editors. Genomics–assisted crop improvement. Volume 1. Netherlands: Springer. pp. 97–119.
57. Yermanos DM, Hemstreet S, Saleeb W, Huszar CK (1972) Oil content and composition of the seed in the world collection of sesame introductions. J Am Oil Chem Soc 49:20–23.
58. Johnson LA, Suleiman TM, Lusas EW (1979) Sesame protein: A review and prospectus. J Am Oil Chem Soc 56:463–468.
59. El Tinay AH, Khattab AH, Khidir MO (1976) Protein and oil compositions of sesame seed. J Am Oil Chem Soc 53:648–653.
60. Uzun B, Arslan Ç, Furat Ş (2008) Variation in fatty acid compositions, oil content and oil yield in a germplasm collection of sesame (*Sesamum indicum* L.). J Am Oil Chem Soc 85:1135–1142.
61. Sarvamangala C, Gowda MVC, Varshney RK (2011) Identification of quantitative trait loci for protein content, oil content and oil quality for groundnut (*Arachis hypogaea* L.). Field Crops Res 122:49–59.
62. Zhao J, Becker HC, Zhang D, Zhang Y, Ecke W (2006) Conditional QTL mapping of oil content in rapeseed with respect to protein content and traits related to plant development and grain yield. Theor Appl Genet 113:33–38.
63. Yu J, Yu S, Fan S, Song M, Zhai H, et al (2012) Mapping quantitative trait loci for cottonseed oil, protein and gossypol content in a *Gossypium hirsutum* ×*Gossypium barbadense* backcross inbred line population. Euphytica 187:191–201.
64. Hwang EY, Song Q, Jia G, Specht JE, Hyten DL, et al (2014) A genome-wide association study of seed protein and oil content in soybean. BMC Genomics 15:1.
65. Zhao JY, Becker HC, Zhang D, Zhang Y, Ecke W (2006) Conditional QTL mapping of oil content in rapeseed with respect to protein content and traits related to plant development and grain yield. Theor Appl Genet 113:33–38.
66. Chen G, Geng J, Rahman M, Liu X, Tu J (2010) Identification of QTL for oil content, seed yield, and flowering time in oilseed rape (*Brassica napus*). Euphytica 175:161–174.
67. Zheng P, Allen WB, Roesler K, Williams ME, Zhang S, et al (2008) A phenylalanine in DGAT is a key determinant of oil content and composition in maize. Nat Genet 40:367–372.
68. Hobbs DH1, Flintham JE, Hills MJ (2004) Genetic control of storage oil synthesis in seeds of Arabidopsis. Plant Physiol 136:3341–3349.
69. Shi A, Chen P, Zhang B, Hou A (2000) Genetic diversity and association analysis of protein and oil content in food-grade soybeans from Asia and the United States. Plant Breeding 129:250–256.
70. Zhang H, Miao HM, Wang L, Qu L, Liu H, et al (2013) Genome sequencing of the important oilseed crop *Sesamum indicum* L. Genome Biol 14:401.
71. Miao HM. The Sesame Genome Project and Sesame Genome Sequencing. Plant and Animal Genome XXII Conference. Plant and Animal Genome, 2014.
72. Holland JB, Nyquist WE, Cervantes-Martnez CT (2003) Estimating and interpreting heritability for plant breeding: An update. Plant breed reviews 22:9–112.
73. Wei LB, Zhang HY, Zheng YZ,Miao HM, Zhang TZ, et al (2009) A Genetic Linkage Map Construction for Sesame (*Sesamum indicum* L.). Genes & genomics 31:199–208.
74. Hedrick PW (1987) Gametic disequilibrium measures: proceed with caution. Genetics 117:331–341.
75. Weir BS (1996) Genetic data analysis 2: methods for discrete population genetic data. Sunderland: Sinauer Associates Inc Publishers. 445 p.
76. Pritchard JK, Stephens M, Donnelly P (2000) Inference of population structure using multilocus genotype data. Genetics 155:945–959.
77. Hardy OJ, Vekemans X (2002) SPAGeDi: a versatile computer program to analyse spatial genetic structure at the individual or population levels. Mol Ecol Notes 2:618–620.
78. Acyl Lipid: pathways Available at: http://aralip.plantbiology.msu.edu

21

First TILLING Platform in *Cucurbita pepo*: A New Mutant Resource for Gene Function and Crop Improvement

Nelly Vicente-Dólera[1], Christelle Troadec[2], Manuel Moya[1], Mercedes del Río-Celestino[1], Teresa Pomares-Viciana[1], Abdelhafid Bendahmane[2], Belén Picó[3], Belén Román[4], Pedro Gómez[1]*

1 IFAPA Centro La Mojonera, Camino de San Nicolás, 1, 04745 La Mojonera, Almería, Spain, **2** INRA-URGV, UMR1165, Unité de Recherche en Génomique Végétale, Saclay Plant Sciences, Evry, France, **3** Institute for the Conservation and Breeding of Agricultural Biodiversity (COMAV-UPV), Universitat Politécnica de Valencia, Camino de Vera s/n, 46022 Valencia, Spain, **4** IFAPA Centro Alameda del Obispo, Avd. Menéndez Pidal s/n, 14004, Córdoba, Spain

Abstract

Although the availability of genetic and genomic resources for *Cucurbita pepo* has increased significantly, functional genomic resources are still limited for this crop. In this direction, we have developed a high throughput reverse genetic tool: the first TILLING (Targeting Induced Local Lesions IN Genomes) resource for this species. Additionally, we have used this resource to demonstrate that the previous EMS mutant population we developed has the highest mutation density compared with other cucurbits mutant populations. The overall mutation density in this first *C. pepo* TILLING platform was estimated to be 1/133 Kb by screening five additional genes. In total, 58 mutations confirmed by sequencing were identified in the five targeted genes, thirteen of which were predicted to have an impact on the function of the protein. The genotype/phenotype correlation was studied in a peroxidase gene, revealing that the phenotype of seedling homozygous for one of the isolated mutant alleles was albino. These results indicate that the TILLING approach in this species was successful at providing new mutations and can address the major challenge of linking sequence information to biological function and also the identification of novel variation for crop breeding.

Editor: Heping Cao, USDA-ARS, United States of America

Funding: Financial support was provided by the Spanish Project INIA (Instituto Nacional de Investigación y Tecnología Agraría y Almentaría) RTA2011-00044-C02-01, the ANR MELODY (ANR-11-BSV7-0024), the European Research Council (ERCSEXYPARTH), FEDER, and FSE funds. NVD has been awarded a grant by the Andalusian Institute of Agronomy Research IFAPA. The funders had no role in study design, data collection and analysis, decision to publish, or preparation of the manuscript.

Competing Interests: The authors have declared that no competing interests exist.

* Email: pedro.gomez.j@juntadeandalucia.es

Introduction

Cucurbita pepo is the main species of the genus *Cucurbita* and represents one of the most important vegetable crops worldwide in terms of food consumption. Of the eight existing morphotypes in this species, the morphotype Zucchini is the most economically valuable. It was recently developed and displays low levels of genetic variation [1] but nevertheless it is commonly cultivated. Nowadays, in this species the development of new cultivars is based on the availability of variable germplasm. However, apart from the intraspecific resources, grouped in three subespecies, only a few crossable *Cucurbita* species can be used as sources of agriculturally useful alleles and new genotypes. In this way, the development of new genetic resources along with the application of new molecular techniques in breeding would facilitate the improvement of characters, mainly those with more complex regulation, allowing breeders to face new challenges in this species.

The high competitiveness of plant breeding has increased the interest in the development of new sources of genetic variation using reverse genetics approaches. The silencing or interruption of individual target genes provide the opportunity to investigate gene function, linking sequence variation to specific phenotypes [2], [3]. Thus, TILLING (Targeting Induced Local Lesions IN Genomes), which combines traditional chemical mutagenesis with high-throughput genome-wide screening for point mutations in desired genes, has been developed in response. In a TILLING project, chemical mutagenesis based on an alkylating agent like EMS (ethyl methane sulphonate) [4] that causes random point mutations at high density, provides an easy and cost-effective way to establish a series of allelic mutations in the whole gene set of any species. The screen of the mutant population with an efficient method to identify point mutations, for example a mismatch specific digestion followed by a Li-COR separation of the resulting fragments, may allow the identification of multiple alleles for a specific gene regardless of the gene size [5], [6]. If the mutation frequency is high enough most, if not all genes, will present mutated alleles in the population.

This method may be preferable to other reverse genetic approaches for various reasons, for example because EMS generates a large spectrum of mutations, including missense and truncation mutations, allowing more flexibility than insertional mutagenesis or transgenesis [5]. The newly identified alleles in TILLING could be used for elucidating gene function. This strategy is invaluable because it generates allelic series often including null alleles, although at low frequency [7]. TILLING also finds application in crop improvement, as the identified

Table 1. Summary of the *Cucurbita pepo* mutant collection development.

EMS concentration	Obtained M1 plants	M2 seed families	Fruits with less than 30 seeds (%)	M3 seed families
80 mM	4500	259	63	95
65 mM	2700	231	56	52
40 mM	2900	974	36	621

mutations can be readily utilized in traditional breeding programs since it is nontransgenic and the novel variations are stably inherited [8–10].

However, to date there are no TILLING-derived crop varieties that have yet been released though there have been many successful examples of TILLING approaches in basic plant science. It has proved to be useful in a large number of agronomically important crops [11] and has allowed to select mutants with allelic variants in virus resistance genes [12], [13], or in genes involved in abiotic stress [14], nutritional properties [15] or shelf life in *Cucumis melo* [16]. Also, some agricultural biotechnology companies have directed this technology towards the enhancement of the tomato shelf life and the reduction of gluten content of wheat for celiacs [2].

Although this reverse genetic approach has been reported in many species, regarding the *Cucurbitaceae* family, only several reports have been published on *Cucumis melo* [13], [16] and recently on *Cucumis sativus* [17]. No TILLING platforms have been reported for other cucurbits species such as *Cucurbita pepo*. The development of a TILLING platform for this species combined with the recent sequence of the genome (a first version available at cucurbigene.net) and other genomic tools such as the first SNP-based genetic map and the first transcriptome [18], [19] could be of great interest, because it could be a way to induce new genetic variation that complements the natural extant variation in this species.

Based on our previously reported method to obtain mutants in *C. pepo* that originated a reference EMS mutant population, with the present study we have established the first TILLING platform developed for this species which could represent an important advance for crop improvement in Zucchini.

Materials and Methods

Mutagenized plant material and growing conditions

Field studies were carried out at the IFAPA research center in Almeria (Spain) (GPS coordinates: 36.78881667, −2.703297222) and no specific permissions were required for these locations/ activities. The field studies did not involve endangered or protected species.

Four to nine seeds of each of the 1464 M2 families from a mutant population previously generated with three different EMS doses: 80 mM, 65 mM and 40 mM [20] were sown and grown in greenhouses following standard local commercial practices for both plant nutrition and pest and disease control. Germination and sterility rates of the M2 families were evaluated and controlled self-pollinations were attempted on each plant. Female flowers were protected the day before anthesis to prevent the transfer of pollen by insects, and early in the morning each plant was self-pollinated by hand. The plants of each family were self-pollinated until one fruit with seeds per each M2 family was obtained. At 60–80 days after self-pollination, M3 seeds were harvested, aired and stored at 4°C. Phenotypic alterations in M2 plants also were regularly monitored and compared with untreated plants that were grown at the same time.

Genomic DNA extraction and pooling

Leaf material from only one plant of each M2 family was collected for DNA isolation. The plant was selected if it produced at least 30 viable seeds. Three leaf discs (diameter 10 mm) were collected and DNA was extracted according to Martin et al. 2009 [21]. Genomic DNAs were quantified on 1% agarose gel using λ DNA (Promega Biotech Iberica) as a concentration reference. DNA concentration was normalized to 3 ng/μl and pooled eightfold in a 96-well format. M3 and M2 seeds of all pooled M2 families are maintained at the genebank of the IFAPA research center.

PCR amplification, mutation detection and validation

Primer sets were designed for five genes (Table S1) and DNA amplification was based on nested-PCR and universal primers to improve the specificity of the amplification [22], [23]. Carotenoid gene structure was established based on the first version of the *C. pepo*, morphotype Zucchini, genomic sequence (available at http://cucurbigene.net) while peroxidase gene structure was derived from NCBI databases. Sequences of the ethylene receptors (CpETR1 and CpERS1) were previously determined in our laboratory from the homology of other cucurbits. Primers were designed to amplify ~1 kb segments based on the selected *C. pepo* genomic sequences. The first PCR amplification was a standard PCR reaction performed in a 25 μl volume consisting of dH2O, 10× PCR buffer, 2.5 mM MgCl2, 5 mM dNTPs, 1 U Taq polymerase, 10 mM forward and reverse target-specific primers and 4.5 ng of genomic DNA. The thermocycling conditions were 95°C for five minutes for initial denaturing, followed by 35 cycles of 94°C for 15 seconds, 60°C for 20 seconds, 72°C from one to two minutes and one cycle of 72°C for five minutes.

One microliter diluted ten times of the first PCR product was employed for the second PCR with a combination of specific primers carrying M13 tail and M13 universal primers, M13F700 (5′-CACGACGTTGTAAAACGAC-3′) and M13R800 (5′-GGATAACATTTCACACAGG-3′), labelled at the 5′end with infrared dyes IRD700 and IRD800 (LI-COR, Lincoln, Nebraska, USA) respectively. This PCR was carried out using 0.125 to 0.25 μM of each primer [23], using the following two step cycling program: 94°C for 5 minutes, 10 cycles at 94°C for 15 seconds, primer-specific annealing temperature for 30 seconds and 72°C for 2 minutes, followed by 25 cycles at 94°C for 15 seconds, 50°C for 30 seconds and 72°C for 1 minute, then a final extension of 5 minutes at 72°C. PCR products were separated in 2% agarose gels.

Mutations were detected in the amplified targets using the mismatch-specific endonuclease ENDO1 as previously described Triques et al. 2007 [24]. Electrophoresis was performed on a

Figure 1. Examples of morphological mutants observed within the *C. pepo* TILLING population. (a) Plant affected in cotyledon colour, albino. (b) Plant affected in cotyledon colour, chlorotic. (c) Plant affected in cotyledon number. (d–e) Female flowers with abnormal stigma. (f–h) Different coloured fruits. (i–k) Semi-dwarf plants with bushy and hyper compact architecture. (l) Albino dwarf plant. (m) Different size and colour leaves.

LICOR 4300 (LI-COR, Lincoln, NE, USA) and gel images were analysed using Adobe Photoshop software (Adobe Systems Inc., San José, CA, USA). After discovery, mutations were validated by sequence analysis. The mutation frequency for each amplicon was calculated as previously described by Dalmais et al. 2008 [23].

Sequence Analysis Tools

The PARSESNP software (Project Aligned Related Sequences and Evaluate SNPs) [25] was used to illustrate the distribution of mutations within the gene and to indicate the nature of each single mutation. The SIFT software (Sorting Intolerant from Tolerant, http://sift.bii.a-star.edu.sg/) [26], was used to predict the impact of the mutation on the protein. Scores below 0.05 are predicted to affect protein function. Multiple sequence alignment of full-length protein sequences was performed with Clustal Omega software (https://www.ebi.ac.uk/Tools/msa/clustalo/)

Mutant protein structure modelling and genotype determination

The APRX three-dimensional structure was generated using the Geno3D server (http://geno3d-pbil.ibcp.fr). Superposition of the wild type APRX structure and the mutant APRX protein was carried out and visualized using Chimera (http://www.cgl.ucsf.edu/chimera).

For each APRX mutant line, homozygous and heterozygous M3 plants were determined by sequencing APRX gene.

How to access the TILLING population

Screening of the *C. pepo* TILLING platform is not established as a service yet but we are currently working on developing a website. For further information about how to access to the platform please contact to: pedro.gomez.j@juntadeandalucia.es

Results

Validation of the *Cucurbita pepo* TILLING platform material

A population of 1464 M2 lines from a previous mutagenesis experiment performed with three different EMS dosages (80 mM, 65 mM and 40 mM) was used and the effect on seed germination, plant development and M3 seed production was evaluated.

All untreated plants used as controls germinated. Nevertheless, germination rate was poor in M2 families obtained with 80 mM and 65 mM dosages (52.6%, and 56.37% respectively) and significantly better in families obtained with 40 mM (70.83%). Additionally, growth of M2 plants was delayed and took longer to germinate at the two higher doses. Moreover, increased EMS concentration led to an increased percentage of M2 fruits in which seeds were non-viable or in which the number of viable seeds was too low to perform TILLING analysis, 63% and 56% for the treatment with 80 mM and 65 mM respectively and 36% for 40 mM treatment. In total 95 and 52 M3 families were collected from the 80 and 65 mM treatments, respectively, and 621 M3 families from the 40 mM treatment (Table 1).

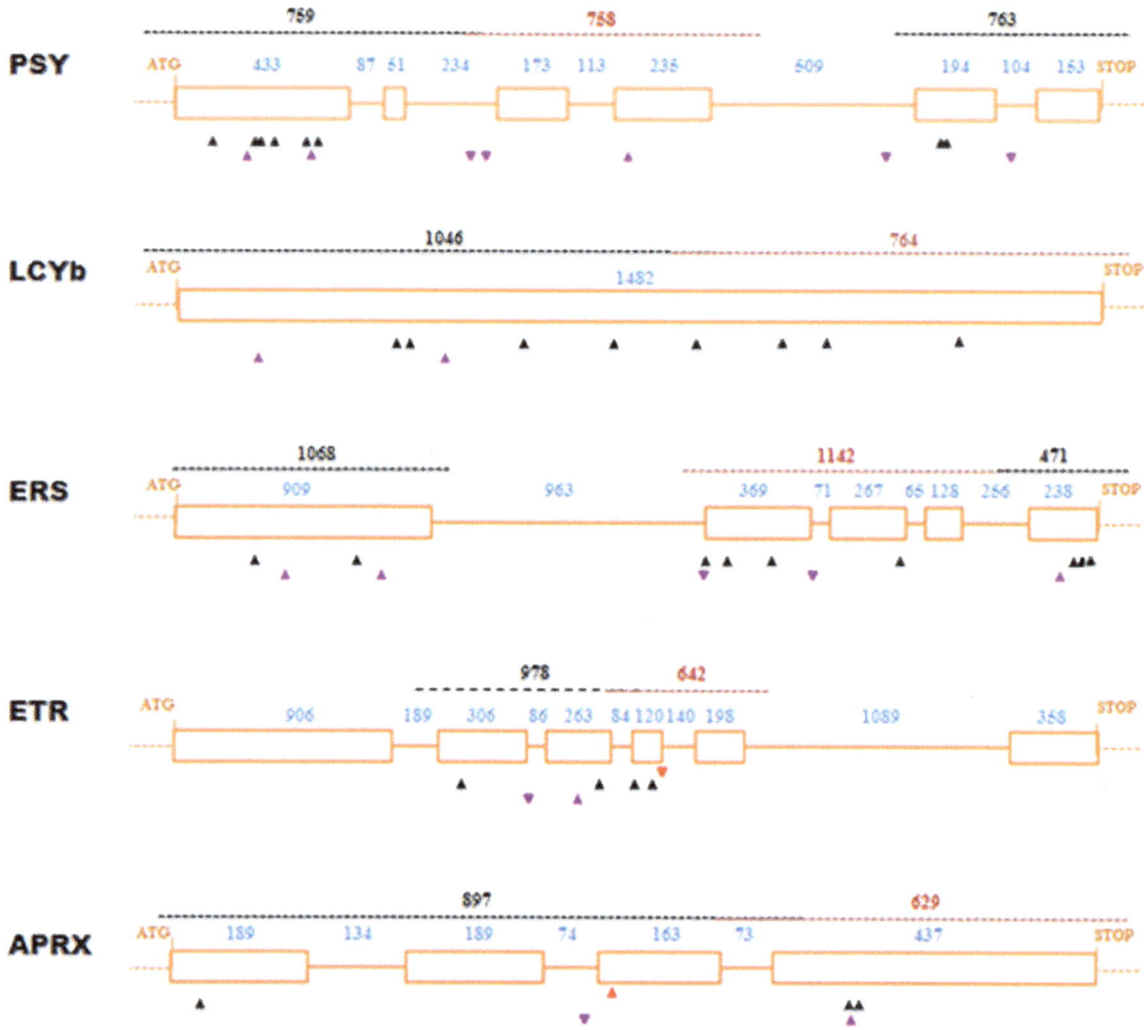

Figure 2. Gene structure of the target genes screened in the *C. pepo* population (PSY, LCYb, ETR1, ERS1, and APRX). This drawing was made using the PARSESNP program, which maps the mutation on a gene model to illustrate the distribution of mutations. Orange boxes represent exons and orange lines introns, the size (pb) is indicated in blue. Dashed lines in red and black indicate amplicons analyzed by TILLING. Purple triangles represent silent mutations and black and red triangles represent missense and truncation mutations, respectively. Mutations before ATG and after STOP codon are not shown.

Compared to untreated plants, each family was evaluated for visual mutations in the phenotypic analysis and a range of different phenotypes was observed in all M2 plants. Albinism and chlorophyll deficiency occurred in 1.3% of the M2 families. About 1.84% showed alterations in cotyledons with variation in their number or shape and 2.2% of the M2 plants segregated individuals displaying "dwarf" or "semidwarf" phenotypes. At flowering and fruit stages the most commonly observed phenotypes were related to the leaf colour and morphology, the growing habit, the plant size or fruit traits. In Figure 1 examples of *C. pepo* mutant phenotypes at key developmental stages are shown confirming the high level of the mutagenesis.

Identification of mutations in target genes by TILLING: mutation efficiency

To set up the *C. pepo* TILLING platform, DNA samples were prepared from 768 M2 lines, each representing an independent family and organized in pools of 8 M2 families. The selection of M2 families was based on the M3 seed disposability.

To validate the TILLING method and to estimate the mutation density of the population, a set of five genes of *C. pepo* involved in diverse metabolic routes were chosen. Two genes were chosen from the ethylene signaling pathway, in particular of the ethylene response such as ERS1 (ethylene response sensor 1) and ETR1 (ethylene receptor 1). LCYb (lycopene beta cyclase) and PSY (phytoene synthase) were also chosen as important enzymes involved in the carotenoid biosynthesis pathway [27]. The last one was the APRX gene, a gene encoding a peroxidase and that seems to have an important function in the growth and development of the plant [28].

PCR products used for TILLING have a maximum size of about 1200 bp and, therefore, longer genes were divided into several amplicons. We focused our efforts on identifying for each candidate gene the most promising regions for TILLING analysis. In this way, new primer pairs flanking this region could be targeted to the intronic sequences, with the aim of improving the screening efficiency on the coding regions in the pilot assay (Figure 2). In total 10 Mb of *C. pepo* genomic sequence were screened in 768

Table 2. Effect of EMS dose on mutation frequency in *C. pepo* mutant collection.

EMS dose	M2 plants	Induced mutations*	Mutation frequency
80 mM	95	8	1/119 Kb
65 mM	52	4	1/130 Kb
40 mM	621	46	1/135 Kb
TOTAL	**768**	**58**	**1/133 Kb**

*Number of identified mutations in all the families for this EMS dose after screening five genes.

M2 plants and 58 induced point mutations were identified of which four and eight mutations were derived from the 65 mM and 80 mM EMS treatment respectively, and 46 mutations from the 40 mM treatment population (Table 2).

Among the 58 identified mutations, we have identified 59% non-synonymous mutations (including nonsense and splicing mutations). Extrapolating from this small set of genes, we calculated an average mutation rate of one in around 133 kb or 6300 mutations per genome for a genome size of 845 Mb [29].

We identified 57 lines, corresponding to 58 different mutations, among which eight missense mutations cause changes in the PSY amino acid sequence, seven in the LCYb amino acid sequence, four and nine in the ETR1 and ERS1 amino acid sequence and four in the APRX amino acid sequence (Table 3). The exonic mutations were mostly present as heterozygotes (36 out of 45 mutations), but 9 lines were homozygous for the mutations. As expected with EMS mutagenesis, these mutations were distributed relatively evenly within the screened amplicons (Figure 2).

The exonic mutations were mostly present as heterozygotes (36 out of 45 mutations) and 9 lines were homozygous for the mutations. Thus, even though these results are higher than expected, the deviation between data is not statistically significant ($P>0.05$). Furthermore, as expected with EMS mutagenesis, these mutations were distributed relatively evenly within the screened amplicons (Figure 2).

All the EMS-induced mutations were detected in single plants, except for the APRX gene, where the same mutation was found in two different plants. Only one induced stop codon mutation was identified for one of the five genes, APRX, also an splice mutant junction was identified in the ETR1 amino acid sequence.

Because of the high number of alleles identified, the possible impact of missense mutations on the function of the protein was assessed before systematic phenotyping of the mutant plants using SIFT (Sorting Intolerant From Tolerant). Out of the 32 missense mutations, a total of 14 mutations (44%) were predicted deleterious for the protein's activity ($p<0.05$): three and two missense mutations for ERS1 and ETR1 respectively, four for each of the two genes in the carotenoid synthesis and one missense mutation for APRX (Table S2).

Genotype/phenotype association: characterization of the APRX mutants

Plant peroxidases are encoded by multigenic families and involved in several important physiological and development processes, although their varied functions are not clearly determined [30]. The appearance of nonsense mutants in these genes has not been described previously for this species so based on the genotype/phenotype relationship we decided to study the function of the only previously peroxidase studied in this species, APRX.

APRX is a globular protein of 330 aa whose structure is highly conserved between distant plant families. Along conserved residues, there are eight cysteine residues bounded by disulphide bridges and essential to stabilise the structure of the protein. The protein also shows thirteen alpha helices with an heme pocket and two distal Ca^{2+} binding sites disposed symmetrically in the center of the heme pocket. Throughout its structure of four exons and three introns we found nine families with mutations that affected diverse structural motifs.

For each APRX mutant family, the genotype/phenotype association was carried out using 18 M3 seeds that were germinated and genotyped by sequencing the APRX gene. No associated phenotype was observed in Cp-34 and Cp-773 families, which agrees with the position of the mutation in an intronic and non-coding region, respectively. Cp-675 and Cp-879 families, which showed a silent mutation, also did not show differences with the wild type.

Among the four mutant families that presented missense mutations, only the Cp-86 did not germinate, so it could not be evaluated. Cp-661 that presented an amino acid change at position R215 and Cp-359 and Cp-579 families with the same mutation at position S14F showed different phenotypes as dwarf plants or plants with deformed leaves. These phenotypes were caused by other mutations since, after genotyping of the lines, we could not associate these genotypes to a particular phenotype.

The induced stop codon mutation indentified in family Cp-161 (W132*) was the only mutation predicted to affect the protein function by SIFT. The nine germinated plants out of the 18 seeds sown were genotyped by sequence the APRX gene and six homozygous mutants, two heterozygous and one homozygous wild type were identified. The phenotype of seedling homozygous for the mutant allele was albino (Figure 3c). The remaining three M3 plants, corresponding to homozygous or heterozygous for the wild type allele, showed normal phenotypes.

Alignment of the homologous proteins showed that the W132* nonsense mutation occurred in the middle of a highly conserved protein motif. Based on the Geno3D derived structure of APRX and X-ray crystallography studies [31] (Figure 3b), the truncated mutant protein is predicted to miss helices A to D and the connected ability to bind heme and Ca^{++}. Consistent with the mutant phenotype, the loss of more than half of the protein should be highly deleterious (Figure 3a).

Discussion

The TILLING platform validated here is the first developed in *Cucurbita pepo*, probably because the low fertility that the mutants of this species have shown in previous attempts, even using different varieties [32], has hampered the perpetuation of the populations. The development of a method for selecting M1 lines with higher fertility used in the first steps of the generation of this

Table 3. Tilled genes and mutation frequency in the *C. pepo* mutant collection with 768 M2 families screened.

Target names	Amplicon size (bp)	GC content (%)	Identified mutants				
			Missense[a]	Nonsense[b]	Splicing[c]	Silent[d]	Non coding[e]
PSY	2263	41	8	0	0	3	5
ERS1	2663	42	9	0	0	3	4
ETR1	1620	39	4	0	1	1	1
APRX	1398	47	4	1	0	2	2
LCYb	2035	42	7	0	0	2	1
TOTAL	**9979**	**42**	**32**	**1**	**1**	**11**	**13**

[a]nucleic acid transition is a non-synonymous mutation and induce amino acid change in the translated protein.
[b]nucleic acid transition produces a stop codon and may induce a truncated protein.
[c]nucleic acid transition is located in splicing motif.
[d]nucleic acid transition induces a synonymous mutation and then no change in the translated protein.
[e]nucleid acid transition is located in an intronic or promotor region.

current platform [20] was successful. Despite the mutagen's toxicity, the analysis of M2 and M3 lines performed in this study demonstrate that this TILLING platform consists of fertile lines with mutations affecting most phenological stages of development.

A primary objective in a mutagenesis project is to generate a saturated resource, where every *locus* is mutated and represented by multiple alleles. In the *C. pepo* TILLING platform more than one mutation was identified per analysed gene. We screened 768 M2 families to obtain a suitable allelic series which averages of 1 mutation per 133 Kb, so we can conclude that our mutant population is sufficiently saturated. Furthermore, by comparing our results with those reported in other TILLING projects with different plant species we can also affirm that our population is suitable for use in high-throughput mutation discovery. The density of mutations that we discovered in this assay appears to be much higher than the obtained for other cucurbit species such as *Cucumis melo* [13], [16] or *Cucumis sativus* [17] and if we compare this with other species, we find only five populations of diploid species with higher mutation densities: *Arabidopsis thaliana* [21], *Brassica rapa* [33], *Oryza sativa* [34], *Linum usitatissimum* [35] and a diploid wheat (*Triticum monococcum*) [36].

The high mutation ratio that we describe here comes from the analysis of one individual plant of each of the 768 M2 families. This mutation ratio is likely to be underestimated as M1 plants that are heterozygous for a new mutation will yield M2 families segregating 1:2:1 for the mutant homozygous, heterozygous and wild type homozygous. The sampling of only one plant per M2, even assuming good viability and fertility of the mutant allele, are likely to cause a bias in this segregation. Therefore, our next step will be to expand the population by generating DNA pools of at least four individuals per M2 family. This fact, maybe, could increase our mutation density. We also found that we had underreported mutations towards both ends of fragments and we attribute these losses primarily to a poor fluorescence signal [4].

A total of 58 mutations per 9 Mb screened were identified in 768 M2 plants and as expected for EMS mutagenesis, single nucleotide substitutions were identified both in coding and non-coding regions [4] and all the induced nucleotide changes by EMS were G/A and C/T substitutions (Table S2) [37].

All the EMS-induced mutations were detected in single families, except for one in the APRX gene, where the same change was found in two different families; this circumstance is expected and observed to occur 4% of the time in *Arabidopsis* based on random distribution of induced mutations, GC content of the genome and distribution of location of mutations discovered in fragments [4]. Finding one coincidence among 58 mutations is not significantly different from our expectations. Furthermore, finding that all 58 mutations are G/C-to-A/T transitions effectively rules out the possibility that they are naturally occurring polymorphisms and although G/C to A/T are the expected transitions, these changes are not always obtained [38]. This indicates that the EMS has had the predicted effect on the background and demonstrates no problems of crossing showing a population with a high homogeneity and inbreeding level.

Despite this higher mutation rate in the population, if we consider that about half of missense mutations are expected to be damaging to a typical protein [39], we expected to have about 16 deleterious mutations in the sequences studied. This fact agrees with the results based on SIFT algorithm which predicts deleterious missense mutations with ~75% overall accuracy. According to SIFT, 44% out of the 32 missense mutations were predicted deleterious for the protein's activity ($p<0.05$) representing an extensive allelic series to address future structure-function studies in this species.

Phenotypic analyses confirm the phenotype/genotype correlation in APRX mutants. The anionic peroxidase analyzed [40], [41] is one of a broad family of Class III Peroxidases with up to 72 members identified in *Arabidopsis* genome [42]. The conserved structure and the low substrate specificity of these proteins have constituted a handicap to determine its specificity of function [28] because redundancy of function prevents phenotypes of overexpressed and antisense strategies. The functional analysis of this *C. pepo* peroxidase has been restricted to heterologous complementation in *A. thaliana*, where overexpression and silencing seems to indicate a role of APRX in auxin catabolism, but conclusive analysis is necessary to assess its specific function [28].

Evaluation of allelic series obtained by TILLING in *C. pepo*, which includes missense, nonsense and null alleles, constitutes an important resource to analyze the function of this gene. Some mutations were situated in significant structures of the protein. One in the signal peptide, and others in the alpha helix F, which conformation could change the position of one cysteine responsible of the structural conformation by covalent forces, however and according to SIFT, these mutations were tolerated and did not show any visual phenotype different from the wild type.

Finally, the truncated protein of line Cp-161, which showed a stop mutation identified at the conserved W132 residue, yielded

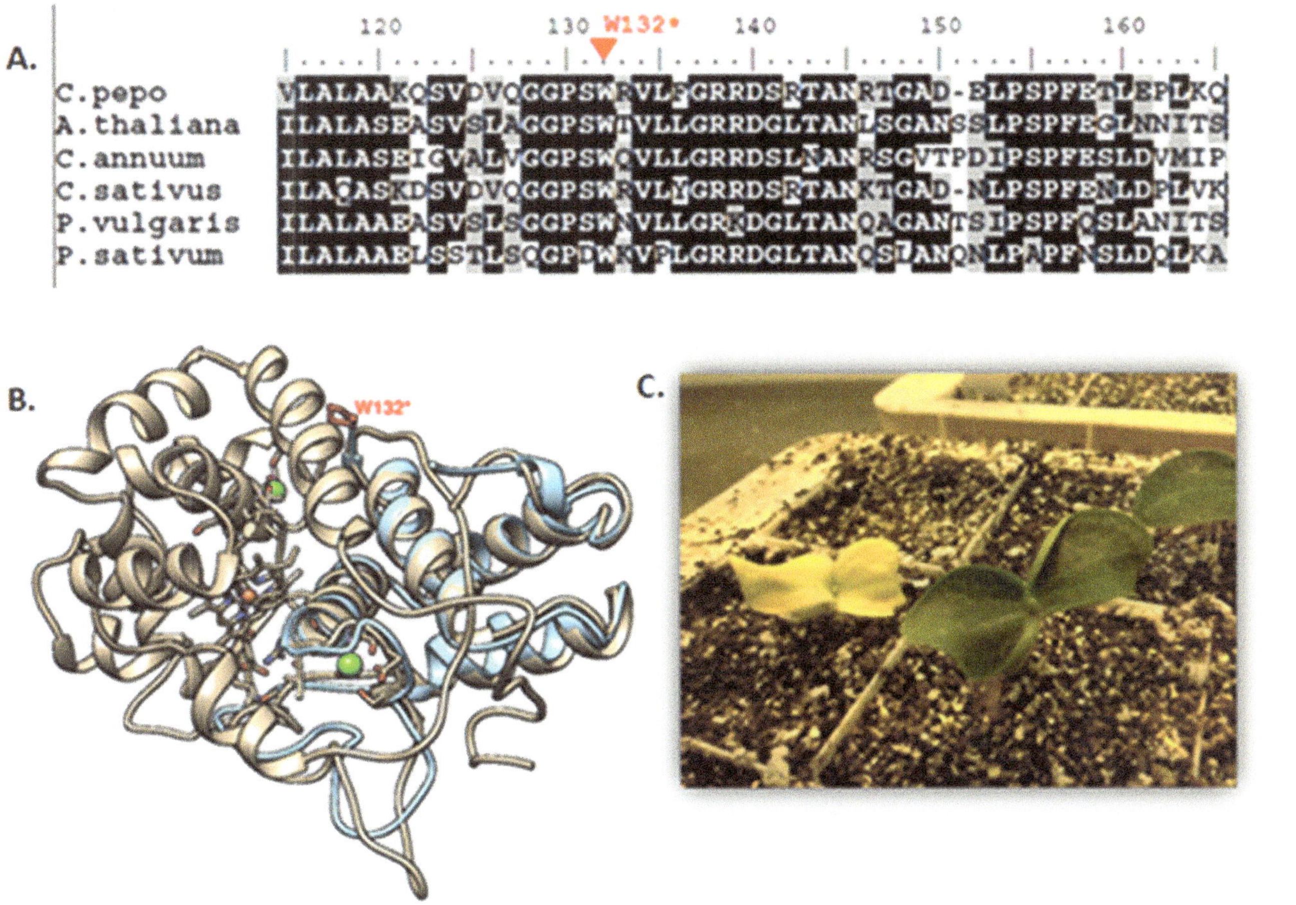

Figure 3. Sequence and structural analysis of APRX. (A) Amino acid alignment of *Cucurbita pepo* APRX and homologous proteins from *Arabidopsis thaliana* (AAM66044.1), *Capsicum annun* (AFU51540.1), *Cucumis sativus* (AAA33128.1), *Phaseolus vulgaris* (AAD37427.1) and *Pytum sativus* (BAD97438.1). Numbers above the alignment indicate the amino acid positions along the APRX protein. The EMS-induced stop mutation in W132 is shown above the alignment in red. (B) Superposition of the predicted 3D structure model of APRX WT protein indicated in grey and the mutant protein indicated in blue. The proximal heme pocket is represented in red and the two calcium ions in green. The position of the induced stop codon mutation indentified in Cp-161 family (W132*) is indicated in red. The APRX model was determined using the Geno3D server (http://geno3d-pbil.ibcp.fr). (C) Wild type and albino phenotypes observed in line Cp-161.

albino seedlings that died in early stages when the mutant allele was homozygous, as was confirmed with the M3 genotyped plants. No differences from the wild type were observed when the mutant allele was heterozygous. Those results provide the first predicted nullimorphic mutation for APRX and a phenotype which is much more severe than that of previously reported mutants [28]. This phenotype could indicate the main role of this peroxidase at early stages of the plant's development and it could be related with the accumulation of hydrogen peroxide that becomes toxic and prevents the development of essential organelles of the cell such as chloroplasts. Although several studies are required to confirm this hypothesis, these mutants constitute a new source to obtain new insights of important and multifunctional proteins as peroxidases.

Conclusions

The establishment of the EMS-mutagenized TILLING population described here represents an important advance in the generation of new genetic variation in *C. pepo*. The high mutation density obtained in this species makes it an attractive genetic system for the identification of new alleles that may be of value for crop improvement. Moreover, TILLING methodology becomes a feasible goal making possible the investigation of the role of key genes for this species bypassing the problems of other functional genomic tools. With the availability of this new resource we hope to fulfill the expectations of both, crop breeders and scientists, who are using this species as their model of study.

Supporting Information

Table S1 Primer pairs used to amplify the five genes tilled in *C. pepo* mutant population. Asterisk indicate that the primer carried M13 tail and M13 universal primers, M13F700 (5′-CACGACGTTGTAAAACGAC-3′) and M13R800 (5′GGA-TAACATTTCACACAGG-3′), labelled at the 5′end with infrared dyes IRD700 (forward primers) and IRD800 (reverse primers).

Acknowledgments

The authors thank Nicholas Davies for his help in grammatical revision of the manuscript and Sara Vázquez for helping with creating figures.

Author Contributions

Conceived and designed the experiments: PG NVD CT AB. Performed the experiments: NVD MM TPV CT. Analyzed the data: NVD CT PG. Contributed reagents/materials/analysis tools: MRC BR BP CT AB. Wrote the paper: NVD PG.

References

1. Paris HS, Yonash N, Portnoy V, Mozes-Daube N, Tzuri G, et al. (2003) Assessment of genetic relationships in *Cucurbita pepo* (Cucurbitaceae) using AFLP, ISSR, and SSR markers. Theor Appl Genet 106: 971–978.
2. Parry MA, Madgwick PJ, Bayon C, Tearall K, Hernandez-Lopez A, et al. (2009) Mutation discovery for crop improvement. J Exp Bot 60(10): 2817–2825.
3. Gilchrist E and Haughn G (2010) Reverse genetics techniques: engineering loss and gain of gene function in plants. Brief Funct Genomics 9: 103–110.
4. Greene EA, Codomo CA, Taylor NE, Henikoff JG, Till BJ, et al. (2003) Spectrum of chemically induced mutations from a largescale reverse-genetic screen in *Arabidopsis*. Genetics 164: 731–740.
5. McCallum CM, Comai L, Greene EA, Henikoff S (2000) Targeting induced local lesions IN genomes (TILLING) for plant functional genomics. Plant Physiol 123(2): 439–442.
6. Colbert T, Till BJ, Tompa R, Reynolds S, Steine MN, et al. (2001) High-throughput screening for induced point mutations. Plant Physiol 126: 480–484.
7. Wang TL, Uauy C, Robson F, Till B (2012) TILLING in extremis. Plant Biotechnol J 10: 761–772.
8. Dong C, Dalton-Morgan J, Vincent K, Sharp P (2009) A Modified TILLING Method for Wheat Breeding. Plant Gen 2: 39.
9. Uauy C, Paraiso F, Colasuonno P, Tran RK, Tsai H, et al. (2009) A modified TILLING approach to detect induced mutations in tetraploid and hexaploid wheat. BMC Plant Biol 9: 115.
10. Kumar AP, Boualem A, Bhattacharya A, Parikh S, Desai N, et al. (2013) SMART- Sunflower mutant population and reverse genetic tool for crop improvement. BMC Plant Biol 13: 38.
11. Kurowska M, Daszkowska-Golec A, Gruszka D, Marzec M, Szurman M, et al. (2011) TILLING: a shortcut in functional genomics. J Appl Genet 52(4): 371–390.
12. Rigola D, van Oeveren J, Janssen A, Bonne A, Schneiders H, et al. (2009) High throughput detection of induced mutations and natural variation using KeyPoint technology. PLoS ONE 4: e4761.
13. González M, Xu M, Esteras C, Roig C, Monforte AJ, et al. (2011) Towards a TILLING platform for functional genomics in Piel de sapo melons. BMC Research Notes 4: 289.
14. Elias R, Till BJ, Al-Safadi B (2009) Optimizing TILLING and Ecotilling techniques for potato (*Solanum tuberosum* L). BMC research notes 2: 141.
15. Gady ALF, Vriezen WH, Van De Wal MHBJ, Huang P, Bovy AG, et al. (2012) Induced point mutations in the phytoene synthase gene cause differences in carotenoid content during tomato fruit ripening. Molecular breeding new strategies in plant improvement 29(3): 801–812.
16. Dahmani-Mardas F, Troadec C, Boualem A, Lévêque S, Alsadon AA, et al. (2010) Engineering Melon Plants with Improved Fruit Shelf Life Using the TILLING Approach. PLoS ONE 5: 12.
17. Boualem A, Fleurier S, Troadec C, Audigier P, Kumar AP, et al. (2014) Development of a *Cucumis sativus* TILLinG Platform for Forward and Reverse Genetics. PLoS One 16; 9(5): e97963.
18. Blanca J, Cañizares J, Roig C, Ziarsolo P, Nuez F, et al. (2011) Transcriptome characterization and high throughput SSRs and SNPs discovery in *Cucurbita pepo* (Cucurbitaceae). BMC Genomics 12: 104.
19. Esteras C, Gómez P, Monforte AJ, Blanca J, Vicente-Dólera N, et al. (2012) High-throughput SNP genotyping in *Cucurbita pepo* for map construction and quantitative trait loci mapping. BMC genomics 13(1): 80.
20. Vicente-Dólera N, Pinillos V, Moya M, Del Río-Celestino M, Pomares-Viciana T, et al. (2014) An improved method to obtain novel mutants in *Cucurbita pepo* by pollen viability. Scientia Horticulturae 169: 14–19.
21. Martín B, Ramiro M, Martínez-Zapater JM, Alonso-Blanco C (2009) A high-density collection of EMS-induced mutations for TILLING in *Landsberg erecta* genetic background of *Arabidopsis*. BMC Plant biology 9: 147.
22. Wienholds E, van Eeden F, Kosters M, Mudde J, Plasterk RHA, et al. (2003) Efficient target-selected mutagenesis in zebrafish. Genome Res 13: 2700–2707.
23. Dalmais M, Schmidt J, Le Signor C, Moussy F, Burstin J, et al. (2008) UTILLdb, a *Pisum sativum* in silico forward and reverse genetics tool. Genome Biol 9(2): R43.
24. Triques K, Sturbois B, Gallais S, Dalmais M, Chauvin S, et al. (2007) Characterization of *Arabidopsis thaliana* mismatch specific endonucleases: application to mutation discovery by TILLING in pea. Plant J 51(6): 1116–1125.
25. Taylor N, Greene EA (2003) PARSESNP: A tool for the analysis of nucleotide polymorphisms. Nucleic Acids Res 31: 3808–3811.
26. Ng PC, Henikoff S (2003) SIFT: predicting amino acid changes that affect protein function. Nucleic Acids Res 31: 3812–3814.
27. Obrero Á, González-Verdejo CI, Die JV, Gómez P, Del Río-Celestino M, et al. (2013) Carotenogenic gene expression and carotenoid accumulation in three varieties of *Cucurbita pepo* during fruit development. J Agric Food Chem 3; 61(26): 6393–403.
28. Cosio C, Vuillemin L, De Meyer M, Kevers C, Penel C, et al. (2009) An anionic class III peroxidase from zucchini may regulate hypocotyls elongation through its auxin oxidase activity. Planta 229 (4), 823–836.
29. Šiško M, Ivančič A, Bohanec B (2003) Genome size analysis in the genus *Cucurbita* and its use for determination of interspecific hybrids obtained using the embryo-rescue technique. Plant Science 165(3): 663–669.
30. Campa A (1991) Biological roles of plant peroxidases: known and potential function. In Peroxidases in Chemistry and Biology Vol. II. (Everse, J., Everse, K.E. & Grisham, M.B., eds), pp. 25–50, CRC Press, Boca Raton, FL.
31. Schuller DJ, Ban N, van Huystee RB, McPherson A, Poulos TL (1996) The crystal structure of peanut peroxidase. Structure 4(3): 311–321.
32. Whitwood WN, Weigle JL (1978) Compact mutations in *Cucurbita pepo* L. induced by ethyl-methanesulfonate. Cucurbit Genet Coop 1: 34.
33. Stephenson P, Baker D, Girin T, Perez A, Amoah S, et al. (2010) A rich TILLING resource for studying gene function in *Brassica rapa*. BMC Plant Biol 10: 62.
34. Suzuki T, Eiguchi M, Kumamaru T, Satoh H, Matsusaka H, et al. (2008) MNU-induced mutant pools and high performance TILLING enable finding of any gene mutation in rice. Mol Genet Genomics 279: 213–223.
35. Chantreau M, Grec S, Gutierrez L, Dalmais M, Pineau C, et al. (2013) PT-Flax (phenotyping and TILLinG of flax): development of a flax (*Linum usitatissimum* L.) mutant population and TILLinG platform for forward and reverse genetics. BMC Plant Biology 13: 159.
36. Rawat N, Sehgal SK, Joshi A, Rothe N, Wilson DL, et al. (2012) A diploid wheat TILLING resource for wheat functional genomics. BMC Plant Biology 12: 205.
37. Till BJ, Colbert T, Tompa R, Enns LC, Codomo CA, et al. (2003) Highthroughput TILLING for functional genomics. Methods in molecular biology 236: 205–220.
38. Minoia S, Petrozza A, D'Onofrio O, Piron F, Mosca G, et al. (2010) A new mutant genetic resource for tomato crop improvement by TILLING technology. BMC Res Notes 3: 69.
39. Markiewicz P, Kleina LG, Cruz C, Ehret S, Miller JH (1994) Genetic studies of the lac repressor. XIV. Analysis of 4000 altered *Escherichia coli* lac repressors reveals essential and nonessential residues, as well as "spacers" which do not require a specific sequence. J Mol Biol 240: 421–433.
40. Carpin S, Crèvecoeur M, Greppin H, Penel C (1999) Molecular cloning and tissue-specific expression of an anionic peroxidase of zucchini. Plant Physiol 120: 1–12.
41. Carpin S, Crevecoeur M, de Meyer M, Simon P, Greppin H, et al. (2001) Identification of a Ca(2+)-pectate binding site on an apoplastic peroxidase. Plant Cell 13: 511–520.
42. Welinder KG, Justesen AF, Kjaersgård IV, Jensen RB, Rasmussen SK, et al. (2002) Structural diversity and transcription of class III peroxidases from *Arabidopsis thaliana*. Eur J Biochem 269(24): 6063–81.

The Population Genetics of Cultivation: Domestication of a Traditional Chinese Medicine, *Scrophularia ningpoensis* Hemsl. (Scrophulariaceae)

Chuan Chen[1,2], Pan Li[1], Rui-Hong Wang[1], Barbara A. Schaal[3]*, Cheng-Xin Fu[1]*

1 The Key Laboratory of Conservation Biology for Endangered Wildlife of the Ministry of Education, College of Life Sciences, Zhejiang University, Hangzhou, China, 2 Hangzhou Botanical Garden, Hangzhou, China, 3 Department of Biology, Washington University, St. Louis, Missouri, United States of America

Abstract

Background: Domestic cultivation of medicinal plants is an important strategy for protecting these species from over harvesting. Some species of medicinal plants have been brought into cultivation for more than hundreds years. Concerns about severe loss of genetic diversity and sustainable cultivation can potentially limit future use of these valuable plants. Genetic studies with comprehensive sampling of multiple medicinal species by molecular markers will allow for assessment and management of these species. Here we examine the population genetic consequences of cultivation and domestication in *Scrophularia ningpoensis* Hemsl. We used chloroplast DNA and genomic AFLP markers to clarify not only the effects of domestication on genetic diversity, but also determine the geographic origins of cultivars and their genetic divergence from native populations. These results will allow both better management of cultivated populations, but also provide insights for crop improvement.

Results: Twenty-one cpDNA haplotypes of *S. ningpoensis* were identified. Wild populations contain all haplotypes, whereas only three haplotypes were found in cultivated populations with wild populations having twice the haplotype diversity of cultivated populations. Genetic differentiation between cultivated populations and wild populations was significant. Genomic AFLP markers revealed similar genetic diversity patterns. Furthermore, Structure analysis grouped all wild populations into two gene pools; two of which shared the same gene pool with cultivated *S. ningpoensis*. The result of Neighbor-Joining analysis was consistent with the structure analysis. In principal coordinate analysis, three cultivated populations from Zhejiang Province grouped together and were separated from other cultivated populations.

Conclusions: These results suggest that cultivated *S. ningpoensis* has experienced dramatic loss of genetic diversity under anthropogenic influence. We postulate that strong artificial selection for medicinal quality has resulted in genetic differentiation between cultivated and wild populations. Furthermore, it appears that wild populations in Jiangxi-Hunan area were involved in the origin of cultivated *S. ningpoensis*.

Editor: Xiao-Yong Chen, East China Normal University, China

Funding: This research was supported by the Key Breeding Project of Zhejiang Province (2012C12912), the National Natural Science Foundation of China (31070205), and Science and Technology Development Project of Hangzhou West Lake Scenic Area Management Committee (2012002). The funders had no role in study design, data collection and analysis, decision to publish, or preparation of the manuscript.

Competing Interests: The authors have declared that no competing interests exist.

* Email: schaal@biology.wustl.edu (BAS); cxfu@zju.edu.cn (CXF)

Introduction

Plant domestication is one of the great milestones in human history. The process of domestication represents a continuum of increasing codependence between plants and people [1]. Crop species have been derived from their wild progenitors as a result of artificial selection for desirable traits by early farmers. In the light of crop domestication, a number of agencies now are recommending that wild medicinal plants be brought into cultivation systems [2,3]. Currently, more than 80% of the world's population in developing countries depends primarily on herbal medicine for basic healthcare needs [4,5]. Given the demand for a reliable and uniform supply of medicinal plants and the accelerating depletion of forest resources, cultivation of medicinal plants species would be an important strategy and a viable alternative to harvesting of natural populations [6,7].

At present some medicinal plants are grown in home farms, and some are cultivated as field crops. But the number of medicinal plants species currently in high intensity cultivation for commercial production are varied world-wide. In India, 20 species are currently under cultivation, in Hungary 40 and in Germany 100 [6,8,9,10]. China, with its history of thousands years of medicinal plant use cultivates only about 250 species [11,12]. Considering the harvest pressure on wild populations and the increasing demands, the most popular medicinal plants are cultivated under the supervision of Chinese government on large scale reaching almost 1 million acres [13].

As the demand for medicinal plant cultivation increased, many studies examined the impact of cultivation on genetic diversity, species such as *Artemisia judaica* [14], *Scutellaria baicalensis* [15], *Corydalis yanhusuo* [16], *Coptis chinensis* [17], *Magnolia officinalis* subsp. *biloba* [18], *Fritillaria cirrhosa* [19]. However, all these studies only used single molecular marker or local sampling which just revealed genetic diversity patterns in limited populations. Since genetic diversity underlies the plasticity of secondary metabolism and hence the production of medicinal compounds [20], understanding the population genetics across the range of wild populations is essential for utilizing native resources for improvement of cultivated species. A study with comprehensive sampling of both cultivated and wild populations of medicinal species, using both chloroplast and nuclear DNA markers, can clarify not only changes in genetic diversity, but also provide information on domestication including the geographic origins, ancestor populations and the impact of domestication on population genetics.

Plant secondary metabolites are responsible for phytochemical diversity and therapeutic efficacy of medicinal species. The production of secondary compounds is influenced by both environmental and genetic factors [20,21]. Thus geographically distinct populations even of the same species may have vastly different medical qualities [22]. Modern methods of phytochemical methods are used to discriminate geographically dispersed cultivars within a medicinal species [23,24,25,26,27]. Yet little is known about the genetic differences among the wild populations of a medicinal species and whether cultivated populations with high pharmaceutical quality have already significantly diverged from wild populations. Additionally, understanding genetic changes associated with cultivation not only helps control the quality of medicinal herbs, but also provides practical information essential for formulating appropriate conservation and cultivation management strategies [28,29].

Scrophularia ningpoensis Hemsl. is a perennial herb native to Southeastern China. Wild populations are distributed in forests along streams, thickets, and tall grasses below alt. 1700 m and reproduce from seed [30]. Roots of this medicinal herb have a long history of widespread use in China to treat inflammation, laryngitis, tonsillitis, abscesses and constipation [31,32,33]. The first recorded use in ancient Chinese literature is from ca. 100 BC [34]. Harpagoside, angroside C, acteoside and cinnamic acid are the main bioactive components [35,36,37,38,39]. According to the Chinese Pharmacopoeia, only the root of *S. ningpoensis* is listed as the Radix Scrophulariae [40]. Our recent field investigations indicated that wild populations of this herb have suffered rapid declines, and the species is extirpated in many locations due to over exploitation and deterioration of habitats. An analysis of the population genetics of wild *S. ningpoensis* populations is important and urgent for appropriate utilization, conservation and preservation of the wild resources.

Today, *S. ningpoensis* is cultivated for its roots on a large scale in several regions of Central and Southeastern China. This species can trace its cultivation history back to Song dynasty (1000 years ago) beginning in backyard gardens [41]. In spite of this long written history, the geographic origin of cultivation has not been recorded. Formal cultivation programs were initiated in China only since the 1950's [42]. Unlike wild populations, cultivated *S. ningpoensis* is propagated vegetatively from rhizome. During the cultivation, large rhizomes are often selected for propagation; flowers are removed before opening so that resources are allocated to vegetative growth, in particular the rhizomes [43,44]. Our previous study used HPLC revealed that among cultivated *S. ningpoensis*, accessions from Zhejiang Province produced higher concentrations of three main bioactive compounds, suggesting that the best quality of cultivated *S. ningpoensis* is from Zhejiang [45]. These results are consistent with the general perspective that *S. ningpoensis* from Zhejiang has the best quality for medical use [46]. Besides the phytochemical differences, the unresolved question of whether cultivated *S. ningpoensis* from Zhejiang is genetically differentiated for others is also important and indispensable to sustainable cultivation of *S. ninpoensis*.

Here we used two chloroplast DNA fragments and AFLP markers to study the population genetics and phylogeography of wild and cultivated *S. ningpoensis* with the aims of: (1) evaluating the genetic erosion of cultivated populations and the change in the pattern of genetic diversity under artificial selection; (2) unraveling the geographical origins and ancestral populations of cultivated *S. ningpoensis*; (3) clarifying the genetic divergence between cultivated populations from Zhejiang Province and the rest of cultivated *S. ningpoensis*; and (4) determining which wild populations are valuable and can be introduced for improvement of cultivation.

Materials and Methods

Ethics Statement

Management Bureau of Mt. Jinggang National Nature Reserve issued the permit for Jinggang Mountain (JGW); Management Bureau of Mt. Tianmu National Nature Reserve issued the permit for Tianmu Mountain (TNW, TM1W, TM2W); Management Bureau of Mt. Dapan National Nature Reserve issued the permit for Dapan Mountain (TWW). No specific permissions were required for other locations which are neither privately owned nor protected and the field study did not involve endangered or protected species.

Sampling and DNA extraction

Field studies were conducted throughout distribution range of *Scrophularia ningpoensis*. Specific locations (GPS coordinates) of our field work are indicated in Table 1. A total of 364 individuals representing 13 cultivated and 14 wild populations were sampled with nine to twenty-two individuals collected randomly for each population (Table 1, Figure 1). Leaf material was dried in silica gel and stored at 4°C. Voucher specimens are deposited in the Herbarium of Zhejiang University (HZU), China. Total genomic DNA was extracted using a modified (CTAB) method [47,48].

Chloroplast (cp) DNA sequences analysis

After preliminary screening for variation in the intergenic spacer (IGS) regions of chloroplast DNA, two regions: *trn*L–*trn*F [49] and *psb*A–*trn*H [50] were chosen for analysis because they contained sufficient levels of informative polymorphic sites. PCR amplification was performed in an ABI 9700 thermocycler (Applied Biosystems). Cycling conditions were 95°C for 2 min followed by 35 cycles of 94°C (45 s), 57°C (45 s), 72°C (1 min), and concluding extension at 72°C (10 min). The reaction mixture (25 μL) contained 2.5 mm/L $MgCl_2$, 0.5 μm/L dNTP, 2.5 μl 10×buffer, 2.5 μm/L primer, 1 U Taq DNA polymerase (*Takara* Bio Inc.) and 20 ng DNA template. PCR products were identified on 1.5% agarose gels in a 0.5×TBE buffer, stained with ethidium bromide (0.5 μg/ml), visualized with ultraviolet light and photographed. DNA Marker DL2000 (*Takara* Bio Inc.) was used. The PCR products with a single band were cleaned by a GENE-CLEAN II Kit (BIO 101, Inc. Carlsbad, USA) for direct sequencing. Samples were sequenced in forward and reverse directions using the dideoxy chain termination method with fluorescent labeling with Big Dye Terminator (version 1.1). Sequences were determined with an ABI 377XL DNA sequencer and edited using Sequencher version 4.0 (Gene Codes Corp., Ann

Table 1. Details of sample locations and sample sizes (N) of *Scrophularia ningpoensis*.

Population code	Locality	Altitude (m)	Lat. (°N) Long. (°E)	N CpDNA	N AFLP	Vouchers
CULTIVATED						
GZC	Guangze, Fujian Province	234	117°20′, 27°32′	15	15	LP0805GZ
JFC	Jinfo Mountain, Chongqin Province	1282	107°12′, 29°00′	14	13	FU0810321
SCC	Guangyuan, Sichuan Province	594	105°47′, 32°27′	14	19	GW20070705
XJC	Xianju, Zhejiang Province	63	120°44′, 28°51′	10	10	CC20060601
*RCC/PAC	Pan'an, Zhejiang Province	356	120°24′, 28°54′	15	4	CC20060602
*YCC/PAC	Pan'an, Zhejiang Province	357	120°33′, 29°05′	11	4	CC20060603
*SHC/PAC	Pan'an, Zhejiang Province	537	120°39′, 29°08′	10	4	CC20060604
*DPC/PAC	Pan'an, Zhejiang Province	581	120°32′, 28°59′	14	4	CC20060605
YLC	Yangling, Shanxi Province	459	108°04′, 34°15′	15	10	LP20080811
ZPC	Zhenping, Shanxi Province	1013	109°33′, 31°45′	15	10	LP20090610
KKC	Suiyan, Guizhou Province	1572	107°09′, 28°13′	10	8	WY20090618
LOC	Long Mountain, Hunan Province	843	109°43′, 29°36′	15	10	ST20090601
NSC	Enshi, Hubei Province	1606	110°12′, 30°44′	13	10	ZZ20080721
WILD						
HNW	Pingjiang, Hunan Province	1428	113°49′, 28°38′	15	15	SS20070821
JGW	Jinggang Mountain, Jiangxi Province	975	114°07′, 26°35′	15	15	CC0708J029
JHW	Jiuhua Mountain, Anhui Province	1246	117°49′, 30°28′	22	22	CC20080507
JMW	Junma, Henan Province	673	111°25′, 33°28′	13	13	SS20070703
LAW	Tiantang zhai, Anhui Province	492	115°47′, 31°11′	15	14	LP0906508
LSW	Lenshui, Jiangxi Province	80	117°11′, 28°15′	15	15	SS20070701
MTW	Matou, Jiangxi Province	101	115°53′, 29°27′	15	15	SS20070702
PTW	Putuo, Zhejiang Province	17	122°23′, 30°00′	9	9	CC20080527
THW	Tiantai Mountain, Zhejiang Province	960	121°05′, 29°14′	15	14	LP0910919
TNW	Tianmu Mountain, Zhejiang Province	1477	119°25′, 30°20′	15	14	LP0906TMS
TTW	Tiantong, Ningbo, Zhejiang	348	121°47′, 29°48′	10	10	CC20071001
TM1W	Tianmu Mountain, Zhejiang Province	315	119°26′, 30°19′	10	10	CC20061001
TM2W	Tianmu Mountain, Zhejiang Province	266	119°29′, 30°19′	15	10	CC20070706
TWW	Dapan Mountain, Zhejiang University	469	120°23′, 28°54′	9	9	CC20060901

*In AFLP analysis, four populations RCC, YCC, SHC and DPC are combined as one population coded PAC, and in Figure 1C these four populations are also indicated as PAC.

Arbor, MI, USA). Sequences of the two IGS regions were combined and aligned by Clustal W version 1.8 [51].

Sequences from all individuals were characterized for their cpDNA haplotypes. Chloroplast DNA haplotype diversity (h) and nucleotide diversity (π) [52] were calculated for each population (h_S, π_S) and overall (h_T, π_T) using DNASP 5.10 [53]. A haplotype network rooted by *Scrophularia buergeriana* that contains all linkages with >95% probability of being most parsimonious was drawn by TCS version 1.21 [54]. In this analysis, indels and one inversion were treated as single mutation events. Phylogenetic relationships among cpDNA haplotypes of *S. nignpoensis* were assessed by Bayesian inference (BI), maximum likelihood (ML), and maximum parsimony (MP) methods with *S. buergeriana*, *S. spicata* and *S. dentate* as the outgroup. For Bayesian inference, the substitution model was determined by MrModeltest 2.2 [55] and the best-fit model (F81+I+G) was selected by Akaike information criterion (AIC). Then Bayesian analysis was performed by MrBayes 3.1.2 [56]. Two independent runs of four Markov chains each starting with a random tree were processed in ten threads simultaneously for 50 million generations, sampling trees

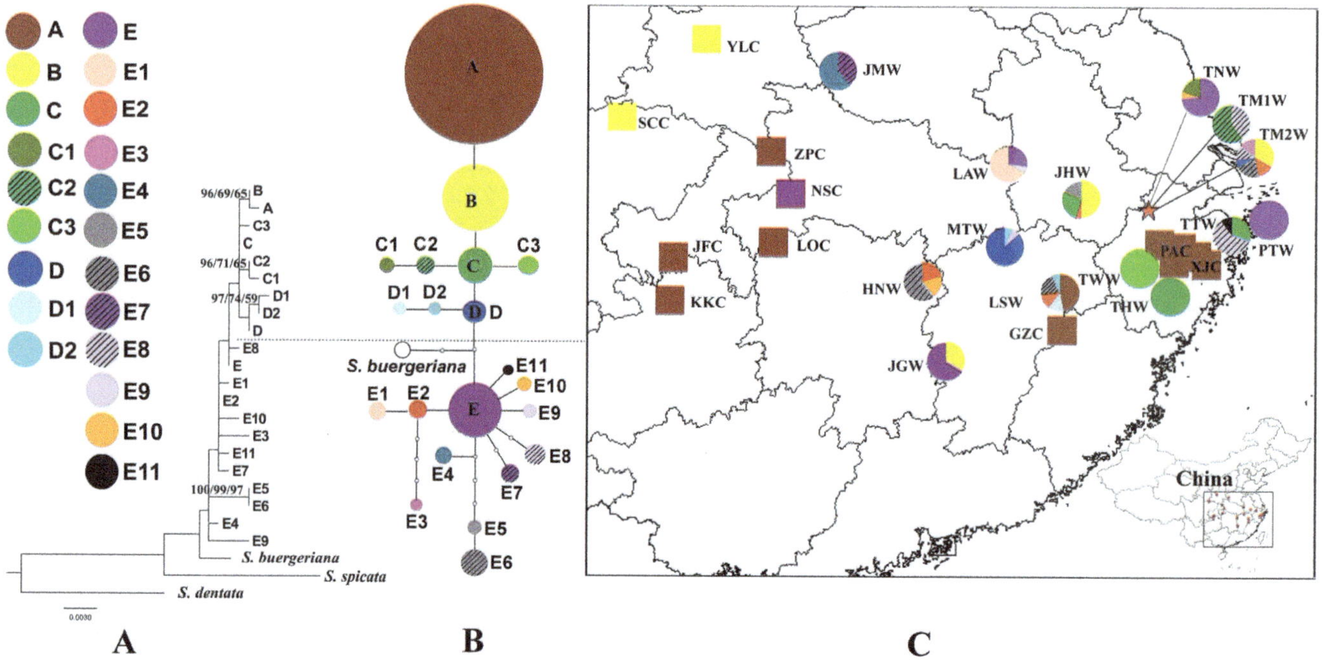

Figure 1. Analyses of 21 chloroplast (cp) DNA haplotypes of *Scrophularia ningpoensis*. (A) The maximum likelihood tree of 21 chloroplast (cp) DNA haplotypes of *S. ningpoensis*, *S. buergeriana*, *S. spicata* and *S. dentata* were used as the outgroup. Statistical supports (Bayesian posterior probability ≥80/maximum likelihood bootstrap value ≥50/maximum parsimony bootstrap value ≥50) are indicated on the branches. (B) 95% plausible network of the 21 cpDNA haplotypes of *S. ningpoensis* (A–E11). The size of circles corresponds to the frequency of each haplotype. Each solid line represents one mutational step that interconnects two haplotypes for which parsimony is supported at the 95% level. The small open circles indicate inferred intermediate haplotypes not detected in this investigation. (C) A geographic distribution of 21 cpDNA haplotypes (A–E11) detected in *Scrophularia ningpoensis*. The populations correspond to those detailed in Table 1. Squares represent cultivated populations and circles represent wild populations.

at every 1000 generation. The first 12500 sampled trees (25%) were discarded as burn-in samples. Maximum likelihood [57] tree and ML bootstrap searches were conducted using RAxML V.7.2.8 [58] on the CIPRES [59]. Final tree was evaluated and optimized under GAMMAGAMMA Model. ML bootstrap values were estimated from 1000 bootstrap replicates. Maximum parsimony analyses were conducted using PAUP* version 4.0b10 [60]. Support values for the relationships were calculated by performing bootstrap analyses of 1000 heuristic search replicates using the TBR branching swapping algorithm with 100 random additions per replicate. The geographical distribution of haplotypes was plotted on a map of China using Mapinfo (Pitney Bowes, Inc.). To test for significant difference in haplotype frequencies between wild and cultivated populations, total diversity (h_T), within-population diversity (h_S), U test (Nst/Gst) and population differentiation (G_{ST}) were calculated in HAPLONST [61]. Hierarchical structure of genetic variation was estimated by an analysis of molecular variance (AMOVA) on three levels: among groups (cultivated and wild groups); among-population within groups; and within-population. All AMOVAs were performed with ARLEQUIN [62]. Sequence variation was tested for deviations from neutrality by Tajima's D statistic [63], and by Fu and Li's D^* and F^* statistics [64] using DNASP 5.10 [65].

AFLP analysis

AFLP method followed the protocol established by Vos [66] with minor modifications. First, selective primer pairs were pre-screened for three individuals. Second, thirty-six primer combinations which had produced well-separated fragments in the pre-screen were used on eight individuals from different populations to determine polymorphisms. Six primer pairs which yielded polymorphic, clear and reproducible fragments were selected for the final AFLP analysis: *Eco*RI-AAC (FAM)/*Mse*I-CTC, *Eco*RI-ACA (FAM)/*Mse*I-CTG, *Eco*RI-ACC (HEX)/*Mse*I-CAC, *Eco*RI-ACC (HEX)/*Mse*I-CTA, *Eco*RI-AGG (TAMAR)/*Mse*I-CAG and *Eco*RI-AGG (TAMAR)/*Mse*I-CTC. FAM, HEX and TAMAR are three differential fluorescence labels for multiplexed electrophoresis.

The protocols are as follows: (i) in the digestion step, 2.5 ul (0.25–0.3 ug) of genomic DNA was restricted in 2.5 ul of a reaction mixture that contained 0.2 μl Eco R I (15 U/μL), 0.15 ul Mse I (10 U/μL), 0.5 μl 10×Buffer and 1.65 μl ddH_2O. The mixture was kept at 37°C for 3 hours, then 65°C for 10 minutes; (ii) the ligation reaction included 5.0 μl digested genomic DNA, 1.0 μl 10× T4 DNA ligase buffer, 1.0 μl Eco R I-adapters (5 μM), 1.0 μl Mse I adapters (50 μM), 0.1 μl T4 DNA ligase (3 U/μL) and 1.9 μl ddH_2O. The ligation reaction was 16°C for 12 h and the product was diluted with 90 μl of ddH_2O; (iii) pre-selective amplifications were carried out in 12.5 μl reactions: 2.5 μl diluted ligation product, 1.25 μl 10×PCR buffer, 1 μl $MgCl_2$ (25 mmol/L), 2 μl dNTPs (2.5 mmol/L), 0.5 μl Eco R I - A primer (5 umol/L), 0.5 μl Mse I - C primer (5 umol/L), 0.05 μl Taq DNA Polymerase (5 U/μL), 4.7 μl ddH_2O. PCR conditions for pre-selective amplification were 72°C for 2 min; 94°C for 20 sec, 56°C for 30 sec, 72°C for 2 min, for 20 cycles, 60°C for 30 sec. The product of the pre-selective PCR was diluted with 150 μl ddH_2O; (iv) selective amplification reactions included 3 μl diluted pre-selective amplification products, 1.25 μl 10×PCR buffer, 1.25 μl $MgCl_2$ (25 mmol/L), 2 μl dNTPs (2.5 mmol/L), 0.8 μl Eco R I primer (0.5 U/μL), 0.5 μl M primer (5 U/μL), 0.5 μl BSA

(20 mg/mL), 0.1 µl Taq DNA Polymerase (5 U/µL), 3.6 µl ddH_2O. PCR conductions were 94°C for 2 min; 13 cycles of 94°C for 30 sec, 65°C–56°C (−0.7°C/cycles) for 1 min, 72°C for 1.5 min; 23 cycles of 94°C for 30 sec, 56°C for 1 min, 72°C for 1.5 min; 72°C for 8 min. The product of the selective amplification was analyzed on ABI3730 (Applied Biosystems) according to the manufacture instructions with GS ROX-labelled 500 size standard (Applied Biosystems). For each primer combination, negative amplification controls without template DNA resulted in no amplification products.

Raw AFLP data were scored and collected as presence/absence matrix firstly by Genemarker v1.7 (SoftGenetics LLC) and then each locus was manually inspected. Only unambiguously detectable fragments in the size range 75–450 bp were scored. To test the reproducibility of the AFLP data, the complete AFLP procedure (digestion, ligation, pre-selective, selective amplification) was repeated for 15 individuals (5% of the samples) and scored independently for each primer pair.

Assuming a set of measures of wild and cultivated populations, population genetic statistics were generated by the program POPGENE version 1.31 [68] including: the number of polymorphic fragments (*F*p), the percentage of polymorphic fragments (*PPF*), Nei's [67] gene diversity (*h*), Shannon's diversity index (*H*), coefficient of gene differentiation (G_{ST} = total genetic diversity - gene diversity within population/total genetic diversity), and the level of gene flow: $N_m = 0.5(1-G_{ST})/G_{ST}$. To explore the relationships among individuals Principal Coordinate Analysis (PCoA) was performed by software MVSP version 3.1 (Kovach Computing Services, Anglesey, Wales) from pairwise Euclidian distances between individuals genotypes. A neighbor-joining (NJ) tree rooted by the populations of *Scrophularia buergeriana* was constructed by the PHYLIP 3.62 package [69] and group support was assessed by a bootstrap analysis with 1000 replicates. Genetic admixture analysis was implemented in STRUCTURE version 2.2.3 [70] with "no admixture model" and assumed "uncorrelated allele frequencies" [71]. The main modeling assumptions of STRUCTURE are Hardy-Weinberg equilibrium within populations [70]. But the cultivated populations are Hardy-Weinberg disequilibrium because of extensively clonal reproduction by rhizome within population. So instead of using all cultivated populations, only one individual from each different cultivated population (total ten individuals) were used for STRUCTURE analysis. The number of clusters (K) was set from 2 to 14 (number of all wild populations) with 10 replicate runs for each K, a burn-in period of 100000, and additional 1000000 replicates of the MCMC chains after burn-in. The structure computations output files were carried out by the freely accessible STRUCTURE HARVESTER [72] to calculate similarity coefficients between the replicate runs and to plot the mean logarithmic likelihood of K values (the means of the estimated log posterior probability of the data over the replicate runs for each K value). All populations were divided into two groups (cultivated and wild) and analysis of molecular variance (AMOVA) [73] was conducted by ARLEQUIN [62] to quantify genetic differentiation at different hierarchical levels. "Frequency- down-weighted marker values" (DW) [74] quantified by the amount of particular AFLP markers in the total dataset were calculated by AFLPdat [75].

Results

CpDNA sequence data

Total 21 different cpDNA haplotypes (A-U) were identified among the 364 individuals from 27 populations of *S. ningpoensis*. A total of 26 polymorphic sites were detected across the two chloroplast regions (Table 2); 21 were single-site mutations, four were length polymorphisms (4 bp, 9 bp, 7 bp and 2 bp, respectively) and one was an inversion (22 bp). Considering that sequencing poly-N regions could easily cause homoplasies due to polymerase error, length variations in mononucleotide repeats (poly A or T stretches) were not treated as polymorphisms. The *trn*H-*psb*A region (21 polymorphic sites detected in 466 aligned positions; 4.5%) was more variable than the *trn*L-*trn*F region (5 polymorphic sites detected in 811 aligned positions; 0.62%). The combined cpDNA sequences of *S. ningpoensis* varied in length from 1255 to 1268 bp, with a consensus length of 1277 bp after alignment. When the two fragments were combined, 2.0% of all sites (26/1277) were polymorphic. Both Tajima's test (*psb*A-*trn*H: $D = -0.42908$, $P>0.10$; *trn*L-*trn*F: $D = 0.03414$, $P>0.10$) and Fu & Li's test (*psb*A-*trn*H: $D^* = 1.26305$, $P>0.10$; $F^* = 0.68518$, $P> 0.10$; *trn*L-*trn*F: $D^* = 1.01550$, $P>0.10$; $F^* = 0.80528$, $P>0.10$) indicated no significant deviation from neutrality for the two regions. The combined sequences of the spacers conformed to the expectation of neutrality as well ($D = -0.33903$, $P>0.10$; $D^* = 1.51081$, $0.10>P>0.05$; $F^* = 0.87503$, $P>0.10$). Sequences of the eighteen *psb*A-*trn*H and six *trn*L-*trn*F haplotypes are available in GenBank database (GenBank accession numbers:KJ194143-KJ194166).

Haplotype diversity and within-population diversity declined under anthropogenic influence during the course of cultivation (Table 3). Wild populations contained all haplotypes, whereas cultivated populations only had three haplotypes (A, B and E). Moreover, in wild *S. ningpoensis*, eleven out of fourteen populations (78.6%) were polymorphic for haplotypes. In contrast, every cultivated population was fixed for a single haplotype. Wild populations had much higher estimates of haplotype diversity ($h_T = 0.919$), the within-population component of diversity ($h_S = 0.444$) and nucleotide diversity ($\pi_T = 0.00301$) than do cultivated populations ($h_T = 0.399$ $h_S = 0$, $\pi_T = 0.00076$). In wild *S. ningpoensis*, haplotype diversity varied: populations LSW, TM2W had the highest haplotype diversity whereas populations TWW, PTW, THW from Zhejiang were each fixed for a single haplotype (Figure 1C). An AMOVA indicated that 41.02% of the total genetic variation occurred between cultivated and wild groups (Table 4), which is consistent with the high observed genetic differentiation ($F_{ST} = 0.71988$) between cultivated and wild groups. And in cultivated *S. ningpoensis* all genetic variation was among populations, but in the wild populations, more variation occurred within populations (58.54%) than among wild populations (41.46%).

Maximum parsimony analysis of these 21 haplotypes, with *S. buergeriana*, *S. spicata* and *S. dentata* as the outgroup, resulted in 501 most parsimonious trees with a length of 87, a consistency index (CI) of 0.828, and a retention index (RI) of 0.583. Maximum likelihood optimization resulted in a final optimization likelihood of −2020.22, with the alpha parameter being 0.020000. The best-scoring ML tree had a length of 0.073087. The Maximum likelihood tree with statistical support indicated above branches was presented in Figure 1A, haplotypes of A, B, C, C1, C2, C3, D, D1 and D2 were monophyletic but with very low bootstrap support. The MP/BI/ML analyses yield mostly congruent topologies. Haplotypes (A, B, E) which cultivated populations harbored occur in different clades. The cpDNA haplotype network for the 27 populations of *S. ningpoensisis* is shown in Figure 1B. The rooted TCS network (Figure 1B) showed a similar structure to the phylogenetic tree (Figure 1A), which grouped 21 haplotypes into two distinct clades. The outgroup *S. buergeriana* is linked with the center of the whole network by two missing haplotypes. Two most common haplotypes A and B are on tip

Table 2. Chloroplast DNA sequence polymorphisms detected in two intergenic spacer (IGS) regions of *Scrophularia ningpoensis* identifying 21 haplotypes (A-E11).

Halpotype	Nucleotide position																									
	*psb*A-*trn*H																					*trn*L-*trn*F				
	5	6	8	1	2	5	7	9	9	1	1	1	1	1	2	2	2	2	3	3	4	8	1	1	2	5
				2	1	3	4	1	2	2	4	5	6	8	4	7	7	8	5	8	0	1	1	6	1	5
										0	5	5	0	3	9	0	6	7	5	0	9		3	2	9	5
A	A	T	1[a]	C	0	0	C	A	C	0	A	T	C	T	G	T	A	T	A	2[f]	T	T	C	C	C	A
B	.	.	.	.	.	.	.	.	T	.	.	.	.	.	.	.	.	.	.	.	.	.	.	.	.	.
C	.	.	.	.	.	.	.	.	T	.	.	.	A	.	.	.	.	.	.	.	.	.	.	.	.	.
D	.	.	.	.	.	.	.	.	T	.	.	.	A	.	.	.	.	.	.	.	.	.	.	.	A	.
E	.	.	.	.	.	.	.	.	T	.	.	.	A	.	.	A	T	.	.	.	.	.	.	.	A	.
C1	.	.	.	.	.	.	.	.	T	.	.	.	A	G	.	.	.	C	.	.	.	.	.	.	.	.
C2	.	.	.	.	.	.	.	.	T	.	.	.	A	G	.	.	.	.	.	.	.	.	.	.	.	.
C3	.	.	.	.	.	.	.	.	T	.	.	G	A	.	.	.	.	.	.	.	.	.	.	.	.	.
D1	.	.	.	.	.	.	.	.	T	.	.	.	A	.	A	.	.	.	.	.	.	.	.	.	A	C
D2	.	.	.	.	.	.	.	.	T	.	.	.	A	.	A	.	.	.	.	.	.	.	.	.	A	.
E1	.	.	.	.	.	.	.	.	T	.	.	.	A	.	.	A	T	.	C	.	.	.	.	.	.	.
E2	.	.	.	.	.	.	.	.	T	.	.	.	A	.	.	A	T	.	.	.	.	.	.	.	.	.
E3	.	.	.	.	.	1[d]	.	.	T	.	.	.	A	.	.	A	T	.	.	2[g]	G	C	.	.	.	.
E4	T	.	0	T	.	.	.	.	T	.	.	.	A	.	.	A	T	.	.	.	.	.	.	.	A	.
E5	.	.	.	.	.	.	.	G	T	.	.	.	A	.	.	A	T	.	.	.	G	.	.	A	A	C
E6	.	.	.	.	.	.	.	G	T	1[e]	.	.	A	.	.	A	T	.	.	.	G	.	.	A	A	C
E7	.	.	.	.	.	.	T	.	T	.	.	.	A	.	.	A	T	.	.	.	G	.	.	.	A	.
E8	.	.	1[b]	.	.	.	.	.	T	.	.	.	A	.	.	A	T	.	.	.	.	.	T	.	A	.
E9	.	.	.	.	.	.	.	.	T	.	C	.	A	.	.	A	T	.	.	.	.	.	.	.	A	.
E10	.	A	.	.	.	.	.	.	T	.	.	.	A	.	.	A	T	.	.	.	.	.	.	.	A	.
E11	.	.	.	.	1[c]	.	.	.	T	.	.	.	A	.	.	A	T	.	.	.	.	.	.	.	A	.

All sequences are compared to the reference haplotype H1. Numbers '0/1' in the sequences indicate absence/presence of four length polymorphisms, and '2' presence of an inversion, whereby superscripts identify corresponding character states. Note that poly-A or poly-T stretches were excluded from analyses.

[a]: AAGC;

[b]: AATC;

[c]:TTTTTATTA;

[d]:AATTTTA;

[e]: AT;

[f]: CCTCTTGATAGAACAAGAAAAA;

[g]:TTTTTCTTGTTCTATCAAGAGG.

Table 3. Comparisons of genetic diversity and genetic structure between wild and cultivated *Scrophularia ningpoensis* populations based on chloroplast DNA sequences.

Parameter	wild	cultivated
Number of haplotype	21	3
haplotype diversity (h_T)	0.919	0.399
Within-population diversity(h_S)	0.444	0
Nucleotide diversity(π_T),	0.00301	0.00076

position linked with haplotype C and D to form one clade, while the majority of rare haplotypes forms a "star" shape dominated by haplotype E. The geographic distribution and haplotype frequencies within populations are shown in Figure 1C. The most frequent haplotypes (A, B and E) were shared by both wild and cultivated groups, while other remaining haplotypes were found only in wild populations (Figure 1C). The most common and widest geographically distributed haplotype in cultivated populations was A which was unique to the wild population LSW located in Lenshui County in northeastern Jiangxi Province. In wild populations, there was no clear geographical pattern of within-population diversity (Figure 1C). Moreover, U test (Nst/Gst) indicated no phylogeographic structure in wild *S. ningpoensis* ($N_{ST} = 0.446$, $G_{ST} = 0.534$, $U = -1.57$, $P>0.05$).

AFLP data

After exclusion of ambiguous or irreproducible fragments, 289 fragments were consistently scored for 24 populations (306 individuals) of which 261 fragments (90.31%) were polymorphic. The test for reproducibility yielded a mean error rate of 1.7% and confirmed the reproducibility of the AFLP fragments. Wild populations had much higher estimates of genetic diversity ($h = 0.0791$–0.1614) than did cultivated populations ($h = 0.0076$–0.0875) (Table 5). In wild *S. ningpoensis*, the percentage of polymorphic fragments varied from 22.15% in population TWW to 50.87% in population LAW; in cultivated *S. ningpoensis*, the percentage of polymorphic fragments varied from 3.11% in population SCC to 27.68% in population NSC (Table 5).

The variation between cultivated and wild groups accounted for 19.62% of the overall genetic variation (Table 6). In the cultivated group, most variation (76.97%) existed in among-population with only 23.03% due to the within-population (Table 6). But the variation distribution in the wild group was opposite that most variation (60.40%) existed in within-population with only 39.60% due to the among-population (Table 6).

Principal coordinate analysis also showed a clear differentiation between cultivated and wild *S. ningpoensis*. All populations were grouped into three main clusters (Figure 2): cluster I included most wild populations; cluster II included most cultivated populations and two wild populations LSW & HNW; and cluster III included three cultivated populations: PAC, XJC and GZC. Populations PAC and XJC were from Zhejiang province. While GZC is in Fujian Province, local farmers indicated that it was directly introduced from Zhejiang cultivated populations. Genetic admixture analysis performed by STRUCTURE divided genetic variation into two gene pools ($k = 2$, Figure 3B): (1) Cultivated individuals, wild population HNW and part of wild population LSW; (2) Other wild populations. In STUCTURE analysis K values ranged from 2 to 14 and the one with the highest likelihood ($k = 2$) was chosen (Figure 4).

The Neighbor-Joining analysis identified two wild populations (HNW & LSW) that had a close relationship with cultivated *S. ningpoensis*. The wild population HNW was sister to cultivated populations with a bootstrap value of 84%; LSW clustered with HNW and all cultivated populations supported by 100% bootstrap value (Figure 3A). Although three cultivated populations, PAC, XJC and GZC, grouped together and are distinct from other *S. ningpoensis* in PCoA analysis, in the NJ tree they located in one clade with hundred percent support and were associated with other cultivated populations.

Table 4. Hierarchical analysis of molecular variance for 28 populations of *Scrophularia ningpoensis* based on chloroplast DNA sequences.

Regional grouping of populations	Source of variation	d.f.	Sum of squares	Variance components	Percentage of variation	F-statistics(P)
Cultivated	Among populations	12	79.731	0.50613	100	$F_{ST} = 1(<0.001)$
	Within populations	158	0	0	0	
Wild	Among populations	13	156.171	0.79368	41.46	$F_{ST} = 0.41462(<0.001)$
	Within populations	179	200.576	1.12053	58.54	
Cultivated vs. wild	Among groups	1	167.858	0.87163	41.02	$F_{CT} = 0.41023(<0.001)$
	Among populations within groups	25	235.902	0.65791	30.96	$F_{SC} = 0.52503(<0.001)$
	Within populations	337	200.576	0.59518	28.01	$F_{ST} = 0.71988(<0.001)$

F_{ST}: genetic differences among populations; F_{CT}: genetic differences among groups defined a priori; F_{SC}: genetic differences among population within groups.

Table 5. Genetic diversity in 24 populations of *S. ningpoensis* based on AFLP.

	N	N_A	N_E	*h*	*H*	F_P	*PPF*	D_W	G_{ST}	N_M
Cultivated										
GZC	15	1.0796	1.0253	0.0177	0.0295	23	7.96	13.26		
JFC	13	1.2491	1.1515	0.0875	0.1307	72	24.91	8.38		
SCC	19	1.0311	1.0114	0.0076	0.0123	9	3.11	8.54		
XJC	10	1.0484	1.0236	0.0149	0.0232	14	4.84	8.68		
PAC	16	1.0657	1.0226	0.0150	0.0247	19	6.57	14.70		
YLC	10	1.0796	1.0326	0.0215	0.0345	23	7.96	4.80		
ZPC	10	1.1211	1.0597	0.0371	0.0576	35	12.11	7.72		
KKC	8	1.1280	1.0821	0.0467	0.0694	37	12.80	3.57		
LOC	10	1.1349	1.0458	0.0324	0.0537	39	13.49	5.79		
NSC	10	1.2768	1.1427	0.0874	0.1346	80	27.68	7.74		
Wild										
HNW	15	1.3322	1.1797	0.1083	0.1649	96	33.22	21.92		
JGW	15	1.4256	1.2273	0.1369	0.2088	123	42.56	24.79		
JHW	22	1.5052	1.2271	0.1405	0.2195	146	50.52	14.90		
JMW	13	1.3668	1.1982	0.1197	0.1825	106	36.68	20.05		
LAW	14	1.5087	1.2677	0.1614	0.2469	147	50.87	12.51		
LSW	15	1.4394	1.2559	0.1503	0.2260	127	43.94	10.81		
MTW	15	1.4221	1.2135	0.1291	0.1985	122	42.21	22.32		
PTW	9	1.2422	1.1340	0.0812	0.1235	70	24.22	5.40		
THW	14	1.4498	1.1980	0.1233	0.1943	130	44.98	9.77		
TNW	14	1.4429	1.2196	0.1345	0.2078	128	44.29	9.33		
TTW	10	1.4118	1.2271	0.1374	0.2092	119	41.18	6.87		
TM1W	10	1.2318	1.1402	0.0821	0.1231	67	23.18	6.33		
TM2W	10	1.3702	1.1929	0.1175	0.1806	107	37.02	5.39		
TWW	9	1.2215	1.1331	0.0791	0.1187	64	22.15	5.04		
Cultivated	121	1.654	1.2591	0.1554	0.2422	189	65.4		0.7568	0.1607
Wild	185	1.8893	1.3385	0.2171	0.3466	257	88.93		0.4352	0.6490
Species	306	1.9031	1.3551	0.2202	0.3481	261	90.31		0.6111	0.3181

N_A: Observed number of alleles; **N_E:** Effective number of alleles [Kimura and Crow (1964)]; ***h*:** Nei's (1973) gene diversity; ***H*:** Shannon's Information index [Lewontin (1972)]; **N_M:** estimate of gene flow from G_{ST} or G_{CS}. E.g., $N_M = 0.5(1 - G_{ST})/G_{ST}$; **N:** the number of individuals in population; **F_P:** the number of polymorphic fragments; ***PPF*:** the percentage of polymorphic fragments; **D_W:** the frequency down-weighed marker value.

Table 6. Hierarchical analysis of molecular variance for 24 populations of *S. ningpoensis* based on AFLP.

Regional grouping of populations	Source of variation	d.f.	Sum of squares	Variance components	Percentage of variation	*F*-statistics
Cultivated	Among populations	9	2251.916	18.72650	76.97	$F_{ST}=0.76974$
	Within populations	111	666.615	5.60181	23.03	
Wild	Among populations	13	2451.695	12.85266	39.60	$F_{ST}=0.39605$
	Within populations	171	3351.516	19.59951	60.40	
Cultivated vs. wild	Among groups	1	1299.331	7.09433	19.62	$F_{CT}=0.19616$
	Among populations within groups	22	4703.610	15.21535	42.07	$F_{SC}=0.52339$
	Within populations	282	4018.131	13.85563	38.31	$F_{ST}=0.61688$

F_{ST}: genetic differences among populations; F_{CT}: genetic differences among groups defined a priori; F_{SC}: genetic differences among population within groups.

Discussion

Change of genetic diversity under domestication

For many crop species the domestication process has left a signature in the pattern of genetic diversity as a result of founder effects, genetic bottlenecks and artificial selection [76]. Many crops show evidence of genome-wide reductions in variation [77,78,79]. In our study, the pattern of genetic diversity between cultivated and wild populations of *Scrophularia ningpoensis* is consistent with other domesticated crop species, even though the domestication history of *S. ningpoensis* is only one thousand years. Out of twenty-one cpDNA haplotypes in the species, only three haplotypes were detected in cultivated populations. Moreover, cultivated populations had little within-population genetic polymorphisms (Table 3). The cpDNA nucleotide diversity (π_T) and haplotype diversity (h_T) of cultivated populations were much less than that of the wild group (Table 3). AFLP data also support these patterns (Table 5). These results are consistence with many crop-relative systems of domestication [80,81] although the degree of decline in diversity of *S. ningpoensis* with such a short history of domestication is surprising. This medicinal plant has lost genetic diversity as a result of domestication which clearly shows that a species can have large genetic alterations even before there are significant changes in morphology, phenology or breeding system. Other species native to China, such as *Corydalis yanhusuo* (traditional Chinese medicine) [16], *Metasequoia glyptostroboides* [82], *Malus pumila* [83] and *Zizania latifolia* [84] also revealed the same genetic pattern.

These large changes in genetic diversity are most likely the result of both founder effects and selection for medicinal quality, and importantly, the clonal mode of reproduction. The extent of the loss of genetic diversity depends in part on population size during domestication and time over which domestication occurred [77]. At the beginning of cultivation, *S. ningpoensis* was gathered from

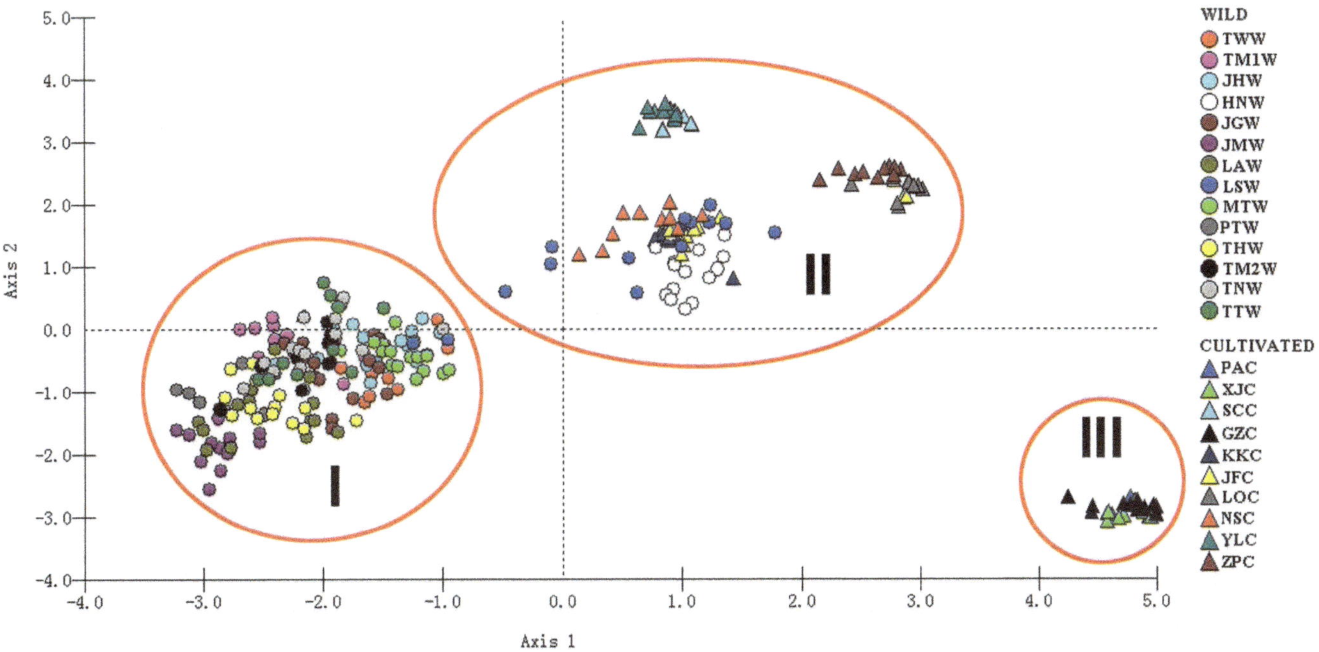

Figure 2. Principal coordinate analysis (PCoA) of 306 individuals from 24 populations of *S. ningpoensis* based on the Euclidean distance generated from AFLP data. To simplify comparison, cultivated *S. ningpoensis* are indicated by triangles, while wild *S. ningpoensis* are indicated by circles.

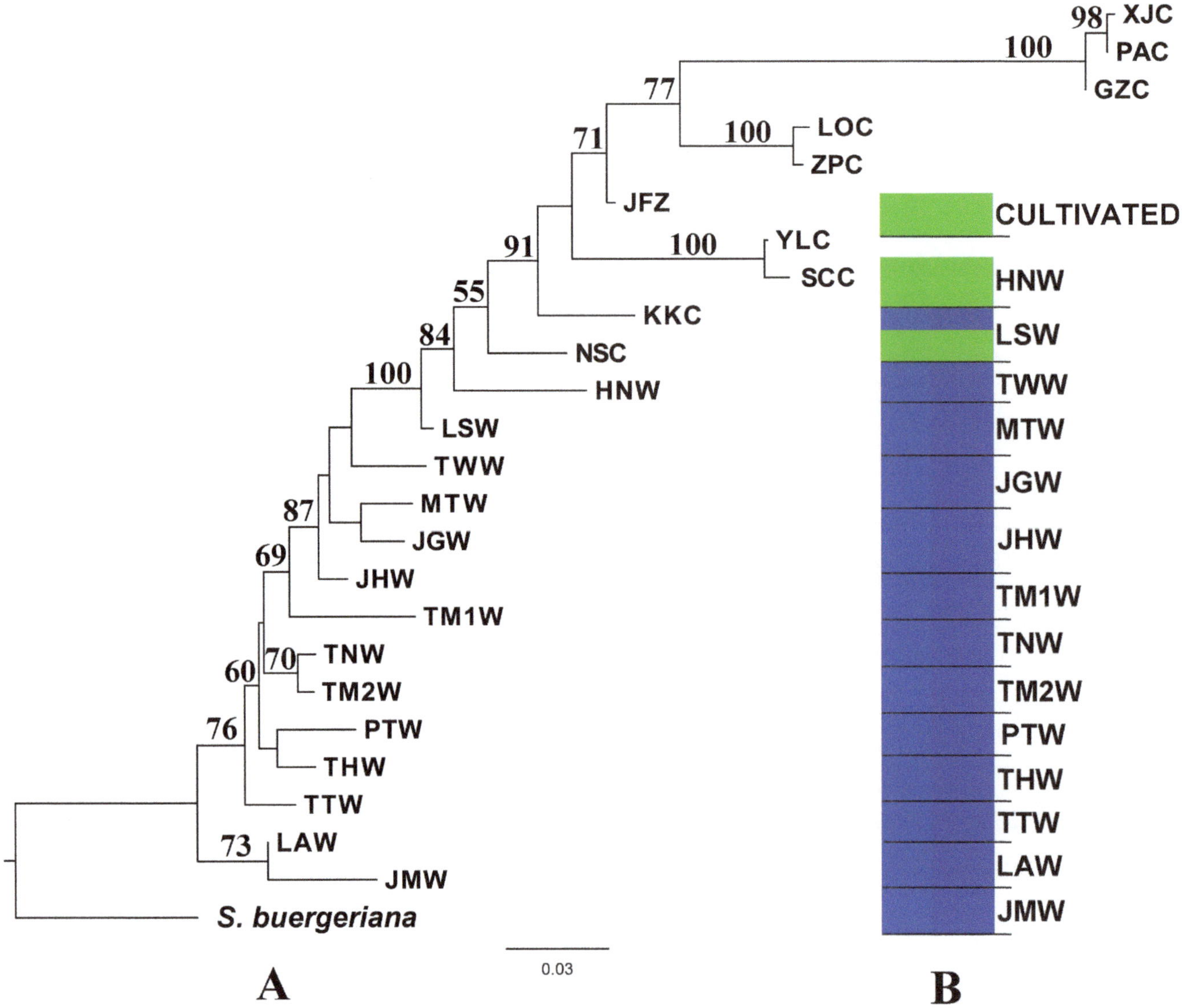

Figure 3. Neighbor-Joining and STRUCTURE analyses of the AFLP data. (A) Neighbor-Joining analysis of the AFLP data for all individuals of *S.ningpoensis* based on Nei's (1979) genetic distances with *S. buergeriana* as the outgroup. (B) Result of the model from clustering (K = 2) of all wild individuals and 10 cultivated individuals of *S. ningpoensis* using STRUCTURE based on the AFLP data set.

the wild. Most likely a limited number of individuals from the wild formed the founding populations of cultivated *S. ningpoensis*, the founders containing only a sample of the genetic diversity of wild populations. Although the domestication history of *S. ningpoensis* is short, under cultivation it is propagated vegetatively from rhizome and flowers are removed. These practices restrict gene flow between wild and cultivated *S. ningpoensis* populations and among individuals within cultivated populations. Moreover, farmers select large and healthy rhizomes to establish new populations. Such selection coupled with clonally propagation could quickly fix alleles carried by selected individuals. The increased F_{ST} values (Table 4, Table 6) for cultivated populations (cpDNA: $F_{ST} = 1$; AFLP: $F_{ST} = 0.76974$) relative to wild populations (cpDNA: $F_{ST} = 0.41462$; AFLP: $F_{ST} = 0.39605$) reflect the different modes of reproduction in cultivated (clonally propagated) and wild (sexually propagated) populations.

Origin of cultivated *Scrophularia ningpoensis*

Locating the geographical site of domestication and assessing the demographic consequences of wild-domesticated species evolution can help to design strategies of use and management of genetic resources [85]. In the current study, three cpDNA haplotypes (A, B and E) found within cultivated populations are scattered on the haplotype network and shared with different wild population gene pools. But the most cultivated populations harbored haplotype A, which is only found in the wild population LSW. Two other haplotypes B & E, distributed in two cultivated populations (YLC and SCC) and one cultivated population (NSC) respectively, are shared with three and four different wild populations respectively. It is interesting that wild populations harboring haplotype B or E didn't have haplotype A. This cpDNA pattern indicates that the wild population LSW in Jiangxi Province contributed substantially to the origin of cultivated *S. ningpoensis*, while other six wild populations (JGW, LAW, JHW, TNW,

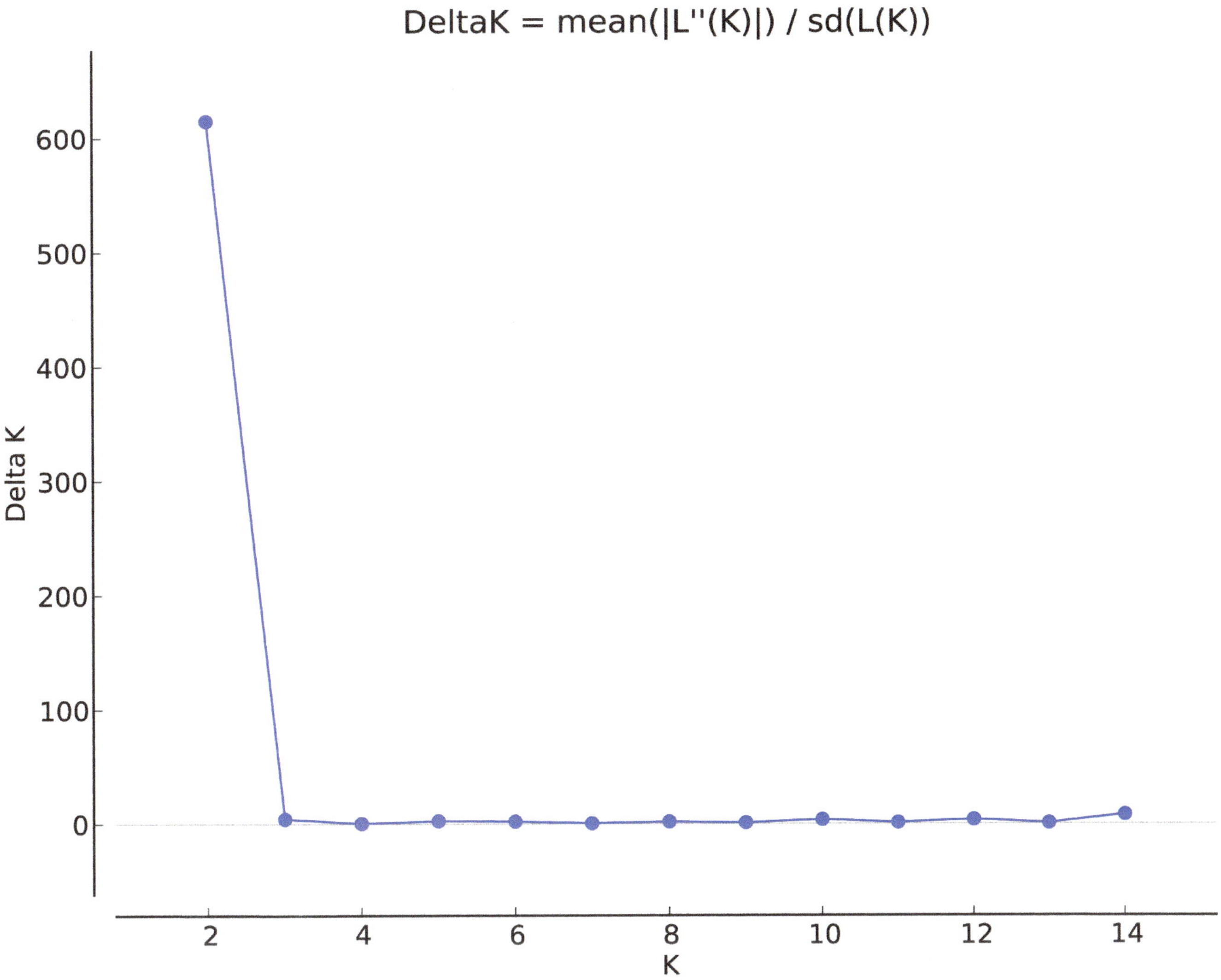

Figure 4. The estimated mean logarithmic likelihood of K values ranging from 2 to 14 with 10 replicates for each K calculated using the R-script Structure-sum.

TM2W, PTW) located in geographically separated places were only involved in origin of three cultivated populations. Moreover, the wild population JGW harboring both haplotype B and E is also in Jiangxi Province as the wild population LSW. Thus, the Jiangxi region is a critical site for the origin of cultivated *S. ningpoensis*.

However, results of AFLP analysis suggested that the origin of cultivated populations is more restricted to only two wild populations, HNW and LSW. In the neighbor-Joining analysis (Figure 3), HNW and LSW clustered with other cultivated populations by 100% bootstrap value and the STRUCTURE analysis was consistent, placing the two wild populations, HNW and LSW, into the same gene pool as the cultivated individuals. In principal coordinate analysis (Figure 2), wild population HNW and LSW grouped with majority of cultivated populations in one cluster. Thus, combined with cpDNA data, we infer that wild populations of the Jiangxi region may have played a major role in the origin of cultivated *S. ningpoensis*. Although three cultivated populations, PAC, XJC and GZC, grouped to another cluster in PCoA analysis (Figure 2), they still located in the cultivated clade identified in the Neighbor-Joining analysis (Figure 3A) and were assigned to the same gene pool as the other cultivated *S. ningpoensis* (Figure 3B). The wild population LSW located in eastern Jiangxi and HNW population is on the border between Hunan and Jiangxi, so AFLP data strongly supported that Jiangxi area of Eastern China is the geographic origin of cultivated *S. ningpoensis*.

Considering the discordance between chloroplast and nuclear markers,the origin of cultivated *S. ningpoensis* must be interpreted with caution. Firstly, patterns of cpDNA indicated broader geographic origin of cultivated *S. ningpoensis* compared with AFLP markers, which can be explained by the characteristics of chloroplast genome evolution. CpDNA evolves slowly and is uniparentally inherited in plants, so it usually only represents the maternal genome and often has low levels of variation making cpDNA less informative for intraspecific studies [86]. Nuclear genomic AFLP markers provided much higher resolution both because of rapid evolution and many independent markers, thus providing a more nuanced interpretation of the origin of cultivated *S. ningpoensis*. Revealed by AFLP markers, cultivated populations originated from Jiangxi area where wild populations were involved in the event of origin. Secondly, if AFLP data has revealed the more resolution, these two wild populations which were involved in the origin of cultivation, HNW & LSW, should harbor all cpDNA haplotypes (A, B and E) found in cultivated populations. But the wild population HNW didn't share any cpDNA haplotype with cultivated populations; only the native population LSW harbored haplotype A. These data suggest that this entire region of China may have served as the source of medicinal *S. ningpoensis* rather than a specific population. The ancestors of cultivated populations YLC, SCC and NSC were probably introduced from wild population similar to the extant HNW; harvesting and cultivation may, by chance, collected all individuals which had the same cpDNA haplotypes with cultivated populations. In addition, the nearest wild population to HNW is JGW, which shared two cpDNA haplotypes, B&E, with cultivated populations. But the very high value of Dw in HNW (21.92) and JGW (24.79) suggests a long-term history of isolation [74]. The long term use, and most likely, trading of rhizomes of *S. ningpoensis* lead to a complicated domestication history for this species.

Crop improvement and sustainable cultivation of *S. ningpoensis*

The therapeutic efficacy of a medicinal plant often varies by geographical location among the populations of a species [22]. Cultivated *S. ningpoensis* from Zhejiang Province is recognized to have the best medicinal quality [87]. Our previous study on HPLC fingerprints of *S. ningpoensis* has also revealed that materials cultivated in Zhejiang Province produce the highest content of bioactive compounds compared to other cultivated populations [45]. In this study, analysis of AFLP markers group cultivated populations from Zhejiang into one cluster which is divergent from other cultivated populations and other wild populations (Figure 2). Moreover, cultivated populations from Zhejiang gather in one clade associated with other cultivated populations in NJ tree (Figure 3A). This evidence strongly supports the genetic distinct nature of cultivated *S. ningpoensis* from Zhejiang (Figure 2). Under strong artificial selection, cultivated *S. ningpoensis* with high pharmaceutical quality has genetically diverged from wild populations. It is generally believed that different medical effects due to secondary compounds are caused by various environmental conditions and habitats where the medicinal plants are grown and harvested [88]. But our study combined with previous phytochemical analysis indicates that under artificial selection, cultivated *S. ningpoensis* from Zhejiang has become genetically differentiated from other cultivated populations and chemical diversity may also be influenced by genetic factors. Thus, this result can help design strategies of crop improvement that consider both genetic and ecological factors for the cultivation of medicinal plants.

Genetic diversity underlies the plasticity of many secondary metabolites [89] and results in phytochemical diversity on which the pharmaceutical quality of medicinal plants relies [90]. Study on *S. ningpoensis* by HPLC and ISSR fingerprinting have already revealed a strong association between chemical and genetic variation of *S. ningpoensis* [45] suggesting that higher genetic diversity can account for higher phytochemical diversity that is critical for the pharmaceutical quality. This study that compares cultivated populations to the range of extant wild populations indicates that wild *S. nignpoensis* has much higher genetic diversity than does the cultivated plant (Figure 1C, Table 5). These wild populations represent an infrequently used source of genetic diversity that is conspecific with the cultivated gene pool. Cultivated and wild plants generally do not exhibit reproductive isolation [79]. Hence, wild resources can be utilized to improve the cultivated *S. ningpoensis* by introducing genetic diversity by sexual reproduction and then selection for desired traits.

The results of this study can be used to select the most diverse natural populations for such a breeding program. Using the AFLP data, the values of Nei's [67] gene diversity (*h*) of most wild *S. ningpoensis* were higher than 0.1 (Table 5), except for PTW, TWW and TM1W. For chloroplast data, three wild populations, TM2W, LSW and JHW, harbored more cpDNA haplotypes than others (Figure 1C). Thus, TM2W, LSW and JHW populations have high genetic diversity for both the cpDNA and nuclear genomes. These native populations have the greatest reservoir of genetic diversity and could be used for further improvement of cultivated *S. ningpoensis* to ensure that it provides an adequate supply of the medicinal plant and the quality of the plant is enhanced. Furthermore, the utilization of wild populations with appropriate guidelines for collection and for regulating management could be an efficient way to find, conserve and deploy desirable agronomic traits which would be beneficial for Radix Scrophulariae cultivation.

Acknowledgments

The authors would like to give special thanks to David Bufford, Joongku Lee, Tetsuo Ohi-Toma, Mingshui Zhao, Shuqing Sun for assistance in sample collection; David Bufford for helpful comments on manuscript; Pu Huang in Schaal lab of Washington University in St. Louis for assistance in DNA sequencing experiment.

Author Contributions

Conceived and designed the experiments: CXF BAS. Performed the experiments: CC PL. Analyzed the data: CC RHW. Wrote the paper: CC CXF BAS.

References

1. Zeder MA (2006) Central questions in the domestication of plants and animals. Evolutionary Anthropology 15: 105–117.
2. Lambert J, Srivastava J, Vietmeyer N (1997) Medicinal plants: Rescuing a global heritage (vol. 355). Washington D. C.: The World Bank. 8–15 p.
3. WHO IUCN, WWF (1993) Guidelines on the Conservation of Medicinal Plants. Gland: The International Union for Conservation of Nature and Natural Resources. 21–25 p.
4. Vines G (2004) Herbal harvests with a future: towards sustainable sources for medicinal plants. Salisbury: Plantlife International. 1–10 p.
5. Canter PH, Thomas H, Ernst E (2005) Bringing medicinal plants into cultivation: opportunities and challenges for biotechnology. Trends Biotechnol 23: 180–185.
6. Uniyal RC, Uniyal MR, Jain P, TRAFFIC I (2000) Cultivation of medicinal plants in India: a reference book. New Delhi: TRAFFIC-India. 20 p.
7. Schippmann U, Leaman DJ, Cunningham AB (2002) Impact of cultivation and gathering of medicinal plants on biodiversity: global trends and issues. Rome: Food and Agriculture Organization of the United Nations.
8. Palevitch D (1991) Agronomy applied to medicinal plant conservation. In: Akerele OH, editor. Conservation of medicinal plants. Cambridge: Cambridge University Press. pp. 167–178.
9. Bernáth J (1999) Biological and economical aspects of utilization and exploitation of wild growing medicinal plants in Middle-and South Europe. Acta Hort (ISHS) 500:31–42.
10. Lange D, Schippmann U (1997) Trade survey of medicinal plants in Germany: a contribution to international plant species conservation. Bonn: German Federal Agency for Nature Conservation.
11. Pei-gen X (1991) The Chinese approach to medicinal plants: Their utilization and conservation. In: Akerele OH, editor. Conservation of medicinal plants. Cambridge: Cambridge University Press. pp. 305–313.
12. He S, Sheng N (1997) Utilization and conservation of medicinal plants in China: Medicinal plants for forest conservation and health care. Rome: Food and Agriculture Organization of the United Nation. 109–115 p.
13. Gao W, Jia W, Gao X, Wang R, Xiao P (2005) In vitro culture and cultivation of Chinese medicinal plants for industrial utilization and genetic resource conservation. Genetic Resour 3: 116.
14. El-Domyati FM, Younis RA, Edris S, Mansour A, Sabir J, et al. (2011) Molecular markers associated with genetic diversity of some medicinal plants in Sinai. Journal of Medicinal Plants Research 5: 1918–1929.

15. Yuan QJ, Zhang ZY, Hu J, Guo LP, Shao AJ, et al. (2010) Impacts of recent cultivation on genetic diversity pattern of a medicinal plant, *Scutellaria baicalensis* (Lamiaceae). BMC Genet 11: 29.
16. Qiu YX, Zong M, Yao Y, Chen BL, Zhou XL, et al. (2009) Genetic variation in wild and cultivated rhizoma corydalis revealed by ISSRs markers. Planta med 75: 94–98.
17. Shi W, Yang CF, Chen JM, Guo YH (2008) Genetic variation among wild and cultivated populations of the Chinese medicinal plant *Coptis chinensis* (Ranunculaceae). Plant Biol 10: 485–491.
18. He J, Chen L, Si Y, Huang B, Ban X, et al. (2009) Population structure and genetic diversity distribution in wild and cultivated populations of the traditional Chinese medicinal plant *Magnolia officinalis* subsp. *biloba* (Magnoliaceae). Genetica 135: 233–243.
19. Zhang DQ, Gao LM, Yang YP (2010) Genetic diversity and structure of a traditional Chinese medicinal plant species, *Fritillaria cirrhosa* (Liliaceae) in southwest China and implications for its conservation. Biochem Syst Ecol 38: 236–242.
20. Hartmann T (2007) From waste products to ecochemicals: Fifty years research of plant secondary metabolism. Phytochemistry 68: 2831–2846.
21. Fernie AR (2007) The future of metabolic phytochemistry: larger numbers of metabolites, higher resolution, greater understanding. Phytochemistry 68: 2861–2880.
22. Woo Y, Kim H, Cho J, Chung H (1999) Discrimination of herbal medicines according to geographical origin with near infrared reflectance spectroscopy and pattern recognition techniques. J Pharm Biomed Anal 21: 407–413.
23. Bagdonaité E, Mártonfi P, Repčák M, Labokas J (2012) Variation in concentrations of major bioactive compounds in *Hypericum perforatum* L. from Lithuania. Ind Crops Prod 35: 302–308.
24. Cirak C, Radusiene J, Stanius Z, Camas N, Caliskan O, et al. (2012) Secondary metabolites of *Hypericum orientale* L. growing in Turkey: variation among populations and plant parts. Acta Physiol Plant 34: 1313–1320.
25. Singh SK, Katoch R, Kapila RK (2013) Chemotypic variation for essential oils in *Valeriana jatamansi* Jones populations from Himachal Pradesh. Journal of Essential Oil Research 25: 154–159.
26. Verma RS, Chauhan A, Padalia RC, Jat SK, Thul S, et al. (2013) Phytochemical Diversity of *Murraya koenigii* (L.) Spreng. from Western Himalaya. Chem Biodivers 10: 628–641.
27. Li Y, Sun S, Zhou Q, Qin Z, Tao J, et al. (2004) Identification of American ginseng from different regions using FT-IR and two-dimensional correlation IR spectroscopy. Vib Spectrosc 36: 227–232.
28. Milligan BG, Leebens-Mack J, Strand AE (1994) Conservation genetics: beyond the maintenance of marker diversity. Mol Ecol 3: 423–435.
29. Francisco-Ortega J, Santos-Guerra A, Kim S, Crawford DJ (2000) Plant genetic diversity in the Canary Islands: a conservation perspective. Am J Bot 87: 909–919.
30. Hong D, Yang H, Jin C, FIscher A M, et al. (1994) Flora of China. China & US: Science Press & Missouri Botanical Garden Press. 13 p.
31. Kajimoto T, Hidaka M, Shoyama K, Nohara T (1989) Iridoids from *Scrophularia ningpoensis*. Phytochemistry 28: 2701–2704.
32. Sagare AP, Kuo CL, Chueh FS, Tsay HS (2001) De novo regeneration of *Scrophularia yoshimurae* Yamazaki (Scrophulariaceae) and quantitative analysis of harpagoside, an iridoid glucoside, formed in aerial and underground parts of in vitro propagated and wild plants by HPLC. Biol Pharm Bull 24: 1311–1315.
33. Yen KY (1992) The Illustrated Chinese Materia Medica Crude and Prepared. Taiwan: SMC Publishing Inc Press. 64 p.
34. Huang S (1982) Shen Nong Ben Cao Jing (The Divine Farmer's Materia Medica). Beijing: Traditional Chinese Medicine Ancient Books Press.
35. Liu L, Hudgins WR, Shack S, Yin MQ, Samid D (1995) Cinnamic acid: a natural product with potential use in cancer intervention. Int J Cancer 62: 345–350.
36. Garcia D, Fernandez A, Saenz T, Ahumada C (1996) Antinflammatory effects of different extracts and harpagoside isolated from *Scrophularia frutescens* L. Farmaco 51: 443–446.
37. de Santos Galíndez J, Díaz Lanza A, Fernandez Matellano L (2002) Biologically active substances from the genus *Scrophularia*. Pharm Biol 40: 45–59.
38. Diáz AM, Abad MJ, Fernández L, Silván AM, De Santos J, et al. (2004) Phenylpropanoid glycosides from *Scrophularia scorodonia*: In vitro anti-inflammatory activity. Life Sci 74: 2515–2526.
39. Miyazawa M, Okuno Y, Nakamura S, Kameoka H (1998) Suppression of SOS-inducing activity of chemical mutagens by cinnamic acid derivatives from *Scrophulia ningpoensis* in the Salmonella typhimurium TA1535/pSK1002 umu test. J Agric Food Chem 46: 904–910.
40. National Pharmacopoeia Committee (2010) Chinese pharmacopoeia (Part I). Beijing: China Medical Science Press. 332 p.
41. Su S (1975) Tu Jing Ben Cao. Beijing: The Science Press.
42. Ran MX, Zhou HQ (1999) Guidelines on Modern Cultivation, Breeding and Processing of Chinese Materia Medica. Beijing: Traditional Chinese Medicine Press.
43. Obeso JR (2002) The costs of reproduction in plants. New Phytol 155: 321–348.
44. McKey D, Elias M, Pujol B, Duputié A (2010) The evolutionary ecology of clonally propagated domesticated plants. New Phytol 186: 318–332.
45. Yang ST, Chen C, Zhao YP, Xi W, Zhou XL, et al. (2011) Association between Chemical and Genetic Variation of Wild and Cultivated Populations of *Scrophularia ningpoensis* Hemsl. Planta Med 77: 865–871.
46. Xu CB (2007) The Textual Research of Common Chinese Materia Medica. Beijing: People's Medical Publishing Press.
47. Doyle JJ (1987) A rapid DNA isolation procedure for small quantities of fresh leaf tissue. Phytochemical bulletin 19: 11–15.
48. Doyle J (1991) DNA protocols for plants-CTAB total DNA isolation. Molecular techniques in taxonomy: 283–293.
49. Taberlet P, Gielly L, Pautou G, Bouvet J (1991) Universal primers for amplification of three non-coding regions of chloroplast DNA. Plant Mol Biol 17: 1105–1109.
50. Sang T, Crawford DJ, Stuessy TF (1997) Chloroplast DNA phylogeny, reticulate evolution, and biogeography of *Paeonia* (Paeoniaceae). Am J Bot 84: 1120.
51. Thompson JD, Higgins DG, Gibson TJ (1994) Clustal W: improving the sensitivity of progressive multiple sequence alignment through sequence weighting, position-specific gap penalties and weight matrix choice. Nucleic Acids Res 22: 4673–4680.
52. Nei M (1987) Molecular evolutionary genetics. New York: Oxford University Press.
53. Rozas J, Rozas R (1999) DnaSP version 3: an integrated program for molecular population genetics and molecular evolution analysis. Bioinformatics 15: 174–175.
54. Clement M, Posada D, Crandall KA (2000) TCS: a computer program to estimate gene genealogies. Mol Ecol 9: 1657–1659.
55. Nylander JAA (2004) MrModeltest v2. Program distributed by the author. Evolutionary Biology Centre, Uppsala University.
56. Ronquist F, Huelsenbeck JP (2003) MrBayes 3: Bayesian phylogenetic inference under mixed models. Bioinformatics 19: 1572–1574.
57. Felsenstein J (1973) Maximum likelihood and minimum steps methods for estimating evolutionary trees from data on discrete characters. Systematic Zoology 22: 240–249.
58. Stamatakis A, Hoover P, Rougemont J (2008) A rapid bootstrap algorithm for the RAxML web servers. Systematic biology 57: 758–771.
59. Miller MA, Pfeiffer W, Schwartz T (2010) Creating the CIPRES Science Gateway for inference of large phylogenetic trees. Proceedings of the Gateway Computing Environments Workshop (GCE). New Orleans, LA: IEEE.
60. Swofford DL (2003) PAUP*: phylogenetic analysis using parsimony (* and other methods), version 4.0 b10. Sunderland: Sinauer Associates.
61. Pons O, Petit RJ (1996) Measuring and testing genetic differentiation with ordered versus unordered alleles. Genetics 144: 1237–1245.
62. Excoffier L, Laval G, Schneider S (2005) Arlequin (version 3.0): an integrated software package for population genetics data analysis. Evol Bioinform Online 1: 47.
63. Tajima F (1989) Statistical-method for testing the neutral mutation hypothesis by DNA polymorphism. Genetics 123: 585–595.
64. Fu YX, Li WH (1993) Statistical tests of neutrality of mutations. Genetics 133: 693–709.
65. Rozas J, Sánchez-DelBarrio JC, Messeguer X, Rozas R (2003) DnaSP, DNA polymorphism analyses by the coalescent and other methods. Bioinformatics 19: 2496–2497.
66. Vos P, Hogers R, Bleeker M, Reijans M, van de Lee T, et al. (1995) AFLP: a new technique for DNA fingerprinting. Nucleic Acids Res 23: 4407–4414.
67. Nei M (1973) Analysis of gene diversity in subdivided populations. Proc Natl Acad Sci U S A 70: 3321–3323.
68. Yeh FC, Yang R, Boyle TB, Ye ZH, Mao JX (1997) Popgene, the user-friendly shareware for population genetic analysis. Molecular biology and biotechnology centre, University of Alberta, Canada.
69. Felsenstein J (1993) Phylip: phylogenetic inference package, version 3.5 c. Department of Genetics, University of Washington, USA.
70. Pritchard JK, Stephens M, Donnelly P (2000) Inference of population structure using multilocus genotype data. Genetics 155: 945–959.
71. Falush D, Stephens M, Pritchard JK (2007) Inference of population structure using multilocus genotype data: dominant markers and null alleles. Mol Ecol Notes 7: 574–578.
72. Earl DA (2012) Structure harvester: a website and program for visualizing Structure output and implementing the evanno method. Conserv Genet Resour 4: 359–361.
73. Excoffier L (1995) Amova: Analysis of molecular variance (version 1.55). University of Geneva, Switzerland.
74. Schönswetter P, Tribsch A (2005) Vicariance and dispersal in the alpine perennial *Bupleurum stellatum* L. (Apiaceae). Taxon: 725–732.
75. Ehrich D (2006) AFLPdat: a collection of R functions for convenient handling of AFLP data. Mol Ecol Notes 6: 603–604.
76. Doebley JF, Gaut BS, Smith BD (2006) The molecular genetics of crop domestication. Cell 127: 1309–1321.
77. Eyre-Walker A, Gaut RL, Hilton H, Feldman DL, Gaut BS (1998) Investigation of the bottleneck leading to the domestication of maize. Proc Natl Acad Sci U S A 95: 4441–4446.
78. Londo JP, Chiang YC, Hung KH, Chiang TY, Schaal BA (2006) Phylogeography of Asian wild rice, *Oryza rufipogon*, reveals multiple independent domestications of cultivated rice,*Oryza sativa*. Proc Natl Acad Sci U S A 103: 9578–9583.
79. Sonnante G, Stockton T, Nodari RO, Becerra Velásquez VL, Gepts P (1994) Evolution of genetic diversity during the domestication of common-bean (*Phaseolus vulgaris* L.). Theor Appl Genet 89: 629–635.

80. Olsen KM, Schaal BA (1999) Evidence on the origin of cassava: phylogeography of *Manihot esculenta*. Proc Natl Acad Sci U S A 96: 5586–5591.
81. Olsen KM, Schaal BA (2007) Insights on the evolution of a vegetatively propagated crop species. Mol Ecol 16: 2838–2840.
82. Li YY, Chen XY, Zhang X, Wu TY, Lu HP, et al. (2005) Genetic differences between wild and artificial populations of Metasequoia glyptostroboides: implications for species recovery. Conservation Biology 19(1): 224–231.
83. Zhang Q, Li J, Zhao Y, Korban S, Han Y (2012) Evaluation of genetic diversity in Chinese wild apple species along with apple cultivars using SSR markers. Plant molecular biology reporter 30(3): 539–546. (PR 7/31)
84. Xu XW, Ke WD, Yu XP, Wen J, Ge S (2008) A preliminary study on population genetic structure and phylogeography of the wild and cultivated *Zizania latifolia* (Poaceae) based on Adh1a sequences. Theor Appl Genet 116:835–843.
85. Frankel OH (1995) The conservation of plant biodiversity. Cambridge: Cambridge University Press.
86. Wolfe KH, Li WH, Sharp PM (1987) Rates of nucleotide substitution vary greatly among plant mitochondrial, chloroplast, and nuclear DNAs. Proc Natl Acad Sci U S A 84: 9054–9058.
87. Wang JK, Wang JL (2007) History and genuine character of traditional Chinese medicine. Beijing: Science and Technology of China Medicine Press.
88. Zhang X, Han T, Zhang Q, Zhang H, Huang B, et al. (2009) Chemical fingerprinting and hierarchical clustering analysis of *Centella asiatica* from different locations in China. Chromatographia 69: 51–57.
89. Hartmann T (2007) From waste products to ecochemicals: Fifty years research of plant secondary metabolism. Phytochemistry 68: 2831–2846.
90. Fernie AR (2007) The future of metabolic phytochemistry: larger numbers of metabolites, higher resolution, greater understanding. Phytochemistry 68: 2861–2880.

PERMISSIONS

All chapters in this book were first published in PLOS ONE, by The Public Library of Science; hereby published with permission under the Creative Commons Attribution License or equivalent. Every chapter published in this book has been scrutinized by our experts. Their significance has been extensively debated. The topics covered herein carry significant findings which will fuel the growth of the discipline. They may even be implemented as practical applications or may be referred to as a beginning point for another development.

The contributors of this book come from diverse backgrounds, making this book a truly international effort. This book will bring forth new frontiers with its revolutionizing research information and detailed analysis of the nascent developments around the world.

We would like to thank all the contributing authors for lending their expertise to make the book truly unique. They have played a crucial role in the development of this book. Without their invaluable contributions this book wouldn't have been possible. They have made vital efforts to compile up to date information on the varied aspects of this subject to make this book a valuable addition to the collection of many professionals and students.

This book was conceptualized with the vision of imparting up-to-date information and advanced data in this field. To ensure the same, a matchless editorial board was set up. Every individual on the board went through rigorous rounds of assessment to prove their worth. After which they invested a large part of their time researching and compiling the most relevant data for our readers.

The editorial board has been involved in producing this book since its inception. They have spent rigorous hours researching and exploring the diverse topics which have resulted in the successful publishing of this book. They have passed on their knowledge of decades through this book. To expedite this challenging task, the publisher supported the team at every step. A small team of assistant editors was also appointed to further simplify the editing procedure and attain best results for the readers.

Apart from the editorial board, the designing team has also invested a significant amount of their time in understanding the subject and creating the most relevant covers. They scrutinized every image to scout for the most suitable representation of the subject and create an appropriate cover for the book.

The publishing team has been an ardent support to the editorial, designing and production team. Their endless efforts to recruit the best for this project, has resulted in the accomplishment of this book. They are a veteran in the field of academics and their pool of knowledge is as vast as their experience in printing. Their expertise and guidance has proved useful at every step. Their uncompromising quality standards have made this book an exceptional effort. Their encouragement from time to time has been an inspiration for everyone.

The publisher and the editorial board hope that this book will prove to be a valuable piece of knowledge for researchers, students, practitioners and scholars across the globe.

LIST OF CONTRIBUTORS

Huaxin Dai, Fangbin Cao, Xianhong Chen, Mian Zhang, Imrul Mosaddek Ahmed, Guoping Zhang and Feibo Wu
Department of Agronomy, College of Agriculture and Biotechnology, Zijingang Campus, Zhejiang University, Hangzhou, P.R. China

Zhong-Hua Chen
School of Science and Health, Hawkesbury Campus, University of Western Sydney, New South Wales, Australia

Chengdao Li
Department of Agriculture, Government of Western Australia, South Perth, Western Australia, Australia

Jie Qiu, Yu Wang, Sanling Wu, Ying-Ying Wang, Chu-Yu Ye, Xuefei Bai, Zefeng Li, Chenghai Yan, Weidi Wang, Ziqiang Wang and Longjiang Fan
Department of Agronomy & James D. Watson Institute of Genome Sciences, Zhejiang University, Hangzhou, China

Qingyao Shu
Institute China, of Nuclear Agricultural Science, Zhejiang University, Hangzhou

Jiahua Xie
Department of Pharmaceutical Sciences, North Carolina Central University, Durham, North Carolina, United States of America

Suk-Ha Lee
Department of Plant Science and Research Institute for Agriculture and Life Sciences, Seoul National University, Seoul, Korea

Geo Coppens d'Eeckenbrugge
CIRAD, UMR 5175 CEFE, Campus du CNRS, Montpellier, France

Jean-Marc Lacape
CIRAD, UMR AGAP, Montpellier, France

Chiara Marchi and Volker Loeschcke
Department of Bioscience, Aarhus University, Aarhus, Denmark

Liselotte Wesley Andersen
Department of Bioscience, Aarhus University, Rønde, Denmark

Vanessa Corby-Harris
United States Department of Agriculture, Carl Hayden Bee Research Center, Tucson, Arizona, United States of America

Patrick Maes
Center for Insect Science, University of Arizona, Tucson, Arizona, United States of America

Kirk E. Anderson
United States Department of Agriculture, Carl Hayden Bee Research Center, Tucson, Arizona, United States of America
Center for Insect Science, University of Arizona, Tucson, Arizona, United States of America

Ola T. Westengen
Nordic Genetic Resource Center (NordGen), Alnarp, Sweden
Centre for Development and the Environment, University of Oslo, Oslo, Norway

Simon Jeppson
Nordic Genetic Resource Center (NordGen), Alnarp, Sweden

Luigi Guarino
The Global Crop Diversity Trust, Bonn, Germany
Kristin L. Mercer, Matthew A. Kost and Brian A. Pace
Ohio State University, Department of Horticulture and Crop Science, Columbus, Ohio, United States of America

D. Jason Emry
University of Kansas, Department of Ecology and Evolutionary Biology, Lawrence, Kansas, United States of America
Washburn University, Topeka, Kansas, United States of America

Allison A. Snow
Ohio State University, Department of Evolution, Ecology, and Organismal Biology, Columbus, Ohio, United States of America

Helen M. Alexander
University of Kansas, Department of Ecology and Evolutionary Biology, Lawrence, Kansas, United States of America

Mian Zhang, Guoping Zhang and Feibo Wu
Department of Agronomy, College of Agriculture and Biotechnology, Zijingang Campus, Zhejiang University, Hangzhou, P.R. China

Weihua Mao
Center of Analysis and Measurement, Zhejiang University, Hangzhou, P.R. China

Qichao Zhao, Minghong Liu, Miaomiao Tan and Zhicheng Shen
State Key Laboratory of Rice Biology, Institute of Insect Sciences, Zhejiang University, Hangzhou, China

Jianhua Gao
College of Life Science, Shanxi Agricultural University, Taigu, China

Mohar Singh, Ishwari Singh Bisht, Sandeep Kumar, Manoranjan Dutta, Kailash Chander Bansal and Moreshwar Karale
National Bureau of Plant Genetic Resources, Pusa, New Delhi, India

Ashutosh Sarker
International Centre for Agricultural Research in Dry Areas, South Asia and China Regional Programme, NASC Complex, DPS Marg, Pusa, New Delhi, India

Ahmad Amri
International Centre for Agricultural Research in Dry Areas, Tunisia, Tunis

Shiv Kumar
International Centre for Agricultural Research in Dry Areas, Rabat, Morocco

Swapan Kumar Datta
Division of Crop Science, Indian Council of Agricultural Research, Krishi Bhawan, New Delhi, India

Von Mark V. Cruz
USDA-ARS National Center for Genetic Resources Preservation, Fort Collins, Colorado, United States of America
Department of Bioagricultural Sciences and Pest Mgt., Colorado State University, Fort Collins, Colorado, United States of America

Andrzej Kilian
Diversity Arrays Technology Pty. Ltd., Yarralumla, ACT, Australia

David A. Dierig
USDA-ARS National Center for Genetic Resources Preservation, Fort Collins, Colorado, United States of America

Wilhelm Klümper, Matin Qaim
Department of Agricultural Economics and Rural Development, Georg-August-University of Goettingen, Goettingen, Germany

Roberto de la Rosa Santamaria, Mon-Ray Shao and Sally A. Mackenzie
Center for Plant Science Innovation, University of Nebraska Lincoln, Lincoln, Nebraska, United States of America
Department of Agronomy and Horticulture, University of Nebraska Lincoln, Lincoln, Nebraska, United States of America

Guomei Wang and David O. Nino-Liu
Monsanto, Chesterfield Village Research Center, Chesterfield, Missouri, United States of America

Hardik Kundariya and Yashitola Wamboldt
Center for Plant Science Innovation, University of Nebraska Lincoln, Lincoln, Nebraska, United States of America

Ismail Dweikat
Department of Agronomy and Horticulture, University of Nebraska Lincoln, Lincoln, Nebraska, United States of America

Gangaprasad Choudhary, Nagireddy Ranjitkumar, Malathi Surapaneni, Dondapati
Annekitty Deborah, Abhilash Vipparla, Ghanta Anuradha, Ebrahimali Abubacker Siddiq and Lakshminarayana Reddy Vemireddy
Institute of Biotechnology, Acharya NG Ranga Agricultural University, Rajendranagar, Hyderabad, Andhra Pradesh, India

Michael Behr, Johanna K. Kaufmann, Dirk M. Nettelbeck, Patrick Ketzer and Sarah Engelhardt
Oncolytic Adenovirus Group, German Cancer Research Center (DKFZ), Heidelberg, Germany

Martin Mück-Häusl
Max von Pettenkofer-Institute, Department of Virology, Ludwig-Maximilians-University Munich, Munich, Germany

Pamela M. Okun
Department of General Pediatrics, Division of Inherited Metabolic Diseases, Heidelberg University Hospital, Heidelberg, Germany

Gabriele Petersen
COS, CellNetworks Deep Sequencing Core Facility, University Heidelberg, Heidelberg, Germany

Frank Neipel
Institute for Clinical and Molecular Virology, Erlangen University Hospital, Erlangen, Germany

Jessica C. Hassel and Alexander H. Enk
Department of Dermatology, Heidelberg University Hospital, Heidelberg, Germany

Anja Ehrhardt
Max von Pettenkofer-Institute, Department of Virology, Ludwig-Maximilians-University Munich, Munich, Germany
Institute of Virology and Microbiology, Center for Biomedical Education and Research, Department of Human Medicine, Faculty of Health, University Witten/Herdecke, Witten, Germany

Bingjun Jiang, Youfei Gao, Liming Ma, Yanlei Yue, Shi Sun, Cunxiang Wu and Tianfu Han
The National Key Facility for Crop Gene Resources and Genetic Improvement and MOA Key Lab of Soybean Biology (Beijing), Institute of Crop Sciences, The Chinese
Academy of Agricultural Sciences, Beijing, China

Haiyang Nan, Sijia Lu
The Key Laboratory of Soybean Molecular Design Breeding, Northeast Institute of Geography and Agroecology, Chinese Academy of Sciences, Harbin, China
University of Chinese Academy of Sciences, Beijing, China

Lili Tang, Dong Cao, Jialin Wang, Xiaohui Yuan, Fanjiang Kong and Baohui Liu
The Key Laboratory of Soybean Molecular Design Breeding, Northeast Institute of Geography and Agroecology, Chinese Academy of Sciences, Harbin, China

Wensheng Hou
The National Key Facility for Crop Gene Resources and Genetic Improvement and MOA Key Lab of Soybean Biology (Beijing), Institute of Crop Sciences, The Chinese
Academy of Agricultural Sciences, Beijing, China
The Key Laboratory of Soybean Molecular Design Breeding, Northeast Institute of Geography and Agroecology, Chinese Academy of Sciences, Harbin, China

Rachel P. Naegele, Samantha Boyle and Mary K. Hausbeck
Department of Plant and Microbial Sciences, Michigan State University, East Lansing, Michigan, United States of America

Lina M. Quesada-Ocampo
Department of Plant Pathology, North Carolina State University, Raleigh, North Carolina, United States of America

Debjani Roy Choudhury, Nivedita Singh, Amit Kumar Singh, Sundeep Kumar and Rakesh Singh
Division of Genomic Resources, National Bureau of Plant Genetic Resources, New Delhi, 110 012, India

Kalyani Srinivasan and R. K. Tyagi
Division of Germplasm Conservation, National Bureau of Plant Genetic Resources, New Delhi, 110 012, India

Altaf Ahmad
Department of Botany, Faculty of Science, Jamia Hamdard (Hamdard University), New Delhi, 110062, India

N. K. Singh
National Research Centre on Plant Biotechnology, Indian Agricultural Research Institute, New Delhi, 110 012, India

Subodh K. Srivastava and Andy Pereira
Crop, Soil and Environmental Sciences, University of Arkansas, Fayetteville, Arkansas, United States of America

Pawel Wolinski
Arkansas High Performance Computing Center, University of Arkansas, Fayetteville, Arkansas, United States of America

Chun Li, Hongmei Miao, Libin Wei, Tide Zhang, Xiuhua Han and Haiyang Zhang
Henan Sesame Research Center, Henan Academy of Agricultural Sciences, Zhengzhou, Henan, People's Republic of China

Nelly Vicente-Dólera, Manuel Moya, Mercedes del Río-Celestino, Teresa Pomares-Viciana and Pedro Gómez
IFAPA Centro La Mojonera, Camino de San Nicolás, 1, 04745 La Mojonera, Almería, Spain

Christelle Troadec and Abdelhafid Bendahmane
INRA-URGV, UMR1165, Unité de Recherche en Génomique Végétale, Saclay
Plant Sciences, Evry, France

Belén Picó
Institute for the Conservation and Breeding of Agricultural Biodiversity (COMAV-UPV), Universitat Politécnica de Valencia, Camino de Vera s/n, 46022 Valencia, Spain

Belén Román
IFAPA Centro Alameda del Obispo, Avd. Mene´ndez Pidal s/n, 14004, Córdoba, Spain

Chuan Chen
The Key Laboratory of Conservation Biology for Endangered Wildlife of the Ministry of Education, College of Life Sciences, Zhejiang University, Hangzhou, China
Hangzhou Botanical Garden, Hangzhou, China

Pan Li, Rui-Hong Wang and Cheng-Xin Fu
The Key Laboratory of Conservation Biology for Endangered Wildlife of the Ministry of Education, College of Life Sciences, Zhejiang University, Hangzhou, China

Barbara A. Schaal
Department of Biology, Washington University, St. Louis, Missouri, United States of America

Index